De Gruyter Series in Nonlinear Analysis and Applications 13

Peter Kosmol
Dieter Müller-Wichards

Optimization in Function Spaces

With Stability Considerations in Orlicz Spaces

De Gruyter

Mathematics Subject Classification 2010: Primary: 46B20, 46E30, 52A41; Secondary: 41-02, 41A17, 41A25, 46B10, 46T99, 47H05, 49-02, 49K05, 49K15, 49K40, 49M15, 90C34.

ISBN 978-3-11-025020-6
e-ISBN 978-3-11-025021-3
ISSN 0941-813X

Library of Congress Cataloging-in-Publication Data

Kosmol, Peter.
Optimization in function spaces with stability considerations on Orlicz spaces / by Peter Kosmol, Dieter Müller-Wichards.
p. cm. – (De Gruyter series in nonliniear analysis and applications ; 13)
Includes bibliographical references and index.
ISBN 978-3-11-025020-6 (alk. paper)
1. Stability – Mathematical models. 2. Mathematical optimization. 3. Orlicz spaces. I. Müller-Wichards, D. (Dieter), 1949– II. Title.
QA871.K8235 2011
515'.392–dc22
2010036123

Bibliographic information published by the Deutsche Nationalbibliothek

The Deutsche Nationalbibliothek lists this publication in the Deutsche Nationalbibliografie; detailed bibliographic data are available in the Internet at http://dnb.d-nb.de.

Typesetting: Da-TeX Gerd Blumenstein, Leipzig, www.da-tex.de
Printing and binding: Hubert & Co. GmbH & Co. KG, Göttingen
∞ Printed on acid-free paper

Printed in Germany

www.degruyter.com

Preface

In this text we present a treatise on optimization in function spaces. A concise but thorough account of convex analysis serves as a basis for the treatment of Orlicz spaces and Variational Calculus as a contribution to non-linear and linear functional analysis.

As examples may serve our discussion of stability questions for families of optimization problems or the equivalence of strong solvability, local uniform convexity, and Fréchet differentiability of the convex conjugate, which do provide new insights and open up new ways of dealing with applications.

A further contribution is a novel approach to the fundamental theorems of Variational Calculus, where pointwise (finite-dimensional) minimization of suitably convexified Lagrangians w.r.t. the state variables x and $\dot{x}$ is performed simultaneously. Convexification in this context is achieved via quadratic supplements of the Lagrangian. These supplements are constant on the restriction set and thus lead to equivalent problems.

It is our aim to present an essentially self-contained book on the theory of convex functions and convex optimization in Banach spaces. We assume that the reader is familiar with the concepts of mathematical analysis and linear algebra. Some awareness of the principles of measure theory will turn out to be helpful.

The methods mentioned above are applied to Orlicz spaces in a twofold sense: not only do we consider optimization and in particular norm approximation problems in Orlicz spaces but we also use these methods to develop a complete theory of Orlicz spaces for σ-finite measures. We also employ the stability principles developed in this text in order to establish optimality for solutions of the Euler–Lagrange equations of certain non-convex variational problems.

Overview

In Chapter 1 we present the classical approximation theory for L^1 and $C(T)$ embedded into the category of Orlicz spaces. In particular we present the theorems of Jackson (zeros of error function) and Bernstein (error estimates) and their extension to Orlicz spaces. For differentiable (1-D) Young functions the non-linear systems of equations that arise from linear modular and norm approximation problems are discussed. For their solution the rapidly convergent numerical methods of Chapter 4 can be used.

Chapter 2 is devoted to Polya-type algorithms in Orlicz spaces to determine a best Chebyshev (or L^1) approximation. The idea is to replace a numerically ill conditioned problem, affected by non-uniqueness and non-differentiability, with a sequence of well-behaved problems. The closedness of the algorithm is already guaranteed by the pointwise convergence of the sequence of Young functions. This is due to the stability principles presented in Chapter 5. Convergence estimates for the corresponding sequence of Luxemburg norms can be gleaned from the Young functions in the discrete and continuous case.

These estimates in turn can be utilized to regularize the Polya algorithm in order to achieve actual convergence to a two-stage solution, a subject that is discussed within a general framework in more detail in Chapter 5. For sequences of Young functions with separation property the convergence of discrete approximations to the strict approximation of Rice is shown.

The last application in this chapter is of a somewhat different nature: it is well known that many applications can be restated as linear programming problems. Semi-infinite optimization problems are a generalization with potentially infinitely many restrictions (for which a fairly large literature exists). We solve the problem by making use of the stability results developed in this book: we approximate the restriction set by a sequence of smooth restriction sets making use of the Lagrange mechanism and the existence of Lagrange multipliers, which in turn is guaranteed by corresponding regularity conditions.

In Chapter 3 we develop the theory of convex functions and convex sets in normed spaces. We stress the central role of the (one-sided) directional derivative and – by using its properties – derive necessary and sufficient conditions for minimal solutions of convex optimization problems.

For later use we consider in particular convex functions on finite dimensional spaces. It turns out that real-valued convex functions are already continuous, and differentiable convex functions are already continuously differentiable. The latter fact is used later to show that the Gâteaux derivative of a continuous convex function is already demi-continuous (a result that is used in Chapter 8 to prove that a reflexive and differentiable Orlicz space is already Fréchet differentiable).

We discuss the relationship between the Gâteaux and Fréchet derivatives of a convex function in normed spaces and give the proof of a not very well-known theorem of Phelps on the equivalence of the continuity of the Gâteaux derivative and the Fréchet differentiability of a convex function.

Based on the Hahn–Banach extension theorem we prove several variants of separation theorems (Mazur, Eidelheit, strict). Using these separation theorems we show the existence of the subdifferential of a continuous convex function, and derive a number of properties of a convex function together with its convex conjugate (in particular Young's equality and the theorem of Fenchel–Moreau).

The separation theorems are then used to prove the Fechel-duality theorem and – in an unconventional way – by use of the latter theorem the existence of minimal solutions of a convex function on a bounded convex subset of a reflexive Banach space (theorem of Mazur–Schauder).

The last section of this chapter is devoted to Lagrange multipliers where we show their existence – again using the separation theorem – and the Lagrange duality theorem which is used in Chapter 7 to prove the (general) Amemiya formula for the Orlicz norm.

In Chapter 5 we turn our attention to stability principles for sequences (and families) of functions. We show the closedness of algorithms that are lower semi-continuously convergent. Using this result we derive stability theorems for monotone convergence. Among the numerous applications of this result is the computation of the right-handed derivative of the maximum norm on $C(T)$ which in turn is used to provide a simple proof for the famous Kolmogoroff criterion for a best Chebyshev approximation (see Chapter 1).

Our main objective in this chapter is to consider pointwise convergent sequences of convex functions. Based on an extension of the uniform boundedness principle of Banach, formulated originally for linear functionals, to families of convex functions, it turns out that pointwise convergent sequences of convex functions are already continuously convergent. We are thus enabled to consider sequences of convex optimization problems (f_n, M_n) where the functions f_n converge pointwise to f and the sets M_n in the sense of Kuratowski to M, while closedness of the algorithm is preserved. In the finite dimensional case we can even guarantee the existence of points of accumulation, provided that the set of minimal solutions of (f, M) is non-empty and bounded.

In the next section of this chapter we consider two-stage solutions where – under certain conditions – we can guarantee the actual convergence of the sequence of minimal solutions and characterize it. Among the methods being considered is differentiation w.r.t. the family parameter: for example, it turns out that the best L^p-approximations converge for $p \to 1$ to the best L^1-approximation of maximal entropy. 'Outer regularizations' in the spirit of Tikhonov are also possible, provided that convergence estimates are available.

In the final section of this chapter we show that a number of stability results for sequences of convex functions carry over to sequences of monotone operators. If P is a non-expansive Féjer contraction then $I - P$ is monotone. Such operators occur in the context of smoothing of linear programming problems where the solution of the latter (non-smooth) problem appears as a second stage solution of the solutions of the smooth operator sequence.

In Chapter 6 we introduce the Orlicz space L^Φ for general measures and arbitrary (not necessarily finite) Young functions Φ. We discuss the properties and relations of Young functions and their conjugates and investigate the structure of Orlicz spaces equipped with the Luxemburg norm. An important subspace is provided by the closure of the space of step functions M^Φ. It turns out that $M^\Phi = L^\Phi$ if and only if Φ satisfies an appropriate Δ_2-condition. Moreover, if Φ does not satisfy a Δ_2-condition then L^Φ contains a closed subspace isomorphic to ℓ^∞, in the case of a finite not purely atomic measure this isomorphy is even isometric. These statements are due to the theorem of Lindenstrauss–Tsafriri for Orlicz sequence spaces and a theorem of Turett for non-atomic measures. These results become important in our discussion of reflexivity in Chapters 7 and 8.

In Chapter 7 we introduce the Orlicz norm which turns out to be equivalent to the Luxemburg norm. Using Jensen's integral inequality we show that norm convergence already implies convergence in measure. We show that the convex conjugate of the modular f^{Φ} is f^{Ψ}, provided that Ψ is the conjugate of the Young function Φ. An important consequence is that the modular is always lower semi-continuous. These facts turn out to be another ingredient in proving the general Amemiya formula for the Orlicz norm.

The main concern of this chapter is to characterize the dual space of an Orlicz space: it turns out that for finite Φ and σ-finite measures we obtain $(M^{\Phi})^* = L^{\Psi}$, provided that Ψ is the conjugate of Φ. Reflexivity of an Orlicz space is then characterized by an appropriate (depending on the measure) Δ_2-condition for Φ and Ψ.

Based on the theorem of Lusin and a theorem of Krasnosielski, establishing that the continuous functions are dense in M^{Φ}, we show that, if Φ and Ψ are finite, and T is a compact subset of $\mathbb{R}^m$, and μ the Lebesgue measure, then M^{Φ} is separable, where separability becomes important in the context of greedy algorithms and the Ritz method (Chapter 8).

We conclude the chapter by stating and proving the general Amemiya formula for the Orlicz norm.

Based on the results of Chapter 6 and 7 we now (in Chapter 8) turn our attention to the geometry of Orlicz spaces. Based on more general considerations in normed spaces we obtain for not purely atomic measures that M^{Φ} is smooth if and only if Φ is differentiable. For purely atomic measures the characterization is somewhat more complicated.

Our main objective is to characterize strong solvability of optimization problems where convergence of the values to the optimum already implies norm convergence of the approximations to the minimal solution. It turns out that strong solvability can be geometrically characterized by the local uniform convexity of the corresponding convex functional (provided that the term local uniform convexity is appropriately defined, which we do). Moreover, we establish that in reflexive Banach spaces strong solvability is characterized by the Fréchet differentiability of the convex conjugate, provided that both are bounded functionals. These results are based in part on a paper of Asplund and Rockafellar on the duality of A-differentiability and B-convexity of conjugate pairs of convex functions, where B is the polar of A.

Before we apply these results to Orlicz spaces, we turn our attention to E-spaces introduced by Fan and Glicksberg, where every weakly closed subset is approximatively compact. A Banach space is an E-space if and only if it is reflexive, strictly convex, and satisfies the Kadec–Klee property. In order to establish reflexivity the theorem of James is applied. Another characterization is given by Anderson: X is an E-space if and only if X^* is Fréchet differentiable. The link between Fréchet differentiability and Kadec–Klee property is then provided through the lemma of Shmulian [42].

With these tools at hand we can show that for finite not purely atomic measures Fréchet differentiability of an Orlicz space already implies its reflexivity. The main theorem gives in 17 equivalent statements a characterization of strong solvability, local uniform convexity, and Fréchet differentiability of the dual, provided that L^Φ is reflexive. It is remarkable that all these properties can also be equivalently expressed by the differentiability of Φ or the strict convexity of Ψ. In particular it turns out that in Orlicz spaces the necessary conditions of strict convexity and reflexivity for an E-space are already sufficient.

We conclude the geometrical part of this chapter by a discussion on the duality of uniform convexity and uniform differentiability of Orlicz spaces based on a corresponding theorem by Lindenstrauss. We restate a characterization by Milne of uniformly convex Orlicz spaces equipped with the Orlicz norm and present an example of a reflexive Orlicz space, also due to Milne, that is not uniformly convex but (according to our results) the square of its norm is locally uniformly convex. Following A. Kaminska we show that uniform convexity of the Luxemburg norm is equivalent to δ-convexity of the defining Young function.

In the last section we discuss a number of underlying principles for certain classes of applications:

- Tikhonov regularization: this method was introduced for the treatment of ill-posed problems (of which there is a whole lot). The convergence of the method was proved by Levitin and Polyak for uniformly convex regularizing functionals. We show here that locally uniformly convex regularizations are sufficient for that purpose. As we have given a complete description of local uniform convexity in Orlicz spaces we can state such regularizing functionals explicitly.
- Ritz method: the Ritz method plays an important role in many applications (e.g. FEM-methods). It is well known that the Ritz procedure generates a minimizing sequence. Actual convergence of the solutions on each subspace is only achieved if the original problem is strongly solvable.
- Greedy algorithms have indeed drawn a growing attention and experienced a rapid development in recent years (see e.g. Temlyakov). The aim is to arrive at a 'compressed' representation of a function in terms of its dominating 'frequencies'. The convergence proof makes use of the Kadec–Klee property of an E-space.

Viewing these 3 'applications' at one glance it turns out that there is an inherent relationship: local uniform convexity, strong solvability, and Kadec–Klee property are 3 facets of the same property which we have completely described in Orlicz spaces.

In the last chapter we describe an approach to variational problems, where the solutions appear as pointwise (finite dimensional) minima for fixed t of the supplemented

Lagrangian. The minimization is performed simultaneously w.r.t. to both the state variable x and $\dot{x}$, different from Pontryagin's maximum principle, where optimization is done only w.r.t. the $\dot{x}$ variable. We use the idea of the Equivalent Problems of Carathéodory employing suitable (and simple) supplements to the original minimization problem. Whereas Carathéodory considers equivalent problems by use of solutions of the Hamilton–Jacobi partial differential equations, we shall demonstrate that quadratic supplements can be constructed, such that the supplemented Lagrangian is convex in the vicinity of the solution. In this way, the fundamental theorems of the Calculus of Variations are obtained. In particular, we avoid any usage of field theory.

We apply the stability principles of Chapter 5 to problems of variational calculus and control theory. As an example, we treat the isoperimetric problem (by us referred to as the Dido problem) as a problem of variational calculus. It turns out that the identification of the circle as a minimal solution can only be accomplished by employing stability principles to appropriately chosen sequences of variational problems.

The principle of pointwise minimization is then applied to the detection of a smooth, monotone trend in time series data in a parameterfree manner. In this context we also employ a Tikhonov-type regularization.

The last part of this chapter is devoted to certain problems in optimal control, where, to begin with, we put our focus on stability questions. In the final section we treat a minimal time problem which turns out to be equivalent to a linear approximation in the mean, and thus closing the circle of our journey through function spaces.

We take this opportunity to express our thanks to the staff of DeGruyter, in particular to Friederike Dittberner, for their friendly and professional cooperation throughout the publication process.

Kiel, November 2010

Peter Kosmol
Dieter Müller-Wichards

Contents

Chapter 1
Approximation in Orlicz Spaces

1.1 Introduction

In the present textbook we want to treat the methods of convex optimization under both practical and theoretical aspects. As a standard example we will consider approximation problems in Orlicz spaces, whose structure and geometry is to a large extent determined by one-dimensional convex functions (so-called Young functions).

Convexity of a one-dimensional function f can be visualized in the following way: if one connects two points of the graph by a chord then the chord will stay above the graph of the function f.

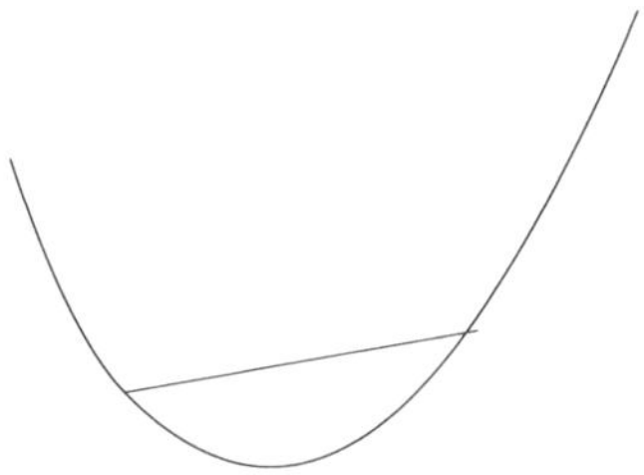

This behavior can be described by Jensen's inequality

$$f(\lambda s + (1-\lambda)t) \leq \lambda f(s) + (1-\lambda) f(t)$$

for all $\lambda \in [0,1]$ and all s, t in the interval, on which f is defined.

The function f is called strictly convex, if Jensen's inequality for $\lambda \in (0,1)$ always holds in the strict sense. If f is differentiable, then an equivalent description of convexity, which is also easily visualized, is available: the tangent at an arbitrary point of the graph always stays below the graph of f:

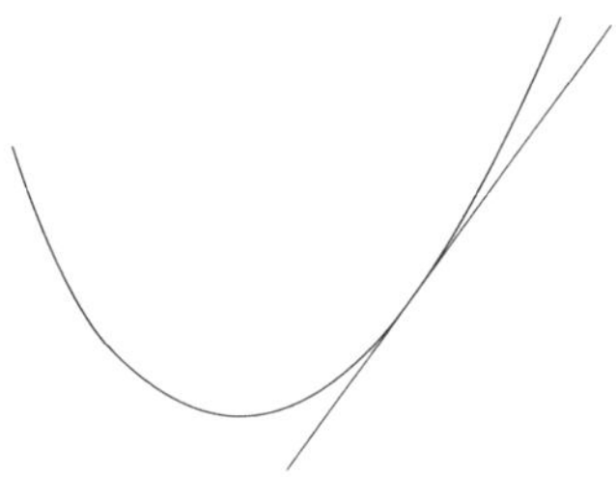

This behavior can be expressed by the following inequality

$$f(t_0) + f'(t_0)(t - t_0) \leq f(t) \tag{1.1}$$

for all t, t_0 in the interval, on which f is defined. Later on we will encounter this inequality in a more general form and environment as *subgradient inequality*.

A Young function Φ is now defined as a non-negative, symmetric, and convex function on $\mathbb{R}$ with $\Phi(0) = 0$.

The problem of the approximation in the mean is well suited to serve as an illustration of the type of questions and methods, which are among the primary objectives of this text.

The approximation in the mean appears as a natural problem in a number of applications. Among them are robust statistics (simplest example: median), where outliers in the data have much smaller influence on the estimate than e.g. in the case of the approximation in the square mean (simplest example: arithmetic mean). A further application can be found in the field of time-optimal controls (see Theorem 9.10.5).

In its simplest setting the problem can be described in the following way: for a given point x of the $\mathbb{R}^m$ we look for a point y in a fixed subset M of $\mathbb{R}^m$, which among all points of M has least distance to x, where the distance is understood in the following sense:

$$\|x - y\|_1 = \sum_{i=1}^{m} |x_i - y_i|. \tag{1.2}$$

In a situation like this we will normally require that the set M is convex, i.e. the chord connecting two arbitrary points of M has to be contained in M. We want to visualize this problem geometrically. For this purpose we first consider the unit ball of the $\|\cdot\|_1$-norm in $\mathbb{R}^2$, and compare this to the unit balls of the Euclidean and the maximum norm.

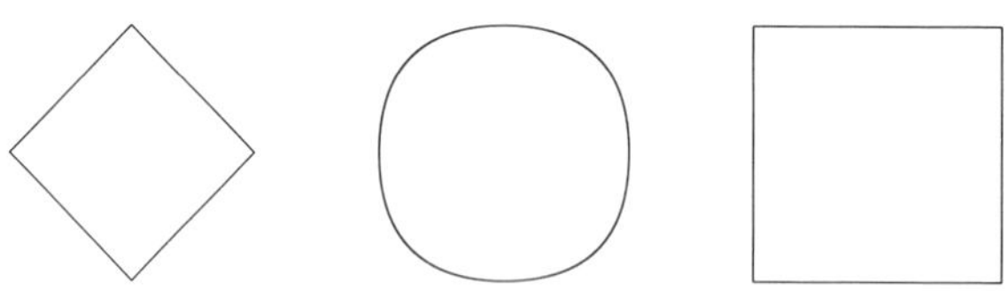

unit balls

We will treat the connection of these norms to the corresponding Young functions in subsequent sections of the current and subsequent chapters, primarily under the aspect of computing linear approximations and in Chapters 6, 7 and 8 under structural points of view.

The shape of the unit balls influences the solvability of the approximation problem: the fact that a ball has vertices leads to non-differentiable problems. Intervals contained in the sphere can result in non-uniqueness of the best approximations.

The best approximation of a point x w.r.t. a convex set M can be illustrated in the following way: the ball with center x is expanded until it touches the set M. The common points of the corresponding sphere and the set M (the 'tangent' points) are then the best approximations to be determined. It is apparent that different types of distances may lead to different best approximations.

Example 1.1.1. Let $x = (2, 2)$ and $M = \{x \in \mathbb{R}^2 \mid x_1 + x_2 = 1\}$ be the straight line through the points $P = (1, 0)$ and $Q = (0, 1)$. The set of best approximations in the mean (i.e. w.r.t. the $\|\cdot\|_1$-norm) is in this case the whole interval $[P, Q]$, whereas the point $(0.5, 0.5)$ is the only best approximation w.r.t. the Euclidean norm and the maximum norm.

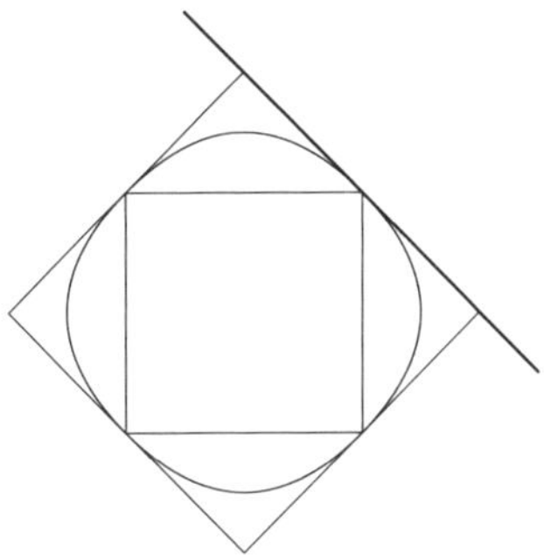

The computation of a best approximation in the mean using differential calculus creates a certain difficulty since the modulus function is not differentiable at zero. Instead we will approximate the modulus function by differentiable Young functions. By a suitable selection of these one-dimensional functions the geometrical properties of the corresponding unit balls can be influenced in a favorable way. This does not only refer to differentiability and uniqueness, but also to the numerical behavior.

In order to solve the original problem of computing the best approximation in the mean, a sequence of approximating problems is considered, whose solutions converge to a solution of the original problem. Each of these approximate problems is determined by a (one-dimensional) Young function. The richness of the available choices provides the opportunity to achieve a stable behavior of the sequence of minimal solutions of the approximating problems in the above sense.

In the sequel we will sketch the framework for this type of objective. The problem of the best approximation in the mean described above will, in spite of its simple structure, give us the opportunity to illustrate the principal phenomena.

For this purpose we adopt a different notation.

Let $\Phi_0 : \mathbb{R} \to \mathbb{R}$ be defined as $\Phi_0(s) := |s|$, then we obtain for the problem of approximation in the mean:

For a given $x \in \mathbb{R}^m$ determine a point $\bar{v} \in M$, such that for all $z \in M$

$$\sum_{i=1}^{m} \Phi_0(x_i - \bar{v}_i) \leq \sum_{i=1}^{m} \Phi_0(x_i - z_i). \tag{1.3}$$

The function Φ_0 is neither strictly convex nor differentiable at zero, resulting in a negative influence on the approximation properties. If we replace the one-dimensional function Φ_0 by approximating functions Φ_k, where k plays the role of a parameter, then we obtain for each k the following approximate problem:

For given $x \in \mathbb{R}^m$ determine a point $v^{(k)} \in M$ such that for all $z \in M$

$$\sum_{i=1}^{m} \Phi_k(x_i - v_i^{(k)}) \leq \sum_{i=1}^{m} \Phi_k(x_i - z_i) \tag{1.4}$$

holds. We consider a few examples of such approximating functions:

(a) $\Phi_k(s) = |s|^{1+k}$

(b) $\Phi_k(s) = \sqrt{s^2 + k^2} - k$

(c) $\Phi_k(s) = |s| - k\log(1 + \frac{1}{k}|s|)$

(d) $\Phi_k(s) = |s| + ks^2$

for $k \in (0, \infty)$.

The first 3 examples represent strictly convex and everywhere differentiable functions, the 4-th is strictly convex, but not differentiable at zero. The type of approximation of Φ_0 can be described in all cases by

$$\lim_{k \to 0} \Phi_k(s) = \Phi_0(s)$$

for all $s \in \mathbb{R}$. The function in the 3rd example seems too complicated at first glance but appears to be particularly well suited for numerical purposes. This is due to the simple form of its derivative

$$\Phi_k'(s) = \frac{s}{k + |s|}.$$

All examples have in common that for each $k \in (0, \infty)$ the solution $v^{(k)}$ is unique.

The mapping $k \mapsto v^{(k)}$ determines a curve in $\mathbb{R}^m$: 'the path to the solution' of the original problem. If one considers this curve as a river and the set of solutions of the original problem as a lake, then a few questions naturally arise:

(a) Does the river flow into the lake? (Question of stability)

(b) How heavily does the river meander? (Question about numerical behavior)

(c) Does the mouth of the river form a delta? (Question about convergence to a particular solution)

We will now raise a number of questions, which we will later on treat in a more general framework. Does the pointwise convergence of the Φ_k to Φ_0 carry over to the convergence of the solutions $v^{(k)}$? In order to establish a corresponding connection we introduce the following functions for $0 \leq k < \infty$:

$$f^{\Phi_k} : \mathbb{R}^m \to \mathbb{R}$$
$$x \mapsto \sum_{i=1}^{m} \Phi_k(x_i).$$

Apparently, the pointwise convergence of the one-dimensional functions Φ_k carries over to the pointwise convergence of the functions f^{Φ_k} on $\mathbb{R}^m$. The stability theorems in Chapter 5 will then yield convergence of the sequence $(v^{(k)})$ in the following sense: each sequence $(v^{(k_n)})_{n\in\mathbb{N}}$ has a convergent subsequence, whose limit is a solution of the original problem (1.2). This behavior we denote as *algorithmic closedness*. If the original problem has a unique solution $\bar{v}$, then actual convergence is achieved

$$\lim_{k\to 0} v^{(k)} = \bar{v}.$$

An implementation of the above process can be sketched in the following way: the original problem of the best approximation in the mean is replaced by a parameter-dependent family of approximate problems, which are numerically easier to handle. These approximate problems, described by the functions Φ_k, will now be chosen in such way that rapidly convergent iteration-methods for the determination of $v^{(k)}$ for fixed k can be employed, requiring normally differentiability of the function f^{Φ_k}. If a minimal solution for fixed k is determined, it is used as a starting point of the iteration for the subsequent approximate problem, because it turns out that better 'initial guesses' are required the more the approximating functions approach their limit. The determination of the subsequent parameter k' depends on the structure of the problem to be solved.

A process of this type we will denote as a *Polya algorithm*. The original application of Polya was the computation of a best Chebyshev approximation by a process similar to the one sketched above using the functions $\Phi_p : s \mapsto |s|^p$ (L^p-approximation) for $p \to \infty$ (compare Example (a) with $p = 1 + k$). This application we will treat in detail in Chapter 2 in a somewhat more general framework. In this context we can raise the following question: under what conditions does the sequence $(\Phi_k)_k$ enforce convergence of the whole sequence $(v^{(k)})$, even if the best L^1-approximations are not uniquely determined. The corresponding question for the continuous Chebyshev-approximation is denoted as the *Polya problem*. Beyond that the question arises, how this limit can be described. This question will be treated in Chapter 2 and – in a more general setting – in Chapter 5 in the section on two-stage optimization.

Through an appropriate selection of Young functions the Orlicz spaces, which we will introduce in the next section, will offer a common platform for the treatment of modular and norm approximation problems. In this context we will adopt the modern perception of functions as points of a vector space and obtain in this manner a framework for geometrical methods and visualizations. A formulation in terms of general measures will enable us to treat discrete and continuous approximation problems in a unified manner. In particular we obtain a common characterization of optimal solutions. For approximations of continuous functions on compact subsets of $\mathbb{R}^r$, in the simplest case on intervals, we merely need the Riemann integral, whereas discrete problems can be described in terms of point measures.

In this chapter we will also treat the classical theorems of Chebyshev approximation by Kolmogoroff, Chebyshev and de la Valleé-Poussin. In the framework of Haar subspaces we discuss the extensions of the also classical theorems of Bernstein and Jackson to linear approximations in Orlicz spaces. For differentiable Young functions we investigate the non-linear systems of equations resulting from the characterization theorem for linear approximations and supply a solution method based on iterative computation of weighted L^2-approximations. In a general setting we will discuss rapidly convergent methods for non-linear systems of equations in Chapter 4.

The geometry of convex sets and the properties of convex functions, in particular with respect to optimization problems, are considered in Chapter 3.

Based on the considerations about the structure of Orlicz spaces and their dual spaces in Chapters 6 and 7, where again the properties of the corresponding Young functions play a decisive role, we turn our attention to the geometry of Orlicz spaces in Chapter 8. Our main interest in this context is the description of strong solvability of optimization problems, where from convergence of the values to the optimum the convergence of the minimizing sequence to the minimal solution already follows. This behavior can be geometrically characterized by local uniform convexity of the corresponding functionals, defined in Chapter 8.

Fréchet differentiability turns out to be a property, which is dual to local uniform convexity.

Local uniform convexity and Fréchet differentiability also play a central role in the applications discussed in the last sections of this chapter. This holds for regularizations of Tychonov type, where ill posed problems are replaced by a sequence of 'regular' optimization problems, for the Ritz method, where sequences of minimal solutions of increasing sequences of finite-dimensional subspaces are considered, but also for Greedy algorithms, i.e. non-linear approximations, which aim at a compressed representation of a given function in terms of dominating basis elements.

1.2 A Brief Framework for Approximation in Orlicz Spaces

In this section we briefly state a framework for our discussion of linear approximation in Orlicz space. A detailed description of the structure of Orlicz spaces, its dual spaces, its differentiability and convexity properties will be given in Chapters 6, 7 and 8.

Let (T, Σ, μ) be an arbitrary measure space. We introduce the following relation on the vector space of μ-measurable real-valued functions defined on T : two functions are called equivalent if they differ only on a set of μ-measure zero. Let E be the vector space of equivalence classes (quotient space).

Let further $\Phi : \mathbb{R} \to \overline{\mathbb{R}}$ be a Young function, i.e. an even, convex function with $\Phi(0) = 0$, continuous at zero and on $\mathbb{R}_+$ continuous from the left (see Section 6.1). It is easily seen that the set

$$L^\Phi(\mu) := \left\{ x \in E \,\middle|\, \text{there is } \alpha > 0 \text{ with } \int_{t\in T} \Phi(\alpha x(t))d\mu \leq \infty \right\}$$

is a subspace of E (see Section 6.2). By use of the Minkowski functional (see Theorem 3.2.7) the corresponding Luxemburg norm is defined on $L^\Phi(\mu)$:

$$\|x\|_{(\Phi)} := \inf\left\{ c > 0 \,\middle|\, \int_{t\in T} \Phi\left(\frac{x(t)}{c}\right) d\mu \leq 1 \right\}.$$

In particular we obtain by choosing $\Phi_\infty : \mathbb{R} \to \overline{\mathbb{R}}$ as

$$\Phi_\infty(t) := \begin{cases} 0 & \text{for } |t| \leq 1 \\ \infty & \text{otherwise} \end{cases}$$

the norm

$$\|x\|_\infty = \inf\left\{ c > 0 \,\middle|\, \int_{t\in T} \Phi_\infty\left(\frac{x(t)}{c}\right) d\mu \leq 1 \right\},$$

and the corresponding space $L^\infty(\mu)$.

For the Young functions $\Phi_p : \mathbb{R} \to \mathbb{R}$ with $\Phi_p(t) := |t|^p$ one obtains the $L^p(\mu)$ spaces ($p \geq 1$).

Furthermore we denote by $M^\Phi(\mu)$ the closed subspace spanned by the step functions of $L^\Phi(\mu)$.

The modular $f^\Phi : L^\Phi(\mu) \to \overline{\mathbb{R}}$ is defined by

$$f^\Phi(x) = \int_{t\in T} \Phi(x(t))d\mu.$$

Remark 1.2.1. The level sets of f^Φ are bounded, because let $f^\Phi(x) \leq M$ (w.l.o.g. $M \geq 1$), then by use of the convexity of f^Φ,

$$f^\Phi\left(\frac{x}{M}\right) = f^\Phi\left(\frac{1}{M}x + \left(1 - \frac{1}{M}\right)0\right) \leq \frac{1}{M} f^\Phi(x) \leq 1,$$

hence by definition: $\|x\|_{(\Phi)} \leq M$.

In this chapter we will consider approximation problems of the following kind:

Let $x \in L^\Phi(\mu)$. We look for a point in a fixed subset K of $L^\Phi(\mu)$, for which the modular f^Φ or the Luxemburg norm $\|\cdot\|_{(\Phi)}$ assumes its least value among all points u in K.

If K is convex we obtain the following characterization theorem for the modular approximation:

Theorem 1.2.2. *Let (T, Σ, μ) be a measure space, Φ a finite Young function and K a convex subset of L^Φ. An element $u_0 \in K$ is a minimal solution of the functional f^Φ, if and only if for all $u \in K$ and $h := u - u_0$ we have*

$$\int_{\{h>0\}} h\Phi'_+(u_0)d\mu + \int_{\{h<0\}} h\Phi'_-(u_0)d\mu \geq 0.$$

Proof. Due to the monotonicity of the difference quotient $t \mapsto \frac{\Phi(s_0+ts)-\Phi(s_0)}{t}$ for all $s_0, s \in \mathbb{R}$ (see Theorem 3.3.1) and the theorem on monotone convergence it follows that

$$(f^\Phi)'_+(x_0, h) = \int_{\{h>0\}} h\Phi'_+(x_0)d\mu + \int_{\{h<0\}} h\Phi'_-(x_0)d\mu.$$

From the Characterization Theorem 3.4.3 we obtain the desired result. □

For linear approximations and differentiable Young functions we obtain the following characterization:

Corollary 1.2.3. *Let (T, Σ, μ) be a measure space, Φ a differentiable Young function, and V a closed subspace of M^Φ. Let $x \in M^\Phi(\mu) \setminus V$, then the element $v_0 \in V$ is a minimal solution of the functional $f^\Phi(x - \cdot)$ on V, if for all $v \in V$ we have*

$$\int_T v\Phi'(x - v_0)d\mu = 0.$$

An element $v_0 \in V$ is best approximation of x w.r.t. V in the Luxemburg norm, if for all $v \in V$ we have

$$\int_T v\Phi'\left(\frac{x - v_0}{\|x - v_0\|_{(\Phi)}}\right)d\mu = 0.$$

Proof. Set $K := x - V$ and $u_0 := x - v_0$, then the assertion follows for the modular approximation. In order to obtain the corresponding statement for the Luxemburg norm, let v_0 be a best approximation w.r.t. the Luxemburg norm, set $\Phi_1(s) := \Phi(\frac{s}{\|x-v_0\|_{(\Phi)}})$, then (see Lemma 6.2.15 together with Remark 6.2.23)

$$\int_T \Phi_1(x-v)d\mu = \int_T \Phi\left(\frac{x-v}{\|x-v_0\|_{(\Phi)}}\right)d\mu \geq \int_T \Phi\left(\frac{x-v}{\|x-v\|_{(\Phi)}}\right)d\mu = 1$$
$$= \int_T \Phi_1(x-v_0)d\mu.$$

i.e. v_0 is a best modular approximation for f^{Φ_1} of x w.r.t. V.

Conversely let

$$\int_T v\Phi'\left(\frac{x-v_0}{\|x-v_0\|_{(\Phi)}}\right)d\mu = 0$$

hold, then v_0 is a minimal solution of $f^{\Phi_1}(x-\cdot)$ on V, hence

$$\int_T \Phi\left(\frac{x-v}{\|x-v_0\|_{(\Phi)}}\right)d\mu = \int_T \Phi_1(x-v)d\mu \geq \int_T \Phi_1(x-v_0)d\mu = 1$$
$$= \int_T \Phi\left(\frac{x-v}{\|x-v\|_{(\Phi)}}\right)d\mu,$$

and thus $\|x-v\|_{(\Phi)} \geq \|x-v_0\|_{(\Phi)}$. □

Linear approximations are used e.g. in parametric regression analysis, in particular in robust statistics (i.e. if Φ' is bounded). An example is the function

$$\Phi(s) = |s| - \ln(1+|s|)$$

with the derivative $\Phi'(s) = \frac{s}{1+|s|}$. To the corresponding non-linear system of equations (see Equation (1.15)) the rapidly convergent algorithms of Chapter 4 can be applied.

Further applications can be found not only in connection with Polya algorithms for the computation of best L^1- and Chebyshev approximations (see Chapter 2), but also in Greedy algorithms.

1.3 Approximation in $C(T)$

Let T be a compact metric space. We consider the space $C(T)$ of continuous functions on T. Based on Borel's σ-algebra (generated from the open sets on T) the elements of $C(T)$ are measurable functions. If we define a finite measure on this σ-algebra then $C(T)$ can be considered as a subspace of $L^\Phi(\mu)$.

1.3.1 Chebyshev Approximations in $C(T)$

In order to prove the criterion of Kolmogoroff for a best Chebyshev approximation we use the right-sided directional derivative of the maximum norm (which is obtained as an example in Theorem 5.2.4 for the Stability Theorem of Monotone Convergence 5.2.3).

Theorem 1.3.1 (Kolmogoroff criterion). *Let V be a convex subset of $C(T)$ and let $x \in C(T) \setminus V$, $v_0 \in V$ and $z := x - v_0$. Let further*

$$E(z) := \{t \in T \mid |z(t)| = \|z\|_\infty\}$$

be the set of extreme values z of the difference function.

The element v_0 is a best Chebyshev approximation of x w.r.t. V, if and only if no $v \in V$ exists, such that

$$z(t)(v(t) - v_0(t)) > 0 \quad \textit{for all } t \in E(z).$$

Proof. Let $f(x) := \|x\|_\infty$ and $K := x - V$. According to Theorem 5.2.4 and the Characterization Theorem 3.4.3 v_0 is a best Chebyshev approximation of x w.r.t. V if and only if for all $v \in V$ we have

$$f'_+(x - v_0, v_0 - v) = \max_{t \in E(z)} \{(v_0(t) - v(t)) \operatorname{sign}(x(t) - v_0(t))\} \geq 0,$$

from which the assertion follows. □

Remark 1.3.2. If V is a subspace one can replace the condition in the previous theorem by: there is no $v \in V$ with $z(t)v(t) > 0$ for all $t \in E(z)$ or more explicitly

$$\begin{aligned} v(t) > 0 &\quad \text{for } t \in E^+(z) := \{t \in T \mid z(t) = \|z\|_\infty\} \\ v(t) < 0 &\quad \text{for } t \in E^-(z) := \{t \in T \mid z(t) = -\|z\|_\infty\}. \end{aligned}$$

As a characterization of a best linear Chebyshev approximation w.r.t. a finite-dimensional subspace of $C(T)$ we obtain

Theorem 1.3.3. *Let $V = \operatorname{span}\{v_1, \ldots, v_n\}$ be a subspace of $C(T)$ and let $x \in C(T) \setminus V$. The element $v_0 \in V$ is a best Chebyshev approximation of x w.r.t. V, if and only if there are k points $\{t_1, \ldots, t_k\} \in E(x - v_0)$ with $1 \leq k \leq n+1$ and k positive numbers $\alpha_1, \ldots, \alpha_k$, such that for all $v \in V$ we have*

$$0 = \sum_{j=1}^{k} \alpha_j (x(t_j) - v_0(t_j)) \cdot v(t_j).$$

Proof. Let the above condition be satisfied and let $\sum_{j=1}^{k} \alpha_j = 1$. Set $z := x - v_0$. From $z(t_j) = \|z\|_\infty$ for $j = 1, \dots, k$ it follows that

$$\|z\|_\infty^2 = \sum_{j=1}^{k} \alpha_j z^2(t_j) = \sum_{j=1}^{k} \alpha_j z(t_j)(z(t_j) - v(t_j))$$

$$\leq \|z\|_\infty \sum_{j=1}^{k} \alpha_j \max_j |z(t_j) - v(t_j)| \leq \|z\|_\infty \|z - v\|_\infty,$$

and hence $\|z\|_\infty \leq \|z - v\|_\infty = \|x - v_0 - v\|_\infty$. As V is a subspace, it follows that $\|x - v_0\|_\infty \leq \|x - v\|_\infty$ for all $v \in V$.

Let now v_0 be a best Chebyshev approximation of x w.r.t. V. We consider the mapping $A : T \to \mathbb{R}^n$, which is defined by $A(t) := z(t)(v_1(t), \dots, v_n(t))$. Let $C := A(E(z))$. We first show: $0 \in \operatorname{conv}(C)$: as a continuous image of the compact set $E(z)$ the set C is also compact. The convex hull of C is according to the theorem of Carathéodory (see Theorem 3.1.19) the image of the compact set

$$C^{n+1} \times \left\{ \alpha \in \mathbb{R}^{n+1} \,\middle|\, \sum_{j=1}^{n+1} \alpha_j = 1, \alpha_j \geq 0, j = 1, \dots, n+1 \right\}$$

under the continuous mapping $(c_1, \dots, c_{n+1}, \alpha) \mapsto \sum_{j=1}^{n+1} \alpha_j c_j$ and hence compact.

Suppose $0 \notin \operatorname{conv}(C)$, then due to the Strict Separation Theorem 3.9.17 there exists a $(a_1, \dots, a_n) \in \mathbb{R}^n$ such that for all $t \in E(z)$ we have

$$\sum_{i=1}^{n} a_i z(t) v_i(t) > 0.$$

For $v_* := \sum_{i=1}^{n} a_i v_i$ and $t \in E(z)$ it follows that $z(t)v_*(t) > 0$, contradicting the Kolmogoroff criterion, i.e. $0 \in \operatorname{conv}(C)$. Due to the theorem of Carathéodory (see Theorem 3.1.19) there exist k numbers $\alpha_1, \dots, \alpha_k \geq 0$ with $1 \leq k \leq n+1$ and $\sum_{j=1}^{k} \alpha_j = 1$, and k points $c_1, \dots, c_k \in C$, thus k points $t_1, \dots, t_k \in E(z)$, such that

$$0 = \sum_{j=1}^{k} \alpha_j c_j = \sum_{j=1}^{k} \alpha_j (x(t_j) - v_0(t_j)) \cdot \begin{pmatrix} v_1(t_j) \\ v_2(t_j) \\ \vdots \\ v_n(t_j) \end{pmatrix}.$$

As $v_1, \dots, v_n$ form a basis of V, the assertion follows. □

Remark 1.3.4. Using the point measures δ_{t_j}, $j = 1, \dots, k$ and the measure $\mu := \sum_{j=1}^{k} \alpha_j \delta_{t_j}$ the above condition can also be written as

$$\int_T (x(t) - v_0(t)) v(t) d\mu(t) = 0.$$

As a consequence of the above theorem we obtain

Theorem 1.3.5 (Theorem of de la Vallée-Poussin I). *Let V be an n-dimensional subspace of $C(T)$, let $x \in C(T) \setminus V$, and let v_0 be a best Chebyshev approximation of x w.r.t. V. Then there is a subset T_0 of $E(x - v_0)$, containing not more than $n+1$ points and for which the restriction $v_0|_{T_0}$ is a best approximation of $x|_{T_0}$ w.r.t. $V|_{T_0}$, i.e. for all $v \in V$ we have*

$$\max_{t\in T} |x(t) - v_0(t)| = \max_{t\in T_0} |x(t) - v_0(t)| \leq \max_{t\in T_0} |x(t) - v(t)|.$$

In the subsequent discussion we want to consider L^Φ-approximations in $C[a, b]$. We achieve particularly far-reaching assertions (uniqueness, behavior of zeros, error estimates) in the case of Haar subspaces.

1.3.2 Approximation in Haar Subspaces

Definition 1.3.6. Let T be a set. An n-dimensional subspace $V = \operatorname{span}\{v_1, \dots, v_n\}$ of the vector space X of the real functions on T is called a *Haar subspace*, if every not identically vanishing function $v \in V$ has at most $n - 1$ zeros in T.

Equivalent statements to the one in the above definition are

(a) For all $\{t_1, \dots, t_n\} \subset T$ with $t_i \neq t_j$ for $i \neq j$ we have

$$\begin{vmatrix} v_1(t_1) & \dots & v_1(t_n) \\ \vdots & \ddots & \vdots \\ v_n(t_1) & \dots & v_n(t_n) \end{vmatrix} \neq 0.$$

(b) To n arbitrary pairwise different points $t_i \in T$, $i = 1, \dots, n$, there is for any set of values s_i, $i = 1, \dots, n$, a unique $v \in V$ with $v(t_i) = s_i$ for $i = 1, \dots, n$, i.e. the interpolation problem for n different points is uniquely solvable.

Proof. From the definition (a) follows, because if $\det(v_i(t_j)) = 0$ there is a non-trivial linear combination $\sum_{i=1}^n a_i z_i$ of the rows of the matrix $(v_i(t_j))$ yielding the zero vector. Hence the function $\sum_{i=1}^n a_i v_i$ has $\{t_1, \dots, t_n\}$ as zeros.

Conversely from (a) the definition follows, because let $\{t_1, \dots, t_n\}$ be zeros of a $v \in V$ with $v \neq 0$, hence $v = \sum_{i=1}^n a_i v_i$ in a non-trivial representation, then we obtain for these $\{t_1, \dots, t_n\}$ a non-trivial representation of the zero-vector: $\sum_{i=1}^n a_i z_i = 0$, a contradiction.

However, (a) and (b) are equivalent, because the linear system of equations

$$\sum_{j=1}^n a_j v_j(t_i) = s_i, \quad i = 1, \dots, n,$$

has a unique solution, if and only if (a) holds. □

Examples for Haar subspaces:

(a) The algebraic polynomials of degree at most n form a Haar subspace of dimension $n+1$ on every real interval $[a,b]$.

(b) The trigonometric polynomials of degree at most n form a Haar subspace of dimension $2n+1$ on the interval $[0,2\pi]$.

(c) Let $\lambda_1,\dots,\lambda_n$ be pairwise different real numbers, then $V := \text{span}\{e^{\lambda_1 t},\dots,e^{\lambda_n t}\}$ is a Haar subspace of dimension n on every interval $[a,b]$.

In the sequel let V be an n-dimensional Haar subspace of $C[a,b]$.

Definition 1.3.7. The zero t_0 of a $v \in V$ is called *double*, if

(a) t_0 is in the interior of $[a,b]$

(b) v is in a neighborhood of t_0 non-negative or non-positive.

In any other case the zero t_0 is called simple.

Lemma 1.3.8. *Every not identically vanishing function $v \in V$ has – taking the multiplicity into account – at most $n-1$ zeros.*

Proof (see [112]). Let $t_1 < \cdots < t_m$ be the zeros of v on (a,b), let r be the number of zeros at the boundary of the interval and let k be the number of double zeros of v, then v has – according to the definition – $m-k+r$ simple zeros and we have: $r+m \le n-1$. Let further be $t_0 := a$ and $t_{m+1} := b$, then we define

$$\mu := \min_{0\le j\le m} \max_{t\in[t_j,t_{j+1}]} |v(t)|.$$

Apparently $\mu > 0$. Furthermore, there is a $v_1 \in V$ that assumes the value 0 in each simple zero of v and in each double zero the value 1 (or -1 respectively), provided that v is in a neighborhood of this zero non-negative (or non-positive respectively). Let $c > 0$ be chosen in such a way that

$$c \cdot \max_{t\in[a,b]} |v_1(t)| < \mu,$$

then the function $v_2 := v - c\cdot v_1$ has the following properties:

(a) each simple zero of v is a zero of v_2

(b) each double zero of v generates two zeros of v_2: to see this let $\bar{t}_j \in [t_j, t_{j+1}]$ be chosen, such that

$$\max_{t\in[t_j,t_{j+1}]} |v(t)| = |v(\bar{t}_j)|$$

for $j = 0,1,\dots,m$, then $|v(\bar{t}_j)| \ge \mu$.

Let now t_j be a double zero of v and let $v_1(t_j) = 1$, then we have

$$\begin{aligned} v_2(t_j) &= v(t_j) - c \cdot v_1(t_j) = -c \\ v_2(\bar{t}_j) &= v(\bar{t}_j) - c \cdot v_1(\bar{t}_j) \geq \mu - c \cdot |v_1(\bar{t}_j)| > 0 \\ v_2(\bar{t}_{j+1}) &= v(\bar{t}_{j+1}) - c \cdot v_1(\bar{t}_{j+1}) \geq \mu - c \cdot |v_1(\bar{t}_{j+1})| > 0. \end{aligned}$$

A corresponding result is obtained for $v_1(t_j) = -1$.

For sufficiently small c all these zeros of v_2 are different. Thus we finally obtain $r + m - k + 2k \leq n - 1$. □

Lemma 1.3.9. *Let $k < n$ and $\{t_1, \ldots, t_k\} \subset (a, b)$. Then there is a $v \in V$ which has precisely in these points a zero with a change of sign.*

Proof. If $k = n - 1$, then – according to the previous lemma – the interpolation problem $v(a) = 1$ and $v(t_i) = 0$ for $i = 1, \ldots, n - 1$ has the desired property, because none of these zeros of v can be a double zero.

For $k < n - 1$ we construct a sequence of points $(t_{k+1}^{(m)}, \ldots, t_{n-1}^{(m)})$ – with pairwise different components – in the open cuboid $(t_k, b)^{n-k-1}$ that converges componentwise to b. In an analogous way as in the case $k = n - 1$ we now determine for the points $a, t_1, \ldots, t_k, t_{k+1}^{(m)}, \ldots, t_{n-1}^{(m)}$ the solution $v^{(m)} = \sum_{i=1}^n a_i^{(m)} v_i$ of the interpolation problem $v(a) = 1$, $v(t_i) = 0$ for $i = 1, \ldots, k$ and $v(t_i^{(m)}) = 0$ for $i = k + 1, \ldots, n - 1$.

Let $\bar{a} := (\bar{a}_1, \ldots, \bar{a}_n)$ be a point of accumulation of the sequence $\frac{a^{(m)}}{\|a^{(m)}\|}$, where $a^{(m)} := (a_1^{(m)}, \ldots, a_n^{(m)})$. Then $\bar{v} := \sum_{i=1}^n \bar{a}_i v_i$ satisfies the requirements. □

Chebyshev's Alternant

We start our discussion with

Theorem 1.3.10 (Theorem of de la Vallée-Poussin II). *Let V be an n-dimensional Haar subspace of $C[a, b]$ and $x \in C[a, b] \setminus V$. If for a $v_0 \in V$ and for the points $t_1 < \cdots < t_{n+1}$ in $[a, b]$ the condition*

$$\operatorname{sign}(x - v_0)(t_i) = -\operatorname{sign}(x - v_0)(t_{i+1}) \quad \textit{for} \in \{1, \ldots, n\}$$

is satisfied, the estimate

$$\|x - v_0\|_\infty \geq \min_{v \in V} \|x - v\|_\infty \geq \min_{1 \leq i \leq n+1} |x(t_i) - v_0(t_i)|$$

follows.

Proof. Suppose for a $v_* \in V$ we have

$$\|x - v_*\|_\infty < \min_{1 \le i \le n+1} |x(t_i) - v_0(t_i)|,$$

then $0 \ne v_* - v_0 = (x - v_0) - (x - v_*)$ and $\operatorname{sign}(v_* - v_0)(t_i) = -\operatorname{sign}(v_* - v_0)(t_{i+1})$ for $i \in \{1, \dots, n\}$ follows. Thus $v_* - v_0$ has at least n zeros in $[a, b]$, a contradiction. □

Theorem 1.3.11 (Theorem of Chebyshev). *Let V be an n-dimensional Haar subspace of $C[a, b]$ and $x \in C[a, b] \setminus V$. Then $v_0 \in V$ is a best Chebyshev approximation of x w.r.t. V, if and only if there are $n + 1$ points $t_1 < \cdots < t_{n+1}$ in $[a, b]$ satisfying the alternant condition, i.e. $|x(t_i) - v_0(t_i)| = \|x - v_0\|_\infty$ and*

$$\operatorname{sign}(x - v_0)(t_i) = -\operatorname{sign}(x - v_0)(t_{i+1})$$

for $i \in \{1, \dots, n\}$.

Proof. Let v_0 be a best Chebyshev approximation. Due to the Characterization Theorem there are $k \le n+1$ points $\{t_1, \dots, t_k\} =: S \subset E(x - v_0)$ and k positive numbers $\alpha_1, \dots, \alpha_k$ such that

$$\sum_{j=1}^{k} \alpha_j (x(t_j) - v_0(t_j)) v(t_j) = 0 \quad \text{for all } v \in V,$$

and $v_0|_S$ is a best approximation of $x|_S$ w.r.t. $V|_S$. As V is a Haar subspace, $k = n + 1$ must hold, because otherwise one could interpolate $x|_S$, contradicting the properties of v_0. We now choose $0 \ne \bar{v} \in V$, having zeros in the $n - 1$ points $t_1, \dots, t_{i-1}, t_{i+2}, \dots, t_{n+1}$ (and no others). Then it follows that

$$\alpha_i (x(t_i) - v_0(t_i)) \bar{v}(t_i) + \alpha_{i+1} (x(t_{i+1}) - v_0(t_{i+1})) \bar{v}(t_{i+1}) = 0,$$

and $\operatorname{sign} \bar{v}(t_i) = \operatorname{sign} \bar{v}(t_{i+1})$, because $t_{i-1} < t_i < t_{i+1} < t_{i+2}$. Conversely, if the alternant condition is satisfied, then due to Theorem II of de la Vallée-Poussin v_0 is a best approximation of x w.r.t. V. □

Corollary 1.3.12. *Let V be an n-dimensional Haar subspace of $C[a, b]$ and $x \in C[a, b] \setminus V$. Then the best Chebyshev approximation v_0 of x w.r.t. V is uniquely determined and the difference function $x - v_0$ has at least n zeros.*

Proof. Let v_1 and v_2 be best Chebyshev approximations of x w.r.t. V. Then $v_0 := \frac{1}{2}(v_1 + v_2)$ is also a best approximation and due to the theorem of Chebyshev there are $n + 1$ points $t_1 < \cdots < t_{n+1}$ in $[a, b]$ with $|x(t_i) - v_0(t_i)| = \|x - v_0\|_\infty$ and $\operatorname{sign}(x - v_0)(t_i) = -\operatorname{sign}(x - v_0)(t_{i+1})$ for $i \in \{1, \dots, n\}$. Then we obtain

$$\|x - v_0\|_\infty \ge \frac{1}{2} |(x - v_1)(t_i)| + \frac{1}{2} |(x - v_2)(t_i)| \ge |(x - v_0)(t_i)| = \|x - v_0\|_\infty.$$

It follows that $|(x - v_j)(t_i)| = \|x - v_0\|_\infty$ for $i = 1, \dots, n+1$ and $j = 1, 2$. Furthermore $(x - v_1)(t_i) = (x - v_2)(t_i)$ for $i = 1, \dots, n+1$, because if for an $i_0 \in \{1, \dots, n+1\}$

$$\operatorname{sign}((x(t_{i_0}) - v_1(t_{i_0})) = -\operatorname{sign}(x(t_{i_0} - v_2(t_{i_0})),$$

then $x(t_{i_0}) - \frac{1}{2}(v_1(t_{i_0}) + v_2(t_{i_0})) = x(t_{i_0}) - v_0(t_{i_0}) = 0$, a contradiction. But from $(x - v_1)(t_i) = (x - v_2)(t_i)$ for $i = 1, \dots, n+1$ it follows that $v_2(t_i) - v_1(t_i) = 0$ for $i = 1, \dots, n+1$, thus $v_1 = v_2 = v_0$, because V is a Haar subspace.

Due to $\operatorname{sign}(x - v_0)(t_i) = -\operatorname{sign}(x - v_0)(t_{i+1})$ for $i \in \{1, \dots, n\}$ and $|x(t_i) - v_0(t_i)| = \|x - v_0\|_\infty$ for $i \in \{1, \dots, n+1\}$ we obtain using the continuity of $x - v_0$ that $x - v_0$ has at least n zeros. □

Remark 1.3.13. The set $\{t_1 < \dots < t_{n+1}\}$ is called Chebyshev's alternant. The determination of a best approximation can now be reduced to the determination of this set and one can pursue the following strategy (see *Remez algorithm* see [22]): choose $n + 1$ points and try by exchanging points to achieve improved bounds in the sense of the theorem of de la Valée-Poussin. This theorem then provides a stop criterion.

Example 1.3.14 (Chebyshev Polynomials). Let $P_n(t) := \cos(n \arccos t)$ on $[-1, 1]$. P_n is a polynomial of degree n, because from the addition theorem $\cos\alpha + \cos\beta = 2\cos\frac{1}{2}(\alpha+\beta)\cos\frac{1}{2}(\alpha-\beta)$ the recursion formula $P_{n+1}(t) = 2tP_n(t) - P_{n-1}(t)$ follows, where $P_0(t) = 1$ and $P_1(t) = t$ for $t \in [-1, 1]$. For $t_i := \cos\frac{i}{n}\pi$ we have $P_n(t_i) = \cos i\pi = (-1)^i$ for $i = 0, 1, \dots, n$. If we define

$$T_n(t) := \frac{1}{2^{n-1}} P_n(t),$$

then the leading coefficient of T_n is equal to 1 and we have $T_n(t_i) = \frac{1}{2^{n-1}} \cdot (-1)^i$, i.e. $t_0, \dots, t_n$ form a Chebyshev's alternant for the approximation of $x(t) = t^n$ by the subspace of the polynomials of degree $n - 1$ on $[-1, 1]$. The difference function $x - v_0 = T_n$ has n zeros.

Uniqueness: The Theorem of Jackson in L^Φ

According to Jackson an alternative holds for the L^1 approximation which is generalized by the following theorem (see [51]).

Theorem 1.3.15 (Jackson's Alternative). *Let Φ be a finite, definite Young function and let v_0 be a best f^Φ or $\|\cdot\|_{(\Phi)}$ approximation of $x \in C[a,b] \setminus V$. Then $x - v_0$ has at least n zeros or the measure of the zeros is positive.*

If Φ is differentiable at 0, then $x - v_0$ has at least n zeros with change of sign.

Proof. Let at first v_0 be a best f^Φ approximation and let Z be the set of zeros of $x-v_0$. If $\mu(Z)=0$ then according to Theorem 1.2.2 for all $h\in V$

$$\int_{\{h>0\}\setminus Z} h\Phi'_+(x-v_0)d\mu + \int_{\{h<0\}\setminus Z} h\Phi'_-(x-v_0)d\mu \geq 0. \tag{1.5}$$

If $x-v_0$ has only $k<n$ zeros with change of sign, say $t_1,\dots,t_k$, we can due to Lemma 1.3.9 find a $v_1\in V$ such that

$$\operatorname{sign} v_1(t) = -\operatorname{sign}(x(t)-v_0(t)) \quad \text{for all } t\in[a,b]\setminus Z.$$

As Φ is symmetrical and definite, we have for $s\in\mathbb{R}\setminus\{0\}$

$$\operatorname{sign}\Phi'_+(s) = \operatorname{sign}\Phi'_-(s) = -\operatorname{sign}\Phi'_+(-s) = -\operatorname{sign}\Phi'_-(-s).$$

Hence for $h=v_1$ both integrands in (1.5) are negative, a contradiction.

If Φ is differentiable at 0, then $\Phi'(0)=0$ and hence also $\int_Z h\Phi'(x-v_0)d\mu=0$. As $x\neq v_0$ and the measure is equivalent to the Lebesgue measure, we can repeat the argument used in the first part of the proof.

Let now v_0 be a best $\|\cdot\|_{(\Phi)}$ approximation of $x\in C[a,b]\setminus V$, and let $c:=\|x-v_0\|_{(\Phi)}$. For $\Phi_1(s):=\Phi(\frac{s}{c})$ apparently v_0 is a best f^{Φ_1} approximation of x, because for all $v\in V$ we have

$$1=\int_a^b \Phi\left(\frac{x-v_0}{c}\right)d\mu \leq \int_a^b \Phi\left(\frac{x-v}{c}\right)d\mu.$$

Repeating the above argument for Φ_1, we obtain the corresponding property for the best approximation in the Luxemburg norm. □

The uniqueness theorem of Jackson for the L^1-approximation (see [1]) can be generalized to the best L^Φ approximation under the following condition:

Theorem 1.3.16. *Let V be a finite-dimensional Haar subspace of $C[a,b]$ and let Φ be a finite, definite Young function. Then every $x\in C[a,b]\setminus V$ has a unique best f^Φ and $\|\cdot\|_{(\Phi)}$ approximation.*

Proof. Let v_1,v_2 be best f^Φ approximations of x w.r.t. V, then also $v_0:=\frac{1}{2}(v_1+v_2)$, and we have

$$\begin{aligned}0&=\frac{1}{2}\int_a^b\Phi(x-v_1)d\mu+\frac{1}{2}\int_a^b\Phi(x-v_2)d\mu-\int_a^b\Phi(x-v_0)d\mu\\&=\int_a^b\left[\frac{1}{2}\Phi(x-v_1)+\frac{1}{2}\Phi(x-v_2)d\mu-\Phi(x-v_0)\right]d\mu.\end{aligned}$$

As Φ is convex, the integrand on the right-hand side is non-negative. As $x - v_0$ is continuous we have for all $t \in [a, b]$

$$\frac{1}{2}\Phi(x - v_1)(t) + \frac{1}{2}\Phi(x - v_2)(t) - \Phi(x - v_0)(t) = 0.$$

Due to Jackson's Alternative 1.3.15 $x - v_0$ has at least n zeros. As Φ is only zero at 0, the functions $x - v_1$ and $x - v_2$ and hence $v_1 - v_2$ have at least the same zeros. As V is a Haar subspace, $v_1 = v_2$ follows. In order to obtain uniqueness for the Luxemburg norm, we choose as we have done above $c := \|x - v_0\|_{(\Phi)}$ and $\Phi_1(s) := \Phi(\frac{s}{c})$. □

The following generalization of the theorem of Bernstein provides an error estimate for linear approximations in $C[a, b]$ w.r.t. the $(n + 1)$-dimensional Haar subspace of the polynomials of degree $\leq n$:

Error Estimate: The Theorem of Bernstein in L^Φ

Lemma 1.3.17. *Let the functions $x, y \in C[a, b]$ have derivatives up to the order $n+1$ on $[a, b]$ and let for the $(n + 1)$-st derivatives $x^{(n+1)}$ and $y^{(n+1)}$ satisfy*

$$|x^{(n+1)}(t)| < y^{(n+1)}(t) \quad \text{for all } t \in [a, b]. \tag{1.6}$$

Then the following inequality holds:

$$|r_x(t)| \leq |r_y(t)| \quad \text{for all } t \in [a, b], \tag{1.7}$$

where $r_x := x - x_n$ and $r_y := y - y_n$ denote the difference functions for the interpolation polynomials x_n and y_n w.r.t. x and y resp. for the same interpolation nodes $t_0 < t_1 < \cdots < t_n$ in $[a, b]$.

Proof. Following the arguments of Tsenov [106]: for $t = t_i$, $i = 0, 1, \ldots, n$, the inequality is trivially satisfied. Let now $t \neq t_i$, $i = 0, 1, \ldots, n$. The auxiliary function

$$z(s) := \begin{vmatrix} x(t) - x_n(t) & x(s) - x_n(s) \\ y(t) - y_n(t) & y(s) - y_n(s) \end{vmatrix}$$

has zeros at $s = t$ and at the $n + 1$ interpolation nodes. Hence, due to the theorem of Rolle there is $s_0 \in [a, b]$, where the $(n + 1)$-st derivative

$$z^{(n+1)}(s) := \begin{vmatrix} x(t) - x_n(t) & x^{(n+1)}(s) \\ y(t) - y_n(t) & y^{(n+1)}(s) \end{vmatrix}$$

is equal to 0, i.e.

$$x^{(n+1)}(s_0)r_y(t) = y^{(n+1)}(s_0)r_x(t).$$

If now $x^{(n+1)}(s_0) = 0$, then $r_x(t) = 0$, thus the Inequality (1.7) is trivially satisfied. If $x^{(n+1)}(s_0) \neq 0$, then using (1.6) the assertion follows. □

In the sequel we will present a generalization of the subsequent theorem of S. N. Bernstein.

Theorem 1.3.18 (Theorem of Bernstein). *Let the functions $x, y \in C[a,b]$ have derivatives of order $n+1$ in $[a,b]$ and let the $(n+1)$-st derivatives $x^{(n+1)}$ and $y^{(n+1)}$ satisfy*

$$|x^{(n+1)}(t)| \leq y^{(n+1)}(t) \quad \text{for all } t \in [a,b].$$

The following inequality holds:

$$E_n(x) \leq E_n(y),$$

where $E_n(z)$ denotes the distance of z to the space of polynomials of degree $\leq n$ w.r.t. the maximum norm.

Definition 1.3.19. A functional $f : C[a,b] \to \mathbb{R}$ is called *monotone*, if $x, y \in C[a,b]$ and $|x(t)| \leq |y(t)|$ for all $t \in [a,b]$ implies $f(x) \leq f(y)$.

Theorem 1.3.20. *Let $f : C[a,b] \to \mathbb{R}$ be monotone. Let the functions $x, y \in C[a,b]$ have derivatives up to order $n+1$ in $[a,b]$ and let the $(n+1)$-st derivatives $x^{(n+1)}$ and $y^{(n+1)}$ satisfy*

$$|x^{(n+1)}(t)| < y^{(n+1)}(t) \quad \text{for all } t \in [a,b].$$

Let further V be the subspace of the polynomials of degree $\leq n$ and let for $v_x, v_y \in V$

(a) $f(x - v_x) = \inf\{f(x-v) \mid v \in V\}$ *and* $f(y - v_y) = \inf\{f(y-v) \mid v \in V\}$

(b) *the functions $x - v_x$ and $y - v_y$ have at least $n+1$ zeros in $[a,b]$.*

Then the inequality $f(x - v_x) \leq f(x - v_y)$ holds.

Proof. If we choose $n+1$ zeros of $y - v_y$ as interpolation nodes and denote by $\bar{v}_x$ the interpolation polynomial belonging to x and these interpolation nodes , then, due to the previous lemma $|x(t) - \bar{v}_x(t)| \leq |y(t) - v_y(t)|$ for all $t \in [a,b]$. The monotonicity of f then implies

$$f(x - v_x) \leq f(x - \bar{v}_x) \leq f(y - v_y).$$ □

Remark 1.3.21. The modular f^Φ as well as the Luxemburg norm $\|\cdot\|_{(\Phi)}$ satisfy the monotonicity requirement of the previous theorem.

Moreover, Jackson's Alternative (see Theorem 1.3.15) provides for Φ finite and definite corresponding information about the set of zeros of the difference function.

Corollary 1.3.22. *Let f be the Luxemburg norm and V be as in the previous theorem and let $y(t) = t^{n+1}$ and let $E_n^f(z) := \inf_{v \in V} f(z - v)$. If for $x \in C[a,b]$ we have*

$$|x^{(n+1)}(t)| < \beta \quad \text{for all } t \in [a,b],$$

then the following inequality holds:

$$E_n^f(x) \leq \frac{\beta E_n^f(y)}{(n+1)!}.$$

Corollary 1.3.23. *Let V be the subspace of polynomials of degree $\leq n$ on $[-1,1]$, let $y(t) = t^{n+1}$, and let $E_n(z) := \inf_{v \in V} \|z - v\|_\infty$. If for $x \in C[a,b]$*

$$|x^{(n+1)}(t)| < \beta \quad \textit{for all } t \in [-1,1]$$

then the following inequality holds:

$$E_n(x) \leq \frac{\beta E_n(y)}{(n+1)!} = \frac{\beta}{2^n(n+1)!},$$

because $E_n(y) = \|T_{n+1}\|_\infty = \frac{1}{2^n}$ (see Example 1.3.14).

1.4 Discrete L^Φ-approximations

Let T be a set containing m elements, $T = \{t_1, \ldots, t_m\}$ and $x : T \to \mathbb{R}$. Hence we can view x as a point in $\mathbb{R}^m$ and consider the approximation problem in $\mathbb{R}^m$. The norms that are used in this section can be understood as norms on the $\mathbb{R}^m$.

For a given point x in $\mathbb{R}^m$ we try to find a point in a certain subset M of $\mathbb{R}^m$ that among all points of M has the least distance to x, where the distance is understood in the sense of the (discrete) Luxemburg norm

$$\|x - y\|_{(\Phi)} = \inf\left\{c > 0 \,\middle|\, \sum_{t \in T} \Phi\left(\frac{x(t) - y(t)}{c}\right) \leq 1\right\}.$$

In the subsequent discussion we will first consider the special case of the discrete Chebyshev approximation which frequently results from the discretization of a continuous problem. On the other hand, due to the Theorems of de la Vallée-Poussin (see above) the continuous problem can always be viewed as a discrete approximation.

1.4.1 Discrete Chebyshev Approximation

For a given point x in $\mathbb{R}^m$ we look for a point in a subset M of $\mathbb{R}^m$ that has the least distance from x where the distance is defined as

$$\|x - y\|_\infty = \max_{t \in T} |x(t) - y(t)|. \tag{1.8}$$

The Strict Approximation

J. R. Rice has introduced the *strict approximation* using the notion of critical points (see Rice [98]). In the sequel the strict approximation is defined using the *universal extreme point* (see also Descloux [25]).

Let $T = \{t_1, \ldots, t_m\}$, V be a subspace of $C(T) = \mathbb{R}^m$ and $x \in C(T) \setminus V$. Let $P_V(x)$ denote the set of all best approximations of x w.r.t. V.

Definition 1.4.1. A point $t \in T$ is called *extreme point* of a best approximation v_0 of x w.r.t. V, if

$$|x(t) - v_0(t)| = \|x - v_0\|_\infty.$$

By $E^T_{v_0}$ we denote the set of all extreme points of a best approximation v_0 of x.

Lemma 1.4.2. *The set*

$$D = \bigcap_{v_0 \in P_V(x)} E^T_{v_0}$$

of the extreme points common to all best approximations (we will call these points universal*) is non-empty.*

Proof. Suppose $D = \emptyset$, then we could for each i, $i = 1, \ldots, m$, find a $v_i \in P_V(x)$ such that

$$|v_i(t_i) - x(t_i)| < \min_{v \in V} \|v - x\|_\infty.$$

But then the function

$$\frac{1}{m} \sum_{i=1}^{m} v_i \in V$$

would approximate the function x better than v_i $(i = 1, 2, \ldots, m)$. As, however $v_i \in P_V(x)$ this would mean a contradiction. □

Let $x \notin V$. A universal extreme point $t \in D$ is called *negative* (or *positive* resp.) w.r.t. a best approximation v_0, if

$$\operatorname{sign}(x(t) - v_0(t)) = -1 \quad (\text{resp.} + 1).$$

Lemma 1.4.3. *Let $x \notin V$. If a universal extreme point is negative (or positive) w.r.t. a best approximation, then it is negative (or positive) w.r.t. every best approximation. Therefore it is simply called negative (or positive).*

Proof. If for $v_1, v_2 \in P_V(x)$ and a $t \in D$

$$x(t) - v_1(t) = -x(t) + v_2(t),$$

then for

$$\frac{1}{2}(v_1 + v_2) \in P_V(x)$$

$$x(t) - \frac{1}{2}(v_1(t) + v_2(t)) = 0,$$

contrary to t being a universal extreme point. □

Let now V be an n-dimensional subspace $(n < m)$ of $C(T)$. Every basis $u_1, \ldots, u_n$ of the subspace V defines a mapping $B : T \to \mathbb{R}^n$,

$$B(t) = \begin{pmatrix} u_1(t) \\ u_2(t) \\ \vdots \\ u_n(t) \end{pmatrix}. \tag{1.9}$$

For every set $H \subset \mathbb{R}^n$ let $[H]$ denote the smallest subspace of $\mathbb{R}^n$, containing H.

Lemma 1.4.4. *Let $P \subset T$, $Q = B^{-1}([B(P)])$, $v_1, v_2 \in P_V(x)$ and*

$$v_1(t) = v_2(t) \quad \textit{for } t \in P,$$

then also

$$v_1(t) = v_2(t) \quad \textit{for } t \in Q.$$

Proof. Let v_1 and v_2 be represented in the above basis:

$$v_k = \sum_{i=1}^{n} a_i^{(k)} u_i, \quad k = 1, 2, \ \ell := |P| \text{ and w.l.o.g.}$$

$$P = \{t_1, t_2, \ldots, t_\ell\}.$$

For $t \in Q$ we have

$$B(t) = \begin{pmatrix} u_1(t) \\ u_2(t) \\ \vdots \\ u_n(t) \end{pmatrix} = \sum_{j=1}^{\ell} c_j \cdot \begin{pmatrix} u_1(t_j) \\ u_2(t_j) \\ \vdots \\ u_n(t_j) \end{pmatrix}$$

and hence

$$u_i(t) = \sum_{j=1}^{\ell} c_j u_i(t_j),$$

where $c_1, c_2, \dots, c_\ell$ are real constants, thus

$$v_1(t) = \sum_{i=1}^{n} a_i^{(1)} u_i(t) = \sum_{i=1}^{n} a_i^{(1)} \sum_{j=1}^{\ell} c_j u_j(t_j) = \sum_{j=1}^{\ell} c_j \sum_{i=1}^{n} a_i^{(1)} u_i(t_j)$$

$$= \sum_{j=1}^{\ell} c_j \sum_{i=1}^{n} a_i^{(2)} u_i(t_j) = \sum_{i=1}^{n} a_i^{(2)} \sum_{j=1}^{\ell} c_j u_i(t_j) = v_2(t). \qquad \square$$

Lemma 1.4.5. *Let T' be a proper subset of T and $q : T' \to \mathbb{R}$. If there is a best approximation of x which extends q, then among these extensions there is at least one, which approximates x best on $T \setminus T'$. Thus, if*

$$W = \{v_0 \in P_V(x) \mid v_0|_{T'} = q\} \neq \emptyset,$$

then also

$$Z = \{v_1 \in W \mid \|v_1 - x\|_{T\setminus T'} \leq \|v_0 - x\|_{T\setminus T'} \text{ for all } v_0 \in W\} \neq \emptyset.$$

Furthermore

$$\bigcap_{v_1 \in Z} E_{v_1}^{T\setminus T'} \neq \emptyset.$$

Proof. The function $N : W \to \mathbb{R}$ with $v_0 \mapsto \|v_0 - x\|_{T\setminus T'}$ assumes its minimum on the convex and compact set W, hence $Z \neq \emptyset$. The proof of the second part can be performed in analogy to the proof of Lemma 1.4.2, because W is convex. $\square$

The *strict approximation* is characterized by the following constructive definition: Let

$$V_0 = P_V(x) = \{v_0 \in V \mid d_0 = \|v_0 - x\|_\infty \leq \|v - x\|_\infty \text{ for all } v \in V\}$$

be the set of all best Chebyshev approximations of x w.r.t. V and let

$$D_0 = \bigcap_{v_0 \in V_0} E_{v_0}^T$$

be the set of the universal extreme points (see Lemma 1.4.2) and

$$T_0 = \{t \in T \mid t \in B^{-1}([B(D_0)])\}.$$

All best approximations v_0 of x are identical in all points of T_0 (see Lemma 1.4.4). Among the $v_0 \in P_V(x)$ we now look for those, which are best approximations of x on $T \setminus T_0$

$$V_1 = \{v_1 \in V_0 \mid d_1 = \|v_1 - x\|_{T\setminus T_0} \leq \|v_0 - x\|_{T\setminus T_0} \text{ for all } v_0 \in V_0\},$$

$$D_1 = \bigcap_{v_1 \in V_1} E_{v_1}^{T\setminus T_0} \quad \text{(see Lemma 1.4.5)}$$

and

$$T_1 = \{t \in T \mid t \in B^{-1}([B(D_0 \cup D_1)])\}.$$

Furthermore we have

$$V_2 = \{v_2 \in V_1 \mid d_2 = \|v_2 - x\|_{T\setminus T_1} \leq \|v_1 - x\|_{T\setminus T_1} \text{ for all } v_1 \in V_1\},$$
$$D_2 = \bigcap_{v_2 \in V_2} E_{v_2}^{T\setminus T_1}$$

and

$$T_2 = \{t \in T \mid t \in B^{-1}([B(D_0 \cup D_1 \cup D_2)])\}.$$

We observe: if $T \setminus T_i$ is non-empty, T_i is a proper subset of T_{i+1}: in fact $D_{i+1} \subset T_{i+1}$, but $D_{i+1} \cap T_i = \emptyset$. The procedure is continued until $T_k = T$. From the construction, using Lemma 1.4.5, it becomes apparent that for $v_i^{(1)}, v_i^{(2)} \in V_i$ we obtain: $v_i^{(1)}|_{T_i} = v_i^{(2)}|_{T_i}$.

Hence V_k consists of exactly one element, and this element we call the *strict approximation* of x w.r.t. V.

1.4.2 Linear L^Φ-approximation for Finite Young Functions

Uniqueness and Differentiability

Before we turn our attention towards the numerical computation, we want to discuss a few fundamental questions:

(a) How can the uniqueness of minimal solutions w.r.t. $\|\cdot\|_{(\Phi)}$ be described?

(b) What can be said about the differentiation properties of the norm $\|\cdot\|_{(\Phi)}$?

Happily, these properties can be easily gleaned from the Young functions:

Strict convexity and differentiability of the Young functions carry over to strict convexity and differentiability of the corresponding norms. The subsequent theorem gives a precise formulation of this statement, which we will – in a more general and enhanced form – treat in Chapter 8:

Theorem 1.4.6. *Let $\Phi : \mathbb{R} \to \mathbb{R}$ be a Young function.*

(a) *If Φ is differentiable, the corresponding Luxemburg norm is differentiable and we have*

$$\nabla \|x\|_{(\Phi)} = \frac{\Phi'\left(\frac{x(\cdot)}{\|x\|_{(\Phi)}}\right)}{\sum_{t \in T} \frac{x(t)}{\|x\|_{(\Phi)}} \Phi'\left(\frac{x(t)}{\|x\|_{(\Phi)}}\right)}. \tag{1.10}$$

(b) *If Φ is strictly convex, so is the corresponding Luxemburg norm.*

Proof. For the function $F : \mathbb{R}^m \times (0, \infty) \to \mathbb{R}$ with

$$(x, c) \mapsto F(x, c) := \sum_{t \in T} \Phi\Big(\frac{x(t)}{c}\Big) - 1$$

we have according to the definition of the Luxemburg norm

$$F(x, \|x\|_{(\Phi)}) = 0.$$

According to the implicit function theorem we only need to show the regularity of the partial derivative w.r.t. c in order to establish the differentiability of the norm. We have

$$\frac{\partial}{\partial c} F(x, c) = \sum_{t \in T} -\frac{x(t)}{c^2} \Phi'\Big(\frac{x(t)}{c}\Big) = -\frac{1}{c} \sum_{t \in T} \frac{x(t)}{c} \Phi'\Big(\frac{x(t)}{c}\Big).$$

Due to the subgradient inequality (see Inequality (3.3)) we have for arbitrary $t \in \mathbb{R}$

$$\Phi'(t)(0 - t) \leq \Phi(0) - \Phi(t),$$

i.e.

$$\Phi(t) \leq \Phi'(t)t. \tag{1.11}$$

As $\sum_{t \in T} \Phi(\frac{x(t)}{\|x\|_{(\Phi)}}) = 1$ we obtain

$$\sum_{t \in T} \frac{x(t)}{c} \Phi'\Big(\frac{x(t)}{c}\Big) \geq 1 \quad \text{for } c = \|x\|_{(\Phi)}. \tag{1.12}$$

Thus the regularity has been established and for the derivative we obtain using the chain rule

$$0 = \frac{d}{dx} F(x, \|x\|_{(\Phi)}) = \frac{\partial}{\partial x} F(x, c) + \frac{\partial}{\partial c} F(x, c) \nabla \|x\|_{(\Phi)},$$

hence

$$-\frac{1}{c}\Big(\sum_{t \in T} \frac{x(t)}{c} \Phi'\Big(\frac{x(t)}{c}\Big)\Big) \nabla \|x\|_{(\Phi)} + \Phi'\Big(\frac{x(\cdot)}{\|x\|_{(\Phi)}}\Big) \cdot \frac{1}{c} = 0$$

for $c = \|x\|_{(\Phi)}$ and hence the first assertion.

In order to show the second part of the theorem we choose $x, y \in \mathbb{R}^m$ with $\|x\|_{(\Phi)} = \|y\|_{(\Phi)} = 1$ and $x \neq y$. Then $\sum_{t \in T} \Phi(x(t)) = 1$ and $\sum_{t \in T} \Phi(y(t)) = 1$ and hence

$$1 = \sum_{t \in T} \Big(\frac{1}{2} \Phi(x(t)) + \frac{1}{2} \Phi(y(t))\Big) > \sum_{t \in T} \Phi\Big(\frac{x(t) + y(t)}{2}\Big),$$

because due to the strict convexity of Φ the strict inequality must hold for at least one of the terms in the sum. Thus $\|x + y\|_{(\Phi)} < 2$. □

Approximation in the Luxemburg Norm

Let Φ now be a twice continuously differentiable Young function. The set w.r.t. which we approximate is chosen to be an n-dimensional subspace V of $\mathbb{R}^m$ with a given basis $\{v_1, \ldots, v_n\}$ (*linear approximation*).

The best approximation y^Φ to be determined is represented as a linear combination $\sum_{i=1}^n a_i v_i$ of the basis vectors, where $\{a_1, \ldots, a_n\}$ denotes the real coefficients to be computed. The determination of the best approximation corresponds to a minimization problem in $\mathbb{R}^n$ for the unknown vector $a = (a_1, \ldots, a_n)$. The function to be minimized in this context is then:

$$\begin{aligned} p : \mathbb{R}^n &\to \mathbb{R} \\ a \mapsto p(a) &:= \left\| x(t) - \sum_{t \in T} a_i v_i(t) \right\|_{(\Phi)}. \end{aligned}$$

This problem we denote as a (linear) *approximation w.r.t. the Luxemburg norm.*

System of Equations for the Coefficients

For a given Φ we choose for reasons of conciseness the following abbreviations:

$$\begin{aligned} z(a,t) &:= \frac{x(t) - \sum_{l=1}^n a_l v_l(t)}{\|x - \sum_{l=1}^n a_l v_l\|_{(\Phi)}} \\ \gamma(a) &:= \sum_{t \in T} z(a,t)\Phi'(z(a,t)). \end{aligned}$$

The number $\gamma(a)$ is because of Inequality (1.11) always greater or equal to 1. Setting the gradient of the function p to zero leads to the following system of equations:

$$\nabla p(a)_i = -\frac{1}{\gamma(a)} \sum_{t \in T} v_i(t)\Phi'(z(a,t)) = 0, \quad \text{for } i = 1, \ldots, n.$$

For the solution of this system of non-linear equations the methods in Chapter 4 can be applied. For the elements of the second derivative matrix (Hessian) of p one obtains by differentiation of the gradient using the chain rule ($\lambda(a) := (\gamma(a) \cdot \|x - \sum_{l=1}^n a_l v_l\|_{(\Phi)})^{-1}$)

$$\frac{\partial^2}{\partial a_i \partial a_j} p(a) = \lambda(a) \sum_{t \in T} (v_i(t) + \nabla p(a)_i z(a,t))(v_j(t) + \nabla p(a)_j z(a,t))\Phi''(z(a,t)).$$

Regularity of the Second Derivative

In order to obtain rapid convergence of numerical algorithms for computing the above best approximation, the regularity of the second derivative (Hessian) at the solution is required. The richness of the available Young functions will enable us to guarantee this property. At the solution the gradient is equal to zero. Hence here the second derivative has the following structure:

$$\frac{\partial^2}{\partial a_i \partial a_j} p(a) = \frac{1}{\gamma(a) \cdot \|x - \sum_{l=1}^n a_l v_l\|_{(\Phi)}} \sum_{t \in T} v_i(t) v_j(t) \Phi''(z(a,t)).$$

It turns out that it is sufficient to choose Φ in such a way that Φ'' has no zeros, in order to guarantee positive definiteness of the second derivative of p everywhere, because the second derivative is a positive multiple of Gram's matrix of the basis vectors $\{v_1, \ldots, v_n\}$ w.r.t. the scalar product

$$\langle u, w \rangle_z := \sum_{t \in T} u(t) w(t) \Phi''(z(a,t)).$$

In particular the unique solvability of the linear systems, occurring during the (damped) Newton method, is guaranteed.

Remark. This property is not provided for the l^p norms with $p > 2$.

In the expression for the gradient as well as in the Hessian of the Luxemburg norm the Luxemburg norm of the difference function occurs $\|x - \sum_{l=1}^n a_l v_l\|_{(\Phi)}$. The computation of this quantity can be performed as the one-dimensional computation of a zero of the function g:

$$c \mapsto g(c) := \sum_{t \in T} \Phi\left(\frac{x(t) - \sum_{i=1}^n a_i v_i(t)}{c}\right) - 1.$$

Instead of decoupling the one-dimensional computation of the zero in every iteration step for the determination of the coefficient vector a, we can by introducing the additional unknown $a_{n+1} := \|x - \sum_{l=1}^n a_l v_l\|_{(\Phi)}$, consider the following system of equations for the unknown coefficient vector $a := (a_1, \ldots, a_n, a_{n+1}) \in \mathbb{R}^{n+1}$:

$$F_i(a) := \sum_{t \in T} v_i(t)\, \Phi'\left(\frac{x(t) - \sum_{j=1}^n a_j v_j(t)}{a_{n+1}}\right) = 0, \quad \text{for } i = 1, \ldots, n \tag{1.13}$$

$$F_{n+1}(a) := \sum_{t \in T} \Phi\left(\frac{x(t) - \sum_{j=1}^n a_j v_j(t)}{a_{n+1}}\right) - 1 = 0. \tag{1.14}$$

For the Jacobian we obtain

$$F_{i,j}(a) = \sum_{t\in T} v_i(t)v_j(t)\,\Phi''\Big(\frac{x(t)-\sum_{j=1}^n a_j v_j(t)}{a_{n+1}}\Big)$$

$$\text{for } i = 1,\dots,n \text{ and } j = 1,\dots,n$$

$$F_{n+1,j}(a) = -\frac{1}{a_{n+1}}\sum_{t\in T} v_j(t)\Phi'\Big(\frac{x(t)-\sum_{j=1}^n a_j v_j(t)}{a_{n+1}}\Big) = -\frac{1}{a_{n+1}}F_j(a)$$

$$\text{for } j = 1,\dots,n$$

$$F_{n+1,n+1}(a) = -\frac{1}{a_{n+1}}\sum_{t\in T}\frac{x(t)-\sum_{j=1}^n a_j v_j(t)}{a_{n+1}}\Phi'\Big(\frac{x(t)-\sum_{j=1}^n a_j v_j(t)}{a_{n+1}}\Big).$$

One obtains the regularity of the corresponding Jacobian at the solution for Φ with $\Phi'' > 0$ in the following way: if we omit in the Jacobian the last row and the last column, the resulting matrix is apparently non-singular. The first n elements of the last row are equal to zero, because $F_i(a) = 0$ for $i = 1,\dots,n$ at the solution, and the $(n+1)$-st element is different from 0, because we have $t\Phi'(t) \geq \Phi(t)$ (see above), hence:

$$\sum_{t\in T}\frac{x(t)-\sum_{j=1}^n a_j v_j(t)}{a_{n+1}}\Phi'\Big(\frac{x(t)-\sum_{j=1}^n a_j v_j(t)}{a_{n+1}}\Big) \geq \sum_{t\in T}\Phi\Big(\frac{x(t)-\sum_{j=1}^n a_j v_j(t)}{a_{n+1}}\Big) = 1.$$

If the determinant of the total Jacobi matrix is expanded using the Laplace expansion w.r.t. the last row, then apparently the determinant is non-zero.

1.5 Determination of the Linear L^Φ-approximation

In the sequel we will generalize our point of view to arbitrary finite measures.

1.5.1 System of Equations for the Coefficients

Let Φ be a differentiable Young function and let V be an n-dimensional subspace of M^Φ with a given basis $\{v_1,\dots,v_n\}$. From the characterization in Corollary 1.2.3 we obtain the following system of equations:

$$\int_T v_i\Phi'\Big(x - \sum_{j=1}^n a_j v_j\Big)d\mu = 0 \quad \text{for } i = 1,\dots,n \tag{1.15}$$

for the modular approximation and

$$\int_T v_i \Phi'\left(\frac{x - \sum_{j=1}^n a_j v_j}{\|x - \sum_{j=1}^n a_j v_j\|_{(\Phi)}}\right) d\mu = 0 \quad \text{for } i = 1, \dots, n$$

for the approximation in the Luxemburg norm. If we introduce as in the case of the discrete approximation $a_{n+1} := \|x - \sum_{j=1}^n a_j v_j\|_{(\Phi)}$ as a further variable, we obtain the additional equation

$$\int_T \Phi\left(\frac{x - \sum_{j=1}^n a_j v_j}{a_{n+1}}\right) d\mu - 1 = 0.$$

The statements about the regularity of the Jacobian for twice continuously differentiable Young functions carry over. Thus the methods in Chapter 4 are applicable (if we are able to compute the integrals). Right now, however, we want to describe a different iterative method, which already makes use of certain stability principles.

1.5.2 The Method of Karlovitz

Karlovitz presented in [47] a method for the computation of best linear L^p-approximations by iterated computation of weighted L^2-approximations. This algorithm carries over to the determination of minimal solutions for modulars and Luxemburg norms, in the discrete as well as in the continuous case:

Let Φ be a strictly convex and differentiable Young function, whose second derivative exists at 0 and is positive there. By $F : \mathbb{R} \to \mathbb{R}$ with $F(s) := \frac{\Phi'(s)}{s}$ we then define a continuous positive function.

Let T, Σ, μ be a finite measure space and V a finite-dimensional subspace of $L^\infty(\mu)$ and $w \in L^\infty(\mu)$.

For the computation of a best approximation of w w.r.t. V in the sense of the modular f^Φ and the Luxemburg norm we attempt a unified treatment. By $z(v)$ we denote in the case of minimization of the modular the vector $w - v$, in the other case the normalized vector $\frac{w-v}{\|w-v\|_{(\Phi)}}$.

Theorem 1.5.1. *Let $v_1 \in V$, then we determine v_{k+1} for given v_k in the following way: let $g_k := F(z(v_k))$ and let the scalar product with the weight function g_k for $x, y \in C(T))$ be given by*

$$(x, y)_{g_k} := \int_T g_k x y \, d\mu$$

then we define u_k as the best approximation of w w.r.t. V in the corresponding scalar product norm: $u_k := M(\|w - \cdot\|_{g_k}, V)$. If we have determined the parameter ρ_k in such a way that for the modular

$$f^\Phi(w - v_k + \rho_k(v_k - u_k)) \le f^\Phi(w - v_k + \rho(v_k - u_k))$$

or for the Luxemburg norm

$$\|w - v_k + \rho_k(v_k - u_k)\|_{(\Phi)} \leq \|w - v_k + \rho(v_k - u_k)\|_{(\Phi)}$$

for all $\rho \in \mathbb{R}$, *then we define* $v_{k+1} := v_k - \rho_k(v_k - u_k)$.

We finally obtain

(a) $\lim_{k\to\infty} v_k = \lim_{k\to\infty} u_k = v_*$, *where* v_* *is the uniquely determined minimal solution of* $f^\Phi(w - \cdot)$ *or* $\|w - \cdot\|_{(\Phi)}$ *respectively.*

(b) *if* $v_k \neq v_*$, *then* $f^\Phi(w - v_{k+1}) < f^\Phi(w - v_k)$ *resp.* $\|w - v_{k+1}\|_{(\Phi)} < \|w - v_k\|_{(\Phi)}$.

For the proof we need the following lemma, where for f^Φ and $\|\cdot\|_{(\Phi)}$ we choose the common notation h:

Lemma 1.5.2. *Let* $v_0 \in V$ *and let* u_0 *be best approximation of* w *w.r.t.* V *in the norm* $\|\cdot\|_{g_0}$. *Then the following statements are equivalent:*

(a) $v_0 \neq u_0$

(b) *there is a* $\rho \neq 0$ *with* $h(w - v_0 + \rho(v_0 - u_0)) < h(w - v_0)$

(c) $v_0 \neq v_*$.

Proof. Due to the requirements for Φ the bilinear form $(\cdot,\cdot)_{g_0}$ is a scalar product and hence the best approximation of w w.r.t. V is uniquely determined.

(a) $\Rightarrow$ (b): Let ρ_0 be chosen such that for all $\rho \in \mathbb{R}$:

$$h(w - v_0 + \rho_0(v_0 - u_0)) \leq h(w - v_0 + \rho(v_0 - u_0))$$

and let $h(w - v_0 + \rho_0(v_0 - u_0)) = h(w - v_0)$. Because of the strict convexity of Φ the function h is strictly convex on its level sets and hence $\rho_0 = 0$. The chain rule then yields:

$$\langle h'(w - v_0), v_0 - u_0\rangle = 0,$$

and hence

$$\int_T \Phi'(z(v_0))(v_0 - u_0)d\mu = 0. \tag{1.16}$$

Furthermore because of $(w - u_0, v)_{g_0} = 0$ for all $v \in V$:

$$\begin{aligned}\|w - u_0\|_{g_0}^2 = (w - u_0, w - v_0)_{g_0} &= \lambda_0 \int_T F(z(v_0))(w - v_0)(w - u_0)d\mu \\ &= \lambda_0 \int_T \Phi'(z(v_0))(w - u_0)d\mu\end{aligned}$$

with $\lambda_0 = 1$ for $h = f^\Phi$ and $\lambda_0 = \|w - v_0\|_{(\Phi)}$ for $h = \|\cdot\|_{(\Phi)}$. Together with (1.16) we then obtain

$$\|w - u_0\|_{g_0}^2 = \lambda_0 \int_T \Phi'(z(v_0))(w - v_0)d\mu = \|w - v_0\|_{g_0}^2,$$

and thus, because of the uniqueness of the best approximation $v_0 = u_0$.

(b) $\Rightarrow$ (a) is obvious.

(c) $\Leftrightarrow$ (a): We have $v_0 = u_0$ if and only if for all $v \in V$:

$$0 = (w - v_0, v)_{g_0} = \lambda_0 \int_T \Phi'(z(v_0)) v d\mu,$$

in other words $\langle h'(w - v_0), v\rangle = 0$ if and only if $v_0 = v_*$ (see Corollary 1.2.3). □

Proof of Theorem 1.5.1. The sequence (v_k) is bounded, because by construction $w - v_k \in S_h(h(w - v_1))$. Suppose now the sequence (u_k) is not bounded, then there is subsequence (v_n) and (u_n) and $\tilde{v} \in V$ such that $\|u_n\|_\infty \geq n$ and $\|v_n - \tilde{v}\|_\infty \to 0$. The continuity of F yields the uniform convergence of $g_n = F(z(v_n))$ to $\tilde{g} = F(z(\tilde{v}))$. Thus the sequence $\|\cdot\|_{g_n}$ converges pointwise to $\|\cdot\|_{\tilde{g}}$ on $C(T)$. According to the stability principle for convex functions (see Remark 5.3.16) also the sequence of the best approximations (u_n) converges, a contradiction.

Let now $\tilde{v}$ be a point of accumulation of the sequence (v_k). Then there are convergent subsequences (v_{k_i}) and (u_{k_i}) with $v_{k_i} \to \tilde{v}$ and $u_{k_i} \to \tilde{u}$. As above the stability principle yields: $\tilde{u}$ is best approximation of w w.r.t. V in the norm $\|\cdot\|_{\tilde{g}}$. Suppose $\tilde{v} \neq \tilde{u}$, then according to the lemma above there is $\tilde{\rho} \neq 0$ with

$$h(w - \tilde{v} + \tilde{\rho}(\tilde{v} - \tilde{u})) < h(w - \tilde{v}).$$

Because of the continuity of h there is a $\delta > 0$ and a $K \in \mathbb{N}$ such that

$$h(w - v_{k_i} + \tilde{\rho}(v_{k_i} - u_{k_i})) < h(w - \tilde{v}) - \delta$$

for $k_i > K$. Furthermore by construction

$$h(w - v_{k_i+1}) \leq h(w - v_{k_i} + \tilde{\rho}(v_{k_i} - u_{k_i})).$$

On the other hand we obtain because of the monotonicity of the sequence $(h(w - v_k))$ the inequality $h(w - \tilde{v}) \leq h(w - v_{k_i+1})$, a contradiction. Using the previous lemma we obtain $\tilde{v} = v_*$. □

Remark 1.5.3. If the elements of V have zeros of at most measure zero, we can drop the requirement $\Phi''(0) > 0$: let $z \in w + V$, then $z \neq 0$, i.e. there is $U \in \Sigma$ with $\mu(U) > 0$ and $z(t) \neq 0$ for all $t \in U$. Hence $F(z(t)) > 0$ on U. Let $v \in V$ arbitrary and N the set of zeros of v, then we obtain

$$\int_T v^2 F(z) d\mu \geq \int_{U \setminus N} v^2 F(z) d\mu > 0.$$

Corollary 1.5.4. *Let $\{b_1, \dots, b_n\}$ be a basis of V, let $A : \mathbb{R}^n \to V$ be defined by $Ax = \sum_{i=1}^n x_i b_i$ and $f : \mathbb{R}^n \to \mathbb{R}$ by $f(x) = h(w - Ax)$. Let $x_k := A^{-1}v_k$ and $y_k := A^{-1}u_k$. Then iterations defined in the previous theorem carry over to the coefficient vectors by the following recursion:*

$$x_{k+1} = x_k - \rho_k \lambda_k \gamma_k (S(x_k))^{-1} f'(x_k),$$

where $S(x_k)$ denotes Gram's matrix for the scalar products $(\cdot,\cdot)_{g_k}$ w.r.t. the basis $\{b_1, \dots, b_n\}$, $\lambda_k = \gamma_k = 1$, if $h = f^\Phi$, and $\lambda_k = \|w - Ax_k\|_{(\Phi)}$ and $\gamma_k = \int_T z(x_k)\Phi'(z(x_k))d\mu \geq 1$ with $z(x_k) := \frac{w - Ax_k}{\lambda_k}$, if $h = \|\cdot\|_{(\Phi)}$.

In particular $\rho_k > 0$ for $f'(x_k) \neq 0$.

Proof. As $\Phi'(s)(0 - s) \leq \Phi(s)$ on $\mathbb{R}$ we have in the case of the Luxemburg norm

$$\gamma_k = \int_T z(x_k)\Phi'(z(x_k))d\mu \geq \int_T \Phi(z(x_k))d\mu = 1.$$

According to the definition of u_k we have: $S(x_k)y_k = r_k$, where

$$\begin{aligned} r_{kj} := (w, b_j)_{g_k} &= (w - Ax_k, b_j)_{g_k} + (Ax_k, b_j)_{g_k} \\ &= \lambda_k \int_T b_j \Phi'(z(x_k))d\mu + \sum_{i=1}^n x_{ki}(b_i, b_j)_{g_k}. \end{aligned}$$

We obtain

$$S(x_k)y_k = -\lambda_k \gamma_k f'(x_k) + S(x_k)x_k,$$

i.e.

$$x_k - y_k = \lambda_k \gamma_k (S(x_k))^{-1} f'(x_k),$$

and due to

$$x_{k+1} = x_k - \rho_k(x_k - y_k)$$

the first part of the assertion follows.

Together with $S(x_k)$ also $(S(x_k))^{-1}$ is positive definite, and hence

$$\lim_{\alpha \to 0} \frac{f(x_k - \alpha(S(x_k))^{-1} f'(x_k)) - f(x_k)}{\alpha} = -\langle f'(x_k), (S(x_k))^{-1} f'(x_k)\rangle < 0,$$

which means $-(S(x_k))^{-1} f'(x_k)$ is a direction of descent, i.e.

$$\begin{aligned} 0 > f(x_{k+1}) - f(x_k) &\geq \langle f'(x_k), x_{k+1} - x_k\rangle = -\rho_k \langle f'(x_k), x_k - y_k\rangle \\ &= -\rho_k \lambda_k \gamma_k \langle f'(x_k), (S(x_k))^{-1} f'(x_k)\rangle, \end{aligned}$$

and hence $\rho_k > 0$, if $f'(x_k) \neq 0$. □

In terms of the coefficient vectors the Karlovitz method can be interpreted as a modified gradient method.

Remark 1.5.5. If Φ' is Lipschitz continuous, one can instead of the exact determination of the step parameter ρ_k use a finite search method like the Armijo Rule (see (4.3) in Chapter 4 and [86]).

Remark 1.5.6. The functions $\Phi_p = |s|^p$ are exactly those Young functions, for which Φ'' and F differ only by a constant factor. For $p \geq 2$ also $\Phi_p''(0)$ is finite. For modulars f^{Φ_p} the method of Karlovitz corresponds to a damped Newton method.

Chapter 2
Polya Algorithms in Orlicz Spaces

2.1 The Classical Polya Algorithm

Let T be a compact subset of $\mathbb{R}^r$ and let $x \in C(T)$. The following theorem was shown by Polya:

Theorem 2.1.1 (Polya Algorithm). *Let V be a finite-dimensional subspace of $C(T)$. Then every point of accumulation of the sequence of best L^p-approximations of x w.r.t. V for $p \to \infty$ is a best Chebyshev approximation of x w.r.t. V.*

In the year 1920 Polya raised the question, whether the above sequence of best approximations actually converges. A negative answer was given in 1963 by Descloux for the continuous case, while he was able to show convergence to the strict approximation for the discrete case (see [25], compare Theorem 2.3.1).

Instead of considering the case $p \to \infty$, one can study the behavior of the sequence of best L^p-approximations for $p \to 1$. We can show that (see Equation (5.7) in Chapter 5) for an arbitrary measure space (T, Σ, μ) the sequence of best approximations converges to the best $L^1(\mu)$-approximation of largest entropy.

2.2 Generalized Polya Algorithm

We will now treat the subject considered by Polya in the framework of Orlicz spaces. In Chapter 1 we have become acquainted with a numerical method which can be used to perform Polya algorithms for the computation of a best approximation in the mean or a best Chebyshev approximation. We will discuss more effective methods in Chapter 4.

An important aim in this context is, to replace the determination of a best approximation for limit problems that are difficult to treat (non-differentiable, non-unique) in an approximate sense by approximation problems that are numerically easier to solve (Polya algorithm). The required differentiability properties of the corresponding function sequence to be minimized can already be provided by properties of their one-dimensional Young functions. The broad variety available in the choice of the Young functions is also used to influence the numerical properties in favorable way. The subsequent discussion will show that the convergence of the sequence of approximative problems to a specific best approximation can even in the case of non-uniqueness of the limit problem frequently be achieved through a proper choice of the sequence of

Young functions or by an outer regularization by use of convergence estimates developed below. The characterization of the limits is treated in the framework of two-stage optimization. The required pointwise convergence of the functions to be optimized is guaranteed through the pointwise convergence of the corresponding Young functions. The best approximation in the mean in this sense was already discussed in the introduction. A related problem is the Chebyshev approximation, to which a considerable part of these principles carries over. In the latter context however the modulars have to be replaced by the corresponding Minkowski functionals.

The following stability principle will play a central role in our subsequent discussion:

Theorem 2.2.1. *Let ρ be a norm on the finite-dimensional vector space X and let $(\rho_k)_{k\in\mathbb{N}}$ be a sequence of semi-norms on X that converges pointwise on X to ρ. Let further V be a subspace of X, $x \in X \setminus V$ and v_k a ρ_k-best approximation of x w.r.t. V. The following statements hold:*

(a) *every subsequence of (v_k) has a ρ-convergent subsequence*

(b) $\lim \rho(x - v_k) = d_\rho(x, V)$

(c) *every ρ-point of accumulation of (v_k) is a ρ-best approximation of x w.r.t. V*

(d) *if x has a unique ρ-best approximation v_ρ w.r.t. V, then*

$$\lim_{k\to\infty} \rho(v_\rho - v_k) = 0.$$

This theorem of Kripke, proved in [73], is a special case of the stability theorem of convex optimization (see Theorem 5.3.21 and Theorem 5.3.25).

2.3 Polya Algorithm for the Discrete Chebyshev Approximation

As was seen above, the maximum norm can be described as a Minkowski functional using the Young function Φ_∞ restated below: $\Phi_\infty : \mathbb{R} \to \overline{\mathbb{R}}$ defined by

$$\Phi_\infty(t) := \begin{cases} 0 & \text{for } |t| \le 1 \\ \infty & \text{otherwise.} \end{cases}$$

Then we obtain

$$\|x\|_\infty = \inf\left\{ c > 0 \,\middle|\, \sum_{t\in T} \Phi_\infty\left(\frac{x(t)}{c}\right) \le 1 \right\}.$$

It is not too far-fetched, to proceed in a similar way as in the introduction of Chapter 1 and to consider sequences $(\Phi_k)_{k\in\mathbb{N}}$ of Young functions with favorable numerical properties (differentiability, strict convexity) that converge pointwise to Φ_∞. For the corresponding Luxemburg norms we obtain

$$\|x\|_{(\Phi_k)} := \inf\left\{c > 0 \,\middle|\, \sum_{t\in T} \Phi_k\left(\frac{x(t)}{c}\right) \le 1\right\},$$

and it turns out that pointwise convergence of the Young functions carries over to pointwise convergence of the Luxemburg norms. We obtain (see below)

$$\lim_{k\to\infty} \|x\|_{(\Phi_k)} = \|x\|_\infty.$$

Thus the preconditions for the application of the stability theorem (see Theorem 2.2.1) are satisfied, i.e. the sequence $(v_k)_{k\in\mathbb{N}}$ of best approximations of x w.r.t. V is bounded and every point of accumulation of this sequence is a best Chebyshev approximation.

Theorem 2.3.1. *Let $(\Phi_k)_{k\in\mathbb{N}}$ be a sequence of Young functions that converges pointwise to the Young function*

$$\Phi_\infty(t) := \begin{cases} 0 & \text{for } |t| \le 1 \\ \infty & \text{otherwise.} \end{cases}$$

Then this carries over to the pointwise convergence of the Luxemburg norms, i.e.

$$\lim_{k\to\infty} \|x\|_{(\Phi_k)} = \|x\|_\infty.$$

Proof. Let $x \neq 0$ and let $\varepsilon > 0$ be given ($\varepsilon \le \frac{1}{2}\|x\|_\infty$). We have:

$$\sum_{t\in T} \Phi_k\left(\frac{x(t)}{\|x\|_\infty + \varepsilon}\right) \le \sum_{t\in T} \Phi_k\left(\frac{\|x\|_\infty}{\|x\|_\infty + \varepsilon}\right) = |T|\Phi_k\left(\frac{\|x\|_\infty}{\|x\|_\infty + \varepsilon}\right) \xrightarrow{k\to\infty} 0$$

because of the pointwise convergence of the sequence of Young functions $(\Phi_k)_{k\in\mathbb{N}}$ to Φ_∞. Thus

$$\|x\|_{(\Phi_k)} \le \|x\|_\infty + \varepsilon \quad \text{for } k \text{ sufficiently large.}$$

On the other hand there is $t_0 \in T$ with $|x(t_0)| = \|x\|_\infty$. We obtain

$$\sum_{t\in T} \Phi_k\left(\frac{x(t)}{\|x\|_\infty - \varepsilon}\right) \ge \Phi_k\left(\frac{x(t_0)}{\|x\|_\infty - \varepsilon}\right) = \Phi_k\left(\frac{\|x\|_\infty}{\|x\|_\infty - \varepsilon}\right) \xrightarrow{k\to\infty} \infty.$$

Hence

$$\|x\|_\infty - \varepsilon \le \|x\|_{(\Phi_k)} \quad \text{for } k \text{ sufficiently large.}$$ □

As a special case we obtain the discrete version of the *classical Polya algorithm*, which has the property that every point of accumulation of the best l^p-approximations ($p > 1$) for p to infinity is a best Chebyshev approximation. For this purpose we define $\Phi_p(t) = |t|^p$. We have $\sum_{t \in T} |\frac{x(t)}{c}|^p = 1$ if and only if $(\sum_{t \in T} |x(t)|^p)^{\frac{1}{p}} = c$. Hence the Luxemburg norm yields the well-known l^p-norm

$$\|x\|_{(\Phi_p)} = \|x\|_p = \left(\sum_{t \in T} |x(t)|^p \right)^{\frac{1}{p}}.$$

For continuous Young functions Φ, we can use the equation

$$\sum_{t \in T} \Phi \left(\frac{x(t)}{\|x\|_{(\Phi)}} \right) = 1 \tag{2.1}$$

for the computation of the unknown Luxemburg norm $\|x\|_{(\Phi)}$.

2.3.1 The Strict Approximation as the Limit of Polya Algorithms

Polya's question, whether even genuine convergence can be shown, was given a positive answer by Descloux in the year 1963 (for the classical Polya algorithm), by proving convergence to the strict approximation.

Theorem 2.3.2. *Let $T = \{t_1, t_2, \ldots, t_m\}$ and $(\Phi_k)_{k \in \mathbb{N}}$ be a sequence of continuous Young functions with the following properties:*

(1) $\lim \Phi_k(s) = \begin{cases} 0 & \textit{for } |s| < 1, \\ \infty & \textit{for } |s| > 1, \end{cases}$

(2) $\Phi_k(r \cdot s) \geq \Phi_k(r) \cdot \Phi_k(s)$ *for* $s, r \geq 0$.

Let further $L^{\Phi_k}(\mu)$ be the Orlicz spaces for the measure μ with $\mu(t_i) = 1$, $i = 1, 2, \ldots, m$, and V a subspace of $C(T)$, $x \in C(T)$, and let v_k be best $L^{\Phi_k}(\mu)$ approximations of x in the corresponding Luxemburg norms and v_* the strict approximation (see Definition 1.4.1*) of x w.r.t. V. Then*

$$\lim_{k \to \infty} v_k = v_*.$$

Proof. We can assume $x \in C(T) \setminus V$, because otherwise $x = v_k = v_*$.

If Φ is continuous and $\|x\|_{(\Phi)} > 0$, then, as already mentioned above

$$\int_T \Phi \left(\frac{x}{\|x\|_{(\Phi)}} \right) d\mu = 1. \tag{2.2}$$

Due to the Stability Theorem 2.2.1 every point of accumulation of the sequence (v_k) is a best Chebyshev approximation. Let (v_{k_λ}) be a convergent subsequence and let v_0 be the limit of (v_{k_λ}).

Since all norms on a finite dimensional linear space are equivalent, v_0 is also the limit of (v_{k_λ}) in the maximum norm.

Now we will show that

$$v_0 = v_*.$$

The assertion is proved via complete induction over j, where we assume

$$v_0(t) = v_*(t) \quad \text{for } t \in T_j, \tag{2.3}$$

(see definition of the Strict Approximation 1.4.1). The start of the induction for $j = 0$ follows directly from the fact that v_0 is a best L_∞-approximation of x.

Suppose (2.3) holds for j. If (2.3) holds for D_{j+1}, then also for T_{j+1} (see definition of the Strict Approximation and Lemma 1.4.4).

If $v_0(\tilde{t}) \neq v_*(\tilde{t})$ for a $\tilde{t} \in D_{j+1}$, then apparently $v_0 \notin V_{j+1}$. Thus there is a $\hat{t} \in T \setminus T_j$ with

$$|x(\hat{t}) - v_0(\hat{t})| > \|x - v_*\|_{T \setminus T_j}.$$

Let

$$z_{k_\lambda} = x - v_{k_\lambda}, \quad z_0 = x - v_0 \quad \text{and} \quad z_* = x - v_*,$$

then we obtain

$$|z_0(\hat{t})| = s_0 > d_0 = \|z_*\|_{T \setminus T_j}.$$

Let $\varepsilon := \frac{s_0 - d_0}{3}$, then there is a k_λ^0 such that for all $k_\lambda > k_\lambda^0$

$$\|v_{k_\lambda} - v_0\|_\infty < \varepsilon, \tag{2.4}$$

and by (1) a $k_\lambda^1 \geq k_\lambda^0$, such that for all $k_\lambda > k_\lambda^1$

$$\Phi_{k_\lambda}\left(\frac{s_0 - \varepsilon}{d_0 + \varepsilon}\right) > m. \tag{2.5}$$

From (2) it follows that

$$\Phi_{k_\lambda}\left(\frac{s_0 - \varepsilon}{d_0 + \varepsilon}\right) = \Phi_{k_\lambda}\left(\frac{\frac{s_0-\varepsilon}{\|z_{k_\lambda}\|_{(\Phi_{k_\lambda})}}}{\frac{d_0+\varepsilon}{\|z_{k_\lambda}\|_{(\Phi_{k_\lambda})}}}\right) \leq \Phi_{k_\lambda}\left(\frac{\frac{z_{k_\lambda}(\hat{t})}{\|z_{k_\lambda}\|_{(\Phi_{k_\lambda})}}}{\frac{d_0+\varepsilon}{\|z_{k_\lambda}\|_{(\Phi_{k_\lambda})}}}\right) \leq \frac{\Phi_{k_\lambda}(\frac{z_{k_\lambda}(\hat{t})}{\|z_{k_\lambda}\|_{(\Phi_{k_\lambda})}})}{\Phi_{k_\lambda}(\frac{d_0+\varepsilon}{\|z_{k_\lambda}\|_{(\Phi_{k_\lambda})}})}.$$

This leads to

$$\begin{aligned}\Phi_{k_\lambda}\left(\frac{z_{k_\lambda}(\hat{t})}{\|z_{k_\lambda}\|_{(\Phi_{k_\lambda})}}\right) &> m\Phi_{k_\lambda}\left(\frac{d_0 + \varepsilon}{\|z_{k_\lambda}\|_{(\Phi_{k_\lambda})}}\right) \geq \sum_{t \in T \setminus T_j} \Phi_{k_\lambda}\left(\frac{z_*(t) + \varepsilon}{\|z_{k_\lambda}\|_{(\Phi_{k_\lambda})}}\right) \\ &\geq \sum_{t \in T \setminus T_j} \Phi_{k_\lambda}\left(\frac{z_*(t) + z_{k_\lambda}(t) - z_0(t)}{\|z_{k_\lambda}\|_{(\Phi_{k_\lambda})}}\right).\end{aligned}$$

By the induction hypothesis $z_*(t) = z_0(t)$ for all $t \in T_j$ and hence

$$1 = \sum_{t \in T} \Phi_{k_\lambda}\left(\frac{z_{k_\lambda}(t)}{\|z_{k_\lambda}\|_{(\Phi_{k_\lambda})}}\right) > \sum_{t \in T} \Phi_{k_\lambda}\left(\frac{z_*(t) + z_{k_\lambda}(t) - z_0(t)}{\|z_{k_\lambda}\|_{(\Phi_{k_\lambda})}}\right).$$

Due to (2.2) we have

$$\|v_{k_\lambda} - v_0 + v - x\|_{(\Phi_{k_\lambda})} < \|v_{k_\lambda} - x\|_{(\Phi_{k_\lambda})},$$

a contradiction to the assumption that v_{k_λ} is a best $L^{\Phi_{k_\lambda}}$-approximation of x. Thus we have shown that every convergent subsequence (v_{k_λ}) of the bounded sequence $\{v_k\}$ converges to v_* so that the total sequence (v_k) converges to v_*. □

2.3.2 About the Choice of the Young Functions

Under numerical aspects, the choice of the sequence of Young functions is of central importance. This is true for the complexity of function evaluation as well as for the numerical stability.

The subsequent remark shows that for pointwise convergence of the Luxemburg norms the pointwise convergence of the Young functions is sufficient in a relaxed sense. The flexibility in the choice of Young functions is enlarged thereby, and plays a considerable role for the actual computation.

Remark. For the pointwise convergence of the Luxemburg norms it is sufficient to require

(a) $\Phi_k(t) \to_{k\to\infty} \infty$ for $t > 1$

(b) $\Phi_k(1) \leq \frac{1}{|T|}$.

Proof. We have

$$\sum_{t \in T} \Phi_k\left(\Phi_k^{-1}\left(\frac{1}{|T|}\right)\frac{x(t)}{\|x\|_\infty}\right) \leq \sum_{t \in T} \Phi_k\left(\Phi_k^{-1}\left(\frac{1}{|T|}\right)\cdot 1\right) = 1,$$

and hence

$$\|x\|_{(\Phi_k)} \leq \frac{\|x\|_\infty}{\Phi_k^{-1}(\frac{1}{|T|})},$$

but from (b) we obtain

$$\Phi_k^{-1}\left(\frac{1}{|T|}\right) \geq 1,$$

whence

$$\|x\|_{(\Phi_k)} \leq \|x\|_\infty.$$

The other direction of the inequality is obtained as in the last part of the proof of Theorem 2.3.1. □

So we can leave the values $\Phi_k(t)$ for $t \in [0,1]$ unchanged when constructing the Young functions $k \in \mathbb{N}$.

Example 2.3.3.

$$\Phi_k(t) := \begin{cases} \frac{1}{|T|}t^2 & \text{for } |t| \leq 1 \\ k^3|t|^3 + (\frac{1}{|T|} - 3k^3)t^2 + 3k^3|t| - k^3 & \text{for } |t| > 1. \end{cases}$$

This example will have prototype character throughout this chapter for the properties of Young functions and corresponding Luxemburg norms.

2.3.3 Numerical Execution of the Polya Algorithm

In the previous chapter we have treated the properties of Young functions which guarantee uniqueness, twice differentiability of the norm, and the regularity of the Hessian. If we choose a sequence of Young functions that converges pointwise in the relaxed sense of the previous section, then the stability theorems guarantee that the sequence of the approximative solutions is bounded and every point of accumulation of this sequence is a best Chebyshev approximation. Thus we can sketch the numerical execution of the Polya algorithm in the following way:

Algorithmic Scheme of the Polya Algorithm

(a) Choose a sequence $(\Phi_k)_{k\in\mathbb{N}}$ in the above sense.

(b) Choose a starting point a_0 for the coefficient vector and set $k = 1$.

(c) Solve the non-linear system

$$\sum_{t\in T} v_i(t)\,\Phi_k'\left(\frac{x - \sum_{l=1}^n a_l v_l}{\|x - \sum_{l=1}^n a_l v_l\|_{(\Phi_k)}}\right) = 0 \quad \text{for } i = 1,\dots,n,$$

using an iterative method from Chapter 1 (or 4), where the computation of the norm as a one-dimensional solution for the quantity c of the equation

$$\sum_{t\in T} \Phi_k\left(\frac{x - \sum_{l=1}^n a_l v_l}{c}\right) - 1 = 0$$

has to be performed in each iteration step.

(d) Take the solution vector computed in this way as a new starting point, set $k \leftarrow k+1$ and go to (c).

Remarks. (a) The sequence of Young functions occurring in the above scheme can as a rule be understood as a subsequence of a parametric family of functions. The change of the parameter then should be done in a way that the current minimal solution is a good starting point for the solution method w.r.t. the subsequent parameter. This is of particular importance, since we want to employ rapidly convergent (Q-superlinear) methods (see Section 4.1).

In the above parametric examples of sequences of Young functions (see e.g. Example 2.3.3) experience has shown that as a parameter the sequence $(c^j)_{j\in\mathbb{N}}$ with $2 \le c \le 8$ can be recommended.

(b) Step (c) of the above algorithmic scheme can also be replaced by the solution of the non-linear system (1.13). As numerical methods those for the solution of non-linear equations have to be employed (see Chapter 4).

(c) The question referring to 'genuine' convergence of the approximate solutions to a best Chebyshev approximation and the description of the limit will be treated in detail below.

(d) The question referring to an appropriate stop criterion for the iteration we will discuss later. As a first choice one can use the change between the k-th and $k+1$-st solution or use the characterisation theorems for a best Chebyshev approximation (see Theorem 1.3.3).

(e) An estimate for the 'quality' of the best approximations w.r.t. $\|\cdot\|_{(\Phi_k)}$ as approximations for a best Chebyshev approximation is obtained by the convergence estimates in Section 2.5.2.

We will now state the Polya algorithm in the framework of general measures.

2.4 Stability of Polya Algorithms in Orlicz Spaces

Let (T, Σ, μ) be a finite measure space. We will now proceed in a similar way as in the previous section and consider sequences $(\Phi_k)_{k\in\mathbb{N}}$ of Young functions with amiable numerical properties (differentiability, strict convexity), which converges pointwise to Φ_∞. If we consider the corresponding Luxemburg norms $\|x\|_{(\Phi_k)}$, it can be shown that also in this situation pointwise convergence of the Young functions carries over to pointwise convergence of the Luxemburg norms, i.e.

$$\lim_{k\to\infty} \|x\|_{(\Phi_k)} = \|x\|_\infty.$$

Thus the requirements for the corresponding stability theorems (see Theorem 2.2.1 or Theorem 5.3.25) are satisfied, i.e. the sequence $(v_k)_{k\in\mathbb{N}}$ of best approximations is bounded and every point of accumulation of this sequence is a best L^∞-approximation, also denoted as Chebyshev approximation. A corresponding statement can be made for best L^1-approximations.

For the following stability considerations we need the fact that $(L^\infty, \|\cdot\|_\infty)$ is a Banach space (see Theorem 6.2.4).

In order to describe the behavior of best approximations related to sequences of modulars (f^{Φ_k}) or of sequences of Luxemburg norms $(\|\cdot\|_{(\Phi_k)})$ in a unified way, the following stability principle that constitutes a broadening of the scope of Theorem 2.2.1 is of central significance (see Theorem 5.3.18):

Theorem 2.4.1. *Let X be a Banach space, K a closed convex subset of X and $(f_n : X \to \mathbb{R})_{n\in\mathbb{N}}$ a sequence of convex continuous functions that converges pointwise to a function $f : X \to \mathbb{R}$. Let x_n be a minimal solution of f_n on K. Then every point of accumulation of the sequence $(x_n)_{n\in\mathbb{N}}$ is a minimal solution of f on K.*

In the finite-dimensional case one has the following situation: if the set of solutions of the limit problem $M(f, K)$ is bounded, then the set of points of accumulation is non-empty and the sequence of the minimal values $(\inf(f_n, K))$ converges to the minimal value $\inf(f, K)$ of the limit problem (see Theorem 5.3.25).

In order to obtain stability of our generalized Polya algorithms, it suffices to establish pointwise convergence of modulars and Luxemburg norms (for an appropriate choice Young functions):

- $f^{\Phi_k} \to \|\cdot\|_1$
- $\|\cdot\|_{(\Phi_k)} \to \|\cdot\|_1$
- $\|\cdot\|_{(\Phi_k)} \to \|\cdot\|_\infty$

on L^∞ or $C(T)$. The theorem of Banach–Steinhaus 5.3.17 for convex functions will permit to reduce the proof of pointwise convergence of the functionals to the pointwise convergence of the (1-dimensional) Young functions (Φ_k), using the fact that the step functions form a dense subset. As a preparation we need the following

Lemma 2.4.2. *Let Φ be a finite Young function, then corresponding modular f^Φ and the corresponding Luxemburg norm $\|\cdot\|_{(\Phi)}$ are continuous on $(L^\infty, \|\cdot\|_\infty)$.*

Proof. Let $\|x_k - x\|_\infty \to_{k\to\infty} 0$, then let $\varepsilon > 0$ and $\delta > 0$ be chosen such that

$$\delta < \varepsilon \Phi^{-1}\left(\frac{1}{\mu(T)}\right),$$

then for k sufficiently large

$$\int_T \Phi\left(\frac{x - x_k}{\varepsilon}\right) d\mu \le \int_T \Phi\left(\frac{\delta}{\varepsilon}\right) d\mu = \mu(T)\Phi\left(\frac{\delta}{\varepsilon}\right) < 1,$$

hence $\|x_k - x\|_{(\Phi)} \le \varepsilon$.

In particular there is a number M such that the modulus of the values of x and of the sequence (x_k) is contained a.e. in the interval $[0, M]$. Since Φ is uniformly continuous on $[0, M]$, there is for given $\varepsilon > 0$ a $\delta > 0$ with $\|x_k - x\|_\infty < \delta$ for k sufficiently large and $|\Phi(x) - \Phi(x_k)| < \varepsilon$ a. e.. It follows that $|f^\Phi(x) - f^\Phi(x_k)| < \varepsilon \cdot \mu(T)$. □

Theorem 2.4.3. *Let (T, Σ, μ) be a finite measure space and $(\Phi_k)_{k\in\mathbb{N}}$ a sequence of finite Young functions, which converges pointwise to the Young function Φ_∞.*

Then this carries over to the pointwise convergence of the Luxemburg norms, i.e. for $x \in L^\infty(\mu)$ we have

$$\lim_{k\to\infty} \|x\|_{(\Phi_k)} = \|x\|_\infty.$$

Proof. $L^\infty(\mu)$ is contained in $L^{\Phi_k}(\mu)$ for all $k \in \mathbb{N}$, since T has finite measure and Φ_k is a finite Young function. The assertion is verified using the theorem of Banach–Steinhaus for convex functions (see Theorem 5.3.17 in Chapter 5). According to the previous lemma $\|\cdot\|_{(\Phi_k)}$ is continuous on $L^\infty(\mu)$ for all $k \in \mathbb{N}$. Furthermore the step functions are dense in the Banach space $(L^\infty(\mu), \|\cdot\|_\infty)$ (see Theorem 6.2.19). We show at first the convergence of the norms for a given step function $x := \sum_{n=1}^m \alpha_n \chi_{T_n}$, where $T_i \cap T_j = \emptyset$ for $i \neq j$. Let $0 < \varepsilon < \|x\|_\infty$ be arbitrary, then we obtain

$$\int_T \Phi_k\left(\frac{x}{\|x\|_\infty + \varepsilon}\right) d\mu = \sum_{n=1}^m \Phi_k\left(\frac{\alpha_n}{\|x\|_\infty + \varepsilon}\right)\mu(T_n) \xrightarrow{k\to\infty} 0,$$

i.e. $\|x\|_{(\Phi_k)} \leq \|x\|_\infty + \varepsilon$ for k sufficiently large. On the other hand there is a n_0 with $\|x\|_\infty = |\alpha_{n_0}|$ and hence

$$\sum_{n=1}^m \Phi_k\left(\frac{\alpha_n}{\|x\|_\infty - \varepsilon}\right)\mu(T_n) \xrightarrow{k\to\infty} \infty,$$

i.e. $\|x\|_{(\Phi_k)} \geq \|x\|_\infty - \varepsilon$ for k sufficiently large.

It is just as simple to verify the pointwise boundedness of the sequence of the norms: let $x \in L^\infty(\mu)$, i.e. there is a $M > 0$ such that $|x(t)|_\infty \leq M$ μ a. e. . Thus we obtain

$$\begin{aligned}\int_T \Phi_k\left(\frac{x(t)}{M+1}\right) d\mu &\leq \int_T \Phi_k\left(\frac{M}{M+1}\right) d\mu \\ &= \Phi_k\left(\frac{M}{M+1}\right)\mu(T) \xrightarrow{k\to\infty} 0,\end{aligned}$$

i.e. there is a $k_0 \in \mathbb{N}$, such that for all $k > k_0$

$$\|x\|_{(\Phi_k)} \leq M + 1,$$

thus the pointwise boundedness is established. □

This theorem is the basis for the Polya algorithm in $C(T)$, if $C(T)$ for an appropriately chosen measure space (T, Σ, μ) is considered as a closed subspace of $L^\infty(\mu)$.

A corresponding statement can be made for approximations of the L^1-norm:

Theorem 2.4.4. *Let (T, Σ, μ) be a finite measure space and $(\Phi_k)_{k\in\mathbb{N}}$ a sequence of finite Young functions, which converges pointwise to the Young function Φ_1 with $\Phi_1(s) = |s|$. Then this carries over to the pointwise convergence of the modulars and the Luxemburg norms, i.e. for $x \in L^\infty(\mu)$ we have*

$$\lim_{k\to\infty} \|x\|_{(\Phi_k)} = \|x\|_1$$

$$\lim_{k\to\infty} f^{\Phi_k}(x) = \|x\|_1.$$

Proof. $L^\infty(\mu)$ is contained in $L^{\Phi_k}(\mu)$ for all $k \in \mathbb{N}$, since T has finite measure and Φ_k is a finite Young function. The proof is again performed by use of the theorem of Banach–Steinhaus for convex functions. By the above lemma, f^{Φ_k} and $\|\cdot\|_{(\Phi_k)}$ are continuous on $L^\infty(\mu)$ for all $k \in \mathbb{N}$. As stated above, the step functions are dense in $L^\infty(\mu)$. First we will show the convergence of the norms for a given step function $x := \sum_{n=1}^m \alpha_n \chi_{T_n}$. Let $0 < \varepsilon < \|x\|_1$ be arbitrary, then we have

$$\begin{aligned}\int_T \Phi_k\left(\frac{x}{\|x\|_1 + \varepsilon}\right) d\mu &= \sum_{n=1}^m \Phi_k\left(\frac{\alpha_n}{\|x\|_1 + \varepsilon}\right)\mu(T_n)\\ &\xrightarrow{k\to\infty} \frac{1}{\|x\|_1 + \varepsilon}\sum_{n=1}^m |\alpha_n|\mu(T_n)\\ &= \frac{\|x\|_1}{\|x\|_1 + \varepsilon} < 1,\end{aligned}$$

i.e. $\|x\|_{(\Phi_k)} \le \|x\|_1 + \varepsilon$ for k sufficiently large. On the other hand we obtain for $\varepsilon < \|x\|_1$

$$\sum_{n=1}^m \Phi_k\left(\frac{\alpha_n}{\|x\|_1 - \varepsilon}\right)\mu(T_n) \xrightarrow{k\to\infty} \frac{1}{\|x\|_1 - \varepsilon}\sum_{n=1}^m |\alpha_n|\mu(T_n) = \frac{\|x\|_1}{\|x\|_1 - \varepsilon} > 1,$$

i.e. $\|x\|_{(\Phi_k)} \ge \|x\|_1 - \varepsilon$ for k sufficiently large.

For modulars we immediately obtain

$$f^{\Phi_k}(x) = \sum_{n=1}^m \Phi_k(\alpha_n)\mu(T_n) \xrightarrow{k\to\infty} \sum_{n=1}^m |\alpha_n|\mu(T_n) = \|x\|_1.$$

It is just as simple to verify the pointwise boundedness of the sequence of norms: let $x \in L^\infty(\mu)$, i.e. there is a $M > 0$ such that $|x(t)| \le M$ μ-a. e. . Hence we obtain

$$\begin{aligned}\int_T \Phi_k\left(\frac{x(t)}{(M+1)\mu(T)}\right) d\mu &\le \int_T \Phi_k\left(\frac{M}{(M+1)\mu(T)}\right) d\mu\\ &= \Phi_k\left(\frac{M}{(M+1)\mu(T)}\right)\mu(T) \xrightarrow{k\to\infty} \frac{M}{M+1} < 1,\end{aligned}$$

i.e. there is a $k_0 \in \mathbb{N}$, such that for all $k > k_0$

$$\|x\|_{(\Phi_k)} \leq (M+1)\mu(T).$$

For the modulars we have

$$|f^{\Phi_k}(x)| \leq |f^{\Phi_k}(M)|,$$

and the pointwise boundedness of the modulars follows from the pointwise convergence of the Young functions □

2.5 Convergence Estimates and Robustness

In a numerical realization of the Polya Algorithm the solutions of the approximating problems are not solved exactly. As a criterion to terminate the iteration one can use the norm of the gradient.

It turns out that stability is essentially retained if the approximating problems are solved only approximately by some $\tilde{x}_k$ provided the sequence $(\|\nabla f_k(\tilde{x}_k)\|)$ goes to zero (robustness).

Theorem 2.5.1 (see Theorem 5.5.1). *Let $(f_k : \mathbb{R}^r \to \mathbb{R})_{k\in\mathbb{N}}$ be a sequence of differentiable convex functions that converges pointwise to a function $f : \mathbb{R}^n \to \mathbb{R}$. Let further the set of minimal solutions of f on $\mathbb{R}^n$ be non-empty and bounded and let $(\tilde{x}_k)_{k\in\mathbb{N}}$ be a sequence in $\mathbb{R}^n$ with the property $\lim_{k\to\infty} \|\nabla f_k(\tilde{x}_k)\| = 0$ then we have*

(a) *The set of limit points of the sequence $(\tilde{x}_k)$ is non-empty and contained in $M(f, \mathbb{R}^n)$.*

(b) $f_k(\tilde{x}_k) \to \inf f(\mathbb{R}^n)$.

(c) $f(\tilde{x}_k) \to \inf f(\mathbb{R}^n)$.

(d) *Let Q be an open bounded superset of $M(f, \mathbb{R}^n)$ and $\varepsilon_k := \sup_{x\in Q} |f(x) - f_k(x)|$, then*

$$|\inf f(\mathbb{R}^n) - f(\tilde{x}_k)| = O(\varepsilon_k) + O(\|\nabla f_k(\tilde{x}_k)\|).$$

The above theorem makes use of the following theorem on convergence estimates (which in turn makes use of the stability theorem):

Theorem 2.5.2 (see Theorem 5.5.2). *Let $(f_k : \mathbb{R}^n \to \mathbb{R})_{k\in\mathbb{N}}$ be a sequence of convex functions that converges pointwise to a function $f : \mathbb{R}^n \to \mathbb{R}$ with bounded level sets. Let further K be a closed convex subset of $\mathbb{R}^n$ and $x_k \in M(f_k, K)$ and let Q be an open bounded superset of $M(f, K)$ where $\varepsilon_k := \sup_{x\in Q} |f(x) - f_k(x)|$. Then $\varepsilon_k \to 0$ and we obtain*

(a) $f(x_k) - \inf f(K) = O(\varepsilon_k)$

(b) $|f_k(x_k) - \inf f(K)| = O(\varepsilon_k)$.

It is now our aim to give estimates of the above kind for the subspace of Hölder–Lipschitz continuous functions on a compact subset of $\mathbb{R}^r$ satisfying a cone condition, and where μ is the Lebesgue measure of $\mathbb{R}^r$ (see Section 2.5.2).

Definition 2.5.3. A subset T of $\mathbb{R}^r$ satisfies a *cone condition* if there is a cone C with interior point and a number $\rho > 0$ such that for all $t \in T$ there exists an orthogonal transformation A_t with the property

$$t + (\rho B \cap A_t C) \subseteq T,$$

where B denotes the unit ball in $\mathbb{R}^r$ w.r.t. the Euclidean norm.

Definition 2.5.4. Let T be a compact subset of $\mathbb{R}^r$ and let $0 < \alpha \leq 1$. We define $\mathrm{Lip}^\alpha(T)$ as the space of all real-valued functions x on T with the property: there is $L \in \mathbb{R}$ with $|x(t) - x(t')| \leq L\|t - t'\|^\alpha$ for all $t, t' \in T$, where $\|\cdot\|$ is the Euclidean norm of $\mathbb{R}^r$.

Theorem 2.5.5. *Let T be a compact subset of $\mathbb{R}^r$ that satisfies a cone condition, μ the Lebesgue measure, and $(\Phi_k)_{k\in\mathbb{N}}$ a sequence of finite Young's functions with $\lim_{k\to\infty} \Phi_k(s) = \Phi_\infty(s)$ for $|s| \neq 1$. Let K be a finite-dimensional subset of $\mathrm{Lip}^\alpha(T)$, then there is a sequence $(\delta_k)_{k\in\mathbb{N}}$ converging to zero with $\delta_k := \max(\varepsilon_k^\alpha, \sigma_k)$, where ε_k is the solution of the equation*

$$s^r \Phi_k(1 + s^\alpha) = 1,$$

and $\sigma_k := \frac{1}{\Phi_k^{-1}(\frac{1}{\mu(T)})} - 1$. There is also a constant $\zeta > 0$ such that for all $x \in K$

$$\big|\|x\|_\infty - \|x\|_{(\Phi_k)}\big| \leq \zeta \|x\|_\infty \delta_k.$$

Remark. If $\Phi_k(1) \leq \frac{1}{\mu(T)}$ then $\|x\|_{(\Phi_k)} \leq \|x\|_\infty$.

Examples. (a) $\Phi_k(s) := as^2 + k(k(|s| - 1)_+)^m$ with $a > 0$ and $n \in \mathbb{N}$. For every $c \in (0, 1)$ there is $k_0 \in \mathbb{N}$ such that for $k \geq k_0$

$$\frac{c}{k^q} < \varepsilon_k^\alpha < \frac{1}{k^q} \quad \text{with } q := \frac{m+1}{m + \frac{r}{\alpha}}.$$

(b) $\Phi_k(s) := |s|^k$. For every $c \in (0, 1)$ there is $k_0 \in \mathbb{N}$ such that for $k \geq k_0$

$$c\frac{r}{\alpha}\frac{\log k}{k} \leq \varepsilon_k^\alpha \leq \frac{r}{\alpha}\frac{\log(k-1)}{k-1}.$$

In the discrete case the situation is somewhat simpler than in the continuous case.

Theorem 2.5.6. *Let* $T := \{t_1, \ldots, t_N\}$ *and* $\mu : T \to (0, \infty)$ *a weight function. Let* $(\Phi_k)_{k\in\mathbb{N}}$ *be a sequence of finite Young's functions with* $\lim_{k\to\infty} \Phi_k(s) = \Phi_\infty(s)$ *for* $|s| \neq 1$ *and* $x : T \to \mathbb{R}$.

Then we obtain the estimate $|\|x\|_{(\Phi_k)} - \|x\|_\infty| \leq \varepsilon_k \cdot \|x\|_\infty$, *where with* $\lambda := \min_{t\in T}\{\mu(t)\}$ *we define* $\varepsilon_k := \max\{\frac{1}{\Phi_k^{-1}(\frac{1}{\mu(T)})} - 1, 1 - \frac{1}{\Phi_k^{-1}(\frac{1}{\lambda})}\}$.

For approximations of the L^1-norm we obtain (see Theorem 2.5.2):

Theorem 2.5.7. *Let* (T, Σ, μ) *be a finite measure space,* $(\Phi_k)_{k\in\mathbb{N}}$ *a sequence of finite, definite Young functions with* $\lim_{k\to\infty} \Phi_k(s) = |s|$ *on* $\mathbb{R}$.

Let further K *be a finite-dimensional subset of* $L^\infty(\mu)$ *and let* ξ *be chosen such that* $\|\cdot\|_\infty \leq \xi\|\cdot\|_1$ *on the span of* K, *then there is a sequence* $(\delta_k)_{k\in\mathbb{N}}$ *tending to zero with* $\delta_k := \max((g_{\Phi_k}(\xi) - 1), \varepsilon_k)$, *where* ε_k *is the larger solution of* $\Phi_k^{-1}(s) = \frac{s}{1-s}$ *and* $g_{\Phi_k}(s) := \frac{s}{\Phi_k^{-1}(s)}$ *for* $s > 0$, *then*

$$|\|x\|_1 - \|x\|_{\Phi_k}| \leq \xi\|x\|_1\delta_k \quad \textit{for all } x \in K.$$

2.5.1 Two-Stage Optimization

Let us return to Polya's original question concerning the convergence of the sequence of minimal solutions. It turns out that a careful choice of the sequence of approximating functions (f_k) (or for Polya algorithms in Orlicz space of the corresponding Young functions) often implies convergence to a second stage solution in the following sense: limit points $\bar{x}$ of the sequence of minimal solutions can – under favorable conditions – be characterized as $\bar{x} \in M(g, M(f, K))$ where g is an implicitly determined or explicitly chosen function. If g is strictly convex convergence of the minimizing sequence is established.

Strategies for this type of regularization are differentiation with respect to the parameter (if the approximating functions are viewed as a parametric family rather than as a sequence) for determination of an implicitly given g or use of the convergence estimates derived above for an explicit regularization. Below we present a finite-dimensional robust version of a theorem pertinent to this question (see Theorem 5.6.12).

Theorem 2.5.8. *Let* $(f_k : \mathbb{R}^n \to \mathbb{R})_{k\in\mathbb{N}}$ *be a sequence of convex functions that converges pointwise to a function* $f : \mathbb{R}^n \to \mathbb{R}$. *Let further the set of minimal solutions of* f *on* $\mathbb{R}^n$ *be non empty and bounded.*

Let $(\alpha_k)_{k\in\mathbb{N}}$ *be a sequence with* $\alpha_k > 0$ *and* $\alpha_k \to 0$ *such that the sequence of functions* $(\frac{f_k - f}{\alpha_k})_{k\in\mathbb{N}}$ *converges lower semi-continuously (see Definition* 5.1.1*) on an open bounded superset* Q *of* $M(f, \mathbb{R}^n)$ *to a function* $g : Q \to \mathbb{R}$.

Let further (δ_k) be $o(\alpha_k)$ and $(\tilde{x}_k)$ be a sequence in $\mathbb{R}^n$ with the property

$$\|\nabla f_k(\tilde{x}_k)\| \leq \delta_k.$$

Then the set of limit points of the sequence $(\tilde{x}_k)$ is non-empty and contained in $M(g, M(f, \mathbb{R}^n))$.

Example 2.5.9. We consider once more the best L^p-approximations ($p > 1$) for $p \to 1$. Let (α_k) be as above. If we define $f_k(x) := \int_T |x(t)|^{1+\alpha_k} d\mu$ and $f(x) := \int_T |x(t)| d\mu$ then the sequence $(\frac{f_k - f}{\alpha_k})$ converges lower semi-continuously to the strictly convex function g with

$$g(x) = \int_T |x(t)| \log |x(t)| d\mu.$$

The reason rests with the fact that the mapping $\alpha \mapsto \int_T |x(t)|^{1+\alpha} d\mu$ is convex and the sequence $(\frac{f_k - f}{\alpha_k})$ represents the difference quotients (for fixed x) that converge to the derivative g (w.r.t. α) at $\alpha = 0$.

If now the best L^{p_k}-solutions with $p_k = 1 + \alpha_k$ are determined only approximately but with growing precision in the sense that the norms of the gradients converge to 0 faster than the sequence (α_k) the sequence of approximate solutions converges the best $L^1(\mu)$-approximation of largest entropy (see Equation (5.7)) (note that $-g$ is the entropy function).

A further consequence of the above theorem is the following

Theorem 2.5.10. *Let $(f_k : \mathbb{R}^n \to \mathbb{R})_{k \in \mathbb{N}}$ be a sequence of convex functions that converges pointwise to a function $f : \mathbb{R}^n \to \mathbb{R}$. Let further $M(f, \mathbb{R}^n)$ be non-empty and bounded.*

Let Q be an open and bounded superset of $M(f, \mathbb{R}^n)$ and $\varepsilon_k := \sup_{x \in Q} |f(x) - f_k(x)|$.

Moreover, let $(\alpha_k)_{k \in \mathbb{N}}$ be chosen such that $\varepsilon_k = o(\alpha_k)$, let the function $g : \mathbb{R}^n \to \mathbb{R}$ be convex, and the sequence $(h_k : \mathbb{R}^n \to \mathbb{R})_{k \in \mathbb{N}}$ be defined by

$$h_k := \alpha_k g + f_k.$$

Let further $(\delta_k)_{k \in \mathbb{N}}$ be chosen such that $\delta_k = o(\alpha_k)$ and let $(\tilde{x}_k)_{k \in \mathbb{N}}$ a sequence in $\mathbb{R}^n$ with the property

$$\|\nabla h_k(\tilde{x}_k)\| \leq \delta_k.$$

Then the set of limit points of the sequence $(\tilde{x}_k)$ is non-empty and contained in $M(g, M(f, \mathbb{R}^n))$.

If g is strictly convex then the sequence $(\tilde{x}_k)$ converges to the uniquely determined second stage solution.

By use of the convergence estimates given above we are now in the position to present a robust regularization of the Polya algorithm for the Chebyshev approximation in Orlicz space using an arbitrarily chosen strictly convex function g (see Theorem 5.6.13).

Theorem 2.5.11. *Let T be a compact metric space and V a finite-dimensional subspace of $C(T)$ and $x \in C(T)$. Let further $(\varepsilon_k)_{k\in\mathbb{N}}$ be a sequence converging to 0 such that for all $y \in x + V$*

$$|\|y\|_{(\Phi_k)} - \|y\|_\infty| \leq \varepsilon_k \|y\|_\infty.$$

We now choose a sequence of positive numbers $(\alpha_k)_{k\in\mathbb{N}}$ converging to zero such that $\lim_{k\to\infty} \frac{\varepsilon_k}{\alpha_k} = 0$ and a strictly convex function $g : x + V \to \mathbb{R}$. An outer regularization of the Luxemburg norms we obtain by defining the sequence $(h_k : x + V \to \mathbb{R})_{k\in\mathbb{N}}$ as

$$h_k := \alpha_k g + \|\cdot\|_{(\Phi_k)}.$$

If the minimal solutions of the functions h_k are determined approximately such that the norms of the gradients go to zero more rapidly than the sequence (α_k) then the sequence of the approximate solutions converges to the best Chebyshev approximation which is minimal w.r.t. g.

In the following section we will give a detailed proof for the estimates stated above.

2.5.2 Convergence Estimates

Discrete Maximum Norm

Lemma 2.5.12. *Let $(\Phi_k)_{k\in\mathbb{N}}$ be a sequence of finite Young functions, which converges pointwise to the Young function*

$$\Phi_\infty(t) := \begin{cases} 0 & \text{for } |t| \leq 1 \\ \infty & \text{otherwise} \end{cases}$$

for $|t| \neq 1$. Then for arbitrary $a > 0$ we obtain

$$\lim_{k\to\infty} \Phi_k^{-1}(a) = 1.$$

Proof. Let $0 < \varepsilon < 1$ be given, then there is a K_0, such that for $k \geq K_0$ we have: $\Phi_k(1+\varepsilon) > a$, i.e. $1+\varepsilon > \Phi_k^{-1}(a)$. Conversely there is a K_1 such that $\Phi_k(1-\varepsilon) < a$ for $k \geq K_1$ and hence $1 - \varepsilon < \Phi_k^{-1}(a)$. □

Theorem 2.5.13. *Let* $T := \{t_1, \dots, t_m\}$ *and* $\mu : T \to (0, \infty)$ *be a weight function. Let* Φ *be a finite Young function and* $x : T \to \mathbb{R}$, *then we obtain for the weighted norm*

$$\|x\|_{(\Phi),T} := \inf\left\{c \,\middle|\, \sum_{t \in T} \mu(t)\Phi\left(\frac{x(t)}{c}\right) \le 1\right\}$$

the following estimate:

$$\big|\|x\|_{(\Phi),T} - \|x\|_{\infty,T}\big| \le \varepsilon_\Phi \cdot \|x\|_{\infty,T}$$

with

$$\varepsilon_\Phi = \max\left\{\frac{1}{\Phi^{-1}(\frac{1}{\mu(T)})} - 1, 1 - \frac{1}{\Phi^{-1}(\frac{1}{\lambda})}\right\},$$

where $\mu(T) := \sum_{t \in T} \mu(t)$ *and* $\lambda := \min_{t \in T}\{\mu(t)\}$.

Proof. For a t_0 we have $|x(t_0)| = \|x\|_{\infty,T}$. Let

$$\bar{x}(t) := \begin{cases} |x(t_0)| & \text{for } t = t_0 \\ 0 & \text{otherwise.} \end{cases}$$

Then we obtain because of the monotonicity of the Luxemburg norm

$$\|x\|_{(\Phi),T} \ge \|\bar{x}\|_{(\Phi),T} \ge \inf\left\{c \,\middle|\, \lambda\Phi\left(\frac{\|x\|_{\infty,T}}{c}\right) \le 1\right\} = \|x\|_{\infty,T} \cdot \frac{1}{\Phi^{-1}(\frac{1}{\lambda})}.$$

On the other hand, if we assume $\tilde{x}$ to be equal to the constant $\|x\|_{\infty,T}$ on T, then we obtain

$$\begin{aligned}\|x\|_{(\Phi),T} \le \|\tilde{x}\|_{(\Phi),T} &= \inf\left\{c \,\middle|\, \mu(T)\Phi\left(\frac{\|x\|_{\infty,T}}{c}\right) \le 1\right\} \\ &= \|x\|_{\infty,T} \cdot \frac{1}{\Phi^{-1}(\frac{1}{\mu(T)})}.\end{aligned}$$

□

Corollary 2.5.14. *Let* $(\Phi_k)_{k \in \mathbb{N}}$ *be a sequence of finite Young functions, which converges pointwise to the die Young function* Φ_∞. *Let*

$$\varepsilon_k = \max\left\{\frac{1}{{\Phi^{-1}}_k(\frac{1}{\mu(T)})} - 1, 1 - \frac{1}{{\Phi^{-1}}_k(\frac{1}{\lambda})}\right\},$$

then $(\varepsilon_k)_{k \in \mathbb{N}}$ *is a sequence tending to zero and we obtain the following estimate:*

$$\big|\|x\|_{(\Phi_k),T} - \|x\|_{\infty,T}\big| \le \varepsilon_k \cdot \|x\|_{\infty,T}.$$

Maximum Norm in $C^\alpha(T)$

It is our aim in this section to present the above mentioned estimates for the subspace of Hölder–Lipschitz continuous functions on a compact subset of $\mathbb{R}^r$ that satisfies a cone condition, where μ is the Lebesgue measure of $\mathbb{R}^r$.

Definition 2.5.15. A subset T of $\mathbb{R}^r$ satisfies a *cone condition*, if there is a cone C with interior point and a number $\rho > 0$, such that for all $t \in T$ there is an orthogonal transformation A_t with the property that

$$t + (\rho B \cap A_t C) \subseteq T,$$

if B denotes the unit ball w.r.t. the Euclidean norm.

Remark 2.5.16. Every finite union of sets which satisfy a cone condition also satisfies this condition. Every compact convex subset T of the $\mathbb{R}^r$ with interior point satisfies a cone condition.

Lemma 2.5.17. *Let μ be the Lebesgue measure on $\mathbb{R}^r$ and T a compact subset of the $\mathbb{R}^r$, which satisfies a cone condition, let further*

$$q_\lambda(t) := \inf\{\delta > 0 \,|\, \mu((t + \delta B) \cap T) \geq \lambda\} \quad \textit{for } 0 < \lambda \leq \mu(T) \textit{ and } t \in T,$$

then there are positive numbers λ_0 and c such that

$$q_\lambda(t) \leq c\lambda^{\frac{1}{r}} \quad \textit{for } 0 < \lambda \leq \lambda_0 \textit{ and } t \in T,$$

and we have

$$\lambda = \mu((t + q_\lambda(t)B) \cap T).$$

Proof. Let $\lambda \leq \rho^r \mu(B \cap C) =: \lambda_0$, and let $Q_{\lambda,t} := \{\delta > 0 \,|\, \mu((t + \delta B) \cap T) \geq \lambda\}$, then we obtain because of the cone condition for T and the invariance of μ w.r.t. translations and orthogonal transformations and the homogeneity of μ

$$\mu((t + \rho B) \cap T) \geq \mu(t + (A_t C \cap \rho B)) = \rho^r \mu(B \cap C) = \lambda_0,$$

hence $\rho \in Q_{\lambda_0,t}$ and thus $q_{\lambda_0}(t) \leq \rho$. For $\lambda \leq \lambda_0$ we have $Q_{\lambda,t} \supseteq Q_{\lambda_0,t}$ and thus $q_\lambda(t) \leq q_{\lambda_0}(t)$. Due to the continuity of μ w.r.t. δ in the definition of $q_\lambda(t)$ the infimum is assumed there and we obtain as above

$$\lambda = \mu((t + q_\lambda(t)B) \cap T) \geq (q_\lambda(t))^r \mu(B \cap C).$$

With the definition $c := (\mu(B \cap C))^{-\frac{1}{r}}$ the assertion follows. □

Definition 2.5.18. Let T be a compact subset of the $\mathbb{R}^r$ and let $0 < \alpha \leq 1$. We define $C^\alpha(T)$ as the space of all real valued functions x on T with the property: there is a number $L \in \mathbb{R}$ with

$$|x(t) - x(t')| \leq L\|t - t'\|^\alpha$$

for all $t, t' \in T$, where $\|\cdot\|$ denotes the Euclidean norm of $\mathbb{R}^r$. Let further be

$$L_\alpha(x) := \inf\{L > 0 \,|\, |x(t) - x(t')| \leq L\|t - t'\|^\alpha \text{ for all } t, t' \in T, t \neq t'\}.$$

Remark 2.5.19. It is easily seen: L_α is a semi-norm on $C^\alpha(T)$, since apparently

$$|(x(t)+y(t)) - (x(t')+y(t'))| \leq (L_\alpha(x)+L_\alpha(y))\|t - t'\|^\alpha \quad \text{for all } t, t' \in T,\ t \neq t',$$

hence $L_\alpha(x + y) \leq L_\alpha(x) + L_\alpha(y)$. The positive homogeneity is obvious.

Lemma 2.5.20. *Let Φ be a finite Young function, then the equation*

$$s^r \Phi(1 + s^\alpha) = 1$$

has a unique positive solution ε_Φ.

Proof. Let $s_0 := \sup\{s \geq 0 \,|\, \Phi(s) = 0\}$, then $0 \leq s_0 < \infty$, because Φ, being a Young function, cannot be identical to zero. Let now $s_0 < s_1 < s_2$, then there is a $\lambda \in (0, 1)$ with $s_1 = (1 - \lambda)s_0 + \lambda s_2$ and thus

$$0 < \Phi(s_1) \leq (1 - \lambda)\Phi(s_0) + \lambda\Phi(s_2) = \lambda\Phi(s_2) < \Phi(s_2),$$

i.e. Φ is strictly increasing on $[s_0, \infty)$. Let now Φ'_+ be the right-sided derivative then the subgradient inequality yields $\Phi(s) \leq s \cdot \Phi'_+(s)$, in particular $\Phi'_+(s) > 0$ for $s > s_0$. Let now $s_1 > s_0$, then we obtain due to the subgradient inequality

$$\Phi(s_1) + \Phi'_+(s_1)(s - s_1) \leq \Phi(s),$$

and hence $\lim_{s\to\infty} \Phi(s) = \infty$. Let now $u(s) := s^r\Phi(1 + s^\alpha)$, then apparently u is strictly increasing on $[s_0, \infty)$ and according to Theorem 5.3.11 continuous there. Furthermore we have: $u(0) = 0$ and $\lim_{s\to\infty} u(s) = \infty$, and the intermediate value theorem yields the assertion. □

Lemma 2.5.21. *Let T be a compact subset of the $\mathbb{R}^r$ that satisfies a cone condition, let μ be the Lebesgue measure, and Φ a finite Young function with $\varepsilon_\Phi \leq \lambda_0^{1/r}$ (see Lemma 2.5.17), if ε_Φ is the solution of the equation $s^r\Phi(1 + s^\alpha) = 1$.*

If $x \in C^\alpha(T)$, then with c as in Lemma 2.5.17 we obtain

$$\|x\|_\infty \leq \|x\|_{(\Phi)} + \varepsilon_\Phi^\alpha(\|x\|_{(\Phi)} + c^\alpha L_\alpha(x)).$$

On the other hand with $\sigma_\Phi := \frac{1}{\Phi^{-1}(\frac{1}{\mu(T)})} - 1$ the inequality

$$\|x\|_{(\Phi)} \leq \|x\|_\infty + \sigma_\Phi\|x\|_\infty$$

holds.

Proof. Let $t_0 \in T$ be chosen such that $|x(t_0)| = \|x\|_\infty$. Let further $I_\Phi := (t_0 + q_{\varepsilon_\Phi{}^r}(t_0)B) \cap T$. For all $t \in T$ we have with $z(t) := \|t_0 - t\|^\alpha$ in the natural order

$$\|x\|_\infty \chi_{I_\Phi} \leq |x|\chi_{I_\Phi} + zL_\alpha(x)\chi_{I_\Phi}.$$

We obtain $\|\chi_{I_\Phi}\|_{(\Phi)} = \frac{1}{\Phi^{-1}(\frac{1}{\mu(I_\Phi)})}$, and due to the monotonicity of the Luxemburg norm $\||x|\chi_{I_\Phi}\|_{(\Phi)} \leq \|x\|_{(\Phi)}$. By Lemma 2.5.17 we have: $q_{\varepsilon_\Phi^r}(t_0) \leq c\varepsilon_\Phi$, hence

$$z\chi_{I_\Phi} \leq (q_{\varepsilon_\Phi^r}(t_0))^\alpha \chi_{I_\Phi} \leq (c\varepsilon_\Phi)^\alpha \chi_{I_\Phi}, \quad \text{i.e. } \|z\chi_{I_\Phi}\|_{(\Phi)} \leq (c\varepsilon_\Phi)^\alpha \|\chi_{I_\Phi}\|_{(\Phi)},$$

using again the monotonicity of the Luxemburg norm. We thus obtain

$$\begin{aligned}
\|x\|_\infty &\leq \Phi^{-1}\left(\frac{1}{\mu(I_\Phi)}\right)\|x\|_{(\Phi)} + (c\varepsilon_\Phi)^\alpha L_\alpha(x) \\
&= \|x\|_{(\Phi)} + \left(\Phi^{-1}\left(\frac{1}{\varepsilon_\Phi^r}\right) - 1\right)\|x\|_{(\Phi)} + (c\varepsilon_\Phi)^\alpha L_\alpha(x) \\
&= \|x\|_{(\Phi)} + \varepsilon_\Phi^\alpha(\|x\|_{(\Phi)} + c^\alpha L_\alpha(x)).
\end{aligned}$$

Apparently

$$\int_T \Phi\left(\frac{x}{\|x\|_\infty}\Phi^{-1}\left(\frac{1}{\mu(T)}\right)\right) d\mu \leq 1,$$

and hence the second part of the assertion. □

Remark 2.5.22. If $\Phi(1) \leq \frac{1}{\mu(T)}$, then $\|x\|_{(\Phi)} \leq \|x\|_\infty$ holds.

Lemma 2.5.23. *Let $(\Phi_k)_{k\in\mathbb{N}}$ be a sequence of finite Young functions with $\lim_{k\to\infty} \Phi_k(s) = \infty$ for $s > 1$.*

Then the solutions ε_k of the equations $s^r\Phi_k(1 + s^\alpha) = 1$ form a sequence tending to zero.

Proof. Suppose this is not the case. Then there is a subsequence (ε_n) and a positive number γ with $\varepsilon_n \geq \gamma$ for $n = 1, 2, \ldots$.

For n large enough the inequality $\Phi_n(1+\gamma^\alpha) \geq \frac{2}{\gamma^r}$ would hold and hence $\varepsilon_n^r\Phi_n(1+\varepsilon_n^\alpha) \geq 2$, contradicting the definition of ε_n. □

Lemma 2.5.24. *Let $(\Phi_k)_{k\in\mathbb{N}}$ as in the previous lemma and let in addition $\lim_{k\to\infty} \Phi_k(s) = 0$ for $s \in (0, 1)$, then*

$$\lim_{k\to\infty} \Phi_k^{-1}(a) = 1 \quad \textit{for } a > 0.$$

Proof. Let $0 < \varepsilon < 1$, then for k large enough: $\Phi_k(1 + \varepsilon) > a$, i.e. $1 + \varepsilon > \Phi_k^{-1}(a)$. On the other hand one obtains for sufficiently large k: $\Phi_k(1 - \varepsilon) < a$ and thus $1 - \varepsilon < \Phi_k^{-1}(a)$. □

We are now in the position to state the convergence estimate in a form suitable for the regularisation method in Theorem 2.5.5.

Theorem 2.5.25. *Let T be a compact subset of $\mathbb{R}^r$, which satisfies a cone condition, μ the Lebesgue measure and $(\Phi_k)_{k\in\mathbb{N}}$ a sequence of finite Young functions with*

$$\lim_{k\to\infty} \Phi_k(s) = \begin{cases} \infty & \text{for } |s| > 1 \\ 0 & \text{for } |s| < 1. \end{cases}$$

Let K be a finite-dimensional subset of $C^\alpha(T)$, then there is a sequence $(\delta_k)_{k\in\mathbb{N}}$ tending to zero, defined by $\delta_k := \max(\varepsilon_k^\alpha, \sigma_k)$, where ε_k is the solution of the equation $s^r\Phi_k(1+s^\alpha) = 1$ and $\sigma_k := \frac{1}{\Phi_k^{-1}(\frac{1}{\mu(T)})} - 1$, and there is a number $\zeta > 0$, such that

$$\big|\|x\|_\infty - \|x\|_{(\Phi_k)}\big| \le \zeta\|x\|_\infty\delta_k$$

for all $x \in K$.

Proof. L_α being a semi-norm is in particular convex and thus according to Theorem 5.3.11 continuous on the span of K, i.e. there is a number κ, such that on the span of K $L_\alpha(x) \le \kappa\|x\|_\infty$.

Lemma 2.5.21 then yields

$$\|x\|_\infty - \|x\|_{(\Phi_k)} \le \varepsilon_k^\alpha(\sigma_k + 1 + c^\alpha\kappa)\|x\|_\infty.$$

The inequality claimed in the assertion of the theorem then follows using the preceding lemmata. □

Example 2.5.26. (a) $\Phi_k(s) := as^2 + k(k(|s|-1)_+)^m$ with $a > 0$ and $m \in \mathbb{N}$.

For every $c \in (0,1)$ there is a $k_0 \in \mathbb{N}$, such that for $k \ge N$

$$\frac{c}{k^q} < \varepsilon_k^\alpha < \frac{1}{k^q} \quad \text{with} \quad q := \frac{m+1}{m+\frac{r}{\alpha}}.$$

Proof. Let $p := q/\alpha$, then because of $p\alpha m - (m+1) = -pr$:

$$\begin{aligned}\left(\frac{1}{k^p}\right)^r \Phi_k\left(1+\frac{1}{k^{p\alpha}}\right) &= \frac{1}{k^{pr}}\left(a\left(1+\frac{1}{k^{p\alpha}}\right)^2 + k\left(k\cdot\frac{1}{k^{p\alpha}}\right)^m\right) \\ &= \frac{1}{k^{pr}}a\left(1+\frac{1}{k^{p\alpha}}\right)^2 + 1 > 1.\end{aligned}$$

On the other hand

$$\begin{aligned}\left(\frac{c}{k^p}\right)^r \Phi_k\left(1+\left(\frac{c}{k^p}\right)^\alpha\right) &= \frac{c^r}{k^{pr}}\left(a\left(1+\frac{c^\alpha}{k^{p\alpha}}\right)^2 + k\left(k\cdot\frac{c^\alpha}{k^{p\alpha}}\right)^m\right) \\ &= \frac{c^r}{k^{pr}}a\left(1+\frac{c^\alpha}{k^{p\alpha}}\right)^2 + c^{\alpha m+r} < 1\end{aligned}$$

for k sufficiently large, hence $c\frac{1}{k^p} < \varepsilon_k < \frac{1}{k^p}$. □

(b) $\Phi_k(s) := |s|^k$. For every $c \in (0,1)$ there is $N \in \mathbb{N}$, such that for $k \geq N$

$$c\frac{r}{\alpha}\frac{\log k}{k} \leq \varepsilon_k^\alpha \leq \frac{r}{\alpha}\frac{\log(k-1)}{k-1}.$$

Proof. For $u > 0$ we have: $(1+\frac{u}{k})^k \leq e^u \leq (1+\frac{u}{k})^{k+1}$. Hence

$$\left(\left(\frac{r}{\alpha}\frac{\log(k-1)}{k-1}\right)^{\frac{1}{\alpha}}\right)^r \left(1+\frac{r}{\alpha}\frac{\log(k-1)}{k-1}\right)^k \geq (k-1)^{\frac{r}{\alpha}}\left(\frac{r}{\alpha}\log(k-1)\right)^{\frac{r}{\alpha}}(k-1)^{-\frac{r}{\alpha}} \xrightarrow{k\to\infty} \infty.$$

On the other hand

$$\left(\left(c\frac{r}{\alpha}\frac{\log k}{k}\right)^{\frac{1}{\alpha}}\right)^r \left(1+c\frac{r}{\alpha}\frac{\log k}{k}\right)^k \leq k^{(c-1)\frac{r}{\alpha}}\left(c\frac{r}{\alpha}\log k\right)^{\frac{r}{\alpha}} \xrightarrow{k\to\infty} 0. \qquad \square$$

This corresponds to the results of Peetre [90] for intervals and Hebden [37] for cuboids in $\mathbb{R}^r$ and continuously differentiable functions.

L^1-norm

Lemma 2.5.27. *Let (T,Σ,μ) be a finite measure space, Φ a finite, definite Young function, then we obtain for all $x \in L^1(\mu) \cap L^\Phi(\mu)$*

$$\|x\|_1 \leq \|x\|_{(\Phi)} + \varepsilon_\Phi(c_\Phi + 1)\|x\|_1,$$

where ε_Φ denotes the larger solution of the equation $\Phi^{-1}(s) = \frac{s}{1-s}$ and c_Φ is defined by $c_\Phi := \frac{1}{\Phi^{-1}(\frac{1}{\mu(T)})}$.

Proof. Let $\Phi'_+(0)$ denote the always existing right-sided derivative of Φ at 0. If $\Phi'_+(0) \geq 1$, then $\Phi(s) \geq |s|$ on $\mathbb{R}$ and hence $\|x\|_1 \leq \|x\|_\Phi$ on $L^\Phi(\mu)$. The equation $\Phi^{-1}(s) = \frac{s}{1-s}$ only has the solution $\varepsilon_\Phi = 0$.

If now $\Phi'_+(0) < 1$, we define $h : \mathbb{R}_+ \to \mathbb{R}$ by $h(s) := \frac{\Phi(s)}{s}$ for $s > 0$ and $h(0) := \Phi'_+(0)$. h is continuous and monotonically increasing on $\mathbb{R}_+$ (see Theorem 3.3.1). Let further be $g(t) := h(\Phi^{-1}(t))$ for $t \geq 0$, then g is also continuous and monotonically increasing on $\mathbb{R}_+$ and we have $g(0) = \Phi'_+(0)$. Let finally $j(t) := t + g(t)$ for $t \geq 0$, then $j(0) < 1$ and $j(1) = 1 + h(\Phi^{-1}(1)) > 1$, because $\Phi^{-1}(1) > 0$ and Φ definite.

Due to the strict monotonicity of j there is a unique ε_Φ with $j(\varepsilon_\Phi) = 1$.

Let now $x \in L^1(\mu) \cap L^\Phi(\mu) \setminus \{0\}$, then we define

$$x_\Phi := |x| + \varepsilon_\Phi\|x\|_1.$$

In the natural order the following inequality holds:

$$x_\Phi \geq \varepsilon_\Phi\|x\|_1;$$

for the integral norm we have

$$\|x_\Phi\|_1 = \|x\|_1(1 + \varepsilon_\Phi \mu(T)).$$

Using the monotonicity of g we then obtain

$$1 + \varepsilon_\Phi \mu(T) = \int_T \frac{x_\Phi}{\|x\|_1} d\mu = \int_T \Phi\left(\frac{x_\Phi}{g(\frac{x_\Phi}{\|x\|_1})\|x\|_1}\right) d\mu$$
$$\leq \int_T \Phi\left(\frac{x}{g(\varepsilon_\Phi)\|x\|_1}\right) d\mu.$$

The monotonicity of the Luxemburg norm then yields

$$g(\varepsilon_\Phi)\|x\|_1 \leq \|x_\Phi\|_{(\Phi)} \leq \|x\|_{(\Phi)} + \varepsilon_\Phi \|x\|_1 \, \|\chi_T\|_{(\Phi)} \quad \text{i.e.}$$
$$\|x\|_1 \leq \|x\|_{(\Phi)} + \|x\|_1(1 - g(\varepsilon_\Phi) + \varepsilon_\Phi c_\Phi).$$

The definition of ε_Φ leads immediately to the assertion. □

Lemma 2.5.28. *Let X be a finite dimensional subspace of $L^\infty(\mu)$, (T, Σ, μ) a finite measure space, Φ a finite Young function and ξ chosen in such a way that $\|\cdot\|_\infty \leq \xi\|\cdot\|_1$ on X, then for $x \in X$*

$$\|x\|_{(\Phi)} \leq \|x\|_1 + (g_\Phi(\xi) - 1)\, \|x\|_1,$$

if $g_\Phi(s) := \frac{s}{\Phi^{-1}(s)}$ for $s > 0$.

Proof. Let $x \in X \setminus \{0\}$, then because of the monotonicity of g_Φ

$$1 = \int_T \frac{|x|}{\|x\|_1} d\mu = \int_T \Phi\left(\frac{x}{g_\Phi(\frac{|x|}{\|x\|_1}\|x\|_1)}\right) d\mu$$
$$\geq \int_T \Phi\left(\frac{x}{g_\Phi(\xi)\|x\|_1}\right) d\mu,$$

i.e. $\|x\|_\Phi \leq g_\Phi(\xi)\|x\|_1$. □

Lemma 2.5.29. *Let $(\Phi_k)_{k\in\mathbb{N}}$ be a sequence of finite, definite Young functions with $\Phi'_{k+}(0) < 1$. If $\lim_{k\to\infty} \Phi_k(s) = |s|$ for $s \in \mathbb{R}$ and if ε_k is the positive solution of the equation $\Phi_k^{-1}(s) = \frac{s}{1-s}$, then $(\varepsilon_k)_{k\in\mathbb{N}}$ and $(g_{\Phi_k}(a) - 1)_{k\in\mathbb{N}}$ for $a > 0$ are sequences tending to zero.*

Proof. Apparently $\lim_{k\to\infty} g_{\Phi_k}(a) = 1$ for $a > 0$. Suppose there is a subsequence (ε_n) and a positive number γ with $\varepsilon_n \geq \gamma$, then due to the monotonicity of g_{Φ_k}

$$\varepsilon_n + g_{\Phi_n}(\varepsilon_n) \geq \gamma + g_{\Phi_n}(\gamma) \geq \frac{\gamma}{2} + 1$$

for n sufficiently large, contradicting the definition of ε_n. □

We are now in the position to formulate the subsequent theorem:

Theorem 2.5.30. *Let (T, Σ, μ) be a finite measure space, $(\Phi_k)_{k\in\mathbb{N}}$ a sequence of finite, definite Young functions with $\lim_{k\to\infty} \Phi_k(s) = |s|$ on $\mathbb{R}$. Let further K be a finite-dimensional subset of $L^\infty(\mu)$ and ξ be chosen such that $\|\cdot\|_\infty \le \xi\|\cdot\|_1$ on the span of K, then there is a sequence $(\delta_k)_{k\in\mathbb{N}}$ tending to zero defined by $\delta_k := \max((g_{\Phi_k}(\xi) - 1), \varepsilon_k)$, where ε_k is the larger solution of $\Phi_k^{-1}(s) = \frac{s}{1-s}$ and $g_{\Phi_k}(s) := \frac{s}{\Phi_k^{-1}(s)}$ for $s > 0$, with the property*

$$|\|x\|_1 - \|x\|_{\Phi_k}| \le \xi\|x\|_1\delta_k \quad \textit{for all } x \in K.$$

Proof. This follows immediately from the preceding lemmata. □

Remark 2.5.31. If in addition $\Phi_k(s) \le |s|$ on $\mathbb{R}$, then

$$|\|x\|_{\Phi_k} - \|x\|_1| \le (1 + \mu(T))\|x\|_1\varepsilon_k$$

for all $x \in L^1(\mu)$.

Examples 2.5.32. (a) $\Phi_k(s) := |s|^{1+\frac{1}{k}}$

Depending on the choice of $c\varepsilon(0, 1)$ we have for sufficiently large k

$$c\frac{\log k}{k} \le \varepsilon_k \le \frac{\log k}{k}.$$

Furthermore we obtain for $\xi > 1$: $g_{\Phi_k}(\xi) - 1 \le \frac{\xi-1}{k+1}$.

(b) $\Phi_k(s) := |s| - \frac{1}{k}\log(1 + k|s|)$ we have for sufficiently large k: $\varepsilon_k \le (\frac{\log k}{k})^{\frac{1}{2}}$.

For modulars the following estimate is available:

Theorem 2.5.33. *Let (T, Σ, μ) be a finite measure space, $(\Phi_k)_{k\in\mathbb{N}}$ a sequence of finite Young functions with $\lim_{k\to\infty} \Phi_k(s) = |s|$ on $\mathbb{R}$.*

Let further K be a finite-dimensional subset of $L^\infty(\mu)$, let $x_0 \in K$ arbitrary and $Q := \{x \in L^\infty \mid \|x\|_1 \le 2\|x_0\|_1\}$.

Then Q is a compact neighborhood of $M(\|\cdot\|_1, K)$ and we have

$$\sup_{x\in Q} |\|x\|_1 - f^{\Phi_k}(x)| \le \mu(T) \sup_{s\in[0,ac]} |s - \Phi_k(s)|$$

with $a := 2\|x_0\|_1$ and c chosen, such that $\|\cdot\|_\infty \le c\|\cdot\|_1$ on the span of K.

Proof. The statement about Q is obvious. Furthermore we have

$$\begin{aligned}\sup_{x\in Q} |\|x\|_1 - f^{\Phi_k}(x)| &\le \sup_{x\in Q} \int_T ||x| - \Phi_k(x)|d\mu \\ &\le \sup_{x\in Q} \mu(T) \sup_{s\in[0,\|x\|_\infty]} |s - \Phi_k(s)| \\ &\le \mu(T) \sup_{s\in[0,ac]} |s - \Phi_k(s)|.\end{aligned}$$

The right-hand side is, according to Theorems 5.3.6 and 5.3.13, a sequence tending to zero. □

Examples 2.5.34. We denote $\sup_{s\in[0,ac]} |s - \Phi_k(s)|$ by ε_k, then we obtain for

(a) $\Phi_k(s) := |s|^{1+\frac{1}{k}}$ the following inequality: $\varepsilon_k \leq \max(1, ac(ac-1))\frac{1}{k}$

(b) for $\Phi_k(s) := |s| - \frac{1}{k}\log(1 + k|s|)$ the inequality

$$\varepsilon_k \leq 2\frac{\log k}{k} \quad \text{for } k \geq ac + 1$$

(c) and for $\Phi_k(s) := \frac{2}{\pi}(s\arctan(ks) - \frac{1}{2k}\log(1 + k^2s^2))$ finally

$$\varepsilon_k \leq 3\frac{\log k}{k} \quad \text{for } k \geq ac + 1.$$

2.6 A Polya–Remez Algorithm in $C(T)$

We will now take up the approximation problems treated in earlier sections for finite-dimensional subspaces of continuous functions on an arbitrary compact metric space T (mostly as a subset of $\mathbb{R}^l$, with $l \in \mathbb{N}$):

For a given point x in $C(T)$ we look for an element in a certain subset M of $C(T)$ that – among all points of M has the least distance from x, where the distance is understood in the following sense:

$$\|x - y\|_{\infty,T} = \max_{t\in T} |x(t) - y(t)|. \tag{2.6}$$

Denoting the norm $\|\cdot\|_{\infty,T}$ with subscript T is meant to express the dependence on the set T, which we intend to vary in the sequel. In analogy to the maximum norm we will use this type of notation also for the Luxemburg norm $\|\cdot\|_{(\Phi),T}$.

The following algorithmic considerations are based on the theorem of de la Vallée-Poussin proved in Chapter 1. For convenience we restate this theorem once again:

Theorem 2.6.1 (Theorem of de la Vallée-Poussin). *Let v_0 be a best Chebyshev approximation of $x \in C(T)$ w.r.t. the n-dimensional subspace V of $C(T)$. Then there is a finite (non-empty) subset T_0 of the extreme points $E(x - v_0, T)$, which does not contain more than $n+1$ points and for which $v_0|_{T_0}$ is a best approximation of $x|_{T_0}$ w.r.t. $V|_{T_0}$, i.e. for all $v \in V$ we have*

$$\|x - v_0\|_{\infty,T} = \|x - v_0\|_{\infty,T_0} \leq \|x - v\|_{\infty,T_0}.$$

This theorem is, as was seen above, a consequence of the Characterization Theorem (see Theorem 1.3.3), which in the sequel will facilitate the constructive choice of a de la Vallée-Poussin set in the sense of the above theorem.

Theorem 2.6.2 (Characterization Theorem). *Let $V = \text{span}\{v_1, \ldots, v_n\}$ be a subspace of $C(T)$ and $x \in C(T)$. v_0 is a best Chebyshev approximation of x w.r.t. V, if and only if q points $t_1, \ldots, t_q \in E(x - v_0, T)$ with $1 \leq q \leq n+1$ and q positive numbers $\alpha_1, \ldots, \alpha_q$ exist, such that for all $v \in V$*

$$\sum_{j=1}^{q} \alpha_j (x(t_j) - v_0(t_j)) v(t_j) = 0.$$

The interpretation of the subsequent theorem is the following: when varying the set T the Kuratowski convergence is responsible for the convergence of the corresponding maximum norms.

Definition 2.6.3. Let X be a metric space and $(M_n)_{n \in \mathbb{N}}$ a sequence of subsets of X. Then

$$\overline{\lim_n} M_n := \Big\{x \in X \mid \text{there is a subsequence } (M_k)_{k \in \mathbb{N}} \text{ of } (M_n)_{n \in \mathbb{N}} \text{ and } x_k \in M_k \text{ with } x = \lim_{k \to \infty} x_k\Big\}$$

$$\underline{\lim_n} M_n := \Big\{x \in X \mid \text{there is a } n_0 \in \mathbb{N} \text{ and } x_n \in M_n \text{ for } n \geq n_0 \text{ with } x = \lim_{n \to \infty} x_n\Big\}.$$

The sequence $(M_n)_{n \in \mathbb{N}}$ is said to be *Kuratowski convergent* to the subset M of X if

$$\overline{\lim_n} M_n = \underline{\lim_n} M_n = M,$$

notation: $M = \lim_n M_n$.

Theorem 2.6.4. *Let T be a compact metric space and $x \in C(T)$. Let further $(T_k)_{k \in \mathbb{N}}$ be a sequence of compact subsets of T, which converges in the sense of Kuratowski to T_0. Then we obtain*

$$\lim_{k \to \infty} \|x\|_{\infty, T_k} = \|x\|_{\infty, T_0},$$

i.e. the sequence of the corresponding maximum norms converges pointwise on $C(T)$.

Proof. Let $\tau_0 \in T_0$ be chosen such that $|x(\tau_0)| = \|x\|_{\infty, T_0}$. From the Kuratowski convergence follows the existence of a sequence $(t_k)_{k \in \mathbb{N}}$ with $t_k \in T_k$ for all $k \geq n_0$, which converges to τ_0. From the continuity of x and $t_k \in T_k \subset T$ it then follows that

$$\|x\|_{\infty, T} \geq \|x\|_{\infty, T_k} \geq |x(t_k)| \xrightarrow{k \to \infty} |x(\tau_0)| = \|x\|_{\infty, T_0}.$$

Let on the other hand $\tau_k \in T_k$ with $|x(\tau_k)| = \|x\|_{\infty, T_k}$ and $s := \sup\{|x(\tau_k)|\}$. Then there is a subsequence $(\tau_{k_m})_{m \in \mathbb{N}}$ converging to a $\tau \in T$ with $\lim_{m \to \infty} |x(\tau_{k_m})| = s$.

From the continuity of x it follows that also $\lim_{m\to\infty}|x(\tau_{k_m})| = |x(\tau)|$. The Kuratowski convergence guarantees $\tau \in T_0$ and thus

$$\|x\|_{\infty,T_0} \geq s \geq \|x\|_{\infty,T_k}$$

for all $k \in \mathbb{N}$. □

This theorem provides the opportunity to develop strategies for the computation of best Chebyshev approximations for functions being defined on a continuous multi-dimensional domain. In addition to an obvious approach to use a successively finer point grid (mesh width to zero), we have the possibility to keep the number of discretization points small, where we just have to take care that the sequence of discretization sets converges in the sense of Kuratowski to a superset of the critical points which result from the theorem of de la Vallée-Poussin.

Using the Stability Theorem 5.3.25 we then obtain

Theorem 2.6.5. *Let M be a closed convex subset of a finite-dimensional subspace of $C(T)$ and $x \in C(T)$. Let T_d be a finite subset of T with the property that the best Chebyshev approximations w.r.t. $\|\cdot\|_{\infty,T_d}$ and $\|\cdot\|_{\infty,T}$ agree as functions on T. Let now $(T_k)_{k\in\mathbb{N}}$ be a sequence of compact subsets of T which converges in the sense of Kuratowski to a superset T_0 of T_d. Then for $K := x - M$ we have*

(a) $\overline{\lim}_k M(\|\cdot\|_{\infty,T_k}, K)$ *is non-empty and* $\bigcup_{k\in\mathbb{N}} M(\|\cdot\|_{\infty,T_k}, K)$ *is bounded*

(b) $\overline{\lim}_k M(\|\cdot\|_{\infty,T_k}, K) \subset M(\|\cdot\|_{\infty,T}, K)$

(c) $\inf_{y\in K}\|y\|_{\infty,T_k} \to_{k\to\infty} \inf_{y\in K}\|y\|_{\infty,T}$

(d) $y_k \in M(\|\cdot\|_{\infty,T_k}, K)$ *implies* $\|y_k\|_{\infty,T} \to \inf_{y\in K}\|y\|_{\infty,T}$.

If one wants to realize the scheme of the previous theorem algorithmically, an obvious approach is to compute each of the best approximations w.r.t. $\|\cdot\|_{\infty,T_k}$ by a corresponding discrete Polya algorithm, at least in an approximative sense. If, beyond that we are able to keep the number of elements in the set sequence (T_k) uniformly bounded, even a *diagonal* procedure is feasible, a process that is put into precise terms in the two subsequent theorems.

Theorem 2.6.6. *Let T be a compact metric space and $x \in C(T)$. Let further $(T_k)_{k\in\mathbb{N}_0}$ be a sequence of finite subsets of T, whose number of elements does not exceed a given constant and converges in the sense of Kuratowski to T_0.*

Let further $(\Phi_k)_{k\in\mathbb{N}}$ be a sequence of Young functions, which converges pointwise to the Young function

$$\Phi_\infty(t) := \begin{cases} 0 & \text{for } |t| \leq 1 \\ \infty & \text{otherwise} \end{cases}$$

for $|t| \neq 1$. Then we obtain

$$\lim_{k\to\infty} \|x\|_{(\Phi_k),T_k} = \|x\|_{\infty,T_0},$$

i.e. the sequence of the corresponding Luxemburg norms converges pointwise on $C(T)$.

Proof. We obtain

$$\left|\|x\|_{(\Phi_k),T_k} - \|x\|_{\infty,T_0}\right| \leq \left|\|x\|_{(\Phi_k),T_k} - \|x\|_{\infty,T_k}\right| + \left|\|x\|_{\infty,T_k} - \|x\|_{\infty,T_0}\right|.$$

According to Theorem 2.5.13 we have

$$\left|\|x\|_{(\Phi_k),T_k} - \|x\|_{\infty,T_k}\right| \leq \varepsilon_k \cdot \|x\|_{\infty,T_k}$$

with

$$\varepsilon_k = \max\left\{\frac{1}{\Phi_k^{-1}(\frac{1}{|T_k|})} - 1, 1 - \frac{1}{\Phi_k^{-1}(1)}\right\}.$$

By assumption there is a $N \in \mathbb{N}$ with the property $|T_k| \leq N$. Then, due to Theorem 2.5.13 the sequence $(\varepsilon_k)_{k\in\mathbb{N}}$ tends to zero. Due to the convergence of the Chebyshev norms the assertion of the theorem follows. □

Under the conditions stated above we can now approximate the best continuous Chebyshev approximation by a sequence of best discrete approximations w.r.t. Luxemburg norms in the sense of Polya.

Theorem 2.6.7. *Let M be a closed convex subset of a finite-dimensional subspace of $C(T)$ and $x \in C(T)$. Let T_d be a finite subset of T with the property that the best Chebyshev approximations w.r.t. $\|\cdot\|_{\infty,T_d}$ and $\|\cdot\|_{\infty,T}$ agree as functions on T. Let now $(T_k)_{k\in\mathbb{N}}$ be a sequence of finite subsets of T, whose number of elements does not exceed a given constant and converges in the sense of Kuratowski to a superset T_0 of T_d. Let further $(\Phi_k)_{k\in\mathbb{N}}$ be a sequence of Young functions, which converges pointwise to the Young function Φ_∞. Then we obtain for $K := x - M$*

(a) $\overline{\lim}_k M(\|\cdot\|_{(\Phi_k),T_k}, K)$ *is non-empty and* $\bigcup_{k\in\mathbb{N}} M(\|\cdot\|_{(\Phi_k),T_k}, K)$ *is bounded*

(b) $\overline{\lim}_k M(\|\cdot\|_{(\Phi_k),T_k}, K) \subset M(\|\cdot\|_{\infty,T}, K)$

(c) $\inf_{y\in K}\|y\|_{(\Phi_k),T_k} \to_{k\to\infty} \inf_{y\in K}\|y\|_{\infty,T}$

(d) *from* $y_n \in M(\|\cdot\|_{(\Phi_k),T_k}, K)$ *it follows that* $\|y_k\|_{\infty,T} \to \inf_{y\in K}\|y\|_{\infty,T}$.

Algorithmic Scheme

Let T be a compact metric space and K a closed convex subset of an n-dimensional subspace of $C(T)$. Let further $0 < \sigma < 1$, $\varepsilon > 0$ and $T_1 \subset T$.

(a) Set $k := 1$, $D_1 := T_1$. Determine $x_1 \in M(\|\cdot\|_{\infty,D_1}, K)$ and set

$$\rho_1 := \frac{\|x_1\|_{\infty,T}}{\|x_1\|_{\infty,D_1}} - 1.$$

(b) If $\rho_k < \varepsilon$, then **Stop**.

(c) *Reduction Step*:

Determine a subset $\tilde{D}_k$ of $E(x_k, D_k)$ where for the number of elements we have $|\tilde{D}_k| \leq n+1$, such that $x_k \in M(\|\cdot\|_{\infty,\tilde{D}_k}, K)$ (i.e. $\tilde{D}_k$ is a de la Vallée-Poussin set relative to D_k)

(d) *Inner Loop*:

i. Choose $t_k \in E(x_k, T)$ and set $i := 1$. Choose $B_{1,k}$ as a superset of $\tilde{D}_k \cup \{t_k\}$.

ii. Determine $y_{i,k} \in M(\|\cdot\|_{\infty,B_{i,k}}, K)$ and set

$$\kappa_{i,k} := \frac{\|y_{i,k}\|_{\infty,T}}{\|y_{i,k}\|_{\infty,B_{i,k}}} - 1.$$

iii. If $\kappa_{i,k} \leq \sigma\rho_k$, then set $D_{k+1} := B_{i,k}$, $x_{k+1} := y_{i,k}$, $\rho_{k+1} := \kappa_{i,k}$.
Set $k := k+1$, **goto (b)**.
Otherwise choose $T_i \subset T$ and set $B_{i+1,k} := B_{i,k} \cup T_i$.
Set $i := i+1$ **goto ii.**

Algorithmic Variants

For the choice of the set T_i in step iii. of the inner loop different strategies are possible:

(a) (static variant) Let $(T_i)_{i\in\mathbb{N}}$ be a sequence of subsets of T that converges in the sense of Kuratowski to T.

(b) (dynamic variant) Choose T_i as a non-empty subset of $E(y_{i,k}, T)$. If one chooses in particular T_i as a set containing just a single point, then this procedure corresponds in the inner loop to the *First Algorithm of Remez* proposed by Cheney (see [22]).

One can view the determination of T_i as a byproduct of the computation of the $\kappa_{i,k}$.

(c) (combined variant) A mixture of both variants in the sense of a union of the corresponding sets.

Theorem 2.6.8. *In variants* (a) *and* (c) *the sequence* $(x_k)_{k\in\mathbb{N}}$ *generated by the algorithm is bounded and every point of accumulation is a best Chebyshev approximation w.r.t.* $\|\cdot\|_{\infty,T}$. *Furthermore the sequence* $(\|x_k\|_{\infty,D_k})_{k\in\mathbb{N}}$ *converges monotonically increasing to* $\inf(\|\cdot\|_{\infty,T}, K)$.

Proof. By construction the inner loop over i ends after a finite number of steps. Furthermore, by construction the sequence (ρ_k) tends to zero and we obtain for k sufficiently large

$$\|x_k\|_{\infty,T} < (1+\varepsilon)\|x_k\|_{\infty,D_k},$$

where $x_k \in M(\|\cdot\|_{\infty,D_k}, K)$. Let $x_0 \in M(\|\cdot\|_{\infty,T}, K)$, then apparently

$$\|x_k\|_{\infty,D_k} \leq \|x_0\|_{\infty,D_k} \leq \|x_0\|_{\infty,T}.$$

Hence the sequence (x_k) is bounded. Let now (x_{k_n}) be a convergent subsequence with $x_{k_n} \to \bar{x}$. Suppose $\bar{x} \notin M(\|\cdot\|_{\infty,T}, K)$, then there is a $\bar{t} \in T$ and $\delta > 0$ with $|\bar{x}(\bar{t})| \geq \|x_0\|_{\infty,T}(1+2\delta)$, due to $x_{k_n}(\bar{t}) \to \bar{x}(\bar{t})$ also $|x_{k_n}(\bar{t})| \geq \|x_0\|_{\infty,T}(1+\delta)$ for n sufficiently large. On the other hand there is a $N \in \mathbb{N}$ with $\rho_k \leq \frac{\delta}{2}$ for $k > N$. Putting these results together we obtain

$$\begin{aligned}\|x_0\|_{\infty,T}(1+\delta) &\leq |\bar{x}_{k_n}(\bar{t})| \leq \|x_{k_n}\|_{\infty,T} \\ &\leq \left(1+\frac{\delta}{2}\right)\|x_{k_n}\|_{\infty,D_{k_n}} \leq \left(1+\frac{\delta}{2}\right)\|x_0\|_{\infty,T},\end{aligned}$$

and hence a contradiction.

The monotonicity follows from $D_{k+1} \supset \tilde{D}_k$, since

$$\|x_k\|_{\infty,D_k} = \|x_k\|_{\infty,\tilde{D}_k} \leq \|x_{k+1}\|_{\infty,D_{k+1}}.$$ □

Algorithm for the Linear Approximation

Let T be a compact metric space and $\{v_1, \ldots, v_n\}$ a basis of a finite-dimensional subspace V of $C(T)$. Let T_0 be a subset of T such that $\{v_1, \ldots, v_n\}$ restricted to T_0 are linearly independent. In order to guarantee uniqueness of the coefficients of the best approximations to be determined, one can in step i. of the inner loop put $B_{1,k} := \tilde{D}_k \cup \{t_k\} \cup T_0$.

The reduction step in (c) can now be realized in the following way: Let $x - v_{0,k}$ be a best approximation w.r.t. $\|\cdot\|_{\infty,D_k}$ and let $\{t_1, \ldots, t_l\} = E(x - v_{0,k}, D_k)$ be the set of extreme points of $x - v_{0,k}$ on D_k. If $l \leq n+1$, put $\tilde{D}_k = E(x - v_{0,k}, D_k)$.

Otherwise determine a solution of the homogeneous linear system of equations resulting from the characterization theorem

$$\sum_{j=1}^{l} \alpha_j (x(t_j) - v_{0,k}(t_j)) v_i(t_j) = 0 \quad \text{for } i = 1, \ldots, n.$$

Here we look for a non-trivial and non-negative solution with at most $n+1$ non-zero components. The non-triviality of the solution can in particular be guaranteed by the requirement

$$\sum_{j=1}^{l} \alpha_j = 1.$$

For non-negative solutions special algorithms are required.

Remark 2.6.9. A further class of algorithms for the computation of a best (continuous) Chebyshev approximation in $C(T)$, also being based on the idea of the Polya algorithm, is obtained in the following way:

After choosing a sequence of Young functions (Φ_k), whose corresponding Luxemburg norms $(\|\cdot\|_{\Phi_k,T})$ converge to $\|\cdot\|_{\infty,T}$, the computation of a best approximation w.r.t. the Luxemburg norm is performed by replacing the integrals by quadrature formulas. In order to keep the number of discretization points low, one can adopt the idea of adaptive quadrature formulas (sparse grids).

2.7 Semi-infinite Optimization Problems

Various problems of optimization theory can be formulated as problems of *linear semi-infinite optimization* (see [71]), where a linear cost function is optimized under (potentially) infinitely many linear restrictions.

Let T be a compact metric space and $C(T)$ be equipped with the natural order, U an n-dimensional subspace of $C(T)$ with basis $\{u_1, \ldots, u_n\}$ and let $u_{n+1} \in C(T)$. Let now $c \in \mathbb{R}^n$ be given, then we denote the following problem

$$\sup_{x\in\mathbb{R}^n} \left\{ \langle c, x\rangle \,\middle|\, \sum_{i=1}^{n} x_i u_i \le u_{n+1} \right\} \tag{2.7}$$

as a *semi-infinite optimization problem.*

If $u_{n+1} + U$ satisfies a Krein condition, i.e. there is a $u \in u_{n+1} + U$ and a $\rho > 0$ with $u(t) \ge \rho$ for all $t \in T$, then we can, without loss of generality, assume that u_{n+1} is positive.

In [66] we have converted the semi-infinite optimization problem into a linear approximation problem w.r.t. the max-function, and have approximated the non-differentiable max-function by a sequence of Minkowski functionals for non-symmetric Young functions, much in the spirit of the Polya algorithms of this chapter. In the next section of the present text we will pursue a different approach: instead of smoothing the function to be minimized we smooth the restriction set and keep the (linear) cost function fixed.

2.7.1 Successive Approximation of the Restriction Set

We will now treat the problem of linear semi-infinite optimization, directly by the method of Lagrange. Let $p : C(T) \to \mathbb{R}$ with $p(u) := \max_{t\in T} u(t)$, then we can formulate the above problem as a restricted convex minimization problem:

$$\inf_{x\in\mathbb{R}^n} \left\{ \langle -c, x\rangle \,\middle|\, p\Big(\sum_{i=1}^{n} x_i u_i - u_{n+1}\Big) \le 0 \right\}. \tag{2.8}$$

The convex function p is not differentiable in general. This also holds for the function $f : \mathbb{R}^n \to \mathbb{R}$, defined by $f(x) := p(\sum_{i=1}^n x_i u_i - u_{n+1})$. Thus the level set $S_f(0)$, which describes the restriction set, is not flat convex (see Theorem 8.1.2).

An equivalent formulation we obtain by using the also convex function $q : C(T) \to \mathbb{R}$ defined by $q(u) := \max(p(u), 0)$

$$\inf_{x \in \mathbb{R}^n} \left\{ \langle -c, x \rangle \,\middle|\, q\Big(\sum_{i=1}^n x_i u_i - u_{n+1} \Big) = 0 \right\}. \tag{2.9}$$

We obtain for the corresponding Lagrange function : $\langle -c, x \rangle + \lambda f(x)$ resp. $\langle -c, x \rangle + \lambda g(x)$, where $g : \mathbb{R}^n \to \mathbb{R}$ is defined by $g(x) := q(\sum_{i=1}^n x_i u_i - u_{n+1})$. The Lagrange functions are, as stated above, not differentiable. By suitable differentiable approximations of the functions f resp. g we obtain differentiable restricted convex optimization problems. If we apply to these the methods of Kuhn–Tucker resp. of Lagrange, one can reduce the determination of minimal solutions to the solution of non-linear equations. In order to apply rapidly convergent numerical methods, the regularity of the corresponding Jacobian matrix at the solution is required. A statement in this direction is provided by the following lemmata:

Lemma 2.7.1. *Let $A \in L(\mathbb{R}^n)$ be symmetric and positive definite and let $b \in \mathbb{R}^n$ be different from zero. Then the matrix*

$$\begin{pmatrix} A & b \\ b^T & 0 \end{pmatrix}$$

is non-singular.

Proof. Suppose there is a $y \in \mathbb{R}^n$ and a $\mu \in \mathbb{R}$, not both of them zero, with $Ay = -\mu b$ and $\langle b, y \rangle = 0$, then

$$\langle Ay, y \rangle = -\mu \langle b, y \rangle = 0$$

holds. Since A is positive definite, $y = 0$ follows, hence $\mu \neq 0$ and therefore $b = 0$ a contradiction. □

Lemma 2.7.2. *Let $f : \mathbb{R}^n \to \mathbb{R}$ be twice continuously differentiable, let $c \in \mathbb{R}^n$ different from zero, let (x^*, λ^*) be a stationary point of the Lagrange function $(x, \lambda) \mapsto \langle -c, x \rangle + \lambda f(x)$. If $f''(x^*)$ is positive definite, the Hessian matrix of the Lagrange function*

$$\begin{pmatrix} \lambda f''(x) & f'(x) \\ f'(x)^T & 0 \end{pmatrix}$$

is non-singular at (x^*, λ^*).

Proof. If (x^*, λ^*) is a stationary point, it satisfies the equations

$$\begin{aligned} -c + \lambda^* f'(x^*) &= 0 \\ f(x^*) &= 0. \end{aligned}$$

Suppose $\lambda^* = 0$ or $f'(x^*) = 0$, then $c = 0$, a contradiction. According to the previous lemma then

$$\begin{pmatrix} f''(x^*) & \frac{1}{\lambda^*} f'(x^*) \\ \frac{1}{\lambda^*} f'(x^*)^T & 0 \end{pmatrix}$$

is non-singular. □

Theorem 2.7.3. *Let $f : \mathbb{R}^n \to \mathbb{R}$ be convex and differentiable, let $c \in \mathbb{R}^n$ be different from zero and let $\beta > \inf f(\mathbb{R}^n)$. Let further*

$$\mu_0 = \inf\{\langle -c, x\rangle \mid f(x) \le \beta\} \tag{2.10}$$

be finite and be attained at x^*. Then there is a $\lambda^* > 0$, such that

$$\mu_0 = \inf\{\langle -c, x\rangle + \lambda^*(f(x) - \beta)\}, \tag{2.11}$$

where the infimum is also attained in (2.11), *and $f(x^*) = \beta$ holds.*

Proof. Apparently there is a $x_1 \in \mathbb{R}^n$ with $f(x_1) < \beta$. Then by Theorem 3.14.4 there is a $\lambda^* \ge 0$ with $\mu_0 = \inf\{\langle -c, x\rangle + \lambda^*(f(x) - \beta)\}$, where the infimum is also attained at x^*. As a necessary (and sufficient) condition we obtain

$$-c + \lambda^* f'(x^*) = 0.$$

Since $c \ne 0$ the gradient $f'(x^*) \ne 0$ and $\lambda^* > 0$. Apparently x^* cannot be an interior point of $S_f(\beta)$. □

Let Φ be a (non-symmetric) twice continuously differentiable Young function. Let further p_Φ be the corresponding Minkowski functional and let $f_\Phi : \mathbb{R}^n \to \mathbb{R}$ be defined by

$$f_\Phi(x) := p_\Phi\bigg(\sum_{i=1}^n x_i u_i - u_{n+1}\bigg).$$

Let $\beta > p_\Phi(-u_{n+1})$. Then we consider the problem

$$\inf_{x\in\mathbb{R}^n} \bigg\{\langle -c, x\rangle \,\bigg|\, p_\Phi\bigg(\sum_{i=1}^n x_i u_i - u_{n+1}\bigg) \le \beta\bigg\} \tag{2.12}$$

resp.

$$\inf_{x\in\mathbb{R}^n} \{\langle -c, x\rangle \mid f_\Phi(x) \le \beta\}. \tag{2.13}$$

For the corresponding Lagrange function we have

$$(x, \lambda) \mapsto \langle -c, x\rangle + \lambda(f_\Phi(x) - \beta).$$

A stationary point (x^*, λ^*) satisfies the following system of equations:

$$-c + \lambda^* f'_\Phi(x^*) = 0 \tag{2.14}$$

$$f_\Phi(x^*) = \beta. \tag{2.15}$$

A more explicit form of the second equation is: $p_\Phi(\sum_{i=1}^n x_i u_i - u_{n+1}) = \beta$. Due to the positive homogeneity of p_Φ

$$\int_T \Phi\left(\frac{\sum_{i=1}^n x_i^* u_i - u_{n+1}}{\beta}\right) d\mu = 1$$

holds. Altogether we obtain the following system of equations:

$$-c + \tilde{\lambda}\left(\int_T u_j \Phi'\left(\frac{\sum_{i=1}^n x_i^* u_i - u_{n+1}}{\beta}\right) d\mu\right)_{j=1}^n = 0 \tag{2.16}$$

$$\int_T \Phi\left(\frac{\sum_{i=1}^n x_i^* u_i - u_{n+1}}{\beta}\right) d\mu = 1. \tag{2.17}$$

Here $\tilde{\lambda} = \lambda^* / \int_T \frac{y(t)}{p_\Phi(y)} \Phi'(\frac{y(t)}{p_\Phi(y)}) d\mu$ with $y(t) := \sum_{i=1}^n x_i^* u_i - u_{n+1}$, where the denominator of the right-hand side corresponds to the expression $\gamma(a)$ in Section 1.4.2. Therefore also $\tilde{\lambda} > 0$.

Let now in particular $\Phi_0 : \mathbb{R} \to \mathbb{R}$ be defined by $\Phi_0(s) := s^2$ and let Φ be a twice continuously differentiable (non-symmetric) Young function with the properties

(a) $\Phi(s) = 0$ for $s \le 1$

(b) $\Phi(s) > 0$ for $s > 1$.

Moreover, let $\alpha > 0$ and let $\Phi_\alpha := \alpha\Phi_0 + \Phi$, and let $p_{\alpha,\Phi}$ be the corresponding Minkowski functional. Let $f_{\alpha,\Phi} : \mathbb{R}^n \to \mathbb{R}$ be defined by

$$f_{\alpha,\Phi}(x) := p_{\alpha,\Phi}\left(\sum_{i=1}^n x_i u_i - u_{n+1}\right).$$

Due to the Krein condition we can assume u_{n+1} to be non-negative, then, due to

$$p_{\alpha,\Phi}(-u_{n+1}) = \sqrt{\alpha}\|u_{n+1}\|_2,$$

(since $\Phi(-u_{n+1}) = 0$) an explicit relation between α and β can be stated as

$$\beta > \sqrt{\alpha}\|u_{n+1}\|_2.$$

A stationary point (x^*, λ^*) of the Lagrange function of the approximating problem

$$\inf_{x\in\mathbb{R}^n}\left\{\langle -c, x\rangle \,\middle|\, p_{\alpha,\Phi}\Big(\sum_{i=1}^n x_i u_i - u_{n+1}\Big) \le \beta\right\} \tag{2.18}$$

satisfies

$$-c + \tilde{\lambda}\bigg(\int_T u_j \Phi'_\alpha\bigg(\frac{\sum_{i=1}^n x_i^* u_i - u_{n+1}}{\beta}\bigg) d\mu\bigg)_{j=1}^n = 0 \tag{2.19}$$

$$\int_T \Phi_\alpha\bigg(\frac{\sum_{i=1}^n x_i^* u_i - u_{n+1}}{\beta}\bigg) d\mu = 1. \tag{2.20}$$

Since $\Phi''_\alpha > 0$ we obtain in analogy to the considerations in Section 1.4.2 that of $f''_{\alpha,\Phi}$ is positive definite.

Let now $(\beta_k)_{\in\mathbb{N}}$ be a sequence of positive numbers tending to zero. At first we will show the Kuratowski convergence of the level sets $S_{p_{0,\Phi}}(\beta_k)$ (here $\alpha = 0$) to $S_p(0)$.

Lemma 2.7.4. *We have*

(a) $S_{p_{0,\Phi}}(0) = S_p(0)$

(b) $S_{p_{0,\Phi}}(\beta_k) \supset S_p(0)$.

Proof. Let $y \in C(T)$, then by definition

$$p_{0,\Phi}(y) = \inf\left\{c \,\middle|\, \int_T \Phi\Big(\frac{y}{c}\Big) d\mu \le 1\right\}.$$

Now for $k \ge 1$ due to the convexity of Φ: $\Phi(s) = \Phi(\frac{1}{k}\cdot ks + (1-\frac{1}{k})\cdot 0) \le \frac{1}{k}\Phi(ks)$ and hence $k\Phi(s) \le \Phi(ks)$. Let now $y(t) > \delta > 0$ for $t \in U$ with $\mu(U) > 0$ and let $c > 0$ be chosen, such that $\delta/2c \ge 1$. Then

$$\int_T \Phi\Big(\frac{y(t)}{c}\Big) d\mu \ge \mu(U)\Phi\Big(\frac{\delta}{c}\Big) = \mu(U)\Phi\bigg(\frac{\delta}{\frac{\delta}{2}c\frac{2}{\delta}}\bigg) \ge \mu(U)\frac{\delta}{2c}\Phi(2)$$

holds. For a $c > 0$ small enough the right-hand side is greater than 1 and hence $p_{0,\Phi}(y) > 0$. If on the other hand $y \le 0$, then apparently $p_{0,\Phi}(y) = 0$. This yields the first part of the assertion. The second part is an immediate consequence of the first part. □

Theorem 2.7.5. *Let $(\alpha_k)_{k\in\mathbb{N}}$ be a sequence of positive numbers tending to zero. Then the sequence of Minkowski functionals $(p_{\alpha_k,\Phi})_{k\in\mathbb{N}}$ converges pointwise to $p_{0,\Phi}$ on $L^\infty(\mu)$.*

Proof. Let $y = \sum_{i=1}^{n} \lambda_i \chi_{T_i}$ be a step function in $L^\infty(\mu)$.

Case 1: $p_{0,\Phi}(y) > 0$.

We have

$$\sum_{i=1}^{n} \Phi_{\alpha_k}\left(\frac{\lambda_i}{p_{0,\Phi}(y)}\right)\mu(T_i) \geq \sum_{i=1}^{n} \Phi\left(\frac{\lambda_i}{p_{0,\Phi}(y)}\right)\mu(T_i) = 1,$$

and hence

$$p_{\Phi_{\alpha_k}}(y) \geq p_{0,\Phi}(y).$$

Suppose, there is a subsequence and a positive ε with $p_{\Phi_{\alpha_k}}(y) \geq p_{0,\Phi}(y) + \varepsilon$, then

$$\sum_{i=1}^{n} \Phi_{\alpha_k}\left(\frac{\lambda_i}{p_{0,\Phi}(y)+\varepsilon}\right)\mu(T_i) \geq \sum_{i=1}^{n} \Phi_{\alpha_k}\left(\frac{\lambda_i}{p_{\alpha_k,\Phi}(y)}\right)\mu(T_i) = 1.$$

Due to the pointwise convergence of the Young functions we then obtain

$$\sum_{i=1}^{n} \Phi\left(\frac{\lambda_i}{p_{0,\Phi}(y)+\varepsilon}\right)\mu(T_i) \geq 1,$$

a contradiction.

Case 2: $p_{0,\Phi}(y) = 0$.

Then by the above lemma $y \leq 0$ and hence $\Phi(\frac{y}{c}) = 0$ for all $c > 0$. Therefore $p_{\alpha_k,\Phi}(y) = \sqrt{\alpha_k}\|y\|_2 \to 0$ for $k \to \infty$.

In a manner similar to Theorem 2.4.2 the continuity of the Minkowski functionals on L^∞ can be established. The remaining part is an application of the theorem of Banach–Steinhaus 5.3.17 for convex functions. □

In order to be able to apply the stability theorems of Chapter 5, we need the Kuratowski convergence of the restriction sets. This can be achieved, if the sequence of the parameters $(\alpha_k)_{k\in\mathbb{N}}$ converges more rapidly to zero than the sequence of the squares of the levels $(\beta_k)_{k\in\mathbb{N}}$.

Theorem 2.7.6. *Let $\alpha_k = o(\beta_k^2)$ and let M be a closed subset of $C(T)$. Then the sequence of the intersects $(M \cap S_{p_{\alpha_k,\Phi}}(\beta_k))_{k\in\mathbb{N}}$ converges in the sense of Kuratowski to the intersect with the negative natural cone $M \cap S_p(0)$.*

Proof. Let $y_k \in M \cap S_{p_{\alpha_k,\Phi}}(\beta_k)$ for $k \in \mathbb{N}$, i.e. $p_{\alpha_k,\Phi}(y_k) \leq \beta_k$ and let $\lim_{k\to\infty} y_k = \bar{y}$. Due to the closedness of M we have $\bar{y} \in M$. We have to show that $\bar{y} \in S_p(0)$.

Pointwise convergence of the Minkowski functionals implies their continuous convergence (see Remark 5.3.16), i.e.

$$p_{\alpha_k,\Phi}(y_k) \xrightarrow{k\to\infty} p_{0,\Phi}(\bar{y}).$$

Let $\varepsilon > 0$ be given, then: $p_{\alpha_k,\Phi}(y_k) \leq \beta_k \leq \varepsilon$ for k sufficiently large, and hence $p_{0,\Phi}(\bar{y}) \leq \varepsilon$.

Conversely, let $\alpha_k = o(\beta_k^2)$ and let $\bar{y} \in M \cap S_p(0)$ (i.e. in particular $\bar{y} \leq 0$), then

$$\begin{aligned} p_{\alpha_k,\Phi}(\bar{y}) &= \inf\left\{ c \,\middle|\, \int_T \Phi_{\alpha_k}\left(\frac{\bar{y}}{c}\right) d\mu \leq 1 \right\} \\ &= \inf\left\{ c \,\middle|\, \alpha_k \int_T \left(\frac{\bar{y}}{c}\right)^2 d\mu \leq 1 \right\} \\ &= \sqrt{\alpha_k}\|\bar{y}\|_2 \end{aligned}$$

holds, but $\sqrt{\alpha_k}\|\bar{y}\|_2 \leq \beta_k$ for k sufficiently large, hence $\bar{y} \in M \cap S_{p_{\alpha_k,\Phi}}(\beta_k)$ for k sufficiently large. □

For the case of semi-infinite optimization we choose $M := -u_{n+1} + U$, where $U := \operatorname{span}\{u_1, \ldots, u_n\}$.

Remark. If $\{u_1, \ldots, u_n\}$ is a basis of U, then the Kuratowski convergence of the level sets of the Minkowski functionals on $C(T)$ carries over to the Kuratowski convergence of the corresponding coefficient sets

$$\lim_{k\to\infty} f_{\alpha_k,\Phi}(\beta_k) = S_f(0).$$

Theorem 2.7.7. *Let $S_k := f_{\alpha_k,\Phi}(\beta_k)$ and $S := S_f(0)$. Let further the set of minimal solutions of $\langle -c, \cdot\rangle$ on S be non-empty and bounded. Then*

(a) *for every $k \in \mathbb{N}$ there exists a uniquely determined minimal solution $x_k \in M(\langle -c, \cdot\rangle, S_k)$. The resulting sequence is bounded and its points of accumulation are elements of $M(\langle -c, \cdot\rangle, S)$.*

(b) $\inf\langle -c, S_k\rangle \to \inf\langle -c, S\rangle$.

(c) *$x_k \in M(\langle -c, \cdot\rangle, S_k)$ for $k \in \mathbb{N}$ implies $\langle -c, x_k\rangle \to \inf\langle -c, S\rangle$.*

Proof. Since for all $k \in \mathbb{N}$ the level sets S_k are strictly convex and bounded, existence and uniqueness of the minimal solutions of $-\langle c, \cdot\rangle$ on S_k follows. The remaining assertions follow from the Stability Theorem of Convex Optimization 5.3.21 in $\mathbb{R}^n$. □

Outer Regularization of the Linear Functional

Theorem 2.7.8. *Let f and g_γ be convex functions in $C^2(\mathbb{R}^n)$ with g_γ'' positive definite. Let $c \in \mathbb{R}^n$ be different from zero and let $\beta > \inf f(\mathbb{R}^n)$. Then:*

(a) *The optimization problem $(g_\gamma(\cdot) - \langle c, \cdot\rangle, S_f(\beta))$ has a unique solution x^*.*

(b) *There is a non-negative Lagrange multiplier* λ^*, *such that the Lagrange function*

$$L_{\lambda^*} = g_\gamma + \langle -c, \cdot \rangle + \lambda^*(f - \beta)$$

has x^* *as minimal solution.*

(c) *If the solution* x^* *is attained at the boundary of the restriction set* $S_f(\beta)$, *then the pair* (x^*, λ^*) *is a solution of the following systems of equations:*

$$\begin{aligned} g'_\gamma(x) - c + \lambda f'(x) &= 0 \\ f(x) &= \beta. \end{aligned}$$

The corresponding Jacobian matrix at (x^*, λ^*)

$$\begin{pmatrix} g''_\gamma(x^*) + \lambda^* f''(x^*) & f'(x^*) \\ f'(x^*)^T & 0 \end{pmatrix}$$

is then non-singular.

Proof. Since $\beta > \inf f(\mathbb{R}^n)$ and $f(x^*) = \beta$ we have $f'(x^*) \neq 0$. As λ^* due to Theorem 3.14.4 is non-negative, the matrix $g''_\gamma(x^*) + \lambda^* f''(x^*)$ is positive definite. The remaining part of the assertion follows by Lemma 2.7.1. □

Remark. The condition required in (c) that x^* should be a boundary point of $S_f(\beta)$, is satisfied, if the global minimal solution of $g_\gamma - \langle c, \cdot \rangle$ is outside of the level set $S_f(\beta)$. This suggests the following strategy: if $g_\gamma(x) = \gamma \|x - x_\gamma\|^2$ with a suitable $x_\gamma \in \mathbb{R}^n$, then the global minimal solution x^*_γ of $g_\gamma - \langle c, \cdot \rangle$ satisfies the equation

$$x^*_\gamma = x_\gamma + \frac{1}{2\gamma} c.$$

If $f(\frac{1}{2\gamma} c) > \beta$, choose $x_\gamma = 0$. If on the other hand a point $\bar{x}$ is known (normally the predecessor in an iteration process) with $f(\bar{x}) > \beta$, put $x_\gamma := \bar{x} - \frac{1}{2\gamma} c$ and hence $x^*_\gamma = \bar{x}$.

Remark (Partial result for second stage). Let $S_k := S_f(\beta_k)$, $S := S_f(0)$ with $f(x) = p_\Phi(\sum_{i=1}^n x_i u_i - u_{n+1})$. Let further be $\bar{x}_k \in M(-\langle c, \cdot \rangle, S_k)$, let (γ_k) be a sequence of positive numbers tending to zero and $h_k := \gamma_k g - \langle c, \cdot \rangle$, then we obtain for x_k minimal solution of h_k on S_k

$$\begin{aligned} h_k(x_k) = \gamma_k g(x_k) - \langle c, x_k \rangle &\leq \gamma_k g(\bar{x}_k) - \langle c, \bar{x}_k \rangle \\ &\leq \gamma_k g(\bar{x}_k) - \langle c, x_k \rangle, \end{aligned}$$

and hence

$$g(x_k) - g(\bar{x}_k) \leq 0.$$

Every point of accumulation $\bar{x}$ of $(\bar{x}_k)$ resp. $\tilde{x}$ of (x_k) is a minimal solution of $-\langle c, \cdot \rangle$ on S and

$$g(\tilde{x}) \leq g(\bar{x}).$$

Chapter 3

Convex Sets and Convex Functions

3.1 Geometry of Convex Sets

In this section we will present a collection of important notions and facts about the geometry of convex sets in vector spaces that we will need for the treatment of optimization problems in Orlicz spaces.

Definition 3.1.1. Let X be a vector space and let $x, y \in X$. Then we call the set

$$[x, y] := \{z \in X \mid z = \lambda x + (1 - \lambda)y \text{ with } 0 \leq \lambda \leq 1\}$$

the closed interval connecting x and y. By the open interval connecting x and y we understand the set

$$(x, y) := \{z \in X \mid z = \lambda x + (1 - \lambda)y \text{ with } 0 < \lambda < 1\}.$$

Half open intervals $(x, y]$ and $[x, y)$ are defined correspondingly.

Definition 3.1.2. A subset K of X is called *convex*, if for all points $x, y \in K$ the interval $[x, y]$ is in K.

Definition 3.1.3. Let A be a subset of X. A point $x \in X$ is called a *convex combination* of elements of A, if there is $n \in \mathbb{N}$ and $x_1, \dots, x_n \in A$, such that $x = \lambda_1 x_1 + \cdots + \lambda_n x_n$, where $\lambda_1, \dots, \lambda_n$ non-negative numbers with $\lambda_1 + \cdots + \lambda_n = 1$.

Using complete induction we obtain the following

Remark 3.1.4. Let K be a convex subset of X, then every convex combination of a finite number of points in K is again an element of K

$$\lambda_1 x_1 + \cdots + \lambda_n x_n \in K.$$

Definition 3.1.5. Let U be a subset of X. A point $x_0 \in U$ is called *algebraically interior point* of U, if for every $y \in X$ there is $\alpha \in \mathbb{R}_{>0}$ with $[x_0 - \alpha y, x_0 + \alpha y] \subset U$. U is called *algebraically open*, if every point in U is an algebraically interior point.

Linearly Bounded Sets

Definition 3.1.6. A subset S of a vector space is called *linearly bounded w.r.t. a point* $x_0 \in S$, if for all $y \in S$ the set

$$\{\alpha \in \mathbb{R}_{\geq 0} \,|\, (1-\alpha)x_0 + \alpha y \in S\}$$

is bounded. S is called *linearly bounded*, if S is linearly bounded w.r.t. a point $x_0 \in S$.

Theorem 3.1.7. *A convex, closed, and linearly bounded subset K of $\mathbb{R}^n$ is bounded.*

Proof. W.l.g. let K be linearly bounded w.r.t. 0. Suppose, K is not bounded. Then there is a sequence $(x_n)_1^\infty$ in K with $\|x_n\| \to \infty$.

For large n the point $s_n := \frac{x_n}{\|x_n\|} = (1 - \frac{1}{\|x_n\|}) \cdot 0 + \frac{1}{\|x_n\|} x_n$ is in K. Let (s_{n_i}) be a subsequence of (s_n) converging to $\bar{s}$. Since K is linearly bounded, there is a $\alpha_0 \in \mathbb{R}_+$ with $\alpha_0 \bar{s} \notin K$. For large n however, $\alpha_0 s_n \in K$ and, because K is closed, a contradiction follows. □

The Interior and the Closure of Convex Sets

In normed spaces the convexity of a subset carries over to the interior and the closure of these subsets.

Theorem 3.1.8. *Let K be a convex subset of X. Then we have:*

(a) *For all x in the interior* $\text{Int}(K)$ *of K and all $y \in K$ the half open interval $[x, y)$ is in* $\text{Int}(K)$.

(b) *The interior of K and the closure of K are convex.*

Proof. (a): Let $x \in \text{Int}(K)$ and let U be a neighborhood of 0, such that $x + U \subset K$. Let $\alpha \in (0, 1)$, then αU is also a neighborhood of 0, and we have

$$\alpha x + (1-\alpha)y + \alpha U = \alpha(x + U) + (1-\alpha)y \subset K.$$

(b): That the interior of K is convex, follows directly from (a), if one observes that for all $x, y \in K$: $[x, y] = \{x\} \cup (x, y) \cup \{y\}$.

That the closure of K is convex, is obtained from the following line of reasoning: let $x, y \in \overline{K}$ and $(x_k)_1^\infty$, $(y_k)_1^\infty$ sequences in K with $x = \lim_k x_k$ and $y = \lim_k y_k$. For a $\lambda \in [0, 1]$ and all $k \in \mathbb{N}$ we have: $(\lambda x_k + (1-\lambda)y_k) \in K$ and thus

$$\lambda x + (1-\lambda)y = \lim_k (\lambda x_k + (1-\lambda)y_k) \in \overline{K}.$$

□

For later use we need the following

Lemma 3.1.9. *Let S be a non-empty convex subset of a finite-dimensional normed space X. Let $x_0 \in S$ and $X_m := \operatorname{span}(S - x_0)$ and let $\dim(X_m) = m > 0$. Then $\operatorname{Int}(S) \neq \emptyset$ w.r.t. X_m.*

Proof. W.l.g. let $0 \in S$. Let $x_1, \dots, x_m \in S$ be linearly independent, then $\{s_i = \frac{x_i}{\|x_i\|}$, $i = 1, \dots, n\}$, is a basis of X_m. Then $t := \frac{1}{2}(\frac{1}{m}\sum_{i=1}^m x_i) + \frac{1}{2}0 = \frac{1}{2m}\sum_{i=1}^m x_i \in S$. We show: $t \in \operatorname{Int}(S)$ w.r.t. X_m: let $x \in X_m$, then we have the following representation: $x = \sum_{i=1}^m \xi_i s_i$. Apparently by $\sum_{i=1}^m |\xi_i|$ a norm is defined on X_m: $\|x\|_m := \sum_{i=1}^m |\xi_i|$. On X_m this norm is equivalent to the given norm. Let now be $\rho > 0$ and $U_\rho := \{u \in X_m \mid \|u\|_m < \rho\}$, then for ρ small enough $t + u \in S$ for all $u \in U_\rho$, because

$$t + u = \frac{1}{2m}\sum_{i=1}^m x_i + \sum_{i=1}^m u_i s_i = \frac{1}{m}\sum_{i=1}^m \left(\frac{1}{2} + m\frac{u_i}{\|x_i\|}\right) x_i.$$

For ρ small enough we have $0 < t_i := \frac{1}{2} + \rho m \frac{u_i}{\|x_i\|} < 1$ and hence $t + \rho u \in S$ since $0 \in S$. □

Extreme and Flat Points of Convex Sets

Let X be a normed space.

Definition 3.1.10. Let K be a convex subset of X. A point $x \in K$ is called *extreme point* of K, if there is no proper open interval in K which contains x, i.e. x is extreme point of K, if $x \in (u, v) \subset K$ implies $u = v$. By $E_p(K)$ we denote the set of extreme points of K.

In finite-dimensional spaces the theorem of Minkowski holds, which we will prove in the section on Separation Theorems (see Theorem 3.9.12).

Theorem 3.1.11 (Theorem of Minkowski). *Let X be an n-dimensional space and S a convex compact subset of X. Then every boundary point (or arbitrary point) of S can be represented as a convex combination of at most n (or $n + 1$ respectively) extreme points.*

Definition 3.1.12. A convex set K is called *convex body*, if the interior of K is non-empty.

Definition 3.1.13. Let K be a convex body in X. K is called *strictly convex*, if every boundary point of K is an extreme point.

Definition 3.1.14. Let K be a convex subset of X. A closed hyperplane H is called *supporting hyperplane* of K, if H contains at least one point of K and K is completely contained in one of the two half-spaces of X.

A notion – corresponding to that of an extreme point in a dual sense – we obtain in the following way:

Definition 3.1.15. A boundary point x of a convex subset K of X is called a *flat point* of K, if at most one supporting hyperplane passes through x.

Definition 3.1.16. A convex body K in X is called *flat convex*, if the boundary of K only consists of flat points.

Convex Hull

Definition 3.1.17. Let A be a subset of X. Then by the term *convex hull* of A we understand the set

$$\operatorname{conv}(A) := \bigcap \{K \mid A \subset K \subset X \text{ where } K \text{ is convex}\}.$$

Remark 3.1.18. Let A be a subset of X. Then the convex hull of A is the set of all convex combinations of elements in A.

Theorem 3.1.19 (Theorem of Carathéodory). *Let A be a subset of a finite-dimensional vector space X and $n = \dim(X)$. Then for every element $x \in \operatorname{conv}(A)$ there are $n+1$ numbers $\lambda_1, \ldots, \lambda_{n+1} \in [0,1]$ and $n+1$ points $x_1, \ldots, x_{n+1} \in A$ with $\lambda_1 + \cdots + \lambda_{n+1} = 1$ and $x = \lambda_1 x_1 + \cdots + \lambda_{n+1} x_{n+1}$.*

The theorem of Carathéodory states that, for the representation of an element of the convex hull of a subset of an n-dimensional vector space, $n+1$ points of this subset are sufficient.

Proof. Let $x \in \operatorname{conv}(A)$. Then there are $m \in \mathbb{N}$, real numbers $\lambda_1 \ldots \lambda_m \in [0,1]$, and points $x_1, \ldots, x_m \in A$ with $\lambda_1 + \cdots + \lambda_m = 1$ and $x = \lambda_1 x_1 + \cdots + \lambda_m x_m$.

If $m \leq n+1$, nothing is left to be shown.

If $m > n+1$, it has to be demonstrated that x can already be represented as a convex combination of $m-1$ points in A, whence the assertion finally follows.

For all $k \in \{1, \ldots, m-1\}$ let $y_k := x_k - x_m$. Since $m > n+1$ and $n = \dim(X)$, the set $\{y_1, \ldots, y_{m-1}\}$ is linearly dependent: hence there is a non-trivial $(m-1)$-tuple of numbers $(\alpha_1, \ldots, \alpha_{m-1})$ such that $\alpha_1 y_1 + \cdots + \alpha_{m-1} y_{m-1} = 0$.

If we put $\alpha_m := -(\alpha_1 + \cdots + \alpha_{m-1})$, then $\alpha_1 + \cdots + \alpha_m = 0$ and

$$\alpha_1 x_1 + \cdots + \alpha_m x_m = 0. \tag{3.1}$$

Since not all numbers $\alpha_1, \dots, \alpha_m$ are zero, there is a subscript k_0, with $\alpha_{k_0} > 0$, which – in addition – can be chosen in such a way that

$$\frac{\lambda_k}{\alpha_k} \geq \frac{\lambda_{k_0}}{\alpha_{k_0}}$$

for all $k \in \{1, \dots, m\}$ with $\alpha_k > 0$. Thus we have for all $k \in \{1, \dots, m\}$

$$\lambda_k - \alpha_k \frac{\lambda_{k_0}}{\alpha_{k_0}} \geq 0,$$

and

$$\sum_{k=1}^{m} \left(\lambda_k - \alpha_k \frac{\lambda_{k_0}}{\alpha_{k_0}} \right) = 1.$$

By multiplying Equation (3.1) by $-\frac{\lambda_{k_0}}{\alpha_{k_0}}$ and adding $x = \sum_{k=1}^{m} \lambda_k x_k$, we obtain

$$x = \sum_{k=1}^{m} \left(\lambda_k - \alpha_k \frac{\lambda_{k_0}}{\alpha_{k_0}} \right) x_k.$$

As $(\lambda_{k_0} - \alpha_{k_0} \frac{\lambda_{k_0}}{\alpha_{k_0}}) = 0$, the point x was represented as a convex combination of $m-1$ elements in A. □

3.2 Convex Functions

Definition 3.2.1. Let K be a convex subset of X and $f : K \to (-\infty, \infty]$ a function. f is called *convex*, if for all $x, y \in K$ and for all $\lambda \in [0, 1]$

$$f(\lambda x + (1 - \lambda) y) \leq \lambda f(x) + (1 - \lambda) f(y).$$

A function $g : K \to [-\infty, \infty)$ is called *concave*, if $-g$ is convex.

In the sequel we will consider such functions f, which are not identical to infinity (proper convex functions).

Definition 3.2.2. Let K be a set and $f : K \to (-\infty, \infty]$ a function. By the *epigraph* of f we understand the set $\text{Epi}(f) := \{(x, r) \in K \times \mathbb{R} \mid f(x) \leq r\}$.

Thus the epigraph contains all points of $K \times \mathbb{R}$, which we find – within the domain of finite values of f – above the graph of f.

Using this notion one obtains the following characterization of convex functions, based on the ideas of J. L. W. V. Jensen.

Theorem 3.2.3. *Let K be a convex subset of X and $f : K \to (-\infty, \infty]$ a function. Then the following statements are equivalent:*

(a) *f is convex*

(b) Epi(f) *is a convex subset of $X \times \mathbb{R}$*

(c) *f satisfies Jensen's inequality, i.e. for all $x_1, \ldots, x_n \in K$ and for all positive $\lambda_1, \ldots, \lambda_n \in \mathbb{R}$ with $\lambda_1 + \cdots + \lambda_n = 1$ the inequality*

$$f\left(\sum_{k=1}^{n} \lambda_k x_k\right) \le \sum_{k=1}^{n} \lambda_k f(x_k)$$

holds.

Proof. (a) $\Rightarrow$ (b): Let $(x, r), (y, s) \in$ Epi(f), and let $\lambda \in [0, 1]$. Since f is convex, we have

$$f(\lambda x + (1 - \lambda)y) \le \lambda f(x) + (1 - \lambda)f(y) \le \lambda r + (1 - \lambda)s,$$

i.e.

$$\lambda(x, r) + (1 - \lambda)(y, s) = (\lambda x + (1 - \lambda)y, \lambda r + (1 - \lambda)s) \in \text{Epi}(f).$$

(b) $\Rightarrow$ (c): Let $n \in \mathbb{N}$, and let $x_1, \ldots, x_n \in K$, furthermore $\lambda_1, \ldots, \lambda_n \in \mathbb{R}_{\ge 0}$ satisfying $\lambda_1 + \cdots + \lambda_n = 1$.

If for a $k \in \{1, \ldots, n\}$ the value of the function $f(x_k)$ is infinite, then the inequality is trivially satisfied.

Let now w.l.o.g. $f(x_1), \ldots, f(x_n) \in \mathbb{R}$. The points $(x_1, f(x_1)), \ldots, (x_n, f(x_n))$ are then in Epi(f). Since Epi(f) is convex by assumption, we have

$$\left(\sum_{k=1}^{n} \lambda_k x_k, \sum_{k=1}^{n} \lambda_k f(x_k)\right) = \sum_{k=1}^{n} \lambda_k (x_k, f(x_k)) \in \text{Epi}(f),$$

hence

$$f\left(\sum_{k=1}^{n} \lambda_k x_k\right) \le \sum_{k=1}^{n} \lambda_k f(x_k).$$

(c) $\Rightarrow$ (a): This implication is obvious. □

Convexity preserving Operations

The following theorem shows, how new convex functions can be constructed from given convex functions. These constructions will play an important role in substantiating the convexity of a function. In this context one tries to represent a 'complicated' function as a composition of simpler functions, whose convexity is 'easily' established.

Theorem 3.2.4. *Let K be a convex subset of X.*

(1) *Let $n \in \mathbb{N}$ and let $f_1, \dots, f_n : K \to \mathbb{R}$ be convex functions. Then for all $\alpha_1, \dots, \alpha_n \in \mathbb{R}_{\geq 0}$ $\alpha_1 f_1 + \cdots + \alpha_n f_n$ is also a convex function.*

(2) *All affine functions, i.e. sums of linear and constant functions, are convex.*

(3) *Let $(X, \|\cdot\|)$ be a normed space. Then the norm $\|\cdot\| : X \to \mathbb{R}$ is a convex function.*

(4) *Let $f : K \to \mathbb{R}$ be a convex function, C a convex superset of $f(K)$ and $g : C \to \mathbb{R}$ a convex and monotone increasing function. Then the composition $g \circ f : K \to \mathbb{R}$ is also a convex function.*

(5) *Let Y be another vector space, $\varphi : X \to Y$ an affine mapping and $f : Y \to \mathbb{R}$ a convex function. Then the composition $f \circ \varphi : K \to \mathbb{R}$ is also a convex function.*

Proof. Is left to the reader. □

Example 3.2.5. Let X be a normed space, then the following functions are convex:

(a) $x \mapsto \|x\|^2$,

(b) for $x_0, v_1, \dots, v_n \in X$ let $f : \mathbb{R}^n \to \mathbb{R}$ with

$$(a_1, \dots, a_n) \mapsto f(a) := \left\| x_0 - \sum_{i=1}^{n} a_i v_i \right\|,$$

(c) more generally: let V be a vector space and $L : V \to X$ affine and $v \mapsto f(v) := \|L(v)\|$.

One obtains a further class of convex functions through the notion of sublinear functionals:

Definition 3.2.6. Let X be a vector space and let $f : X \to \mathbb{R}$ a function.

(a) f is called *positive homogeneous*, if for all $\alpha \in \mathbb{R}_{\geq 0}$ and for all $x \in X$ we have

$$f(\alpha x) = \alpha f(x)$$

(b) f is called *subadditive*, if for all $x, y \in X$ the inequality

$$f(x + y) \leq f(x) + f(y)$$

holds.

(c) f is called *sublinear*, if f is positive homogeneous and subadditive.

(d) f is called a *semi-norm*, if f is sublinear and symmetric, i.e. for all $x \in X$ we have $f(-x) = f(x)$.

Minkowski Functional

In this section we want to become familiar with a method that relates convex sets to sublinear functions. The method was introduced by Minkowski. This relation will play a central role in our treatise, because the definition of an Orlicz space is based on it.

Theorem 3.2.7. *Let K be a convex subset of X, which contains* 0 *as an algebraically interior point. Then the* Minkowski functional $q : X \to \mathbb{R}$, *defined by*

$$q(x) := \inf\{\alpha > 0 \,|\, x \in \alpha K\}$$

is a sublinear function on X. If K is symmetrical (i.e. $x \in K$ implies $-x \in K$), then q is a semi-norm. If in addition K is linearly bounded w.r.t. 0, then q is a norm on X.

Proof. (1) q_K is positive homogeneous, because for $\beta > 0$ we have $x \in \alpha K$ if and only if $\beta x \in \beta\alpha K$.

(2) Since K is convex, we have for $\alpha, \beta > 0$

$$\alpha K + \beta K = (\alpha+\beta)\left(\frac{\alpha}{\alpha+\beta}K + \frac{\beta}{\alpha+\beta}K\right) \subset (\alpha+\beta)K,$$

hence for $x, y \in X$ we have

$$\begin{aligned} q_K(x) + q_K(y) &= \inf\{\alpha > 0 \,|\, x \in \alpha K\} + \inf\{\beta > 0 \,|\, y \in \beta K\} \\ &= \inf\{\alpha + \beta \,|\, x \in \alpha K \text{ and } y \in \beta K\} \\ &\ge \inf\{\alpha+\beta \,|\, x + y \in (\alpha+\beta)K\} = q_K(x+y). \end{aligned}$$

(3) If K is symmetrical, then for $\alpha > 0$ we have: $\alpha x \in K$, if and only if $-\alpha x = \alpha(-x) \in K$. But then $q_K(x) = q_K(-x)$.

(4) Let $x \neq 0$ and K linearly bounded w.r.t. 0. Then there is $\alpha_0 > 0$ such that for all $\alpha \ge \alpha_0$ we obtain: $\alpha x \notin K$, i.e. $q_K(x) \ge \frac{1}{\alpha_0} > 0$. □

As an illustration of the above theorem we want to introduce the l^p-norms in $\mathbb{R}^n$ $(p > 1)$. The triangle inequality for this norm is the well-known Minkowski inequality, which in this context is obtained as a straightforward conclusion.

The function $f : \mathbb{R}^n \to \mathbb{R}$ with

$$x \mapsto f(x) := \sum_{i=1}^{n} |x_i|^p$$

is convex. Thus the level set

$$K := \{x \,|\, f(x) \le 1\}$$

is a convex subset of the $\mathbb{R}^n$, which contains 0 as an algebraically interior point and is obviously linearly bounded w.r.t. 0.

For $c_0 := \sqrt[p]{\sum_{i=1}^n |x_i|^p}$ we have $f(\frac{x}{c_0}) = 1$ and for the Minkowski functional q_K of K we thus obtain

$$q_K(x) = \inf\left\{c > 0 \,\middle|\, f\left(\frac{x}{c}\right) \le 1\right\} = \sqrt[p]{\sum_{i=1}^n |x_i|^p}.$$

According to the above theorem the function $x \mapsto \|x\|_p := \sqrt[p]{\sum_{i=1}^n |x_i|^p}$ is a norm.

3.3 Difference Quotient and Directional Derivative

Theorem 3.3.1 (Monotonicity of the difference quotient). *Let f be convex, let $z \in X$ $x_0 \in \mathrm{Dom}(f)$. Then the mapping $u : \mathbb{R}_{>0} \to (-\infty, \infty]$ defined by*

$$t \mapsto u(t) := \frac{f(x_0 + tz) - f(x_0)}{t}$$

is monotonically increasing.

Proof. Let $h : \mathbb{R}_{\ge 0} \to (-\infty, \infty]$ define by

$$t \mapsto h(t) := f(x_0 + tz) - f(x_0).$$

Apparently h is convex and we have for $0 < s \le t$

$$h(s) = h\left(\frac{s}{t}t + \frac{t-s}{t}0\right) \le \frac{s}{t}h(t) + \frac{t-s}{t}h(0) = \frac{s}{t}h(t).$$

Thus we obtain

$$\frac{f(x_0 + sz) - f(x_0)}{s} \le \frac{f(x_0 + tz) - f(x_0)}{t}.$$

In particular the difference quotient $\frac{f(x_0+tz)-f(x_0)}{t}$ is monotonically decreasing for $t \downarrow 0$. □

Corollary 3.3.2. *The* right-sided directional derivative

$$f'_+(x_0, z) := \lim_{t \downarrow 0} \frac{f(x_0 + tz) - f(x_0)}{t}$$

and left-sided directional derivative

$$f'_-(x_0, z) := \lim_{t \uparrow 0} \frac{f(x_0 + tz) - f(x_0)}{t}$$

exist in $\overline{\mathbb{R}}$. The latter is due to

$$f'_-(x_0, z) = -f'_+(x_0, -z). \tag{3.2}$$

3.3.1 Geometry of the Right-sided Directional Derivative

The monotonicity of the difference quotient, we have just proved, yields an inequality for $t = 1$ that can be interpreted as follows: the growth of a convex function is at least as large as the growth of the directional derivative, since we have

$$f'_+(x, y - x) \leq f(y) - f(x), \tag{3.3}$$

where $x \in \mathrm{Dom}(f)$ and $y \in X$ arbitrary.

This inequality will serve as a basis for the introduction of the notion of the subgradient in the description of a supporting hyperplanes of the epigraph of f (subgradient inequality).

Sublinearity of the Directional Derivative

A particularly elegant statement about the geometry of the directional derivative is obtained, if it is formed at an algebraically interior point of $\mathrm{Dom}(f)$.

Theorem 3.3.3. *Let X be a vector space, $f : X \to \mathbb{R} \cup \infty$ a convex function, x_0 an algebraically interior point of* $\mathrm{Dom}(f)$. *Then for all directions $z \in X$ right-sided and left-sided derivatives are finite and we obtain:*

(a) *the mapping $f'_+(x_0, \cdot) : X \to \mathbb{R}$ is sublinear*

(b) *the mapping $f'_-(x_0, \cdot) : X \to \mathbb{R}$ is superlinear*

(c) *for all $z \in X$ we have: $f'_-(x_0, z) \leq f'_+(x_0, z)$.*

Proof. At first we have to show that for all $z \in X$ the one-sided derivatives $f'_+(x_0, z)$ and $f'_-(x_0, z)$ are finite. Let $z \in X$. Since x_0 is an algebraically interior point of $\mathrm{Dom}(f)$, there is a $\varepsilon > 0$ with $[x_0 - \varepsilon z, x_0 + \varepsilon z] \subset \mathrm{Dom}(f)$. As the difference quotient is monotonically increasing, we have

$$f'_+(x_0, z) \leq \frac{f(x_0 + \varepsilon z) - f(x_0)}{\varepsilon} < \infty.$$

Since f is convex, we obtain for all $t \in (0, 1]$

$$\begin{aligned} f(x_0) &= f\left(\frac{1}{1+t}(x_0 + t\varepsilon z) + \frac{t}{1+t}(x_0 - \varepsilon z)\right) \\ &\leq \frac{1}{1+t} f(x_0 + t\varepsilon z) + \frac{t}{1+t} f(x_0 - \varepsilon z). \end{aligned}$$

hence

$$f(x_0) + tf(x_0) = (1+t)f(x_0) \leq f(x_0 + t\varepsilon z) + tf(x_0 - \varepsilon z),$$

and thus

$$tf(x_0) - tf(x_0 - \varepsilon z) \le f(x_0 + t\varepsilon z) - f(x_0).$$

We obtain

$$-\infty < \frac{f(x_0) - tf(x_0 - \varepsilon z)}{\varepsilon} \le \frac{f(x_0 + t\varepsilon z) - f(x_0)}{t\varepsilon} \to f'_+(x_0, z),$$

which means $f'_+(x_0, z) \in \mathbb{R}$.

(a) We have to show the positive homogeneity and the subadditivity of $f'_+(x_0, \cdot)$.

Concerning the positive homogeneity we observe: let $z \in X$ and $\alpha \ge 0$. If $\alpha = 0$, then $f'_+(x_0, 0 \cdot z) = 0 = 0 \cdot f'_+(x_0, z)$. If $\alpha > 0$, then

$$\begin{aligned} f'_+(x_0, \alpha z) &= \lim_{\lambda \downarrow 0} \frac{f(x_0 + \lambda \alpha z) - f(x_0)}{\lambda} \\ &= \alpha \cdot \lim_{\lambda \downarrow 0} \frac{f(x_0 + \lambda \alpha z) - f(x_0)}{\lambda \alpha} = \alpha f'_+(x_0, z). \end{aligned}$$

We now turn our attention to the subadditivity: let $z_1, z_2 \in X$, then, because of the convexity of f

$$\begin{aligned} f'_+(x_0, z_1 + z_2) &= \lim_{\lambda \downarrow 0} \frac{f(x_0 + \lambda(z_1 + z_2)) - f(x_0)}{\lambda} \\ &= \lim_{\lambda \downarrow 0} \frac{1}{\lambda} \left(f\left(\frac{1}{2}(x_0 + 2\lambda z_1) + \frac{1}{2}(x_0 + 2\lambda z_2) \right) - f(x_0) \right) \\ &\le \lim_{\lambda \downarrow 0} \frac{1}{\lambda} \left(\frac{1}{2} f(x_0 + 2\lambda z_1) + \frac{1}{2} f(x_0 + 2\lambda z_2) - f(x_0) \right) \\ &= \lim_{\lambda \downarrow 0} \frac{f(x_0 + 2\lambda z_1) - f(x_0)}{2\lambda} + \lim_{\lambda \downarrow 0} \frac{f(x_0 + 2\lambda z_2) - f(x_0)}{2\lambda} \\ &= f'_+(x_0, z_1) + f'_+(x_0, z_2). \end{aligned}$$

(b) Follows from (a) due to Equation (3.2).

(c) Let $z \in X$, then

$$0 = f'_+(x_0, z - z) \le f'_+(x_0, z) + f'_+(x_0, -z),$$

hence

$$f'_-(x_0, z) = -f'_+(x_0, -z) \le f'_+(x_0, z).$$

□

Definition 3.3.4. Let X be a vector space, U a subset of X, Y a normed space, $F : U \to Y$ a mapping, $x_0 \in U$ and $z \in X$. Then F is called *differentiable* (or *Gâteaux differentiable*) at x_0 in direction z, if there is a $\varepsilon > 0$ with $[x_0 - \varepsilon z, x_0 + \varepsilon z] \subset U$ and the limit

$$F'(x_0, z) := \lim_{t \to 0} \frac{F(x_0 + tz) - F(x_0)}{t}$$

in Y exists. $F'(x_0, z)$ is called the *Gâteaux derivative* of F at x_0 in direction z. F is called *Gâteaux differentiable at* x_0, if F is differentiable at x_0 in every direction $z \in X$. The mapping $F'(x_0, \cdot) : X \to Y$ is called the *Gâteaux derivative of* F *at* x_0.

Linearity of the Gâteaux Derivative of a Convex Function

Theorem 3.3.5. *Let X be a vector space, $f : X \to \mathbb{R} \cup \infty$ a convex function, x_0 an algebraically interior point of* $\mathrm{Dom}(f)$. *Then we have: f is Gâteaux differentiable at x_0, if and only if the right-sided derivative $f'_+(x_0, \cdot) : X \to \mathbb{R}$ is linear. In particular the right-sided derivative is then equal to the Gâteaux derivative.*

Proof. Let f be Gâteaux differentiable, then one obtains the homogeneity for $\alpha < 0$ from

$$f'(x_0, \alpha z) = f'_-(x_0, \alpha z) = -f'_+(x_0, -\alpha z) = (-\alpha)(-f'_+(x_0, z)) = \alpha f'(x_0, \alpha z).$$

Conversely, if the right-sided derivative is linear, then using Equation (3.2):

$$f'_-(x_0, h) = -f'_+(x, -h) = f'(x, h).$$ □

Theorem 3.3.6. *Let X be a vector space, U an algebraically open convex subset of X, and $f : U \to \mathbb{R} \cup \infty$ a function, which is Gâteaux differentiable at each point of U, then the following statements are equivalent:*

(a) *f is convex*

(b) *for all $x \in U$ the derivative $f'(x, \cdot) : X \to \mathbb{R}$ is linear, and for all $x_0, x \in U$ the following inequality holds*

$$f'(x_0, x - x_0) \le f(x) - f(x_0).$$

Proof. (b) follows from (a) using Inequality (3.3) and Theorem 3.3.5.

Conversely, let $x_1, x_2 \in U$ and $\lambda \in [0, 1]$. Then $x_0 := \lambda x_1 + (1 - \lambda)x_2 \in U$ and we have: $\lambda(x_1 - x_0) + (1 - \lambda)(x_2 - x_0) = 0$ and hence

$$\begin{aligned}
0 &= f'(x_0, \lambda(x_1 - x_0) + (1 - \lambda)(x_2 - x_0)) \\
&= \lambda f'(x_0, x_1 - x_0) + (1 - \lambda) f'(x_0, x_2 - x_0) \\
&\le \lambda(f(x_1) - f(x_0)) + (1 - \lambda)(f(x_2) - f(x_0)) \\
&= \lambda f(x_1) + (1 - \lambda) f(x_2) - f(x_0),
\end{aligned}$$

thus we obtain

$$f(x_0) = f(\lambda x_1 + (1 - \lambda)x_2) \le \lambda f(x_1) + (1 - \lambda) f(x_2).$$ □

Monotonicity of the Gâteaux Derivative of a Convex Function

Let X be a vector space and $U \subset X$ algebraically open and X' the vector space of all linear functionals on X.

Definition 3.3.7. A mapping $A : U \to X'$ is called *monotone* on U, if for all $x_1, x_2 \in U$ the subsequent inequality holds:

$$\langle A(x_1) - A(x_2), x_1 - x_2 \rangle \geq 0.$$

Then we obtain the following characterization of convex functions:

Theorem 3.3.8. *Let U be an algebraically open convex set and $f : U \to \mathbb{R}$ Gâteaux differentiable. f is convex, if and only if for all $x \in U$ the Gâteaux derivative $f'(x, \cdot)$ at x is linear and the Gâteaux derivative is $f' : U \to X'$ monotone.*

Proof. Let f be convex. Due to the previous theorem we have for all $x, y \in U$

$$\begin{aligned} -\langle f'(x,\cdot) - f'(y,\cdot), x - y \rangle &= f'(x, y - x) + f'(y, x - y) \\ &\leq f(y) - f(x) + f(x) - f(y) = 0, \end{aligned}$$

i.e. f' is monotone on U.

Conversely, in order to establish convexity of f we again make use of the previous theorem: let $x, y \in U$, then for $h := y - x$ and a $\beta > 0$ the interval $[x - \beta h, x + \beta h] \subset U$. Let now $\varphi : (-\beta, \beta) \to \mathbb{R}$ be defined by $\varphi(t) := f(x + th)$, then: φ is differentiable and $\varphi'(t) = f'(x + th, h)$, since

$$\begin{aligned} \varphi'(t) = \lim_{\tau \to 0} \frac{\varphi(t + \tau) - \varphi(t)}{\tau} &= \lim_{\tau \to 0} \frac{f(x + th + \tau h) - f(x + th)}{\tau} \\ &= f'(x + th, h). \end{aligned}$$

Due to the mean value theorem for φ there exists a $\alpha \in (0, 1)$ such that $f(y) - f(x) = f'(x + \alpha h, h)$. From the monotonicity of f' and the linearity of $f'(x, \cdot)$ and $f'(x + \alpha h, \cdot)$ we obtain

$$\begin{aligned} \alpha(f(y) - f(x)) &= f'(x + \alpha h, \alpha h) - f'(x, \alpha h) + f'(x, \alpha h) \\ &= \langle f'(x + \alpha h, \cdot) - f'(x, \cdot), x + \alpha h - x \rangle + f'(x, \alpha h) \\ &\geq f'(x, \alpha h) = \alpha f'(x, h), \end{aligned}$$

i.e. $f(y) - f(x) \geq f'(x, y - x)$. The assertion then follows by the previous theorem.

□

3.4 Necessary and Sufficient Optimality Conditions

3.4.1 Necessary Optimality Conditions

For the minimal solution of a Gâteaux differentiable function we can state the following necessary condition, which is a complete analogy to the well-known condition from real analysis:

Theorem 3.4.1. *Let U be a subset of a vector spaces X and let $f : U \to \mathbb{R}$ have a minimal solution at $x_0 \in U$.*

If for a $z \in X$ and a $\varepsilon > 0$ the interval $(x_0 - \varepsilon z, x_0 + \varepsilon z)$ is contained in U and if f is differentiable at x_0 in direction z, then: $f'(x_0, z) = 0$.

Proof. The function $g : (-\varepsilon, \varepsilon) \to \mathbb{R}$ with $t \mapsto g(t) := f(x_0 + tz)$ has a minimal solution at 0. Thus $0 = g'(0) = f'(x_0, z)$. □

Corollary 3.4.2. *Let V be a subspace of the vector space X, $y_0 \in X$ and $f : y_0 + V \to \mathbb{R}$ be a Gâteaux differentiable function. If $x_0 \in M(f, y_0 + V)$, then for all $z \in V$: $f'(x_0, z) = 0$.*

3.4.2 Sufficient Condition: Characterization Theorem of Convex Optimization

Theorem 3.4.3. *Let K be a convex subset of the vector space X and $f : X \to \mathbb{R}$ a convex function. f has a minimal solution $x_0 \in K$ if and only if for all $x \in K$*

$$f'_+(x_0, x - x_0) \geq 0. \tag{3.4}$$

Proof. Let x_0 be a minimal solution of f on K. For $x \in K$ and $t \in (0, 1]$ we have $x_0 + t(x - x_0) = tx + (1 - t)x_0 \in K$ and hence

$$\frac{f(x_0 + t(x - x_0)) - f(x_0)}{t} \geq 0.$$

The limit for t to zero yields the above Inequality (3.4). Conversely from Inequalities (3.3) and (3.4) it follows for all $x \in K$

$$f(x) - f(x_0) \geq f'_+(x_0, x - x_0) \geq 0,$$

i.e. $x_0 \in M(f, K)$. □

Corollary 3.4.4. *Let V be a subspace of the vector space X, $y_0 \in X$ and $f : y_0 + V \to \mathbb{R}$ a Gâteaux differentiable convex function. Then the following equivalence holds:*

$$x_0 \in M(f, y_0 + V), \quad \textit{if and only if for all } z \in V\textit{: } f'(x_0, z) = 0.$$

Proof. According to Corollary 3.4.2 the condition is necessary and due to the characterization theorem sufficient. □

3.5 Continuity of Convex Functions

We will see below that from the boundedness of a convex function on an open set its continuity already follows:

Definition 3.5.1. Let X be a metric space and $f : X \to \overline{\mathbb{R}}$. f is called *lower semi-continuous* (resp. *upper semi-continuous*), if for every $r \in \mathbb{R}$ the level set $S_f(r) = \{x \in X \mid f(x) \le r\}$ (resp. $\{x \in X \mid f(x) \ge r\}$) is closed.

Remark 3.5.2. The supremum of lower semi-continuous functions is lower semi-continuous.

Proof. Let M be a set of lower semi-continuous functions on X and $g(x) := \sup\{f(x) \mid f \in M\}$. From $S_g(r) = \bigcap_{f \in M} S_f(r)$ closedness of $S_g(r)$ follows. □

Theorem 3.5.3. *Let X be a metric space and $f : X \to \overline{\mathbb{R}}$. The following statements are equivalent:*

(a) *f is lower semi-continuous (resp. upper semi-continuous)*

(b) *for every $x_0 \in X$ and every $\varepsilon > 0$ there exists a $\delta > 0$, such that for all x in the ball $K(x_0, \delta)$ we have*

$$f(x) \ge f(x_0) - \varepsilon \quad (\textit{resp. } f(x) \le f(x_0) + \varepsilon)$$

(c) *for every $x \in X$ and every sequence $(x_k)_{k \in \mathbb{N}}$ in X convergent to x we have*

$$\varliminf_{k \to \infty} f(x_k) \ge f(x) \quad (\textit{resp. } \varlimsup_{k \to \infty} f(x_k) \le f(x)),$$

(d) *the epigraph of f is closed in $X \times \mathbb{R}$.*

Proof. Is left to the reader. □

Theorem 3.5.4. *Let X be a normed space, U an open and convex subset of X and let $f : U \to \mathbb{R}$ be a convex function. Then the following statements are equivalent:*

(a) *f is continuous on U*

(b) *f is upper semi-continuous on U*

(c) *f is bounded from above on an open subset U_0 of U*

(d) *f is continuous at a point in U.*

The proof is given in the chapter on stability within the framework of families of convex functions (see Theorem 5.3.8).

There we also demonstrate that in Banach spaces continuity, upper semi-continuity and lower semi-continuity coincide (see Theorem 5.3.12).

Theorem 3.5.5. *Let U be an open and convex subset of a normed space X and let $f : U \to \mathbb{R}$ convex. Then the following statements are equivalent:*

(a) $\text{Epi}(f)$ *has an interior point*

(b) f *is continuous*

(c) $\text{Graph}(f)$ *is closed in* $U \times \mathbb{R}$

(d) *every point in* $\text{Epi}(f) \setminus \text{Graph}(f)$ *is an interior point of* $\text{Epi}(f)$.

Proof. (a) $\Rightarrow$ (b): Let (x_0, r) be an interior point of $\text{Epi}(f)$. Then for a neighborhood V of x_0

$$V \times \{r\} \subset \text{Epi}(f),$$

i.e. $f(x) \leq r$ for all $x \in V$. Due to Theorem 3.5.4 f is continuous.

(b) $\Rightarrow$ (c): we show that the complement of $\text{Graph}(f)$ in $U \times \mathbb{R}$ is open: let $f(x_0) \neq r$ and let $I(f(x_0))$ and $I(r)$ be disjoint open intervals. Then

$$f^{-1}(I(f(x_0)) \times I(r)) \cap \text{Graph}(f) = \emptyset.$$

(c) $\Rightarrow$ (d): let $f(x_0) < r$ and $K = K((x_0, r), \rho)$ a ball in $X \times \mathbb{R}$ contained in $U \times \mathbb{R}$, such that $k \cap \text{Graph}(f) = \emptyset$. For $(y, s) \in K$ we have $f(y) < s$.

(d) $\Rightarrow$ (a) holds trivially. □

3.6 Fréchet Differentiability

In the sequel the following notion will play an important role, in particular in the context of function spaces:

Definition 3.6.1. Let X, Y be normed spaces, U an open subset of X, $F : U \to Y$ a mapping, $x_0 \in U$. Then F is called *Fréchet differentiable* at x_0, if there is a linear and continuous mapping $A : X \to Y$ such that

$$\lim_{\|h\| \to 0} \frac{F(x_0 + h) - F(x_0) - A(h)}{\|h\|} = 0.$$

A is called the *Fréchet derivative* of F at x_0 and denoted by $F'(x_0)$. F is called *Fréchet differentiable on* U, if F is Fréchet differentiable at every point of U. The mapping $F' : X \to L(X, Y)$ is called the *Fréchet derivative of* F *at* x_0.

Remark 3.6.2. Let X, Y be normed spaces, U an open subset of X, $F : U \to Y$ Fréchet differentiable at $x \in U$ then

(a) the Fréchet derivative is uniquely determined

(b) F is Gâteaux differentiable and for all $h \in X$ we have

$$F(x, h) = F'(x)(h).$$

Proof. Is left to the reader. □

3.7 Convex Functions in $\mathbb{R}^n$

The results of this section will again demonstrate the exceptional properties of convex functions: in finite-dimensional spaces finite convex functions on open sets are already continuous, and partial, Gâteaux, Fréchet and continuous differentiability coincide.

Theorem 3.7.1. *Every real-valued convex function f on an open subset of $\mathbb{R}^n$ is continuous.*

Proof. See Theorem 5.3.11. □

We will turn our attention to the differentiability of convex functions in $\mathbb{R}^n$. It will turn out that partial, total and Gâteaux differentiability coincide. In order to show this equivalence we need the following

Lemma 3.7.2. *Let g be a finite sublinear functional on a finite-dimensional vector space V and let $v_1, \dots, v_n$ be a basis of V, for which: $g(-v_i) = -g(v_i)$, $i \in \{1, \dots, n\}$. Then g is linear on V.*

Proof. Apparently then also $g(\lambda(-v_i)) = -g(\lambda v_i)$ for $i \in \{1, \dots, n\}$ and $\lambda \in \mathbb{R}$ arbitrary holds.

Let now $x \in V$, then the sublinearity of g implies

$$0 = g(0) = g\left(\frac{1}{2}(x - x)\right) \leq \frac{1}{2}g(x) + \frac{1}{2}g(-x),$$

i.e. $g(x) \geq -g(-x)$. Let furthermore x have the representation $x = \sum_{i=1}^n \lambda_i v_i$, then the sublinearity implies

$$\sum_{i=1}^n g(\lambda_i v_i) \geq g(x) \geq -g(-x) \geq -\sum_{i=1}^n g(-\lambda_i v_i) = \sum_{i=1}^n g(\lambda_i v_i).$$

Therefore

$$g\left(\sum_{i=1}^n \lambda_i v_i\right) = \sum_{i=1}^n \lambda_i g(v_i).$$

□

From the existence of the partial derivatives we obtain for convex functions already the Gâteaux differentiability, which – as we already know – implies the linearity of the Gâteaux differential.

Theorem 3.7.3. *Let U be an open subset of $\mathbb{R}^n$ and $f : U \to \mathbb{R}$ convex. If f is partially differentiable at $x \in U$, then f is Gâteaux differentiable at x and we have*

$$f'(x,h) = \sum_{i=1}^{n} h_i f'(x, e_i) = \langle \nabla f(x), h \rangle,$$

if $e_1, \ldots, e_n$ denote the unit-vectors in $\mathbb{R}^n$.

Proof. For $i = 1, \ldots, n$ we have

$$f'_+(x, e_i) = f_{x_i}(x) = f'_-(x, e_i) = -f'_+(x, -e_i).$$

According to the above lemma $f'_+(x, \cdot)$ is linear and thus by Theorem 3.3.5 f Gâteaux differentiable at x. The linearity yields in particular

$$f'(x,h) = f'\Bigl(x, \sum_{i=1}^{n} h_i e_i\Bigr) = \sum_{i=1}^{n} h_i f'(x, e_i) = \langle \nabla f(x), h \rangle. \qquad \square$$

Theorem 3.7.4. *Let f be convex on an open subset U of $\mathbb{R}^n$ and partially differentiable at $x \in U$. Then f is Fréchet differentiable at x and the Fréchet derivative is the gradient.*

Proof. For the linear mapping, that we have to identify, we take the multiplication with the 1-row matrix A of the partial derivatives. Let $x \in U$. Since U is open, we can find a ball $K(0, \varepsilon)$ with center 0 and radius ε, such that $x + nK(0, \varepsilon)$ is contained in U. We now consider the convex function $g : K(0, \varepsilon) \to \mathbb{R}$, defined by

$$z \mapsto g(z) = f(x + z) - f(x) - A \cdot z. \tag{3.5}$$

The vector $z = (z_1, \ldots, z_n)$ can be rewritten as the sum $z = \sum_{i=1}^{n} z_i e_i$. The convexity of g yields

$$g(z) = g\Bigl(\sum_{i=1}^{n} \frac{n}{n} z_i e_i\Bigr) \le \frac{1}{n} \sum_{i=1}^{n} g(n z_i e_i), \tag{3.6}$$

and the definition of g

$$g(n z_i e_i) = f(x + z_i n e_i) - f(x) - \frac{\partial f}{\partial x_i}(x) n z_i. \tag{3.7}$$

For $z_i = 0$ we have $g(n z_i e_i) = 0$ and for $z_i \neq 0$ (3.7) implies

$$\lim_{z_i \to 0} \frac{g(n z_i e_i)}{n z_i} = 0. \tag{3.8}$$

Using (3.6) and the Cauchy–Schwarz inequality we obtain after omitting the terms with $z_i = 0$

$$g(z) \leq \sum z_i \frac{g(nz_i e_i)}{z_i n} \leq \|z\| \sqrt{\sum \left(\frac{g(nz_i e_i)}{z_i n}\right)^2} \leq \|z\| \sum \left|\frac{g(nz_i e_i)}{z_i n}\right|. \tag{3.9}$$

In a similar manner we obtain

$$g(-z) \leq \|z\| \sum \left|\frac{g(-nz_i e_i)}{z_i n}\right|. \tag{3.10}$$

Due to $g(0) = 0$ we have

$$0 = g\left(\frac{z + (-z)}{2}\right) \leq \frac{1}{2} g(z) + \frac{1}{2} g(-z),$$

therefore

$$g(z) \geq -g(-z). \tag{3.11}$$

Using (3.9), (3.10), and (3.11) we have

$$-\|z\| \sum \left|\frac{g(-z_i n e_i)}{z_i n}\right| \leq -g(-z) \leq g(z) \leq \|z\| \sum \left|\frac{g(z_i n e_i)}{z_i n}\right|, \tag{3.12}$$

(3.8) then implies

$$\lim_{\|z\| \to 0} \frac{g(z)}{\|z\|} = 0.$$

□

Theorem 3.7.5. *Let U be an open subset of $\mathbb{R}^n$ and let $f : U \to \mathbb{R}$ be a convex function differentiable in U. Then $f' : U \to \mathbb{R}^n$ is continuous.*

Proof. Let w.l.o.g. $0 \in U$ and let $g : U \to \mathbb{R}$ be defined by

$$x \mapsto g(x) := f(x) - f(0) - \langle f'(0), x \rangle,$$

then $g(0) = 0$ and $g'(0) = 0$. The subgradient inequality yields $g(x) = g(x) - g(0) \geq \langle g'(0), x - 0 \rangle = 0$. Apparently it is sufficient to show the continuity of g' at 0.

Let Q be the unit ball w.r.t. the maximum norm on $\mathbb{R}^n$ and let z_i, $i = 1, \ldots, m$, the extreme points of Q. Since U is open, there is a $k > 0$ with $kQ \subset U$. Let now $0 < h < k$ and let x be chosen such that $\|x\|_\infty \leq \frac{1}{2} h$.

Let further $y \in \mathbb{R}^n \setminus \{0\}$ be arbitrary, then because of the continuity of the norm there is $\lambda > 0$ with $\|x + \lambda y\|_\infty = h$. Let $z := x + \lambda y$, then we obtain using the subgradient inequality

$$g(z) - g(x) \geq \langle g'(x), z - x \rangle = \lambda \langle g'(x), y \rangle. \tag{3.13}$$

The theorem of Minkowski 3.9.12 provides a representation of z as a convex combination of the extreme points of hQ, i.e. $z = \sum_{i=1}^{m} \lambda_i h z_i$. Therefore

$$g(z) \leq \sum_{i=1}^{m} \lambda_i g(hz_i) \leq \max_i g(hz_i) \sum_{i=1}^{m} \lambda_i = \max_i g(hz_i).$$

From $g(0) = 0$ and the convexity of g it follows that $g(hz_i) = g(k\frac{h}{k}z_i) \leq \frac{h}{k}g(kz_i)$ and hence $g(z) \leq \frac{h}{k}\max_i g(kz_i)$.

Furthermore: $h = \|z\|_\infty \leq \|x\|_\infty + \lambda\|y\|_\infty \leq \frac{1}{2}h + \lambda\|y\|_\infty$ and hence $h \leq 2\lambda\|y\|_\infty$.

Using (3.13) we obtain

$$\langle g'(x), y\rangle \leq \frac{1}{\lambda}(g(z) - \underbrace{g(x)}_{\geq 0}) \leq \frac{1}{\lambda}\frac{2\lambda\|y\|_\infty}{k}\max_i g(kz_i) = 2\|y\|_\infty \max_i \frac{g(kz_i)}{k}.$$

In the same way we obtain

$$-\langle g'(x), y\rangle = \langle g'(x), -y\rangle \leq 2\|y\|_\infty \max_i \frac{g(kz_i)}{k},$$

in total therefore

$$|\langle g'(x), y\rangle| \leq 2\|y\|_\infty \max_i \frac{g(kz_i)}{k}.$$

But by definition: $\lim_{k\to 0} \frac{g(kz_i)}{k} = \langle g'(0), z_i\rangle = 0$. Let now $\varepsilon > 0$ be given, choose k, such that $2\max_i \frac{g(kz_i)}{k} < \varepsilon$, then we obtain

$$|\langle g'(x), y\rangle| \leq \|y\|_\infty \varepsilon,$$

and thus the continuity of g' at 0. □

3.8 Continuity of the Derivative

Definition 3.8.1. Let X a normed space and U an open subset of X. A mapping $G : U \to X^*$ is called *hemi-continuous*, if for arbitrary $x \in U$ and $u \in X$ and every real sequence $t_n \to t_0$ with $x + t_n u \in U$ for $n = 0, 1, 2, \ldots$ we have

$$G(x + t_n u) \rightharpoonup G(x + t_0 u).$$

Lemma 3.8.2. *Let U be an open subset of X and let $f : U \to \mathbb{R}$ convex and Gâteaux differentiable, then the Gâteaux derivative of f is hemi-continuous.*

Proof. Let $x \in U$ and $u, w \in X$ be arbitrary. Let $W \subset \mathbb{R}^2$ be a suitable open neighborhood of the origin, such that $x + tu + sw \in U$ for all $t, s \in W$. Then $g : W \to \mathbb{R}$ with $g(t, s) := f(x + tu + sw)$ is convex and differentiable and for the partial derivatives (which at the same time are the directional derivatives of f at $x + tu + sw$ in direction u resp. w) we have

$$\frac{\partial}{\partial t} g(t, s) = \langle f'(x + tu + sw), u\rangle$$
$$\frac{\partial}{\partial s} g(t, s) = \langle f'(x + tu + sw), w\rangle.$$

In particular: $\frac{\partial}{\partial s} g(t, 0) = \langle f'(x + tu), w\rangle$.

According to Theorems 3.7.4 and 3.7.5 the partial derivatives of g are continuous, hence in particular for $t_n \to t_0$ we obtain

$$\frac{\partial}{\partial s} g(t_n, 0) = \langle f'(x + t_n u), w\rangle \to \frac{\partial}{\partial s} g(t_0, 0) = \langle f'(x + t_0 u), w\rangle,$$

and thus the assertion. □

Theorem 3.8.3. *Let U be an open subset of the Banach space X and let $f : U \to \mathbb{R}$ be continuous, convex, and Gâteaux differentiable. Let $x_0 \in U$ be arbitrary, then there is a $\varepsilon > 0$ such that the Gâteaux derivative of f is norm-bounded on the ball $K(x_0, \varepsilon)$ i.e. there is a constant K with $\|f'(x)\| \leq K$ for all $x \in K(x_0, \varepsilon)$.*

Proof. Since f is continuous, there is $\varepsilon > 0$, such that $|f|$ is bounded on the ball $K(x_0, 2\varepsilon)$. Let $x \in K(x_0, \varepsilon)$ and $z \in K(0, \varepsilon)$ arbitrary, then apparently $x + z \in K(x_0, 2\varepsilon)$. Using the subgradient inequality we obtain for a suitable constant M

$$M \geq f(x + z) - f(x) \geq \langle f'(x), z\rangle$$
$$M \geq f(x - z) - f(x) \geq \langle f'(x), -z\rangle,$$

therefore $|\langle f'(x), z\rangle| \leq M$ for all $z \in K(0, \varepsilon)$ and all $x \in K(x_0, \varepsilon)$. Let now $y \in U$ arbitrary, then there is a $z \in K(0, \varepsilon)$ and a $\lambda \geq 0$ with $y = \lambda z$, hence

$$\lambda M \geq |\langle f'(x), y\rangle|$$

for all $x \in K(x_0, \varepsilon)$. Thus the family of linear continuous functionals $\{f'(x) \mid x \in K(x_0, \varepsilon)\}$ is pointwise bounded and hence according to the theorem of Banach on Uniform Boundedness (see 5.3.14) norm-bounded, i.e. there is a constant K with $\|f'(x)\| \leq K$ for all $x \in K(x_0, \varepsilon)$. □

Definition 3.8.4. Let X be a normed space and U an open subset of X. A mapping $G : U \to X^*$ is called *demi-continuous*, if for arbitrary sequences $x_n \to x_0 \in U$ we have

$$G(x_n) \rightharpoonup G(x_0).$$

The line of reasoning in the proof of the subsequent theorem – aside from the substantiation of local boundedness – follows from that in Vainberg [109] for monotone operators.

Theorem 3.8.5. *Let U be an open subset of a Banach space X and let $f : U \to \mathbb{R}$ be convex, continuous, and Gâteaux differentiable, then the Gâteaux derivative of f is demi-continuous.*

Proof. Let (x_n) be a sequence with $x_n \to x_0$ and let $u \in U$ arbitrary. According to Theorem 3.8.3 the sequence $\|f'(x_n)\|$ is bounded. Let $t_n := \|x_n - x_0\|^{\frac{1}{2}}$ and $u_n := x_0 + t_n u$. Due to the hemi-continuity of f' by Lemma 3.8.2 it follows that $f'(u_n) \rightharpoonup f'(x_0)$. In particular $\|f'(u_n)\|$ is bounded. The monotonicity of f' yields $t_n^{-1}\langle f'(u_n) - f'(x_n), u_n - x_n\rangle \geq 0$ and hence

$$\langle f'(x_n), u\rangle \leq t_n^{-1}\langle f'(u_n), u_n - x_n\rangle + t_n^{-1}\langle f'(x_n), t_n u - (u_n - x_n)\rangle. \qquad (3.14)$$

We now investigate the first term on the right-hand side

$$t_n^{-1}\langle f'(u_n), u_n - x_n\rangle = \underbrace{\left\langle f'(u_n), t_n \frac{x_0 - x_n}{\|x_0 - x_n\|}\right\rangle}_{\to 0} + \underbrace{\langle f'(u_n), u\rangle}_{\to \langle f'(x_0), u\rangle}.$$

For the 2nd term of the right-hand side of (3.14) we obtain

$$t_n^{-1}\langle f'(x_n), t_n u - (u_n - x_n)\rangle = \left\langle f'(x_n), \frac{x_n - x_0}{\|x_n - x_0\|^{\frac{1}{2}}}\right\rangle \leq \|f'(x_n)\| t_n.$$

Thus there is a sequence tending to zero (c_n) with

$$\langle f'(x_n) - f'(x_0), u\rangle \leq c_n.$$

In the same way we obtain another sequence (d_n) tending to zero with

$$\langle f'(x_n) - f'(x_0), -u\rangle \leq d_n,$$

in total

$$-d_n \leq \langle f'(x_n) - f'(x_0), u\rangle \leq c_n,$$

and thus the assertion, since u was arbitrary. □

The subsequent theorem can be found in Phelps [91] and is a generalization of a corollary of the lemma of Shmulian 8.4.20 (see [42], p. 145) for norms.

Theorem 3.8.6. *Let U be an open subset of a Banach space X and let $f : U \to \mathbb{R}$ be convex, continuous, and Gâteaux differentiable. Then f is Fréchet differentiable, if and only if the Gâteaux derivative of f is continuous.*

Proof. Let f Fréchet differentiable and $x_0 \in U$. We have to show: for every $\varepsilon > 0$ there is a $\delta > 0$, such that for all $x \in K(x_0, \delta)$ we have $f'(x) \in K(f'(x_0), \varepsilon)$. Suppose this is not the case, then there is a $\rho > 0$, and a sequence $(x_n) \subset U$, such that $\|x_n - x_0\| \to 0$ but $\|f'(x_n) - f'(x_0)\| > 2\rho$. Consequently there is a sequence (z_n) in X with $\|z_n\| = 1$ and the property $\langle f'(x_n) - f'(x_0), z_n\rangle > 2\rho$. Due to the Fréchet differentiability of f at x_0 there is $\alpha > 0$ with

$$f(x_0 + y) - f(x_0) - \langle f'(x_0), y\rangle \leq \rho\|y\|,$$

provided that $\|y\| \leq \alpha$. The subgradient inequality implies

$$\langle f'(x_n), (x_0 + y) - x_n\rangle \leq f(x_0 + y) - f(x_n),$$

and hence

$$\langle f'(x_n), y\rangle \leq f(x_0 + y) - f(x_0) + \langle f'(x_n), x_n - x_0\rangle + f(x_0) - f(x_n).$$

Let now $y_n = \alpha \cdot z_n$, i.e. $\|y_n\| = \alpha$, then we obtain

$$\begin{aligned} 2\rho\alpha &< \langle f'(x_n) - f'(x_0), y_n\rangle \\ &\leq \{f(x_0 + y_n) - f(x_0) - \langle f'(x_0), y_n\rangle\} \\ &\quad + \langle f'(x_n), x_n - x_0\rangle + f(x_0) - f(x_n) \\ &\leq \rho\alpha + \langle f'(x_n), x_n - x_0\rangle + f(x_0) - f(x_n), \end{aligned}$$

because of the local boundedness of f' we obtain

$$|\langle f'(x_n), x_n - x_0\rangle| \leq \|f'(x_n)\|\|x_n - x_0\| \to 0,$$

and because of the continuity of f we have $f(x_0) - f(x_n) \to 0$, hence $2\rho\alpha < \rho\alpha$, a contradiction.

Conversely let the Gâteaux derivative be continuous. By the subgradient inequality we have for $x, y \in U$: $\langle f'(x), y-x\rangle \leq f(y)-f(x)$ and $\langle f'(y), x-y\rangle \leq f(x)-f(y)$. Thus we obtain

$$0 \leq f(y)-f(x)-\langle f'(x), y-x\rangle \leq \langle f'(y)-f'(x), y-x\rangle \leq \|f'(y)-f'(x)\|\|y-x\|.$$

Putting $y = x + z$ we obtain

$$0 \leq \frac{f(x + z) - f(x) - \langle f'(x), z\rangle}{\|z\|} \leq \|f'(x + z) - f'(x)\|,$$

and by the continuity of f' the assertion. □

3.9 Separation Theorems

The basis for our further investigations in this section is the following theorem on the extension of a linear functional (see e.g. [59])

Theorem 3.9.1 (Theorem of Hahn–Banach). *Let X be a vector space and $f : X \to \mathbb{R}$ a convex function. Let V be a subspace of X and $\ell : V \to \mathbb{R}$ a linear functional with $\ell(x) \leq f(x)$ for all $x \in V$. Then there is a linear functional $u : X \to \mathbb{R}$ with $u_{|V} = \ell$ and $u(x) \leq f(x)$ for all $x \in X$.*

Remark 3.9.2. In the theorem of Hahn–Banach it suffices, instead of the finiteness of f to only require that 0 *is an algebraically interior point of* Dom f.

Corollary 3.9.3. *Let X be a normed space and $f : X \to \mathbb{R}$ a continuous convex function. Let V be a subspace of X and $\ell : V \to \mathbb{R}$ a linear functional with $\ell(x) \leq f(x)$ for all $x \in V$. Then there is a continuous linear functional $u : X \to \mathbb{R}$ with $u_{|V} = \ell$ and $u(x) \leq f(x)$ for all $x \in X$.*

A geometrical version of the theorem of Hahn–Banach is provided by the theorem of Mazur.

Definition 3.9.4. Let X be a vector space. A subset H of X is called *hyperplane* in X, if there is a linear functional $u : X \to \mathbb{R}$ different from the zero functional and an $\alpha \in \mathbb{R}$ such that $H = \{x \in X \mid u(x) = \alpha\}$. H is called a *zero hyperplane*, if $0 \in H$. A subset R of X is called a *half-space*, if there is a linear functional $u : X \to \mathbb{R}$ and an $\alpha \in \mathbb{R}$ such that $R = \{x \in X \mid u(x) \leq \alpha\}$.

Remark 3.9.5. Let K be a convex subset of a normed space with $0 \in \text{Int}(K)$. For the corresponding Minkowski functional $q : X \to \mathbb{R}$ we have: $q(x) < 1$ holds, if and only if $x \in \text{Int}(K)$.

Proof. Let $q(x) < 1$, then there is a $\lambda \in [0, 1)$ with $x \in \lambda K$. Putting $\alpha := 1 - \lambda$ then αK is a neighborhood of the origin, and we have: $x + \alpha K \subset \lambda K + \alpha K \subset K$, i.e. $x \in \text{Int}\, K$.

Conversely, let $x \in \text{Int}\, K$, then there is a $\mu > 1$ with $\mu x \in K$. Hence $x \in \frac{1}{\mu} K$ and $q(x) \leq \frac{1}{\mu} < 1$. □

Theorem 3.9.6 (Mazur). *Let X be a normed space, K a convex subset of X with non-empty interior and V a subspace of X with $V \cap \text{Int}(K) = \emptyset$. Then there is a closed zero hyperplane H in X with $V \subset H$ and $H \cap \text{Int}(K) = \emptyset$.*

Proof. Let $x_0 \in \text{Int}(K)$ and $q : X \to K$ the Minkowski functional of $K - x_0$. According to the above remark $q(x - x_0) < 1$ if and only if $x \in \text{Int}(K)$. Put $f : X \to \mathbb{R}$ with $x \mapsto f(x) := q(x - x_0) - 1$, then f is convex and $f(x) < 0$ if and only if

$x \in \mathrm{Int}(K)$. Due to $V \cap \mathrm{Int}(K) = \emptyset$ we have $f(x) \geq 0$ for all $x \in V$. Let ℓ be the zero functional on V. According to the corollary of the theorem of Hahn–Banach there is a continuous linear functional $u : X \to \mathbb{R}$ with $u_{|V} = \ell$ and $u(x) \leq f(x)$ for all $x \in X$. Let $H = \{x \in X \mid u(x) = 0\}$ the closed zero hyperplane defined by u. Then $V \subset H$, and for all $x \in \mathrm{Int}(K)$ we have $u(x) \leq f(x) < 0$, i.e. $H \cap \mathrm{Int}(K) = \emptyset$. □

In the sequel we need a shifted version of the theorem of Mazur:

Corollary 3.9.7. *Let X be a normed space, K a convex subset of X with non-empty interior and $0 \in \mathrm{Int}(K)$. Let V be a subspace of X, $x_0 \in X$ with $x_0 + V \cap \mathrm{Int}(K) = \emptyset$. Then there is a closed hyperplane H in X with $x_0 + V \subset H$ and $H \cap \mathrm{Int}(K) = \emptyset$.*

The proof is (almost) identical.

An application of the theorem of Mazur is the theorem of Minkowski. In the proof we need the notion of an extreme set of a convex set:

Definition 3.9.8. Let S be a closed convex subset of a normed space. Then a non-empty subset M of S is called *extreme set* of S, if

(a) M is convex and closed

(b) each open interval in S, which contains a point of M, is entirely contained in M.

Example 3.9.9. For a cube in $\mathbb{R}^3$ the vertices are extreme points (see Definition 3.1.10). The edges and faces are extreme sets.

Remark 3.9.10. Let M be an extreme set of a closed convex set S then

$$E_p(M) \subset E_p(S).$$

To see this, let x be an extreme point of M, suppose x is not extreme point of S, then there is an open interval in S, which contains x. However, since M is an extreme set, this interval is already contained in M, a contradiction.

As a preparation we need the following

Lemma 3.9.11. *Let X be a normed space, $S \neq \emptyset$ a compact convex subset of X, let $f \in X^*$ and let $\gamma := \sup\{f(x) \mid x \in S\}$. Then the set $f^{-1}(\gamma) \cap S$ is an extreme set of S, i.e. the hyperplane $H := \{x \in X \mid f(x) = \gamma\}$ has a non-empty intersection with S, and $H \cap S$ is an extreme set of S.*

Proof. By the theorem of Weierstraß the set $f^{-1}(\gamma) \cap S$ is – due to the compactness of S – non-empty, apparently also compact and convex. Let (x_1, x_2) be an open interval in S, which contains a point $x_0 \in f^{-1}(\gamma) \cap S$, i.e. $x_0 = \lambda x_1 + (1 - \lambda) x_2$ for a $\lambda \in (0, 1)$. Since $f(x_0) = \gamma = \lambda f(x_1) + (1 - \lambda) f(x_2)$, it follows by the definition of γ that $f(x_1) = \gamma = f(x_2)$, i.e. $(x_1, x_2) \subset f^{-1}(\gamma) \cap S$. □

Theorem 3.9.12 (Theorem of Minkowski). *Let X be an n-dimensional space and S a convex compact subset of X. Then every boundary point (resp. arbitrary point) of S can be represented as a convex combination of at most n (resp. $n+1$) extreme points of S.*

Proof. We perform induction over the dimension of S, where by the dimension of S we understand the dimension of the affine hull $\bigcap\{A \mid S \subset A \subset X \text{ and } A \text{ affine}\}$ of S.

If $\dim S = 0$, then S consists of at most one point and therefore $S = E_p(S)$.

Assume, the assertion holds for $\dim S \leq m-1$. Let now $\dim S = m$. W.l.g. let $0 \in S$. Let $X_m := \operatorname{span}\{S\}$, then the interior $\operatorname{Int}(S) \neq \emptyset$ w.r.t. X_m (see Lemma 3.1.9) and convex.

a) Let x_0 be a boundary point of S w.r.t. X_m. Due to the corollary of the separation theorem of Mazur there exists a closed hyperplane H in X_m with $H \cap \operatorname{Int}(S) = \emptyset$ and $x_0 \in H$, i.e. H is supporting hyperplane of S in x_0. The set $H \cap S$ is compact and by the previous lemma an extreme set of S with dimension $\leq m-1$. According to the induction assumption x_0 can be represented as a convex combination of at most $(m-1)+1 = m$ extreme points from $H \cap S$. Since by Remark 3.9.10 we have $E_p(H \cap S) \subset E_p(S)$, the first part of the assertion follows.

b) Let now $x_0 \in \operatorname{Int}(S)$ (w.r.t. X_m) arbitrary and $z \in E_p(S)$ (exists due to a)) be arbitrarily chosen. The set $S \cap \overline{z, x_0}$ with $\overline{z, x_0} := \{x \in X \mid x = \lambda z + (1 - \lambda)x_0, \lambda \in \mathbb{R}\}$ is because of the boundedness of S an interval $[z, y]$, whose endpoints are boundary points of S, and which contains x_0 as an interior point. Since y can, due to a), be represented as a convex combination of at most m extreme points and $z \in E_p(S)$, the assertion follows. □

The separation of convex sets is given a precise meaning by the following definition:

Definition 3.9.13. Let X be a vector space, A, B subsets of X and $H = \{x \in X \mid u(x) = \alpha\}$ a hyperplane in X. H separates A and B, if

$$\sup\{u(x) | x \in A\} \leq \alpha \leq \inf\{u(x) | x \in B\}.$$

H strictly separates A and B, if H separates the sets A and B and if one of the two inequalities is strict.

A consequence of the theorem of Mazur is the

Theorem 3.9.14 (Eidelheit). *Let X be a normed vector space, let A, B disjoint, convex subsets of X. Let further* $\operatorname{Int}(A) \neq \emptyset$. *Then there is a closed hyperplane H in X, which separates A and B.*

Proof. Let $K := A - B = \{a - b \,|\, a \in A, b \in B\}$. Then $A - B$ is a convex subset of X. Since $\text{Int}(A) \neq \emptyset$ we have $\text{Int}(K) \neq \emptyset$. Since A, B are disjoint, it follows that $0 \notin K$. With $V := \{0\}$ the theorem of Mazur implies the existence of a closed zero hyperplane, which separates $\{0\}$ and K, i.e. there exists $u \in X^*$, such that for all $x_1 \in A$, $x_2 \in B$: $u(x_1 - x_2) \leq u(0) = 0$ or $u(x_1) \leq u(x_2)$. □

Remark 3.9.15. Instead of $A \cap B = \emptyset$ it is enough to require $\text{Int}(A) \cap B = \emptyset$.

Proof. By Theorem 3.1.8 we have $A \subset \overline{\text{Int}(A)}$. According to the theorem of Eidelheit there is a closed hyperplane, which separates $\text{Int}(A)$ and B. □

In order to obtain a statement about strict separation of convex sets, we will prove the subsequent lemma:

Lemma 3.9.16. *Let X be a normed space, A a closed and B a compact subset of X. Then $A + B := \{a + b \,|\, a \in A, b \in B\}$ is closed.*

Proof. For $n \in \mathbb{N}$ let $a_n \in A$, $b_n \in B$ with $\lim_n(a_n + b_n) = z$. Since B is compact, the sequence (b_n) has a subsequence (b_{n_i}), that converges to $b \in B$, and hence $\lim_i a_{n_i} = z - b \in A$ due to the closedness of A. Thus $z = (z - b) + b \in A + B$. □

Theorem 3.9.17. *Let X be a normed vector space, let A, B be disjoint, convex subsets of X. Furthermore let A be closed and B compact. Then there is a closed hyperplane H in X, which strictly separates A and B.*

Proof. Let B at first consist of a single point, i.e. $B = \{x_0\} \subset X$. The complement of A is open and contains x_0. Therefore there is an open ball V with center 0, such that $x_0 + V$ is contained in the complement of A. According to the separation theorem of Eidelheit there is a closed hyperplane in X that separates $x_0 + V$ and A, i.e. there exists a $u \in X^* \setminus \{0\}$ with $\sup u(x_0 + V) \leq \inf u(A)$. Since $u \neq 0$ there is $v_0 \in V$ with $u(v_0) > 0$. Hence

$$u(x_0) < u(x_0 + v_0) \leq \sup u(x_0 + V) \leq \inf u(A).$$

Let now B be an arbitrary compact convex set with $A \cap B = \emptyset$. By the previous lemma, $A - B$ is closed. Since A, B are disjoint, we have $0 \notin A - B$. Due to the first part we can strictly separate 0 and $A - B$, corresponding to the assertion of the theorem. □

An immediate application of the strict separation theorem is the

Theorem 3.9.18. *Let K be a closed convex subset of a normed space X, then K is weakly closed.*

Proof. Every closed half-space contained in X is a weakly closed set. If one forms the intersection of all half-spaces containing K, we again obtain a weakly closed set M, which apparently contains K. Suppose M contains a point x_0, which does not belong to K. Then – due to the strict separation theorem – there is a closed hyperplane, which strictly separates x_0 and K. As one of the two corresponding closed half-spaces contains K, we obtain a contradiction. □

This is a theorem we will use in the sequel in various situations.

Another consequence of the separation theorem is the following result on extreme points:

Theorem 3.9.19 (Krein–Milman). *Every closed convex subset S of a normed space is the closed convex hull of its extreme points, i.e.*

$$S = \overline{\text{conv}(E_p(S))}.$$

Proof. Let $B := \overline{\text{conv}(E_p(S))}$, then we have to show $B = S$. Apparently $B \subset S$, since the closure of a convex set is convex (see Theorem 3.1.8). Suppose there is $x_0 \in S$ such that $x_0 \notin B$. Then according to the Strict Separation Theorem 3.9.17 there is a continuous linear functional f such that

$$f(x) < f(x_0) \quad \text{for all } x \in B.$$

Let now $\gamma := \sup\{f(x) \mid x \in S\}$ then $f^{-1}(\gamma) \cap S$ contains an extreme point y of S according to Remark 3.9.10 and Lemma 3.9.11, but this is a contradiction to $f(B) < f(x_0)$ since $\gamma = f(y) < f(x_0) \leq \gamma$. □

3.10 Subgradients

Definition 3.10.1. Let X be a normed space, $f : X \to \mathbb{R} \cup \{\infty\}$. A $u \in X^*$ is called *subgradient* of f at $x_0 \in \text{Dom}\, f$, if for all $x \in X$ the *subgradient inequality*

$$f(x) - f(x_0) \geq \langle u, x - x_0 \rangle$$

holds. The set $\partial f(x_0) := \{u \in X^* \mid u \text{ is subgradient of } f \text{ at } x_0\}$ is called *subdifferential* of f at x_0.

The existence of subgradients is described by the following theorem:

Theorem 3.10.2. *Let X be a normed space, $f : X \to \mathbb{R} \cup \{\infty\}$ convex. If f is continuous at $x_0 \in \text{Dom}\, f$, then $\partial f(x_0)$ is non-empty.*

Proof. Let $g : X \to \mathbb{R} \cup \{\infty\}$ with $x \mapsto f(x + x_0) - f(x_0)$, then g is continuous at 0. Put $V = \{0\}$ and $\ell(0) = 0$. Then there is by the theorem of Hahn–Banach an extension $u \in X^*$ of ℓ with $\langle u, x \rangle \leq g(x) = f(x + x_0) - f(x_0)$ for all $x \in X$. Choose $x = x_1 - x_0$, then $\langle u, x_1 - x_0 \rangle \leq f(x_1) - f(x_0)$, i.e. $u \in \partial f(x_0)$. □

A direct consequence of the definition is the monotonicity of the subdifferential in the following sense:

Theorem 3.10.3. *Let X be a normed space, $f : X \to \mathbb{R} \cup \{\infty\}$ convex. Let $x, y \in$ Dom f, $\phi_x \in \partial f(x)$ and $\phi_y \in \partial f(x_0)$, then*

$$\langle \phi_x - \phi_y, x - y \rangle \geq 0.$$

Proof. By the subgradient inequality we have:

$$\langle \phi_x, y - x \rangle \leq f(y) - f(x)$$
$$\langle \phi_y, x - y \rangle \leq f(x) - f(y).$$

We obtain

$$\langle \phi_x - \phi_y, x - y \rangle \geq f(x) - f(y) - (f(x) - f(y)) = 0. \qquad \square$$

The connection between one-sided derivatives and the subdifferential is provided by (see [59])

Theorem 3.10.4 (Theorem of Moreau–Pschenitschnii). *Let X be a normed space, $f : X \to \mathbb{R} \cup \{\infty\}$ convex. Let f be continuous at $x_0 \in$ Dom f, then for all $h \in X$ we obtain*

$$f'_+(x_0, h) = \max\{\langle u, h \rangle \mid u \in \partial f(x_0)\} \tag{$*$}$$

and

$$f'_-(x_0, h) = \min\{\langle u, h \rangle \mid u \in \partial f(x_0)\}. \tag{$**$}$$

Proof. For all $u \in \partial f(x_0)$ and all $t \in \mathbb{R}_{>0}$ we have by definition

$$\langle u, th \rangle \leq f(x_0 + th) - f(x_0),$$

and hence

$$\langle u, h \rangle \leq \lim_{t \downarrow 0} \frac{f(x_0 + th) - f(x_0)}{t} = f'_+(x_0, h).$$

On the other hand, let for a $h \in X \setminus \{0\}$ the linear functional

$$l(th) := t f'_+(x_0, h)$$

be defined on $V := \operatorname{span}\{h\}$. Then for all $x \in V$ we have

$$l(x) \leq f(x_0 + x) - f(x_0) =: q(x),$$

since for $t \in \mathbb{R}_{>0}$ by the positive homogeneity (see Theorem 3.3.3)

$$tf'_+(x_0, h) = f'_+(x_0, th) \leq f(x_0 + th) - f(x_0) = q(th).$$

The subadditivity

$$0 = f'_+(x_0, th - th) \leq f'_+(x_0, th) + f'_+(x_0, -th)$$

implies

$$-tf'_+(x_0, h) = -f'_+(x_0, th) \leq f'_+(x_0, -th) \leq f(x_0 - th) - f(x_0) = q(-th).$$

The function $q\colon X \to \mathbb{R} \cup \{\infty\}$ is convex and $0 \in \text{Int}(\text{Dom}\, q)$. Due to the theorem of Hahn–Banach l has a continuous linear extension u such that for all $x \in X$

$$\langle u, x \rangle \leq q(x) = f(x_0 + x) - f(x)$$

holds.

Thus $u \in \partial f(x_0)$, and for h we obtain

$$\langle u, h \rangle = l(h) = f'_+(x_0, h).$$

hence $(*)$. The relation

$$f'_-(x_0, h) = -f'_+(x_0, -h) = -\max\{\langle u, -h \rangle \mid u \in \partial f(x_0)\}$$

yields $(**)$. □

Minimal solutions of a convex function on a subspace can be characterized in the following way:

Theorem 3.10.5. *Let X be a normed space, $f : X \to \mathbb{R}$ a continuous convex function and U a subspace of X. Then $u_0 \in M(f, U)$, if and only if there is a $x_0^* \in \partial f(u_0) \subset X^*$ such that*

$$\langle x_0^*, u - u_0 \rangle = 0 \quad \textit{for all } u \in U.$$

Proof. Let $x_0^* \in \partial f(u_0)$ with $\langle x_0^*, u - u_0 \rangle = 0$ for all $u \in U$, then

$$0 = \langle x_0^*, u - u_0 \rangle \leq f(u) - f(u_0) \quad \text{for all } u \in U.$$

Conversely let $u_0 \in M(f, U)$, then for the zero functional θ on U

$$\langle \theta, u - u_0 \rangle = 0 \leq f(u) - f(u_0) \quad \text{for all } u \in U.$$

Due to the theorem of Hahn–Banach θ can be extended to all of X as a continuous linear functional x_0^* in such a way that

$$\langle x_0^*, x - u_0 \rangle \leq f(x) - f(u_0) \quad \text{for all } x \in X.$$

By definition then $x_0^* \in \partial f(u_0)$. □

3.11 Conjugate Functions

Definition 3.11.1. Let X be a normed space, $f : X \to \mathbb{R} \cup \{\infty\}$ with $\operatorname{Dom} f \neq \emptyset$. The (convex) conjugate function $f^* : X^* \to \mathbb{R} \cup \{\infty\}$ of f is defined by

$$f^*(y) := \sup\{\langle y, x\rangle - f(x) | x \in X\}.$$

We will also call f^* the dual of f.

Being the supremum of convex (affine) functions f^* is also convex.

For all $x \in X$ and all $y \in X^*$ we have by definition

$$\langle x, y\rangle \leq f(x) + f^*(y). \tag{3.15}$$

This relation we will denote as *Young's inequality.*

The connection between subgradients and conjugate functions is described by the following

Theorem 3.11.2. *Let X be a normed space, $f : X \to \mathbb{R} \cup \{\infty\}$ and $\partial f(x) \neq \emptyset$. For a $y \in X^*$* Young's equality

$$f(x) + f^*(y) = \langle x, y\rangle$$

holds, if and only if $y \in \partial f(x)$.

Proof. Let $f(x) + f^*(y) = \langle x, y\rangle$. For all $u \in X$ we have $\langle u, y\rangle \leq f(u) + f^*(y)$ and hence $f(u) - f(x) \geq \langle u - x, y\rangle$ i.e. $y \in \partial f(x)$.

Conversely let $y \in \partial f(x)$, i.e. for all $z \in X$ we have: $f(z) - f(x) \geq \langle z - x, y\rangle$, hence $\langle x, y\rangle - f(x) \geq \langle z, y\rangle - f(z)$ and therefore

$$\langle x, y\rangle - f(x) = \sup\{\langle z, y\rangle - f(z) \,|\, z \in X\} = f^*(y). \qquad \square$$

In the sequel we will make use of the following statement:

Lemma 3.11.3. *Let X be a normed space. Then every lower semi-continuous convex function $f : X \to \overline{\mathbb{R}}$ has an affine minorant, more precisely: for each $x \in \operatorname{Dom} f$ and $d > 0$ there is $z \in X^*$ such that for all $y \in X$*

$$f(y) > f(x) + \langle y - x, z\rangle - d.$$

Proof. Let w.l.o.g. f not identical ∞. Since f is lower semi-continuous and convex, $\operatorname{Epi} f$ is closed, convex and non-empty. Apparently $(x, f(x) - d) \notin \operatorname{Epi} f$. By the Strict Separation Theorem 3.9.17 there is a $(z, \alpha) \in X^* \times \mathbb{R}$ with

$$\langle x, z\rangle + \alpha(f(x) - d) > \sup\{\langle y, z\rangle + r\alpha \,|\, (y, r) \in \operatorname{Epi} f\}.$$

Since $(x, f(x)) \in \operatorname{Epi} f$, it follows that $-\alpha d > 0$, i.e. $\alpha < 0$. We can w.l.g assume $\alpha = -1$. For all $y \in \operatorname{Dom} f$ we have $(y, f(y)) \in \operatorname{Epi} f$, hence

$$\langle x, z\rangle - (f(x) - d) > \langle y, z\rangle - f(y),$$

and thus the assertion holds. □

Definition 3.11.4. A function $f : X \to \overline{\mathbb{R}}$, not identical to ∞, is called *proper*.

Lemma 3.11.5. *A convex lower semi-continuous function $f : X \to \overline{\mathbb{R}}$ is proper, if and only if f^* is proper.*

Proof. If f is proper, then there is $x_0 \in X$ with $f(x_0) < \infty$ and hence $f^*(y) > -\infty$ for all $y \in X$. Due to the previous lemma there is a $z \in X^*$ with

$$-f(x_0) + \langle x_0, z\rangle + d > \sup\{-f(y) + \langle y, z\rangle \mid y \in X\} = f^*(z),$$

hence f^* is proper. Conversely let $f^*(z_0) < \infty$ for a $z_0 \in X^*$. From Young's inequality it follows that $f(x) > -\infty$ for all $x \in X$ and from $f^*(z_0) < \infty$ follows the existence of a $x_0 \in X$ with $f(x_0) < \infty$. □

Definition 3.11.6. Let X be a normed space and $f^* : X^* \to \overline{\mathbb{R}}$ the conjugate function of $f : X \to \overline{\mathbb{R}}$. Then we call the function

$$f^{**} : X \to \overline{\mathbb{R}}, \quad x \mapsto f^{**}(x) := \sup\{\langle x, x^*\rangle - f^*(x^*) \mid x^* \in X^*\}$$

the *biconjugate function* of f.

Theorem 3.11.7 (Fenchel–Moreau). *Let X be a normed space and $f : X \to \overline{\mathbb{R}}$ a proper convex function. Then the following statements are equivalent:*

(a) *f is lower semi-continuous*

(b) $f = f^{**}$.

Proof. (a) ⇒ (b): We show at first $\operatorname{Dom} f^{**} \subset \overline{\operatorname{Dom} f}$. Let $x \notin \overline{\operatorname{Dom} f}$. Then according to the Strict Separation Theorem 3.9.17 there exists a $u \in X^*$ with

$$\langle x, u\rangle > \sup\{\langle y, u\rangle \mid y \in \overline{\operatorname{Dom} f}\}.$$

By the above lemma f^* is proper. Hence there exists a $v \in X^*$ with $f^*(v) < \infty$. For all $t > 0$ we then have

$$f^*(v+tu) = \sup\{\langle y, v+tu\rangle - f(y) \mid y \in X\} \le f^*(v)+t\cdot\sup\{\langle y, u\rangle \mid y \in \overline{\operatorname{Dom} f}\}.$$

Thus we obtain

$$\begin{aligned} f^{**}(x) &\geq \langle x, v+tu\rangle - f^*(v+tu) \\ &\geq \langle x, v\rangle - f^*(v) + t(\langle x, u\rangle - \sup\{\langle y, u\rangle \mid y \in \overline{\operatorname{Dom} f}\}) \xrightarrow{t\to\infty} \infty, \end{aligned}$$

i.e. $x \notin \operatorname{Dom} f^{**}$.

Let $y \in X$. From the definition of f^* it follows for all $x^* \in X^*$ that: $f(y) \geq \langle y, x^*\rangle - f^*(x^*)$ and hence $f^{**}(y) \leq f(y)$ for arbitrary $y \in X$.

For $y \notin \operatorname{Dom} f^{**}$ we apparently have $f^{**}(y) \geq f(y)$. Suppose, for a $x \in \operatorname{Dom} f^{**}$ we have $f(x) > f^{**}(x)$, then $(x, f^{**}(x)) \notin \operatorname{Epi} f$. Since f is lower semi-continuous, $\operatorname{Epi} f$ is closed. Due to the Strict Separation Theorem 3.9.17 there exists a $(x^*, \alpha) \in X^* \times \mathbb{R}$ with

$$\langle x, x^*\rangle + af^{**}(x) > \sup\{\langle y, x^*\rangle + ar \mid (y,r) \in \operatorname{Epi} f\}.$$

Here $a < 0$ must hold, because $a = 0$ implies

$$\langle x, x^*\rangle > \sup\{\langle y, x^*\rangle \mid y \in \operatorname{Dom} f\} =: \alpha,$$

i.e. $x \notin \{y \in X \mid \langle y, x^*\rangle \leq \alpha\} \supset \operatorname{Dom} f$ contradicting $x \in \operatorname{Dom} f^{**}$.

The assumption $a > 0$ would imply $\langle x, x^*\rangle + af^{**}(x) = \infty$, contradicting $x \in \operatorname{Dom} f^{**}$. Let w.l.o.g. $a = -1$, i.e.

$$\begin{aligned} \langle x, x^*\rangle - f^{**}(x) &> \sup\{\langle y, x^*\rangle - r \mid (y,r) \in \operatorname{Epi} f\} \\ &\geq \sup\{\langle y, x^*\rangle - f(y) \mid y \in \operatorname{Dom} f\} = f^*(x^*), \end{aligned}$$

contradicting the definition of f^{**}.

(b) $\Rightarrow$ (a): Being the supremum of the family $\{\langle \cdot, x^*\rangle - f^*(x^*) \mid x^* \in X^*\}$ of continuous functions, f^{**} is lower semi-continuous. □

Remark 3.11.8. If $f : X \to \overline{\mathbb{R}}$ is a proper convex and lower semi-continuous function, then by the theorem of Fenchel–Moreau we have $f(x) = \sup\{\langle x, x^*\rangle - f^*(x^*) \mid x^* \in X^*\}$. But each functional $\langle \cdot, x^*\rangle - f^*(x^*)$ is by definition weakly lower semi-continuous and hence also f weakly lower semi-continuous.

As a consequence of Theorem 3.11.7 we obtain an extension of Theorem 3.11.2 and thereby a further interpretation of Young's equality:

Theorem 3.11.9. *Let X be a normed space and $f : X \to \overline{\mathbb{R}}$ a proper convex and lower semi-continuous function. Then the following statements are equivalent:*

(a) $x^* \in \partial f(x)$

(b) $x \in \partial f^*(x^*)$

(c) $f(x) + f^*(x^*) = \langle x, x^* \rangle$.

Proof. Let $f(x) + f^*(x^*) = \langle x, x^* \rangle$. Apparently $f^*(x^*) < \infty$. Using Young's inequality we have: $f(x) + f^*(u) \geq \langle x, u \rangle$ for all $u \in X^*$. Therefore $f^*(u) - f^*(x^*) \geq \langle x, u - x^* \rangle$, i.e. $x \in \partial f^*(x^*)$.

Let now $x \in \partial f^*(x^*)$, then for all $u \in X^*$: $f^*(u) - f^*(x^*) \geq \langle x, u - x^* \rangle$, hence $\langle x, x^* \rangle - f^*(x^*) \geq \langle x, u \rangle - f^*(u)$ and using the theorem of Fenchel–Moreau we obtain

$$\langle x, x^* \rangle - f^*(x^*) = \sup\{\langle x, u \rangle - f^*(u) \,|\, u \in X^*\} = f^{**}(x) = f(x). \qquad \square$$

We will take up the subsequent example for a conjugate function in Chapter 8:

Example 3.11.10. Let $\Phi : \mathbb{R} \to \mathbb{R}$ be a Young function and let $(X, \|\cdot\|)$ be a normed space. We consider the convex function $f : X \to \mathbb{R}$, defined by $f(x) := \Phi(\|x\|)$. For the conjugate $f^* : (X^*, \|\cdot\|_d) \to \overline{\mathbb{R}}$ we have

$$f^*(y) = \Phi^*(\|y\|_d).$$

Proof. Using the definition of the norm $\|\cdot\|_d$ and Young's inequality it follows that

$$\langle x, y \rangle - \Phi(\|x\|) \leq \|x\| \cdot \|y\|_d - \Phi(\|x\|) \leq \Phi^*(\|y\|_d).$$

Due to the definition of Φ^* and $\|\cdot\|_d$ there is a sequence (s_n) in $\mathbb{R}$ and a sequence (x_n) in X with

(a) $s_n \cdot \|y\|_d - \Phi(s_n) \to_{n\to\infty} \Phi^*(\|y\|_d)$

(b) $\|x_n\| = 1$ and $|\langle x_n, y \rangle - \|y\|_d| < \frac{1}{n(1+|s_n|)}$.

Then we obtain

$$\langle s_n x_n, y \rangle - \Phi(\|s_n x_n\|) = s_n \|y\|_d - \Phi(s_n) + s_n(\langle x_n, y \rangle - \|y\|_d) \xrightarrow{n\to\infty} \Phi^*(\|y\|_d). \quad \square$$

In particular we have

$$\left(\frac{\|\cdot\|^2}{2}\right)^* = \frac{\|\cdot\|_d^2}{2}.$$

3.12 Theorem of Fenchel

Let X be a normed space.

Definition 3.12.1. For a concave function $g : X \to [-\infty, \infty)$ with $\mathrm{Dom}\, g := \{x \in X \,|\, g(x) > -\infty\} \neq \emptyset$ the *concave conjugate function* $g^+ : X^* \to [-\infty, \infty)$ of g is defined by

$$g^+(y) := \inf\{\langle y, x \rangle - g(x) \,|\, x \in \mathrm{Dom}\, g\}.$$

By definition we have the subsequent relation between the concave and the convex conjugate function

$$g^+(y) = -(-g)^*(-y). \tag{3.16}$$

We now consider the optimization problem $\inf(f - g)(X)$, where $f - g : X \to \mathbb{R} \cup \{\infty\}$ are assumed to be convex. Directly from the definitions it follows that for all $x \in X$ and $y \in X^*$ the inequality

$$f(x) + f^*(y) \geq \langle y, x \rangle \geq g(x) + g^+(y)$$

holds, i.e.

$$\inf\{(f - g)(x) \,|\, x \in X\} \geq \sup\{(g^+ - f^*)(y) \,|\, y \in X^*\}. \tag{3.17}$$

As a direct application of the Separation Theorem of Eidelheit 3.9.14 we obtain the

Theorem 3.12.2 (Theorem of Fenchel). *Let $f, -g : X \to \mathbb{R} \cup \{\infty\}$ be convex functions and let a $x_0 \in \operatorname{Dom} f \cap \operatorname{Dom} g$ exist such that f or g are continuous at x_0. Then we obtain*

$$\inf\{(f - g)(x) \,|\, x \in X\} = \sup\{(g^+ - f^*)(y) \,|\, y \in X^*\}.$$

If in addition $\inf\{(f - g)(x) \,|\, x \in X\}$ is finite, then on the right-hand side the supremum is attained at a $y_0 \in X^*$.

Proof. W.l.g. let f be continuous at x_0. Then we have $x_0 \in \operatorname{Dom} f$ and

$$f(x_0) - g(x_0) \geq \inf\{(f - g)(x) \,|\, x \in X\} =: \alpha.$$

The assertion of the theorem is for $\alpha = -\infty$ a direct consequence of Inequality (3.17). Under the assumption $-\infty < \alpha < \infty$ we consider the sets

$$A := \{(x, t) \in X \times \mathbb{R} \,|\, x \in \operatorname{Dom} f,\ t > f(x)\}$$
$$B := \{(x, t) \in X \times \mathbb{R} \,|\, t \leq g(x) + \alpha\}.$$

The above sets are convex and disjoint. From the continuity of f at x_0 it follows that $\operatorname{Int}(A) \neq \emptyset$. Due to the separation theorem of Eidelheit there is a hyperplane H separating A and B, i.e. there is a $(0, 0) \neq (y, \beta) \in X^* \times \mathbb{R}$ and $r \in \mathbb{R}$, such that for all $(x_1, t_1) \in A$ and all $(x_2, t_2) \in B$ we have

$$\langle y, x_1 \rangle + \beta t_1 \leq r \leq \langle y, x_2 \rangle + \beta t_2. \tag{3.18}$$

It turns out that $\beta \neq 0$, because otherwise for all $x_1 \in \operatorname{Dom} f$ and all $x_2 \in \operatorname{Dom} g$

$$\langle y, x_1 \rangle \leq r \leq \langle y, x_2 \rangle.$$

Since $x_0 \in \operatorname{Dom} f \cap \operatorname{Dom} g$ and $x_0 \in \operatorname{Int}(\operatorname{Dom} f)$ it follows on a neighborhood U of the origin

$$\langle y, x_0 + U\rangle \le r = \langle y, x_0\rangle,$$

hence $\langle y, U\rangle \le 0$, contradicting $(0,0) \neq (y, \beta)$.

Moreover $\beta < 0$, because otherwise $\sup\{\beta t \mid (x_0, t) \in A\} = \infty$ in contradiction to Inequality (3.18).

Let now $\varepsilon > 0$. For $\lambda := \frac{r}{-\beta}$ and $w := \frac{y}{-\beta}$ and all $x \in \operatorname{Dom} f$ it follows from $(x, f(x) + \varepsilon) \in A$ that

$$\langle w, x\rangle - (f(x) + \varepsilon) \le \lambda.$$

Since ε is arbitrary, we have $f^*(w) \le \lambda$. Let $z \in \operatorname{Dom} g$. Apparently $(z, g(z) + \alpha) \in B$ and hence by Inequality (3.18)

$$\langle w, z\rangle - (g(z) + \alpha) \ge \lambda$$

yielding $g^+(w) \ge \lambda + \alpha$. Together with Inequality (3.17) it follows that

$$\begin{aligned}\alpha \le g^+(w) - \lambda \le g^+(w) - f^*(w) &\le \sup\{g^+(y) - f^*(y) \mid y \in X^*\}\\ &\le \inf\{f(x) - g(x) \mid x \in X\} = \alpha,\end{aligned}$$

and thereby the assertion. □

As a consequence of the theorem of Fenchel we obtain the

Theorem 3.12.3. *Let X be a normed space, $K \subset X$ convex, $f : X \to \mathbb{R} \cup \{\infty\}$ convex and continuous at a point $\bar{k} \in K$. Then $k_0 \in M(f, K)$, if and only if for a $u \in X^*$*

(a) $f^*(u) + f(k_0) = \langle u, k_0\rangle$

(b) $\langle u, k_0\rangle = \min(u, K)$

holds.

Proof. Let $k_0 \in M(f, K)$ and let at first $k_0 \notin M(f, X)$. Let

$$g(x) = \begin{cases} 0 & \text{for } x \in K \\ -\infty & \text{otherwise.}\end{cases}$$

For $y \in X^*$ we have $g^+(y) = \inf\{\langle y, k\rangle \mid k \in K\}$. By the theorem of Fenchel there exists a $u \in X^*$ with

$$\inf\{\langle u, k\rangle \mid k \in K\} - f^*(u) = f(k_0).$$

It turns out that $u \neq 0$, because otherwise due to $f^*(0) = -f(k_0)$ we have by Theorem 3.11.2 $0 \in \partial f(k_0)$, hence apparently $k_0 \in M(f, X)$, a contradiction to our assumption. By Young's inequality we then obtain

$$\inf\{\langle u, k\rangle \mid k \in K\} = f^*(u) + f(k_0) \geq \langle u, k_0\rangle \geq \inf\{\langle u, k\rangle \mid k \in K\},$$

hence (a) and (b).

If $k_0 \in M(f, X)$, then for $u = 0$ we have (a) by definition of f^* and apparently also (b).

Conversely (a) implies – using Theorem 3.11.2 – that $u \in \partial f(k_0)$. Hence $\langle u, k - k_0\rangle \leq f(k) - f(k_0)$ for all $k \in K$. By (b) we then have $f(k) \geq f(k_0)$ for all $k \in K$. □

As another application of the theorem of Fenchel we obtain the duality theorem of linear approximation theory in normed spaces:

Theorem 3.12.4. *Let X be a normed space, V a subspace of X and $z \in X$, then*

$$\inf\{\|z - v\| \mid v \in V\} = \max\{\langle u, z\rangle \mid \|u\| \leq 1 \text{ and } u \in V^{\perp}\}.$$

Proof. Let $f(x) := \|x\|$ and

$$g(x) = \begin{cases} 0 & \text{for } x \in z - V \\ -\infty & \text{otherwise.} \end{cases}$$

Then we have

$$f^*(u) = \begin{cases} 0 & \text{for } \|u\| \leq 1 \\ \infty & \text{otherwise} \end{cases}$$

and

$$g^+(u) = \inf\{\langle u, x\rangle \mid x \in z - V\} = \begin{cases} \langle u, z\rangle & \text{for } u \in V^{\perp} \\ -\infty & \text{otherwise.} \end{cases}$$ □

Application 3.12.5 (Formula of Ascoli). We treat the following problem:

In a normed space X the distance of a point $z \in X$ to a given hyperplane has to be computed. If for a $\bar{u} \in X^*$ and a $x_0 \in X$ the hyperplane is represented by $H = \{x \in X \mid \langle \bar{u}, x\rangle = 0\} + x_0$, then one obtains the distance via the *formula of Ascoli*

$$\frac{|\langle \bar{u}, z\rangle - \langle \bar{u}, x_0\rangle|}{\|\bar{u}\|}, \tag{3.19}$$

because due to the previous theorem we have for $V =: \{x \in X \mid \langle \bar{u}, x\rangle = 0\}$:

$$\inf\{\|z - x_0 - v\| \mid v \in V\} = \max\{\langle u, z - x_0\rangle \mid \|u\| \leq 1 \text{ and } \langle u, v\rangle = 0\, \forall v \in V\}.$$

As the restriction set of the maximum problem only consists of $\frac{\bar{u}}{\|\bar{u}\|}$, the assertion follows.

Theorem 3.12.3 leads to the following characterization of a best linear approximation:

Theorem 3.12.6 (Singer). *Let X be a normed space, V a subspace of X and $x \in X$. The point $v_0 \in V$ is a best approximation of x w.r.t. V, if and only if for a $u \in X^* \setminus \{0\}$ we have*

(a) $\|u\| = 1$

(b) $\langle u, v \rangle = 0 \,\forall v \in V$

(c) $\langle u, x - v_0 \rangle = \|x - v_0\|$.

Proof. Let $h \mapsto f(h) := \|h\|$, $K := x - V$, then

$$f^*(u) = \begin{cases} 0 & \text{for } \|u\| \leq 1 \\ \infty & \text{otherwise.} \end{cases}$$

From (a) in Theorem 3.12.3 both (a) and (c) follow, from (b) in Theorem 3.12.3 finally (b). □

3.13 Existence of Minimal Solutions for Convex Optimization

Definition 3.13.1. The problem '*maximize* $(g^+ - f^*)$ *on* X^*' we call the Fenchel-dual problem to '*minimize* $(f - g)$ *on* X'.

Using the theorem of Fenchel we obtain the following principle for existence proofs:

If an optimization problem with finite values is the Fenchel dual of another problem, then it always has a minimal solution (resp. maximal solution).

A natural procedure for establishing existence for a given optimization problem is to form the dual twice, hoping in this way to return to the original problem. In doing so we encounter a formal difficulty that, by twice forming the dual, a problem in $(X^*)^*$ results, and not in X. One can always view X as a subset of $(X^*)^*$, by considering the following mapping $E : X^* \to (X^*)^*$, where for a $x \in X$ the functional $E(x) : X^* \to \mathbb{R}$ is defined by

$$x^* \mapsto E(x)(x^*) = \langle E(x), x^* \rangle := \langle x^*, x \rangle.$$

Of particular interest are – due to what has just been said – those normed spaces X, for which all elements in $(X^*)^*$ can be obtained in this way, because then $(X^*)^*$ can be identified with X. This leads to the

Definition 3.13.2. A normed space X is called *reflexive*, if $E(X) = (X^*)^*$ holds.

Remark 3.13.3. Let X be a reflexive Banach space and $f : X \to \overline{\mathbb{R}}$. For the second conjugate $(f^*)^* : X^{**} \to \overline{\mathbb{R}}$ and the biconjugate $f^{**} : X \to \overline{\mathbb{R}}$ we have for all $x \in X$

$$\begin{aligned}(f^*)^*(E(x)) &= \sup\{\langle E(x), x^*\rangle - f^*(x^*) \,|\, x^* \in X^*\}\\ &= \sup\{\langle x^*, x\rangle - f^*(x^*) \,|\, x^* \in X^*\} = f^{**}(x).\end{aligned}$$

Lemma 3.13.4. *Let K be a convex subset of a normed space X and let $f : K \to \mathbb{R}$ be a continuous convex function. Then f is bounded from below on every bounded subset B of K.*

Proof. Let $a > 0$ and $x_0 \in K$. As f is continuous, there is an open ball $K(x_0, r)$ with radius $1 > r > 0$ such that

$$f(x) > f(x_0) - a$$

holds for all $x \in K(x_0, r)$. Let $M > 1$ be chosen such that $K(x_0, M) \supset B$. Let $y \in B$ be arbitrary and $z := (1 - \frac{r}{M})x_0 + \frac{r}{M}y$. Then it follows that

$$\|z - x_0\| = \frac{r}{M}\|y - x_0\| < r,$$

i.e. $z \in K(x_0, r)$. Since f is convex, we have

$$f(z) \le \left(1 - \frac{r}{M}\right) f(x_0) + \frac{r}{M} f(y),$$

and therefore

$$f(y) \ge -\frac{M}{r}\left(1-\frac{r}{M}\right) f(x_0)+\frac{M}{r} f(z) \ge \left(1-\frac{M}{r}\right) f(x_0)+\frac{M}{r}(f(x_0)-a) =: c. \quad \square$$

Theorem 3.13.5 (Mazur–Schauder). *Let X be a reflexive Banach space, K a non-empty, closed, bounded and convex subset of X. Then every continuous convex function $h : K \to \mathbb{R}$ has a minimal solution.*

Proof. Let $f, -g : X \to (-\infty, \infty]$ be defined by

$$f(x) = \begin{cases} h(x) & \text{for } x \in K \\ \infty & \text{otherwise,} \end{cases}$$

and

$$g(x) = \begin{cases} 0 & \text{for } x \in K \\ -\infty & \text{otherwise.} \end{cases}$$

Then for all $y \in X^*$ it follows that

$$g^+(y) = \inf\{\langle y, x\rangle - g(x) \,|\, x \in X\} = \inf\{\langle y, x\rangle \,|\, x \in K\}.$$

Since K is bounded, there is a $r > 0$, such that we have for all $x \in K$: $\|x\| \leq r$ and hence $|\langle y, x\rangle| \leq \|y\|\|x\| \leq r\|y\|$. Therefore $g^+(y) \geq -r\|y\| > -\infty$. Being the infimum of continuous linear functions $\{\langle \cdot, x\rangle \mid x \in K\}$ g^+ is also continuous on the Banach space X^* (see Theorem 5.3.13). Due to Lemma 3.11.5 f^* is a proper convex function, i.e. Dom $f^* \neq \emptyset$. Therefore the preconditions for treating the problem

$$\inf\{f^*(y) - g^+(y) \mid y \in X^*\} =: \alpha$$

are satisfied. Using the theorem of Fenchel and the reflexivity of X it follows that

$$\alpha = \sup\{(g^+)^+(E(x)) - (f^*)^*(E(x)) \mid x \in X\}. \tag{3.20}$$

For all $x \in X$ we have by Remark 3.13.3 $(f^*)^*(E(x)) = f^{**}(x)$. The continuity of h on K implies the lower semi-continuity of f on X and by the theorem of Fenchel–Moreau 3.11.7 we obtain $f^{**}(x) = f(x)$. Putting $z = E(x)$ we obtain (see Equation (3.16))

$$(g^+)^+(z) = [-(-g)^*]^+(-z) = -[(-g)^*]^*(z) = -(-g)^{**}(z),$$

and due to the lower semi-continuity of $-g$ we obtain – again using the theorem of Fenchel–Moreau:

$$(g^+)^+(E(x)) = -(-g)^{**}(E(x)) = g(x).$$

By Equation (3.20) we obtain

$$\alpha = \sup\{g(x) - f(x) \mid x \in X\} = \sup\{-f(x) \mid x \in K\} = -\inf\{f(x) \mid x \in K\}.$$

By Lemma 3.13.4 the number α is finite. The theorem of Fenchel then implies the existence of a $\bar{x}$, where the supremum in (3.20) is attained, i.e. $f(\bar{x}) = -\alpha = \inf\{f(x) \mid x \in K\}$. □

3.14 Lagrange Multipliers

In this section we consider convex optimization problems, for which the restrictions are given as convex inequalities (see [79]).

Definition 3.14.1. Let P be a convex cone in a vector space X. For $x, y \in X$ we write $x \geq y$, if $x - y \in P$. The cone P defining this relation is called *positive cone* in X.

In a normed space X we define the convex cone dual to P by

$$P^* := \{x^* \in X^* \mid \langle x, x^*\rangle \geq 0 \text{ for all } x \in P\}.$$

Definition 3.14.2. Let Ω be a convex subset of a vector space X and (Z, P) an ordered vector space. Then $G : \Omega \to (Z, P)$ is called *convex*, if

$$G(\lambda x_1 + (1-\lambda)x_2) \leq \lambda G(x_1) + (1-\lambda)G(x_2)$$

for all $x_1, x_2 \in \Omega$ and all $\lambda \in [0, 1]$.

We now consider the following type of problem:

Let Ω be a convex subset of the vector space X, $f : \Omega \to \mathbb{R}$ a convex function, $(Z, \leq)$ an ordered vector space and $G : \Omega \to (Z, \leq)$ convex.

Then minimize $f(x)$ under the restriction

$$x \in \Omega, \quad G(x) \leq 0.$$

In the sequel we want to develop the concept of *Lagrange multipliers*. For this purpose we need a number of preparations.

Let $\Gamma := \{z \mid \exists x \in \Omega\, G(x) \leq z\}$. It is easily seen that Γ is convex. On Γ we define a function $w : \Gamma \to \mathbb{R} \cap \{\infty\}$ by $z \mapsto \inf\{f(x) \mid x \in \Omega, G(x) \leq z\}$. Then $w(0)$ is the minimal value of f on $\{x \in \Omega, G(x) \leq 0\}$.

Theorem 3.14.3. *The function w is convex and monotonically decreasing, i.e. for all $z_1, z_2 \in \Gamma$ we have*

$$z_1 \leq z_2 \Rightarrow w(z_1) \geq w(z_2).$$

Proof. Let $\alpha \in [0, 1]$ and let $z_1, z_2 \in \Gamma$, then

$$\begin{aligned}
&w(\alpha z_1 + (1-\alpha)z_2) = \inf\{f(x) \mid x \in \Omega, G(x) \leq \alpha z_1 + (1-\alpha)z_2\} \\
&\leq \inf\{f(x) \mid \exists x_1, x_2 \in \Omega : x = \alpha x_1 + (1-\alpha)x_2, G(x_1) \leq z_1, G(x_2) \leq z_2\} \\
&= \inf\{f(\alpha x_1 + (1-\alpha)x_2) \mid x_1, x_2 \in \Omega : G(x_1) \leq z_1, G(x_2) \leq z_2\} \\
&\leq \inf\{\alpha f(x_1) + (1-\alpha)f(x_2) \mid x_1, x_2 \in \Omega : G(x_1) \leq z_1, G(x_2) \leq z_2\} \\
&= \alpha \inf\{f(x_1) \mid x_1 \in \Omega : G(x_1) \leq z_1\} + (1-\alpha) \inf\{f(x_2) \mid x_2 \in \Omega : G(x_2) \leq z_2\} \\
&= \alpha w(z_1) + (1-\alpha)w(z_2).
\end{aligned}$$

This proves the convexity of w, the monotonicity is obvious. □

We start our consideration with the following assumption: there is a non-vertical supporting hyperplane of $\operatorname{Epi}(w)$ at $(0, w(0))$, i.e. there is a $z_0^* \in Z^*$ with

$$\langle -z_0^*, z\rangle + w(0) \leq w(z)$$

for all $z \in \Gamma$. Then apparently $w(0) \leq \langle z_0^*, z\rangle + w(z)$ holds. For all $x \in \Omega$ also $G(x) \in \Gamma$ holds. Therefore we have for all $x \in \Omega$

$$w(0) \leq w(G(x)) + \langle z_0^*, G(x)\rangle \leq f(x) + \langle z_0^*, G(x)\rangle.$$

Hence

$$w(0) \leq \inf\{f(x) + \langle z_0^*, G(x)\rangle \mid x \in \Omega\}.$$

If z_0^* can be chosen, such that $z_0^* \geq 0$, then we have for all $x \in \Omega$

$$G(x) \leq 0 \Rightarrow \langle z_0^*, G(x)\rangle \leq 0,$$

and thus

$$\begin{aligned}\inf\{f(x) + \langle z_0^*, G(x)\rangle \mid x \in \Omega\} &\leq \inf\{f(x) + \langle z_0^*, G(x)\rangle \mid x \in \Omega, G(x) \leq 0\}\\ &\leq \inf\{f(x) \mid x \in \Omega, G(x) \leq 0\}\\ &= w(0).\end{aligned}$$

We finally obtain

$$w(0) = \inf\{f(x) + \langle z_0^*, G(x)\rangle \mid x \in \Omega\},$$

and conclude that we have thus converted an optimization problem with restriction $G(x) \leq 0$ into an optimization problem without restrictions.

Theorem 3.14.4. *Let X be a vector space, (Z, P) an ordered normed space with* $\text{Int}(P) \neq \emptyset$ *and let Ω be a convex subset of X. Let $f : \Omega \to \mathbb{R}$ and $G : \Omega \to Z$ convex mappings. Let further the following regularity conditions be satisfied:*

(a) *there exists a $x_1 \in \Omega$ with $-G(x_1) \in \text{Int}(P)$*

(b) *$\mu_0 := \inf\{f(x) \mid x \in \Omega, G(x) \leq 0\}$ is finite.*

Then there is $z_0^* \in P^*$ with

$$\mu_0 = \inf\{f(x) + \langle z_0^*, G(x)\rangle \mid x \in \Omega\}. \tag{3.21}$$

If the infimum in (b) *is attained at a $x_0 \in \Omega$ with $G(x_0) \leq 0$, the infimum in* (3.21) *is also attained and the equation $\langle z_0^*, G(x_0)\rangle = 0$ holds.*

Proof. In the space $W := \mathbb{R} \times Z$ we define the following sets:

$$\begin{aligned}A &:= \{(r, z) \mid \text{there is } x \in \Omega : \ r \geq f(x),\ z \geq G(x)\}\\ B &:= \{(r, z) \mid r \leq \mu_0,\ z \leq 0\}.\end{aligned}$$

Since P, f and G are convex, the sets A and B are convex sets. From the definition of μ_0 it follows that A contains no interior point of B. Since P does contain interior points, the interior of B is non-empty. Then by the Separation Theorem of Eidelheit 3.9.14 there is a non-zero functional $w_0^* = (r_0, z_0^*) \in W^*$ such that

$$r_0 r_1 + \langle z_1, z_0^*\rangle \geq r_0 r_2 + \langle z_2, z_0^*\rangle$$

for all $(r_1, z_1) \in A$ and all $(r_2, z_2) \in B$. The properties of B immediately imply $w_0^* \geq 0$ i.e. $r_0 \geq 0$ and $z_0^* \geq 0$. At first we show: $r_0 > 0$. Apparently we have $(\mu_0, 0) \in B$, hence

$$r_0 r + \langle z, z_0^* \rangle \geq r_0 \mu_0$$

for all $(r, z) \in A$. Suppose $r_0 = 0$, then in particular $\langle G(x_1), z_0^* \rangle \geq 0$ and $z_0^* \neq 0$. Since however $-G(x_1)$ is an interior point of P and $z_0^* \geq 0$ it then follows that $\langle G(x_1), z_0^* \rangle < 0$, a contradiction. Thus $r_0 > 0$ and we can w.l.o.g. assume $r_0 = 1$.

Let now (x_n) be a minimizing sequence of f on $\{x \in \Omega \mid G(x) \leq 0\}$, i.e. $f(x_n) \to \mu_0$. We then obtain because of $(f(x), G(x)) \in A$ for all $x \in \Omega$

$$\mu_0 \leq \inf\{r + \langle z_0^*, z \rangle\} \mid (r, z) \in A\} \leq \inf\{f(x) + \langle z_0^*, G(x) \rangle \mid x \in \Omega\} \leq f(x_n) \to \mu_0,$$

and thus the first part of the assertion.

If now there is a $x_0 \in \Omega$ with $G(x_0) \leq 0$ and $f(x_0) = \mu_0$, then

$$\mu_0 \leq f(x_0) + \langle G(x_0), z_0^* \rangle \leq f(x_0),$$

and hence $\langle G(x_0), z_0^* \rangle = 0$. □

The following theorem about Lagrange duality is a consequence of the previous theorem:

Theorem 3.14.5. *Let X be a vector space, (Z, P) an ordered normed space with* $\text{Int}(P) \neq \emptyset$ *and let Ω be a convex subset of X. Let $f : \Omega \to \mathbb{R}$ and $G : \Omega \to Z$ convex mappings. Moreover, let the following regularity conditions be satisfied:*

(a) *there exists a $x_1 \in \Omega$ with* $-G(x_1) \in \text{Int}(P)$

(b) $\mu_0 := \inf\{f(x) \mid x \in \Omega, G(x) \leq 0\}$ *is finite.*

Then

$$\inf\{f(x) \mid x \in \Omega, G(x) \leq 0\} = \max\{\varphi(z^*) \mid z^* \in Z^*, z^* \geq 0\},$$

where $\varphi(z^*) := \inf\{f(x) + \langle z^*, G(x) \rangle \mid x \in \Omega\}$ *for* $z^* \in Z^*, z^* \geq 0$.

If the maximum on the right-hand side is attained at $z_0^* \in Z^*$ with $z_0^* \geq 0$ and also the infimum at a $x_0 \in \Omega$, then $\langle z_0^*, G(x_0) \rangle = 0$ holds, and x_0 minimizes $x \mapsto f(x) + \langle z_0^*, G(x) \rangle : \Omega \to \mathbb{R}$ on Ω.

Proof. Let $z^* \in Z^*$ with $z^* \geq 0$. Then we have

$$\begin{aligned}\varphi(z^*) &= \inf\{f(x) + \langle z^*, G(x) \rangle \mid x \in \Omega\} \\ &\leq \inf\{f(x) + \langle z^*, G(x) \rangle \mid x \in \Omega, G(x) \leq 0\} \\ &\leq \inf\{f(x) \mid x \in \Omega, G(x) \leq 0\} = \mu_0.\end{aligned}$$

Hence

$$\sup\{\varphi(z^*) \mid z^* \in Z^*, z^* \geq 0\} \leq \mu_0.$$

The previous theorem implies the existence of an element z_0^*, for which equality holds. The remaining part of the assertion follows also by the previous theorem. □

Chapter 4

Numerical Treatment of Non-linear Equations and Optimization Problems

The treatment of numerical methods in this chapter follows to a large extent that in [60]. Due to its theoretical and practical significance we will start with the *Newton method*, which can be considered as a prototype of all rapidly convergent methods. The Newton method is primarily a method of determining the solution of a non-linear equation. For minimization problems this equation results from setting the derivative of the function to be minimized to zero. For n-dimensional minimization problems this yields a system of n non-linear equations with n unknowns. The idea of the Newton method can be described in the following way: starting with an initial guess the successor is determined as the solution of the linearized equation at the present position. In one dimension we obtain the following figure:

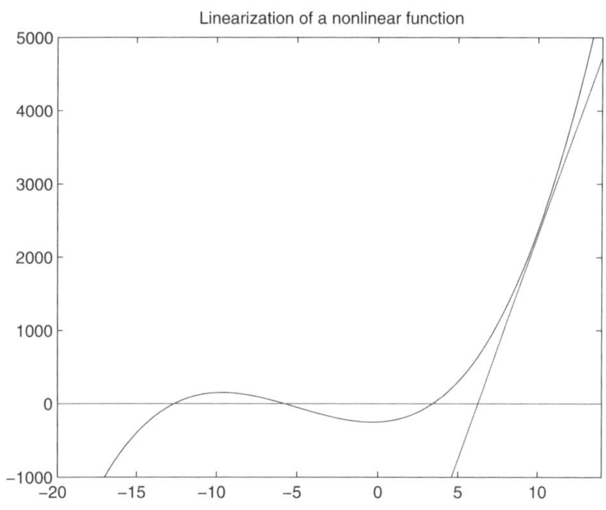

Let

$$F : \mathbb{R}^n \to \mathbb{R}^n$$

be a continuously differentiable function, for which we want to determine a $x^\star \in \mathbb{R}^n$ with $F(x^\star) = 0$. Then for the linearized equation at $\bar{x}$ we obtain

$$F(\bar{x}) + F'(\bar{x})(x - \bar{x}) = 0.$$

If one wants to minimize a function $f : \mathbb{R}^n \to \mathbb{R}$, then for the corresponding equation one has

$$f'(\bar{x}) + f''(\bar{x})(x - \bar{x}) = 0.$$

For the general case one obtains the following algorithm:

4.1 Newton Method

Let

$$F : \mathbb{R}^n \to \mathbb{R}^n$$

be a continuously differentiable function.

(a) Choose a starting point x_0, a stopping criterion, and put $k = 0$.

(b) Compute x_{k+1} as a solution of the *Newton equation*

$$F(x_k) + F'(x_k)(x - x_k) = 0.$$

(c) Test the stopping criterion for x_{k+1} and – if not satisfied – increment k by 1 and continue with (b).

Concerning the convergence of the Newton method we mention the following: if the function F has a zero $x^\star$, and if the Jacobian matrix $F'(x^\star)$ is invertible, then there is a neighborhood U of $x^\star$, such that for every choice of a starting point in U the Newton equation is solvable for each k, the total iteration sequence $(x_k)_{k\in\mathbb{N}_0}$ remains in U and converges to $x^\star$. A characteristic property of the Newton method is its rapid convergence. Under the assumptions mentioned above the convergence is *Q-superlinear*, i.e. the sequence of the ratios

$$\frac{\|x_{k+1} - x^\star\|}{\|x_k - x^\star\|} \tag{4.1}$$

converges to zero [60]. For twice differentiable F (Lipschitz-continuity of F' in $x^\star$) one even has quadratic convergence, i.e. for a constant $C > 0$ one has for all $k \in \mathbb{N}_0$

$$\|x_{k+1} - x^\star\| \le C\|x_k - x^\star\|^2.$$

The Newton method can be considered to be the prototype of a rapidly convergent method in the following sense: a convergent sequence in $\mathbb{R}^n$, for which F' is regular at the limit, is Q-superlinearly convergent to a zero of F, if and only if it is *Newton-like*, i.e.

$$\frac{\|F(x_k) + F'(x_k)(x_{k+1} - x_k)\|}{\|x_{k+1} - x_k\|} \xrightarrow{k\to\infty} 0.$$

In other words: all rapidly convergent methods satisfy the Newton equation up to $o(\|x_{k+1} - x_k\|)$. However, Newton-like sequences can also be generated without computing the Jacobian matrix.

Disadvantages of the Newton method consist in the fact that the method is only locally convergent and at each iteration step one has to

- compute the Jacobian matrix
- solve a linear system of equations.

We will – as we proceed – encounter rapidly convergent (Newton-like) methods, which do not display these disadvantages. The two latter disadvantages mean – in particular for large dimensions – a high computational effort. This is also the case if the Jacobian matrix is approximated by a suitable matrix of difference quotients, where in addition rounding errors create problems.

Application to Linear Modular Approximation

Let $\Phi : \mathbb{R} \to \mathbb{R}_{\geq 0}$ be a twice continuously differentiable Young function. Let $x \in \mathbb{R}^m$ and let V be an n-dimensional subspace of $\mathbb{R}^m$ with a given basis $\{v_1, \dots, v_n\}$. The approximation y^Φ to be determined is a linear combination $\sum_{i=1}^n a_i v_i$ of the basis vectors, where $\{a_1, \dots, a_n\}$ are the real coefficients to be computed. The search for a best approximation corresponds thus to a minimization problem in $\mathbb{R}^n$ with the unknown vector $a = (a_1, \dots, a_n)$. The function to be minimized is

$$\begin{aligned} f &: \mathbb{R}^n \to \mathbb{R} \\ a &\mapsto f(a) := f^\Phi\Big(x - \sum_{i=1}^n a_i v_i\Big). \end{aligned}$$

The minimization of this function we denote as (linear) modular approximation. Setting the first derivative of the function f to zero leads to the following system of equations (compare Equation (1.15)):

$$\sum_{t \in T} v_i(t)\, \Phi'\Big(x(t) - \sum_{l=1}^n a_l v_l(t)\Big) = 0, \quad \text{for } i = 1, \dots, n.$$

For the elements of the matrix of the second derivative of f one obtains

$$\frac{\partial^2}{\partial a_i \partial a_j} f(a) = \sum_{t \in T} v_i(t) v_j(t)\, \Phi''\Big(x(t) - \sum_{l=1}^n a_l v_l(t)\Big).$$

The computational complexity of determining the Hessian matrix is apparently n^2 times as high as the effort for evaluating the function f. In particular for large n it is urgent to decrease the computational effort.

4.2 Secant Methods

The above difficulties can be avoided by employing the so-called secant methods (Quasi-Newton methods). Here the Jacobian matrix at x_k (for minimization the Hessian matrix) is replaced by a matrix B_k, which is easily determined. If B_k is given, then x_{k+1} is determined as the solution of the equation

$$F(x_k) + B_k(x - x_k) = 0.$$

If the inverse of B_k is available we have

$$x_{k+1} = x_k - B_k^{-1}F(x_k).$$

For the successor B_{k+1} one requires that the *secant equation*

$$B_{k+1}(x_{k+1} - x_k) = F(x_{k+1}) - F(x_k)$$

is satisfied. The solutions of the secant equation form an affine subspace of the space of $n \times n$-matrices. For a concrete algorithmic realization of the computation of B_{k+1} an *update formula* is chosen, which determines, how B_{k+1} is obtained from B_k. The sequence of the matrices $(B_k)_{k\in\mathbb{N}_0}$ are called *update matrices*. A particular advantage of this scheme is that even *update formulas for the inverse* B_{k+1}^{-1} are available and thus the solution of the linear system of equations can be replaced by a matrix-vector multiplication.

Newton-likeness and hence rapid convergence of the secant method is obtained if

$$(B_{k+1} - B_k) \xrightarrow{k\to\infty} 0$$

holds, due to the subsequent

Theorem 4.2.1. *A secant method converging to $x^\star \in \mathbb{R}^n$ with invertible matrix $F'(x^\star)$ and*

$$B_{k+1} - B_k \xrightarrow{k\to\infty} 0$$

converges Q-superlinearly and $F(x^\star) = 0$ holds.

Before we begin with the proof we want to draw a connection between the secant and the Newton method: according to the mean value theorem in integral form we have for

$$Y_{k+1} := \int_0^1 F'(x_k + t(x_{k+1} - x_k))dt$$

the equation

$$Y_{k+1}(x_{k+1} - x_k) = F(x_{k+1}) - F(x_k),$$

i.e. Y_{k+1} satisfies the secant equation. One can verify directly that convergence of a sequence $(x_k)_{k\in\mathbb{N}_0}$ and the Lipschitz continuity of F' implies the convergence of the sequence $(Y_k)_{k\in\mathbb{N}_0}$ to the Jacobian matrix at the limit $x^\star$.

This leads to the following

Proof. Let $s_k = x_{k+1} - x_k \neq 0$ for all $k \in \mathbb{N}_0$. Due to $(B_{k+1} - Y_{k+1})s_k = 0$ and $F(x_k) = -B_k s_k$ the Newton-likeness follows:

$$\begin{aligned}\frac{\|F(x_k) + F'(x_k)s_k\|}{\|s_k\|} &= \frac{\| - B_k s_k + F'(x_k)s_k\|}{\|s_k\|} \\ &= \frac{\|(F'(x_k) - Y_{k+1} + B_{k+1} - B_k)s_k\|}{\|s_k\|} \\ &\le \|F'(x_k) - Y_{k+1}\| + \|B_{k+1} - B_k\| \xrightarrow{k\to\infty} 0,\end{aligned}$$

provided that F' is Lipschitz continuous. □

Update Formula of Broyden

A particularly successful update formula in practice is the following formula introduced by C. G. Broyden

$$B_{k+1} = B_k + \frac{(y_k - B_k s_k)s_k^T}{s_k^T s_k}$$

where $s_k := x_{k+1} - x_k$ and $y_k := F(x_{k+1}) - F(x_k)$. The success of the *Broyden formula* can be understood in view of the previous theorem, because it has the following minimality property: among all matrices B that satisfy the secant equation

$$Bs_k = y_k$$

the Broyden matrix B_{k+1} has the least distance to its predecessor B_k in the Frobenius norm. For a practical realization the 'inverse' update formula

$$H_{k+1} = H_k + \frac{(s_k - H_k y_k)s_k^T H_k}{s_k^T H_k y_k}$$

can be used, which allows the computation of $B_{k+1}^{-1} = H_{k+1}$ from $B_k^{-1} = H_k$. This formula is obtained from the general *Sherman–Morrison–Woodbury Lemma*, which guarantees the existence of the inverse and the validity of the above formula for non-vanishing denominator.

Geometry of Update Matrices

In order to understand the Broyden method and the subsequent algorithms, we want to formulate a general geometric statement about successive orthogonal projections on affine subspaces.

Theorem 4.2.2. *Let $(V_n)_{n\in\mathbb{N}_0}$ be a sequence of affine subspaces of a finite-dimensional vector space X, equipped with a scalar product and the corresponding norm. Let there exist a $H_k \in V_k$ for all $k \in \mathbb{N}$ such that*

$$\sum_{k=0}^{\infty} \|H_{k+1} - H_k\| < \infty.$$

Then for every starting point $B_0 \in X$ the sequence defined by the recursion

$$B_{k+1} \text{ is best approximation of } B_k \text{ w.r.t. } V_k$$

is bounded and we have $B_{k+1} - B_k \to_{k\to\infty} 0$.

In order to describe the geometry of the Broyden method, one chooses X as the space of matrices, equipped with the Frobenius norm. As affine subspace in this context we define $V_k := \{B \mid Bs_k = y_k\}$ and for H_k the mean value matrix Y_{k+1} for $k \in \mathbb{N}$ is chosen.

4.3 Global Convergence

The class of rapidly convergent methods corresponds, as we have seen, to the class of the convergent Newton-like methods. In order to ensure global convergence, we take up the idea of the classical *gradient methods*. The gradient method is a method for determining the minimal solution of functions, which displays global convergence for a large class of functions. The main idea of the gradient method relies on the following *descent property*: starting from the present position one can – in the direction of the negative gradient – find a point with a lower function value. The following iteration

$$x_{k+1} = x_k - \alpha_k \nabla f(x_k) \quad \text{for } \nabla f(x_k) \neq 0$$

with suitable *step size* α_k is based on this property. The determination of a step size is an important issue and is provided by use of a *step size rule*. Instead of the direction of the negative gradient $\nabla f(x_k)$ one can use a modified direction d_k with descent property. In this way we will be enabled to construct a rapidly as well as globally convergent method. The pure gradient method, however, is only linearly convergent. This leads to the *generalized gradient methods*, which after choice of a starting point x_0 are defined by the following iteration

$$x_{k+1} = x_k - \alpha_k d_k$$

as soon as step size α_k and *search direction* d_k are specified. Damped versions of the Newton method will also belong to this class, which we will treat below. At first we want to turn our attention to the determination of the step size.

Some Step Size Rules

The most straightforward way of determining a step size is, to search for the point with least function value in the chosen direction. This leads to the following *minimization rule (Rule of optimal step size)*:

Choose α_k, such that

$$f(x_k - \alpha_k d_k) = \min_{\alpha \geq 0} f(x_k - \alpha d_k).$$

This rule has a crucial disadvantage: the step size can in general be determined only by an infinite process. Therefore this rule can only be realized on a computer in an approximate sense. Experience shows that *infinite rules* can be replaced by constructive *finite* rules. A common property of these rules is, to keep the ratio of the slopes of secant and tangent

$$\left(\frac{f(x_k) - f(x_k - \alpha_k d_k)}{\alpha_k}\right) / \nabla f(x_k)^T d_k$$

above a fixed constant $\sigma > 0$. The simplest version is obtained by the following:

Armijo Rule (AR)

Let $\beta \in (0,1)$ and $\sigma \in (0, \frac{1}{2})$. Choose $\alpha_k = \beta^{m_k}$, such that m_k is the smallest number $\mathbb{N}_0$, for which

$$f(x_k) - f(x_k - \beta^{m_k} d_k) \geq \sigma \beta^{m_k} \nabla f(x_k)^T d_k,$$

i.e. one starts with step size 1 and decreases the step size by multiplication with β until the ratio of the slopes of secant and tangent exceeds σ.

In the early phase of the iteration, the step size 1 accepted by the Armijo rule can be unnecessarily small. This problem can be eliminated by the following modification:

Armijo Rule with Expansion (ARE)

Let $\beta \in (0, \frac{1}{2})$ and $\sigma \in (0, \frac{1}{2})$. Choose $\alpha_k = \beta^{m_k}$ such that m_k is the smallest integer for which

$$f(x_k) - f(x_k - \beta^{m_k} d_k) \geq \sigma \beta^{m_k} \nabla f(x_k)^T d_k.$$

Algorithmically one proceeds in the following way: if the required inequality is already satisfied for $\beta = 1$, then, different from the Armijo rule, the search is not discontinued. Through multiplication by $1/\beta$ the expansion is continued, until the inequality is violated. The predecessor then determines the step size.

A different procedure prevents the step size from becoming too small. This can be achieved by the requirement, that the ratio of the slopes of secant and tangent should not come too close to the number 1.

Goldstein Rule (G)

Let $\sigma \in (0, \frac{1}{2})$. Choose α_k, such that

$$1 - \sigma \geq \frac{f(x_k) - f(x_k - \alpha_k d_k)}{\alpha_k \nabla f(x_k)^T d_k} \geq \sigma.$$

For a concrete realization of this rule one can proceed in a similar way as in the Armijo rule, using an interval partitioning method for the second inequality. In any case the acceptance of the full step size ($\alpha_k = 1$) should be tested.

Powell–Wolfe Rule (PW)

Let $\sigma \in (0, \frac{1}{2})$ and $\beta \in (0, 1)$. Choose α_k, such that

(a) $\frac{f(x_k) - f(x_k - \alpha_k d_k)}{\alpha_k \nabla f(x_k)^T d_k} \geq \sigma$

(b) $\nabla f(x_k - \alpha_k d_k)^T d_k \leq \beta \nabla f(x_k)^T d_k$.

A concrete realization of this rule provides an expansion phase similar in spirit to ARE.

These constructive rules have the additional advantage that in the subsequent globalization of the Newton method the damped method automatically will merge eventually into an undamped method. Here (and also for Newton-like directions (see [60], p. 120, 128) for sufficiently large k the step size $\alpha_k = 1$ will be accepted. One can show that the ratio of the slopes of secant and tangent converges to $1/2$.

Remark 4.3.1. For the determination of a zero of a function

$$F : \mathbb{R}^n \to \mathbb{R}^n,$$

which is *not* derivative of a function $f : \mathbb{R}^n \to \mathbb{R}$ (or where f is unknown), the determination of the step size can be performed by use of an alternative function as e.g. $x \mapsto h(x) = \frac{1}{2}\|F(x)\|^2$.

4.3.1 Damped Newton Method

The pure Newton method is, as mentioned above, only locally, but on the other hand rapidly convergent. Damped Newton methods, whose algorithms we will discuss below, combine the advantages of rapid convergence of the Newton method and the global convergence of the gradient method. As indicated above we have to distinguish the determination of minimal solutions of a function $f : \mathbb{R}^n \to \mathbb{R}$ and the determination of a zero of a function $F : \mathbb{R}^n \to \mathbb{R}^n$ which is not necessarily the derivative of a real function on the $\mathbb{R}^n$. For the first of the two cases we have the following algorithm:

Damped Newton Method for Minimization

Let

$$f : \mathbb{R}^n \to \mathbb{R}$$

be a twice continuously differentiable function.

(a) Choose a starting point x_0, a stop criterion and put $k \leftarrow 0$.

(b) Test the stop criterion w.r.t. x_k. Compute a new direction d_k as solution of the linear equation systems

$$f''(x_k)d = f'(x_k).$$

(c) Determine $x_{k+1} := x_k - \alpha_k d_k$, where α_k is chosen according to one of the finite step size rules described above.

(d) Put $k \leftarrow k + 1$ and continue with step (b).

For this method the following convergence statement holds:

Theorem 4.3.2. *Let $f : \mathbb{R}^n \to \mathbb{R}$ be a twice continuously differentiable function and $x_0 \in \mathbb{R}^n$ a starting point. Let the level set belonging to the starting point*

$$\{y \in \mathbb{R}^n \mid f(y) \le f(x_0)\}$$

be convex, bounded, and the second derivative be positive definite there. Then the damped Newton method converges to the uniquely determined minimal solution of f and the damped Newton method merges into the (undamped) Newton method, i.e. there is a $k_0 \in \mathbb{N}$ such that for all $k \ge k_0$ the step size $\alpha_k = 1$ is accepted by the rule chosen.

For the determination of a zero of an arbitrary function $F : \mathbb{R}^n \to \mathbb{R}^n$ we introduce a function $x \mapsto h(x) := \frac{1}{2}\|F(x)\|^2$ as descent function. This leads to following algorithm:

Damped Newton Method for Equations

(a) Choose a starting point x_0, a stop criterion and put $k \leftarrow 0$.

(b) Test the stop criterion for x_k. Compute a new direction d_k as solution of the linear equation systems

$$F'(x_k)d = F(x_k).$$

(c) Determine $x_{k+1} := x_k - \alpha_k d_k$, where α_k is chosen according to one of the finite step size rules described above.

(d) Put $k \leftarrow k + 1$ and continue with step (b).

Remark 4.3.3. The finite step size rules test the ratio of the slopes of secant and tangent. For the direction $d_k = F'(x_k)^{-1}F(x_k)$ the slope of the tangent can be expressed in terms of $h(x_k)$, since we have

$$\langle h'(x_k), d_k\rangle = \langle F'(x_k)^T F(x_k), F'(x_k)^{-1}F(x_k)\rangle = \langle F(x_k), F(x_k)\rangle = 2h(x_k).$$

For the step size α_k we obtain the following simple condition:

$$\frac{h(x_k - \alpha_k d_k)}{h(x_k)} \leq 1 - 2\sigma\alpha_k. \tag{4.2}$$

The convergence is described by the following theorem (see [60], p. 131):

Theorem 4.3.4. *Let $F : \mathbb{R}^n \to \mathbb{R}^n$ be a continuously differentiable function and let $h : \mathbb{R}^n \to \mathbb{R}$ be defined by $h(x) = \frac{1}{2}\|F(x)\|^2$. Let x_0 be a starting point with bounded level set*

$$S_h(x_0) := \{x \in \mathbb{R}^n \mid h(x) \leq h(x_0)\}$$

and let $F'(x)$ be invertible for all $x \in S_h(x_0)$. Let further F' be Lipschitz-continuous in a ball $K \supset S_h(x_0)$.

Then the damped Newton method for equations converges quadratically to a zero of F. Furthermore this method merges into the (proper) Newton method.

4.3.2 Globalization of Secant Methods for Equations

The central issue of the secant methods was to avoid the computation of the Jacobian matrix. Global convergence in the previous section was achieved by use of a step size rule. These rules employ the derivative of the corresponding descent function. In the case of the determination of a zero of a function $F : \mathbb{R}^n \to \mathbb{R}$ and the use of $h : \mathbb{R}^n \to \mathbb{R}$ where $h(x) = \frac{1}{2}\|F(x)\|^2$ as descent function the derivative $h'(x) = F'(x)^T F(x)$ contains the Jacobian matrix $F'(x)$. Thus for secant methods the step size rules w.r.t. h are not available. As a substitute one can use the ratio of subsequent values of h (compare Equation (4.2) for the Newton direction) as a criterion. At first the full step size is tested. If not successful, one can try a few smaller step sizes. If in this way the direction of the secant is not found to be a direction of descent, an alternative direction is constructed. We will now describe a number of methods, which differ by the choice of the alternative direction.

SJN method

Here the Newton direction is chosen as a substitute. Since here the Jacobian matrix is computed, we can use it as a new start for update matrices. We denote the resulting method as *Secant method with Jacobian matrix New start (SJN)*.

The SJN-method

(a) Choose a regular $n \times n$-matrix B_0, $x_0 \in \mathbb{R}^n$, $\beta \in (0,1)$, $\sigma \in (0, 1/4)$ and a formula for the determination of update matrices for a secant method (resp. for its inverse). Put $k = 0$ and choose a stop criterion.

(b) If the stop criterion is satisfied, then **Stop**.

(c) Put $d_k := B_k^{-1} F(x_k)$ and search for $\alpha_k \in \mathbb{R}$, such that $h(x_k - \alpha_k d_k) \leq (1-\sigma)h(x_k)$, where at first $\alpha_k = 1$ is tested. If the search is successful, put $x_{k+1} = x_k - \alpha_k d_k$ and goto (e).

(d) *New start*: Compute $F'(x_k)^{-1}$, put $B_k^{-1} = F'(x_k)^{-1}$ and $d_k = B_k^{-1} F(x_k)$. Find the smallest $m \in \mathbb{N}_0$ (Armijo rule) with

$$h(x_k - \beta^m d_k) \leq (1 - \beta^m \sigma) h(x_k) \tag{4.3}$$

and put $x_{k+1} = x_k - \beta^m d_k$.

(e) Put $s_k = x_{k+1} - x_k$ and $y_k = F(x_{k+1}) - F(x_k)$. Compute B_{k+1} resp. B_{k+1}^{-1} by the update formula chosen above.

(f) Put $k \leftarrow k+1$ and goto (b).

Theorem 4.3.5. *Let $F : \mathbb{R}^n \to \mathbb{R}^n$ be continuously differentiable, F' be locally Lipschitz continuous, $x_0 \in \mathbb{R}^n$, and let the level set $S_h(x_0)$ belonging to the starting point be bounded. Further let x^* be the only zero of F and let $F'(x)$ be regular for all $x \in S_h(x_0)$. Then the sequence $(x_k)_{k \in \mathbb{N}_0}$ generated by the SJN-method is at least Q-superlinearly convergent. Moreover, from a certain index $k_0 \in \mathbb{N}$ on, the step size $\alpha_k = 1$ is accepted.*

Derivative-free Computation of Zeros

The disadvantage of the choice of the alternative direction described in the previous section is that the Jacobian matrix has to be computed. On the other hand by making the Jacobian matrix available we have achieved two goals:

(a) Necessary and sufficient for rapid convergence is the Newton-likeness of the iteration sequence (see [60], p. 71). In this sense the New start provides an improvement of the current update matrix.

(b) The Newton direction is always a direction of descent w.r.t. h.

These two goals we will now pursue using derivative-free methods. The geometry of the update matrices will simplify our understanding of the nature of the problem and will lead to the construction of concrete algorithms.

A possibility for exploiting the geometry, can be obtained in the following way: in order to construct a new update matrix, which – if possible – should approximate the Jacobian matrix, we introduce as a substitute an 'artificial' subspace and determine a new update matrix by projection of the current matrix onto this subspace. A straightforward strategy is to choose the substitute to be orthogonal to the current subspace V_k. Instead of the orthogonality of the subspaces in the subsequent algorithmic realization we construct mutually orthogonal vectors.

The OS method

(a) Choose a regular $n \times n$-matrix B_0, $x_0 \in \mathbb{R}^n$, $C, \tilde{C} \in (0,1)$, $\delta > 0$ and a formula for the determination of update matrices for a secant method (resp. for its inverse). Put $k = 0$ and choose a stop criterion.

(b) Put $N = 0$. If the stop criterion is satisfied, then **Stop**.

(c) Put $d_k := B_k^{-1} F(x_k)$ and search for $\alpha_k \in \mathbb{R}$ satisfying $h(x_k - \alpha_k d_k) \leq Ch(x_k)$, where at first $\alpha_k = 1$ is tested. If the search is successful, put

$$\begin{aligned} x_{k+1} &:= x_k - \alpha_k d_k \\ \tilde{s}_k &:= -\alpha_k d_k \\ \tilde{y}_k &:= F(x_{k+1}) - F(x_k) \end{aligned}$$

and goto (e).

(d) *New start*: Put $x_{k+1} := x_k$ and $N \leftarrow N + 1$. Let α be the last step size tested in (c) and let $d := \frac{\alpha}{N^2} d_k$. If $k = 0$, we put $\tilde{y}_k := F(x_k - d) - F(x_k)$. Otherwise compute $t := d^T \tilde{s}_{k-1} / \tilde{s}_{k-1}^T (d - \tilde{s}_{k-1})$ and put $\tilde{s}_k := t\tilde{s}_{k-1} + (1-t)d$, $\tilde{y}_k := F(x_k - \tilde{s}_k) - F(s_k)$.

(e) Compute B_{k+1} resp. B_{k+1}^{-1} by the update formula chosen above by use of $\tilde{s}_k$ and $\tilde{y}_k$.

(f) Put $k \leftarrow k + 1$. In case a New start was necessary, goto (c), otherwise to (b).

Remark. In case of a multiple New start an inner loop with loop index N comes into play. The desired effect is the asymptotic convergence of the update matrices. In order to achieve this, the division by N^2 is performed in (d).

Further methods for the solution of equations can be found in [60].

4.3.3 Secant Method for Minimization

When looking for a solution of a non-restricted optimization problem, setting the gradient to zero leads to a non-linear equation. In contrast to the preceding section we have in this situation the opportunity to take the function to be minimized as the descent function for the step size rule. The direction of the secant is always a direction of descent, provided that the update matrix is positive definite. Those update formulas will be of particular significance in this context, which generate positive definite matrices. In theory as well as in practice the BFGS method will play an eminent role.

BFGS method

(a) Choose a symmetrical and positive definite $n \times n$-matrix H_0, $x_0 \in \mathbb{R}^n$, and a step size rule (G), (ARE) or (PW). Put $k = 0$ and choose a stop criterion (e.g. test of the norm of the gradient).

(b) If stop criterion is satisfied, then **Stop**.

(c) Put $d_k := H_k \nabla f(x_k)$ and determine a step size α_k with f as descent function. Put $x_{k+1} := x_k - \alpha_k d_k$. Put $s_k := x_{k+1} - x_k$ and $y_k := \nabla f(x_{k+1}) - \nabla f(x_k)$. Compute H_{k+1} according to the following formula:

$$H_{k+1} := H_k + \frac{v_k s_k^T + s_k v_k^T}{\langle y_k, s_k \rangle} - \frac{\langle v_k, y_k \rangle s_k s_k^T}{\langle y_k, s_k \rangle^2}$$

with $v_k := s_k - H_k y_k$.

(d) Put $k \leftarrow k + 1$ and goto (b).

Theorem 4.3.6. *Let $f : \mathbb{R}^n \to \mathbb{R}$ be twice continuously differentiable and $x_0 \in \mathbb{R}^n$. Let the level set $S_f(x_0)$ be bounded and f'' positive definite on $S_f(x_0)$. Then the BFGS-method can be performed with every symmetrical positive definite start matrix and is at least Q-superlinearly convergent to the minimal solution. The matrices of the sequence $(H_k)_{k \in \mathbb{N}_0}$ are uniformly positive definite, i.e. there are positive numbers m, M, such that for all $z \in \mathbb{R}^n$ with $\|z\| = 1$*

$$m \leq \langle H_k z, z \rangle \leq M$$

holds. Furthermore, the damped method merges into the undamped method.

Remark. If one wants to apply the BFGS-method to non-convex functions, it turns out that the Powell–Wolfe rule assumes a particular significance, because it generates also in this situation positive definite update matrices.

Testing the Angle

The BFGS-method preserves positive definiteness of the update matrices and guarantees that the direction of the secant is a direction of descent. Even then the angle between the current gradient and the search direction d_k can become unfavorable at times. As an additional measure, one can test in the context of secant methods the cosine $\beta_k := \frac{\langle \nabla f(x_k), d_k \rangle}{\|\nabla f(x_k)\| \|d_k\|}$ of the angle in question. If β_k falls below a given threshold, a new direction of descent is constructed, e.g. a convex combination of gradient- and direction of the secant. The BFGS-method does have a self correcting property w.r.t. the sequence $(\beta_k)_{k \in \mathbb{N}_0}$, but also here an improvement of the convergence can be achieved by such a test for the angle. An algorithmic suggestion for the whole class of secant methods could be the following:

Secant Method with Angle Check

(a) Choose a symmetrical and positive definite $n \times n$-Matrix H_0, $x_0 \in \mathbb{R}^n$, a precision $\gamma > 0$ and a step size rule (G), (AR), (ARE) or (PW). Put $k = 0$ and choose a stop criterion (e.g. test of the norm of the gradient).

(b) If stop criterion is satisfied, then **Stop**.

(c) Compute $p_k := H_k \nabla f(x_k)$ and $\beta_k := \frac{\langle \nabla f(x_k), p_k \rangle}{\|\nabla f(x_k)\| \|p_k\|}$. If $\beta_k < \gamma$, choose a d_k, such that $\frac{\langle \nabla f(x_k), d_k \rangle}{\|\nabla f(x_k)\| \|d_k\|} \geq \gamma$, otherwise put $d_k := p_k$.

(d) Determine a step size α_k with f as descent function. Put $x_{k+1} := x_k - \alpha_k d_k$. Put $s_k := x_{k+1} - x_k$ and $y_k := \nabla f(x_{k+1}) - \nabla f(x_k)$. Compute H_{k+1} using the update formula of a secant method for the inverse matrix.

(e) Put $k \leftarrow k + 1$ and goto (b).

For the theoretical background see [60] the theorem in Section 11, p. 185.

4.4 A Matrix-free Newton Method

See [64]: Let U be an open subset of $\mathbb{R}^n$ and $F : U \to \mathbb{R}^n$ differentiable. In order to determine x_{k+1} the Newton equation

$$F'(x_k)(x - x_k) = F(x_k)$$

has to be solved. Putting $A_k := F'(x_k)$, $b_k := -F(x_k)$, and $y := x - x_k$ the above equation amounts to $A_k y = b_k$. Its solution y_k yields $x_{k+1} = x_k + y_k$. For large n storing the Jacobian $F'(x_k)$ becomes prohibitive. A matrix-free variant of the Newton method is obtained by replacing $F'(x_k)(x - x_k)$ by the directional derivative

$$F'(x_k, x - x_k) = \lim_{t \to 0} \frac{F(x_k + t(x - x_k)) - F(x_k)}{t}.$$

For the solution of $A_k y = b_k$ a method is employed that does not require the knowledge of A_k, but only the result of the multiplication $A_k u$ for a $u \mathbb{R}^n$, where u denotes a vector appearing in an intermediate step of the corresponding so-called MVM-method (Matrix-Vector-Multiply-method). Here $A_k \cdot u$ is replaced by the directional derivative $F(x_k, u)$. If not available analytically, it is replaced by the approximation

$$\frac{F(x_k + \varepsilon u) - F(x_k)}{\varepsilon}$$

for small $\varepsilon > 0$.

A suitable class of (iterative) solution methods for linear equations is represented by the Krylov space methods. Also the RA-method in [60] can be used. For the important special cases of symmetrical and positive definite matrices the conjugate gradient and conjugate residual methods, and for non-symmetrical matrices A_k the CGS-method can be employed.

Chapter 5
Stability and Two-stage Optimization Problems

In our discussion of Polya algorithms in Chapter 2 the question referring to the closedness of the algorithm, i.e. if every point of accumulation of the sequence of solutions of the approximating problems is a solution of the limit function, is of central importance.

In this chapter we will treat this question in a more general framework. It turns out that lower semi-continuous convergence is responsible for the closedness of the algorithm. Monotone sequences of functions, which converge pointwise, will turn out to be lower semi-continuously convergent and pointwise convergent sequences of convex functions even continuously convergent. The latter statement is based on an extension of the principle of uniform boundedness of Banach to families of convex functions. For the verification of pointwise convergence for sequences of convex functions, a generalized version of the theorem of Banach–Steinhaus for convex functionals turns out to be a useful tool: in the case of Orlicz spaces e.g. it suffices to consider the (dense) subspace of step functions, where pointwise convergence of the functionals is essentially reduced to that of the corresponding Young functions.

We will investigate the actual convergence of the approximate solutions – beyond the closedness of the algorithm – in the section on two-stage optimization. In this context we will also discuss the question of robustness of the algorithms against ε-solutions.

In the last section of this chapter we will see that a number of results about sequences of convex functions carry over to sequences of monotone operators.

Much of the material presented in this chapter is contained in [59].

5.1 Lower Semi-continuous Convergence and Stability

Definition 5.1.1. Let X be a metric space. A sequence $(f_n : X \to \overline{\mathbb{R}})_{n\in\mathbb{N}}$ is called *lower semi-continuously convergent* resp. *upper semi-continuously convergent* to $f : X \to \overline{\mathbb{R}}$, if $(f_n)_{n\in\mathbb{N}}$ converges pointwise to f and for every convergent sequence $(x_n)_{n\in\mathbb{N}}$ in X with $\lim x_n = x$

$$\underline{\lim}_{n\to\infty} f_n(x_n) \geq f(x)$$

resp.

$$\overline{\lim}_{n\to\infty} f_n(x_n) \leq f(x)$$

holds.

Remark. For the practical verification of lower semi-continuous convergence the following criterion turns out to be useful: f is lower semi-continuous and there exists a sequence (α_n) tending to zero, such that $f_n + \alpha_n \geq f$.

An immediate consequence of the definition is the

Theorem 5.1.2. *Let X be a metric space and let the function sequence $(f_n : X \to \overline{\mathbb{R}})_{n\in\mathbb{N}}$ converge lower semi-continuously to $f : X \to \overline{\mathbb{R}}$, then every point of accumulation of the minimal solutions $M(f_n, X)$ is an element of $M(f, X)$.*

Proof. Let $\bar{x}$ be such a point of accumulation, i.e. $\bar{x} = \lim_{i\to\infty} x_{n_i}$ with $x_{n_i} \in M(f_{n_i}, X)$. For an arbitrary $x \in X$ we have

$$f_{n_i}(x_{n_i}) \leq f_{n_i}(x).$$

The lower semi-continuous convergence then implies

$$f(\bar{x}) \leq \varliminf_{i\to\infty} f_{n_i}(x_{n_i}) \leq \varliminf_{i\to\infty} f_{n_i}(x) = f(x). \qquad \square$$

The principle of lower semi-continuous convergence provides a framework for further investigations: in particular the question arises, how we can establish that lower semi-continuous convergence holds.

A simple criterion is provided by *monotone convergence* (see below).

In the case of pointwise convergence of real valued, convex functions even continuous convergence is obtained (see Theorem 5.3.6 and Theorem 5.3.8).

In the context of convex functions one obtains stability results even if the sets, on which the minimization is performed, are varied. The *Kuratowski convergence*, also referred to as topological convergence (see Definition 5.1.5), is the appropriate notion of set convergence in this situation.

A stability theorem for sequences of convex functions with values in $\overline{\mathbb{R}}$ is obtained, if their *epigraphs* converge in the sense of Kuratowski (see Theorem 5.1.7).

5.1.1 Lower Semi-equicontinuity and Lower Semi-continuous Convergence

Definition 5.1.3. Let X be a metric space. A sequence of functions $(f_n : X \to \overline{\mathbb{R}})_{n\in\mathbb{N}}$ is called *lower semi-equicontinuous* at $x \in X$, if for all $\varepsilon > 0$ there is a neighborhood $U(x)$ of x, such that for all $n \in \mathbb{N}$ and all $y \in U(x)$ we have

$$f_n(x) - \varepsilon \leq f_n(y).$$

Theorem 5.1.4. *Let $(f_n : X \to \overline{\mathbb{R}})_{n\in\mathbb{N}}$ be a sequence of functions, which converges pointwise to f on X and let $\{f_n\}_{n\in\mathbb{N}}$ be lower semi-equicontinuous on X, then for all $x \in X$ we obtain: f is lower semi-continuous at x and the convergence is lower semi-continuous.*

Proof. Let $x \in X$, then there is a neighborhood $U(x)$, such that for all $n \in \mathbb{N}$ and all $y \in U(x)$ we have: $f_n(x) - \varepsilon \leq f_n(y)$. From the pointwise convergence it follows that $f(x) - \varepsilon \leq f(y)$, i.e. f is lower semi-continuous at x.

If $x_n \to x$, then $x_n \in U(x)$ for $n \geq N \in \mathbb{N}$, i.e. for $n \geq N$: $f_n(x) - \varepsilon \leq f_n(x_n)$.

(a) Let $x \in \operatorname{Dom} f$, then there is $N_1 \geq N$, such that $f(x) - \varepsilon \leq f_n(x)$ for all $n \geq N_1$, hence

$$f(x) - 2\varepsilon \leq f_n(x) - \varepsilon \leq f_n(x_n)$$

for all $n \geq N_1$. Therefore $\underline{\lim} f_n(x_n) \geq f(x)$.

(b) $x \notin \operatorname{Dom} f$, i.e. $f(x) = \infty$ and $f_n(x) \to f(x)$. Due to $f_n(x) - \varepsilon \leq f_n(x_n)$ it follows that $f_n(x_n) \to \infty$. □

5.1.2 Lower Semi-continuous Convergence and Convergence of Epigraphs

Definition 5.1.5. Let X be a metric space and $(M_n)_{n \in \mathbb{N}}$ a sequence of subsets of X. Then we introduce the following notation

$$\overline{\lim_n} M_n := \Big\{ x \in X \mid \text{there is a subsequence } (M_k)_{k \in \mathbb{N}} \text{ of } (M_n)_{n \in \mathbb{N}} \text{ and } x_k \in M_k, \text{ such that } x = \lim_{k \to \infty} x_k \Big\}$$

$$\underline{\lim_n} M_n := \Big\{ x \in X \mid \text{there is } n_0 \in \mathbb{N} \text{ and } x_n \in M_n \text{ for } n \geq n_0, \text{ such that } x = \lim_{n \to \infty} x_n \Big\}.$$

The sequence $(M_n)_{n \in \mathbb{N}}$ is called *Kuratowski convergent* to the subset M of X, if

$$\overline{\lim_n} M_n = \underline{\lim_n} M_n = M,$$

notation: $M = \lim_n M_n$.

Theorem 5.1.6. *Let the sequence $(f_n)_{n \in \mathbb{N}}$ converge to f lower semi-continuously, then* Epi $f_n \to$ Epi f *converges in the sense of Kuratowski.*

Proof. Let $(x_{n_k}, r_{n_k}) \in \operatorname{Epi} f_{n_k}$ be a convergent subsequence with $(x_{n_k}, r_{n_k}) \to (x, r)$, then $r = \lim r_{n_k} \geq \underline{\lim} f_{n_k}(x_{n_k}) \geq f(x)$, i.e. $(x, r) \in \operatorname{Epi} f$, hence $\overline{\lim} \operatorname{Epi} f_n \subset \operatorname{Epi} f$.

Let now $x \in \operatorname{Dom} f$ and $(x, r) \in \operatorname{Epi} f$, i.e. $f(x) \leq r < \infty$. Let further $\varepsilon > 0$ be given, then $f_n(x) \leq f(x) + \varepsilon \leq r + \varepsilon$ for n sufficiently large. Then $(x, f_n(x) + r - f(x)) \in \operatorname{Epi} f_n$ and we have $(x, f_n(x) + r - f(x)) \to (x, f(x) + r - f(x)) = (x, r)$. If $x \notin \operatorname{Dom} f$, $(x, r) \notin \operatorname{Epi} f$ for all $r \in \mathbb{R}$. □

A weak version of the converse of the previous theorem we obtain by

Theorem 5.1.7. *Let the sequence* $(f_n)_{n\in\mathbb{N}}$ *converge to* f *in the pointwise sense and let* $\overline{\lim}$ Epi $f_n \subset$ Epi f. *Then* $(f_n)_{n\in\mathbb{N}}$ *converges to* f *lower semi-continuously.*

Proof. Let (x_n) converge to x.

(a) $x \in \text{Dom } f$: let $(f_{n_k}(x_{n_k})) \to r$ be a convergent subsequence, then

$$(x_{n_k}, f_{n_k}(x_{n_k})) \to (x, r) \in \text{Epi } f,$$

and hence $r \geq f(x)$. In this way we obtain: $\underline{\lim}\, f_n(x_n) \geq f(x)$.

(b) $x \notin \text{Dom } f$: let $(x_{n_k}, f_{n_k}(x_{n_k})) \to (x, r)$, then $r = \infty$, because if $r < \infty$, then $(x, r) \in \text{Epi } f$, contradicting $x \notin \text{Dom } f$. Therefore $f_n(x_n) \to \infty$. □

In the context of sequences of convex functions we will apply the convergence of the corresponding epigraphs for further stability results.

5.2 Stability for Monotone Convergence

A particularly simple access to lower semi-continuous convergence and hence to stability results is obtained in the case of monotone convergence, because in this case lower semi-continuous convergence follows from pointwise convergence.

Theorem 5.2.1. *Let* X *be a metric space and* $(f_n : X \to \overline{\mathbb{R}})_{n\in\mathbb{N}}$ *a monotone sequence of lower semi-continuous functions, which converges pointwise to a lower semi-continuous function* $f : X \to \overline{\mathbb{R}}$. *Then the convergence is lower semi-continuous.*

Proof. Let $x_n \to x_0$ and (f_n) monotonically decreasing, then $f_n(x_n) \geq f(x_n)$ and therefore

$$\underline{\lim}\, f_n(x_n) \geq \underline{\lim}\, f(x_n) \geq f(x_0).$$

Let now (f_n) monotonically increasing and $k \in \mathbb{N}$. Then for $n \geq k$

$$f_n(x_n) \geq f_k(x_n),$$

and hence, due to the lower semi-continuity of f_k

$$\underline{\lim}_n\, f_n(x_n) \geq \underline{\lim}_n\, f_k(x_n) \geq f_k(x_0).$$

Therefore for all $k \in \mathbb{N}$

$$\underline{\lim}_n\, f_n(x_n) \geq f_k(x_0).$$

From the pointwise convergence of the sequence $(f_k)_{k\in\mathbb{N}}$ to f the assertion follows. □

Remark. If the convergence of (f_n) is monotonically decreasing, apparently just the lower semi-continuity of the limit function is needed. If however the convergence is monotonically increasing, then the lower semi-continuity of the limit function f follows from the lower semi-continuity of the elements of the sequence. A corresponding theorem and proof is available for upper semi-continuous functions and upper semi-continuous convergence.

The previous theorem and the preceding remark lead to the following

Theorem 5.2.2 (Theorem of Dini). *Let X be a metric space and $(f_n : X \to \mathbb{R})_{n\in\mathbb{N}}$ a monotone sequence of continuous functions, which converges pointwise to a continuous function $f : X \to \mathbb{R}$. Then the convergence is continuous.*

Proof. Upper semi-continuous and lower semi-continuous convergence together imply continuous convergence according to Definition 5.3.1. □

As a consequence of Theorem 5.1.2 and Theorem 5.2.1 we obtain

Theorem 5.2.3 (Stability Theorem of Monotone Convergence). *Let X be a metric space and $(f_n : X \to \overline{\mathbb{R}})_{n\in\mathbb{N}}$ a monotone sequence of lower semi-continuous functions, which converges pointwise to a lower semi-continuous function $f : X \to \overline{\mathbb{R}}$. Then every point of accumulation of the minimal solutions $M(f_n, X)$ is an element of $M(f, X)$.*

If the function sequence is monotonically decreasing, the values of the optimization problems (f_n, X) converge to the value of the optimization problem (f, X), i.e.

$$\inf f_n(X) \xrightarrow{n\to\infty} \inf f(X). \tag{5.1}$$

If in addition X is a compact set, then $\overline{\lim}_{n\to\infty} M(f_n, X) \neq \emptyset$ and the convergence of the values is guaranteed for monotonically increasing function sequences.

Proof. Only (5.1) remains to be shown. Let (f_n) be monotonically decreasing. The sequence $r_n := \inf f_n(X)$ is also decreasing and hence convergent to a $r_0 \in \overline{\mathbb{R}}$.

If $r_0 = -\infty$ or $r_0 = \infty$ (5.1) holds. Suppose $r_0 \in \mathbb{R}$ and $r_0 > \inf f(X)$. Then there is $x_0 \in X$ such that $f(x_0) < r_0$. From pointwise convergence we obtain $f_n(x_0) \to f(x_0)$ and thus the contradiction $r_0 = \lim_{n\to\infty} \inf f(X) \le f(x_0) < r_0$.

Let now (f_n) be increasing and X compact. Let $x_n \in M(f_n, X)$ and (x_{n_i}) a subsequence of (x_n) converging to $\bar{x}$. By the first part of the theorem $\bar{x} \in M(f, X)$. Pointwise and lower semi-continuous convergence then yield

$$f(\bar{x}) = \lim f_{n_i}(\bar{x}) \ge \lim f_{n_i}(x_{n_i}) \ge f(\bar{x}) = \inf f(X). \qquad \square$$

As an application of the above stability theorem we consider the directional derivative of the maximum norm:

Theorem 5.2.4. *Let T be a compact metric space and $C(T)$ the space of the continuous functions on T with values in $\mathbb{R}$. For the function $f : C(T) \to \mathbb{R}$ with $f(x) := \max_{t\in T} |x(t)|$ and $h \in C(T)$, $x \in C(T) \setminus \{0\}$ we obtain*

$$f'_+(x,h) = \max\{h(t)\,\mathrm{sign}(x(t)) \,|\, t \in T\}.$$

Proof. From the monotonicity of the difference quotient of convex functions (see Theorem 3.3.1) it follows that the sequence (g_{α_n}) with

$$g_{\alpha_n}(t) := \frac{|x(t)| - |x(t) + \alpha_n h(t)| + f(x) - |x(t)|}{\alpha_n}$$

for $\alpha_n \downarrow 0$ is a monotonically increasing sequence, which converges pointwise for $|x(t)| = f(x)$ to $g(t) := -h(t)\,\mathrm{sign}(x(t))$, for $|x(t)| < f(x)$ however to ∞. Since T is compact, the sequence of the values $\min_{t\in T} g_{\alpha_n}(t)$ converges due to the above theorem to $\min_{t\in T} g(t)$, thus

$$\begin{aligned}
\lim_{\alpha\downarrow 0} \max_{t\in T} -g_\alpha(t) &= \lim_{\alpha\downarrow 0} \max_{t\in T} \frac{|x(t) + \alpha h(t)| - |x(t)| + |x(t)| - f(x)}{\alpha} \\
&= \max\left\{ \lim_{\alpha\downarrow 0} \frac{|x(t) + \alpha h(t)| - |x(t)|}{\alpha} \,\middle|\, t \in T \text{ and } |x(t)| = f(x) \right\} \\
&= \max\{h(t)\,\mathrm{sign}(x(t)) \,|\, t \in T \text{ and } |x(t)| = f(x)\}.
\end{aligned}$$

□

5.3 Continuous Convergence and Stability for Convex Functions

It will turn out that the following notion is of central significance in this context:

Definition 5.3.1. Let X and Y be metric spaces. A sequence of real functions $(f_n)_{n\in\mathbb{N}}$ with $f_n : X \to Y$ is called *continuously convergent* to $f : X \to Y$, if for every sequence $(x_n)_{n\in\mathbb{N}}$ converging to a x_0

$$f_n(x_n) \to f(x_0)$$

holds.

The significance of continuous convergence for stability assertions is immediately apparent from the following simple principle:

Theorem 5.3.2. *Let X be a metric space and $(f_n : X \to \mathbb{R})_{n\in\mathbb{N}}$ a sequence of functions, which converges continuously to $f : X \to \mathbb{R}$. If for each $n \in \mathbb{N}$ the point x_n is a minimal solution of f_n on X, then every point of accumulation of the sequence $(x_n)_{n\in\mathbb{N}}$ is a minimal solution of f on X.*

Proof. Since x_n is a minimal solution of f_n, we have for all $x \in X$

$$f_n(x_n) \leq f_n(x). \tag{5.2}$$

Let x_0 be a point of accumulation of $(x_n)_n \in \mathbb{N}$, whose subsequence $(x_k)_{k \in \mathbb{N}}$ converges to x_0. Then the continuous convergence implies

$$f_k(x_k) \to f(x_0) \quad \text{and} \quad f_n(x) \to f(x).$$

Going to the limit in Inequality (5.2) preserves its validity and we obtain

$$f(x_0) \leq f(x).$$

□

We will see that for certain classes of functions pointwise convergence already implies continuous convergence. Among them are the convex functions. This will also hold for sums, products, differences and compositions of convex functions, because continuous convergence is inherited according to the following general principle.

Theorem 5.3.3. *Let X, Y, Z be metric spaces and $(F_n)_{n \in \mathbb{N}}$ with $F_n : X \to Y$ and $(G_n)_{n \in \mathbb{N}}$ with $G_n : Y \to Z$ sequences of continuously convergent functions. Then the sequence of compositions $(F_n \circ G_n)_{n \in \mathbb{N}}$ is continuously convergent.*

Proof. Let (F_n) to F and (G_n) to G continuously convergent. Let $(x_n)_n$ be a sequence in X convergent to x, then we have: $F_n(x_n) \to F(x)$ in Y and hence $G_n(F_n(x_n)) \to G(F(x))$ in Z. □

In order to be responsive to the above announcement about products and differences, we will mention the following special case:

Theorem 5.3.4. *Let $Y = \mathbb{R}^m$, $Z = \mathbb{R}$, $G : \mathbb{R}^m \to \mathbb{R}$ be continuous and $(F_n)_{n \in \mathbb{N}} = (f_n^{(1)}, \ldots, f_n^{(m)})_{n \in \mathbb{N}}$ continuously convergent. Then the sequence $(G(f_n^{(1)}, \ldots, f_n^{(m)}))_{n \in \mathbb{N}}$ is continuously convergent.*

At first we will explicate that the notion of continuous convergence is closely related to the notion of equicontinuity.

Continuous Convergence and Equicontinuity

Definition 5.3.5. Let X, Y be metric spaces and F a family of functions $f : X \to Y$.

(a) Let $x_0 \in X$. F is called *equicontinuous at* x_0, if for every neighborhood V of $f(x_0)$ there is a neighborhood U of x_0, such that for all $f \in F$ and all $x \in U$ we have

$$f(x) \in V.$$

(b) F is called equicontinuous, if F is equicontinuous at each $x_0 \in X$.

Theorem 5.3.6. *Let X, Y be metric spaces and $(f_n : X \to Y)_{n\in\mathbb{N}}$ a sequence of continuous functions, which converges pointwise to the function $f : X \to Y$. Then the following statements are equivalent:*

(a) *$\{f_n\}_{n\in\mathbb{N}}$ is equicontinuous,*

(b) *f is continuous and $\{f_n\}_{n\in\mathbb{N}}$ converges continuously to f,*

(c) *$\{f_n\}_{n\in\mathbb{N}}$ converges uniformly on compact subsets to f.*

Proof. (a) $\Rightarrow$ (b): Let $x_0 \in X$ and $\varepsilon > 0$. Then there is a $\alpha > 0$, such that for all $x \in K(x_0, \alpha)$ and all $n \in \mathbb{N}$ we have: $d(f_n(x), f_n(x_0)) \leq \varepsilon$.

The pointwise convergence implies for all $x \in K(x_0, \alpha)$: $d(f(x), f(x_0)) \leq \varepsilon$ and thus the continuity of f.

Let $x_n \to x_0$. For $n \geq n_0$ we have $x_n \in K(x_0, \alpha)$ and $d(f_n(x_0), f(x_0)) \leq \varepsilon$. Therefore

$$d(f_n(x_n), f(x_0)) \leq d(f_n(x_n), f_n(x_0)) + d(f_n(x_0), f(x_0)) \leq 2\varepsilon.$$

(b) $\Rightarrow$ (c): Suppose there is a compact subset K of X, on which the sequence (f_n) does not converge uniformly to f. Then we have

$$\exists \varepsilon > 0\ \forall n \in \mathbb{N}\ \exists k_n \in \mathbb{N},\ x_{k_n} \in K : d(f_{k_n}(x_{k_n}), f(x_{k_n})) \geq \varepsilon.$$

The sequence (k_n) has a strictly monotonically increasing subsequence (i_n), such that (x_{i_n}) converges to $\bar{x} \in K$. This leads to the contradiction

$$\varepsilon \leq d(f_{i_n}(x_{i_n}), f(x_{i_n})) \leq d(f_{i_n}(x_{i_n}), f(\bar{x})) + d(f(x_{i_n}), f(\bar{x})) \to 0.$$

(c) $\Rightarrow$ (a): Suppose the family $\{f_n\}$ is not equicontinuous at a point $x_0 \in X$, then

$$\exists \varepsilon > 0\ \forall n \in \mathbb{N}\ \exists k_n \in \mathbb{N},\ x_{k_n} \in K\left(x_0, \frac{1}{n}\right) :\ d(f_{k_n}(x_{k_n}), f_{k_n}(x_0)) \geq \varepsilon.$$

Since $x_{k_n} \to x_0$ and finitely many continuous functions are equicontinuous, the set $J = \{k_n\}$ contains infinitely many elements. Hence there exists in J a strictly monotonically increasing sequence (i_n). By assumption the sequence (f_n) converges uniformly on the compact set $\{x_{k_n}\}_{n\in\mathbb{N}} \cup \{x_0\}$ to f. Hence there is a $\bar{n} \in \mathbb{N}$, such that for all $n \geq \bar{n}$:

$$d(f_{i_n}(x_{i_n}), f(x_{i_n})) < \frac{\varepsilon}{4} \quad \text{and} \quad d(f_{i_n}(x_0), f(x_0)) < \frac{\varepsilon}{4}.$$

The function f is the uniform limit of continuous functions and hence continuous. Therefore

$$d(f(x_{i_n}), f(x_0)) < \frac{\varepsilon}{4},$$

provided that $\bar{n}$ was chosen large enough. From the triangle inequality it follows that for $n \geq \bar{n}$

$$\begin{aligned}\varepsilon &\leq d(f_{i_n}(x_{i_n}), f_{i_n}(x_0)) \\ &\leq d(f_{i_n}(x_{i_n}), f(x_{i_n})) + d(f(x_{i_n}), f(x_0) + d(f_{i_n}(x_0), f(x_0)) \\ &< \frac{\varepsilon}{4} + \frac{\varepsilon}{4} + \frac{\varepsilon}{4},\end{aligned}$$

a contradiction. □

For stability assertions of convex optimization we will show the following non-linear extension of the *uniform boundedness principle of Banach* about continuous linear operators to convex functions:

A pointwise bounded family of continuous convex functions is equicontinuous.

Definition 5.3.7. Let Y be a metric space. A family F of real functions on Y is called *lower semi-equicontinuous* (resp. *upper semi-equicontinuous*) at the point y_0, if for every $\varepsilon > 0$ there is a neighborhood V of y_0, such that for all $y \in V$ and all $f \in F$

$$f(y) - f(y_0) \geq -\varepsilon \quad \text{(resp. } f(y) - f(y_0) \leq \varepsilon)$$

holds.

We remark that the family F is *equicontinuous*, if F is lower semi-equicontinuous and upper semi-equicontinuous.

Theorem 5.3.8. *Let X be a normed space, U an open and convex subset of X and let F be a family of convex functions on U, which is pointwise bounded. Then the following statements are equivalent:*

(a) *F is equicontinuous on U*

(b) *F is upper semi-equicontinuous on U*

(c) *F is uniformly bounded from above on an open subset U_0 of U*

(d) *F is equicontinuous at a point in U.*

Proof. (a) $\Rightarrow$ (b) and (b) $\Rightarrow$ (c) follow from the definitions.

(c) $\Rightarrow$ (d): Let $x_0 \in U_0$ and $a > 0$, such that $K(x_0, a) \subset U_0$. Let $0 < \varepsilon < 1$ be given. If $y \in K(x_0, \varepsilon a)$, then there exists a $x \in K(x_0, a)$ with $y = (1 - \varepsilon)x_0 + \varepsilon x$. The convexity implies for all $f \in F$

$$f(y) \leq \varepsilon f(x) + (1 - \varepsilon) f(x_0) = f(x_0) + \varepsilon (f(x) - f(x_0)).$$

By assumption there is a $M > 0$ such that for all $x \in U_0$ and all $f \in F$: $f(x) \leq M$. If we put, using the pointwise boundedness, $\lambda := M - \inf\{f(x_0) \mid f \subset F\}$, then we obtain for all $y \in K(x_0, \varepsilon a)$

$$f(y) - f(x_0) \leq \varepsilon\lambda.$$

On the other hand for each $z \in K(x_0, \varepsilon a)$ there is a $x \in K(x_0, a)$ with

$$z = x_0 - \varepsilon(x - x_0) \quad \text{resp. } x_0 = \frac{1}{1+\varepsilon} z + \frac{\varepsilon}{1+\varepsilon} x.$$

The convexity implies for all $f \in F$

$$f(x_0) \leq \frac{1}{1+\varepsilon} f(z) + \frac{\varepsilon}{1+\varepsilon} f(x),$$

hence via multiplication of both sides by $1 + \varepsilon$:

$$f(z) - f(x_0) \geq \varepsilon(f(x_0) - f(x)) \geq \varepsilon(\inf\{f(x_0) \mid f \in F\} - M) = -\varepsilon\lambda.$$

Thus we obtain the equicontinuity of F at x_0.

(d) $\Rightarrow$ (a): Due to the above considerations it suffices to show that each point $x \in U$ has a neighborhood, on which F is uniformly bounded from above. Let F be equicontinuous at $y \in U$ and let $a > 0$, such that for all $z \in K(y, a) =: V$ and all $f \in F$

$$f(z) \leq f(y) + 1 \leq \sup\{f(y) \mid f \in F\} + 1 =: r$$

holds. (Here we have, of course, reused the pointwise boundedness). Let now $x \in U$ and $0 < \alpha < 1$ be chosen, such that $(1 + \alpha)x \in U$ and $V_\alpha := x + \frac{\alpha}{1+\alpha} V \subset U$ holds. For each $s \in V_\alpha$ there is apparently a $z \in V$ with $s = x + \frac{\alpha}{1+\alpha} z$ and it follows that

$$\begin{aligned} f(s) = f\left(x + \frac{\alpha}{1+\alpha} z\right) &\leq \frac{1}{1+\alpha} f((1+\alpha)x) + \frac{\alpha}{1+\alpha} f(z) \\ &\leq \frac{1}{1+\alpha} \sup\{f((1+\alpha)x) \mid f \in F\} + \frac{\alpha}{1+\alpha} r =: r' < \infty. \end{aligned}$$

□

For a family consisting of a single function, we immediately obtain

Corollary 5.3.9. *Let X be a normed space, U an open and convex subset of X and let $f : U \to \mathbb{R}$ be a convex function. Then the following statements are equivalent:*

(a) *f is continuous on U*

(b) *f is upper semi-continuous on U*

(c) *f is bounded from above on an open subset U_0 of U*

(d) *f is continuous at a point in U.*

Corollary 5.3.10. *Let X be a Banach space, U an open and convex subset of X. Let $(f_k : U \to \mathbb{R})_{k \in \mathbb{N}}$ be a sequence of continuous convex functions, which converges pointwise to the function $f : U \to \mathbb{R}$. Then f is a continuous convex function.*

Proof. From $f \leq \bar{f} := \sup_{k \in \mathbb{N}} f_k$ continuity of f follows using Theorems 5.3.13 and 5.3.9. □

As a consequence from Corollary 5.3.9 we obtain the following important and beautiful assertion that every convex function on $\mathbb{R}^n$ is continuous.

Theorem 5.3.11. *Let U be an open convex subset of $\mathbb{R}^n$. Then every convex function $f : U \to \mathbb{R}$ is continuous.*

Proof. Let w.l.o.g. $0 \in U$ and $0 < \alpha < 1$, such that the ℓ^1-ball

$$V := \left\{ x \in \mathbb{R}^n \,\middle|\, \sum_{i=1}^n |x_i| < \alpha \right\}$$

is contained in U. For $x \in V$ we have

$$x = \sum_{i=1}^n x_i e_i = \sum_{i=1}^n \frac{|x_i|}{\alpha} \operatorname{sign}(x_i)\alpha e_i + \left(1 - \sum_{i=1}^n \frac{|x_i|}{\alpha}\right) \cdot 0.$$

Then for all $x \in V$ we obtain the following inequality:

$$\begin{aligned} f(x) &\le \sum_{i=1}^n \frac{|x_i|}{\alpha} f(\operatorname{sign}(x_i)\alpha e_i) + \left(1 - \sum_{i=1}^n \frac{|x_i|}{\alpha}\right) f(0) \\ &\le \max(\{|f(\alpha e_i)|\}_1^n, \{|f(-\alpha e_i)|\}_1^n, f(0)). \end{aligned}$$

□

Equicontinuity of Convex Functions in Banach Space

In Banach spaces we have the

Theorem 5.3.12. *Let U be an open and convex subset of a Banach space X and $f : U \to \mathbb{R}$ a convex function. Then the following statements are equivalent:*

(a) *f is continuous*

(b) *f is lower semi-continuous*

(c) *f is upper semi-continuous.*

Proof. (b) $\Rightarrow$ (a): Suppose f is not continuous. Then f is by Theorem 5.3.8 unbounded from above on any open subset of U. Then for every $k \in \mathbb{N}$ the set $B_k := \{x \in U \mid f(x) > k\}$ is non-empty and also open, because f is lower semi-continuous.

Now we determine iteratively a sequence of closed non-empty balls: in B_1 we choose a ball U_1 with radius ≤ 1. If the k-th ball U_k with radius $\le \frac{1}{k}$ has been determined, we choose a non-empty closed ball U_{k+1} with radius $\le \frac{1}{k+1}$ in $B_{k+1} \cap \operatorname{Int}(U_k)$.

The sequence of centers (x_k) is a Cauchy sequence, since for all $k, p \in \mathbb{N}$ we have: $\|x_{k+p} - x_k\| \le \frac{1}{k}$. Since X is a Banach space, the sequence (x_k) converges to a $x^* \in X$. For $p \to \infty$ we then obtain

$$\|x^* - x_k\| \le \frac{1}{k},$$

i.e. for all $k \in \mathbb{N}$ we have $x^* \in U_k$. Since $U_k \subset B_k$ it follows that $x^* \in U$ and $f(x^*) = \infty$ in contradiction to f being finite on U. □

We now approach the central assertion for our stability considerations: the following theorem yields an extension of the theorem of Banach on uniform boundedness to families of convex functions.

Theorem 5.3.13 (Uniform boundedness for convex functions). *Let X be a Banach space, U an open convex subset of X and F a family of continuous convex functions $f : U \to \mathbb{R}$, which is pointwise bounded. Then F is equicontinuous. Moreover, the functions $x \mapsto \sup\{f(x) \mid f \in F\}$ and $x \mapsto \inf\{f(x) \mid f \in F\}$ are continuous.*

Proof. The function $\bar{f} := \sup_{f \in F} f$ being the supremum of continuous convex functions is lower semi-continuous and convex and according to Theorem 5.3.12 continuous. In particular $\bar{f}$ is bounded from above on a neighborhood. Due to Theorem 5.3.8 the family F is then equicontinuous.

Let $x_0 \in U$. Then for every $\varepsilon > 0$ there is a neighborhood V of x_0, such that for all $f \in F$ and all $x \in V$ we have

$$f(x) \geq f(x_0) - \varepsilon \geq \inf_{f \in F} f(x_0) - \varepsilon,$$

hence

$$\inf_{f \in F} f(x) \geq \inf_{f \in F} f(x_0) - \varepsilon,$$

i.e. the function $\inf_{f \in F} f$ is lower semi-continuous and being the infimum of continuous functions also upper semi-continuous.

In particular every point in U has a neighborhood, on which F is uniformly bounded. □

Remark 5.3.14. We obtain the theorem of Banach on uniform boundedness, if one assigns to a linear operator $A : X \to Y$ (X Banach space, Y normed space) the function $f_A : X \to \mathbb{R}$ with $f_A(x) := \|Ax\|$.

For the discussion of strong solvability in Chapter 8 we need the following

Theorem 5.3.15. *Let X be a Banach space and M a weakly bounded subset, then M is bounded.*

Proof. Let $x^* \in X^*$, then there is a number α_{x^*}, such that $|\langle x, x^* \rangle| \leq \alpha_{x^*}$ holds for all $x \in M$. If we consider M as a subset of X^{**}, then M is pointwise bounded on X^*. According to the theorem on uniform boundedness of Banach M is bounded in X^{**}. It is well known that the canonical mapping $\phi : X \to X^{**}$ with $x \mapsto \langle \cdot, x \rangle$ is an isometry, since

$$\|\phi(x)\| = \sup\{\langle x^*, x \rangle \mid \|x^*\| \leq 1\} \leq \|x\|.$$

On the other hand there is according to the theorem of Hahn–Banach for $x \in X$ a $x^* \in X^*$ with $\|x^*\| = 1$ and $\langle x^*, x \rangle = \|x\|$. Thus M is also bounded in X. □

Remark 5.3.16. For convex functions the above Theorem 5.3.13 permits to reduce the assertion of continuous convergence to pointwise convergence. In function spaces this assertion can often be performed successfully using the theorem of Banach–Steinhaus, where essentially the pointwise convergence on a dense subset is required. Due to the equicontinuity shown above this carries over to sequences of continuous convex functions.

Theorem 5.3.17 (Banach–Steinhaus for convex functions). *Let U be an open and convex subset of a Banach space. A sequence of continuous convex functions $(f_n : U \to \mathbb{R})_{n\in\mathbb{N}}$ converges pointwise to a continuous convex function f, if and only if the following two conditions are satisfied:*

(a) *the sequence $(f_n)_{n\in\mathbb{N}}$ is pointwise bounded*

(b) *the sequence $(f_n)_{n\in\mathbb{N}}$ converges pointwise on a dense subset D of U to f.*

Proof. The necessity is obvious. Let now be $x \in U$. By Theorem 5.3.13 the family $\{f_n\}_{n=1}^{\infty}$ is equicontinuous, i.e. there exists a neighborhood V of x, such that for all $n \in \mathbb{N}$ and $y \in V$

$$|f_n(x) - f_n(y)| \leq \varepsilon.$$

Since D is dense in U, we have $D \cap V \neq \emptyset$. Let $x' \in D \cap V$, then there is a $N \in \mathbb{N}$ with $|f_n(x') - f_m(x')| \leq \varepsilon$ for all $n, m \geq N$, hence for $n, m \geq N$

$$|f_n(x) - f_m(x)| \leq |f_n(x) - f_n(x')| + |f_n(x') - f_m(x')| + |f_m(x') - f_m(x)| \leq 3\varepsilon.$$

Therefore the limit $\lim_{n\to\infty} f_n(x) =: f(x)$ exists. But f is due to Corollary 5.3.10 convex and continuous. □

5.3.1 Stability Theorems

Pointwise convergence leads – according to Theorem 5.3.13 – to equicontinuity and hence by Theorem 5.3.6 to continuous convergence, which in turn implies the following stability assertion:

Theorem 5.3.18. *Let X be a Banach space, K a closed convex subset of X and $(f_n : X \to \mathbb{R})_{n\in\mathbb{N}}$ a sequence of continuous convex functions, which converges pointwise to a function $f : X \to \mathbb{R}$. Let x_n be a minimal solution of f_n on K. Then every point of accumulation of the sequence $(x_n)_{n\in\mathbb{N}}$ is a minimal solution of f on K.*

The essential message of the above stability Theorem 5.3.18 remains valid, if one admits that in each step the optimization is done on different domains, provided that these converge to the domain of the limit problem in the sense of Kuratowski (see Definition 5.1.5). This notion for the convergence of sets displays an analogy to the pointwise convergence of functions.

The continuous convergence of functions leads together with the Kuratowski convergence of the restriction sets to the following general stability principle:

Theorem 5.3.19. *Let X be a metric space, $(S_n)_{n\in\mathbb{N}}$ a sequence of subsets of X, which converges to $S \subset X$ in the sense of Kuratowski. Let $(f_n : X \to \mathbb{R})_{n\in\mathbb{N}}$ be continuously convergent to $f : X \to \mathbb{R}$. Then it follows that*

$$\overline{\lim}\, M(f_n, S_n) \subset M(f, S).$$

Proof. Let $x_{n_i} \in M(f_{n_i}, S_{n_i})$ and $x_{n_i} \to x$. Let $y \in S$. Then there exists a $(y_n \in S_n)_{n\in\mathbb{N}}$ with $y = \lim y_n$. Since $f_n \to f$ continuously, we have

$$f(x) = \lim_i f_{n_i}(x_{n_i}) \leq \lim_i f_{n_i}(y_{n_i}) = f(y). \qquad \square$$

The following variant of the stability theorem admits a weaker version of topological convergence:

Theorem 5.3.20. *Let $(f_n : X \to \mathbb{R})_{n\in\mathbb{N}}$ a sequence of functions, which converges continuously to $f : X \to \mathbb{R}$. Let $S \subset X$ and $(S_n)_{n\in\mathbb{N}}$ a sequence of subsets of X with*

$$\underline{\lim}_n S_n \supset S. \tag{5.3}$$

Let for each $n \in \mathbb{N}$ x_n^* be a minimal solution of f_n on S_n.

Then every point of accumulation of the sequence $(x_n^*)_{n\in\mathbb{N}}$ lying in S is a minimal solution of f on S.

Proof. Let $x_{n_i}^* \in M(f_{n_i}, S_{n_i})$ and $x_{n_i}^* \to x^* \in S$.

By (5.3) for each $x \in S$ there is a $n_0 \in \mathbb{N}$, such that for all $n \geq n_0$

$$x_n \in S_n \quad \text{and} \quad x_n \to x.$$

By definition of a minimal solution we have

$$f_{n_i}(x_{n_i}^*) \leq f_{n_i}(x_{n_i}).$$

The continuous convergence of $(f_n)_{n\in\mathbb{N}}$ yields

$$f(x^*) = \lim_{i\to\infty} f_{n_i}(x_{n_i}^*) \leq \lim_{i\to\infty} f_{n_i}(x_{n_i}) = f(x). \qquad \square$$

Theorem 5.3.21 (Stability Theorem of Convex Optimization). *Let X be a Banach space, U an open and convex subset of X and $(f_n : U \to \mathbb{R})_{n\in\mathbb{N}}$ a sequence of convex continuous functions, which converges pointwise to a function $f : U \to \mathbb{R}$. Furthermore let $(M_n)_{n\in\mathbb{N}}$ be a sequence of subsets of U with $U \supset M = \lim_n M_n$. Let x_n be a minimal solution of f_n on M_n. Then every point of accumulation of the sequence $(x_n)_{n\in\mathbb{N}}$ is a minimal solution of f on M.*

Corollary 5.3.22. *Let in addition*

(a) M_n *for all* $n \in \mathbb{N}$ *be closed*

(b) *exist an in* X *compact subset* K *of* U *such that* $M_n \subset K$ *for all* $n \in \mathbb{N}$.

Then the additional statements hold:

(a) *the sets* $\overline{\lim}_n M(f_n, M_n)$ *and* $M(f_n, M_n)$ *are non-empty*

(b) *from* $x_n \in M(f_n, M_n)$ *it follows that* $f(x_n) \to \inf f(M)$

(c) $\inf f_n(M_n) \to \inf f(M)$.

Corollary 5.3.23. *If in the above stability theorem we require instead of* $M = \lim_n M_n$ *only* $M \subset \overline{\lim}_n M_n$, *then still*

$$M \cap \overline{\lim} M(f_n, M_n) \subset M(f, M). \tag{5.4}$$

Proof of Corollaries 5.3.22 *and* 5.3.23. Let $x = \lim_i x_{n_i}$ with $x_{n_i} \in M(f_{n_i}, M_{n_i})$ and $x \in M$. Let $y \in M$ be arbitrarily chosen and $y = \lim_n y_n$ with $y_n \in M_n$. The sequence f_n converges by Theorem 5.3.6 and Theorem 5.3.8 continuously to f and hence

$$f(x) = \lim_i f_{n_i}(x_{n_i}) \le \lim_i f_{n_i}(y_{n_i}) = f(y).$$

The compactness of the sets M_n and K yields (a).

Let $x_n \in M(f_n, M_n)$ and let (x_k) be a convergent subsequence converging to $x_0 \in M$, then the continuous convergence implies $\lim_k f_k(x_k) = f(x_0) = \inf f(M)$. Thus every subsequence of $(f_n(x_n))$ has a subsequence converging to $\inf f(M)$ and hence (c). Using the continuity of f we obtain (b) in a similar manner. □

The requirement $M \subset U$ cannot be omitted, as the following example shows:

Example 5.3.24. Let $U := (0, 1)$, let $f_n := |\cdot - (1 - \frac{2}{n}|$ and $f := |\cdot - 1|$. Let further $M_n := [\frac{1}{n}, 1 - \frac{1}{n}]$, then $\lim M_n = [0, 1] = M$. The sequence of minimal solutions $x_n = 1 - \frac{2}{n}$ of f_n on M_n converges to 1, outside of the domain of definition of f.

Theorem 5.3.25 (Stability Theorem of Convex Optimization in $\mathbb{R}^n$). *Let* U *be an open and convex subset of a finite dimensional normed space* X *and let* $(f_k : U \to \mathbb{R})_{k\in\mathbb{N}}$ *be a sequence of convex functions, which converges pointwise to a function* $f : U \to \mathbb{R}$. *Let further* $(S_k)_{k\in\mathbb{N}}$ *be a sequence of closed convex subsets in* X *with* $S = \lim_k S_k$. *For* $\tilde{S} := U \cap S$ *let the set of minimal solutions of* f *on* $\tilde{S}$ *be non-empty and bounded. Then for* $\tilde{S}_k := U \cap S_k$ *the following statements hold:*

(a) *for large* $k \in \mathbb{N}$ *the set of minimal solutions* $M(f_k, \tilde{S}_k)$ *is non-empty*

(b) $\overline{\lim}_k M(f_k, \tilde{S}_k)$ *is non-empty and from a* $k_0 \in \mathbb{N}$ *on uniformly bounded, i.e.* $\bigcup_{k \ge k_0} M(f_k, \tilde{S}_k)$ *is bounded*

(c) $\overline{\lim}_k M(f_k, \tilde{S}_k) \subset M(f, \tilde{S})$

(d) $\inf f_k(\tilde{S}_k) \to \inf f(\tilde{S})$

(e) *from* $x_k \in M(f_k, \tilde{S}_k)$ *for* $k \in \mathbb{N}$ *it follows that* $f(x_k) \to \inf f(\tilde{S})$

(f) *if* $M(f, \tilde{S})$ *consists of a single point* $\{x_0\}$, *then* $x_k \in M(f_k, \tilde{S}_k)$ *implies for* $k \in \mathbb{N}$ *the convergence* $x_k \to x_0$.

Proof. Let $x \in M(f, \tilde{S})$ and $r = f(x) = \inf f(\tilde{S})$. Since $M(f, \tilde{S})$ is compact, there is a ball $K(0, d)$ with $d > 0$ such that

$$A := M(f, \tilde{S}) + \overline{K}(0, d) \subset U$$

holds. Furthermore there exists an $\alpha > 0$, such that for a $n_0 \in \mathbb{N}$ and all $n \geq n_0$

$$H_n := \{u \,|\, u \in \tilde{S}_n,\ f_n(u) \leq r + \alpha\} \subset A$$

holds, because otherwise there is a strictly monotone sequence (k_n) in $\mathbb{N}$ and $u_{k_n} \in \tilde{S}_{k_n}$ with $f_{k_n}(u_{k_n}) \leq r + \frac{1}{k_n}$ but $u_{k_n} \notin A$. Since (S_n) converges to S, there is a sequence $(y_n \in S_n)$ with $y_n \to x$. For large k_n there exists an intersection z_{k_n} of the interval $[y_{k_n}, u_{k_n}]$ with the set

$$D := \Big\{w \in A \,|\, \inf_{v \in M(f, \tilde{S})} \|w - v\| = d\Big\}.$$

For $k_n \geq n_0$ we have $y_{k_n} \in A \subset U$ and hence $y_{k_n} \in \tilde{S}_{k_n}$. Due to the convexity of $\tilde{S}_{k_n}$ we obtain $z_{k_n} \in \tilde{S}_{k_n}$. Since D is non-empty and compact, the sequence (z_{k_n}) contains a subsequence $(z_l)_{l \in L}$ convergent to $\bar{z} \in D$. As S_n is convex and $\lim S_n = S$, we have $\bar{z} \in S$, on the other hand $\bar{z} \in A \subset U$, i.e. $\bar{z} \in \tilde{S}$. For each $l \in L$ there is $\alpha_l \in [0, 1]$ with $z_l = \alpha_l u_l + (1 - \alpha_l) y_l$, and we obtain

$$f_l(z_l) = \alpha_l f_l(u_l) + (1 - \alpha_l) f_l(y_l) \leq \alpha_l \left(r + \frac{1}{l} \right) + (1 - \alpha_l) f_l(y_l).$$

The continuous convergence implies $f_l(y_l) \to f(x)$ and $f_l(z_l) \to f(\bar{z})$, hence $f(\bar{z}) \leq r$. Thus we have $\bar{z} \in M(f, \tilde{S}) \cap D$, contradicting the definition of D.

For large n the sets H_n are non-empty, because for every sequence $(v_n \in \tilde{S}_n)$ convergent to x we have: $f_n(v_n) \to f(x) = r$. Therefore minimization of f_n on $\tilde{S}_n$ can be restricted to H_n.

H_n – being a subset of A – is apparently bounded but also closed: for let (h_k) be a sequence in H_n with $h_k \to \bar{h}$, then $\bar{h} \in A \subset U$ and $\bar{h} \in S_n$, since S_n is closed, i.e. $\bar{h} \in \tilde{S}_n$. Due to the closedness of the level sets of f_n it follows that $\bar{h} \in H_n$. For $K := A$ we obtain with Corollary 5.3.22 to Stability Theorem 5.3.21 (a) to (e). Assertion (f) follows from the fact that (x_n) is bounded and every point of accumulation of (x_n) is equal to x_0. □

Remark. The limit $S = \lim_k S_k$ is a closed subset of X. If, instead of the above limit, one would consider $\lim_k \tilde{S}_k$, this limit is – if it exists – not necessarily a subset of U. (Hence the careful wording in the above theorem.)

Example 5.3.26. Let $U := \{(x,y) \in \mathbb{R}^2 \,|\, x > 0, y > 0\}$, let $f : U \to \mathbb{R}$ be defined by $f(x,y) = \frac{1}{x} + \frac{1}{y}$, let further $f_k := f + \frac{1}{k}$ and $S_k := \{(x,y) \in \mathbb{R}^2 \,|\, x^2 + y^2 \leq (1 - \frac{1}{k})^2\}$. Then $\lim S_k = \{(x,y) \in \mathbb{R}^2 \,|\, x^2 + y^2 \leq 1\} = S$. However $\lim \tilde{S}_k = \tilde{S}$ does not hold, since the limit set is necessarily closed.

A consequence of the above theorem is the following

Theorem 5.3.27. *Let X be a finite dimensional normed space and let $(f_k : X \to \overline{\mathbb{R}})_{k\in\mathbb{N}}$ a sequence of lower semi-continuous convex functions, whose epigraphs converge in the sense of Kuratowski to the epigraph of the function $f_0 : X \to \overline{\mathbb{R}}$. Let further K be a closed convex subset of X and the set of minimal solutions of f_0 on K be non-empty and bounded. Then we have*

(a) *for large $k \in \mathbb{N}$ the set of minimal solutions $M(f_k, \tilde{K})$ is non-empty*

(b) *$\overline{\lim}_k M(f_k, K)$ is non-empty and from a $k_0 \in \mathbb{N}$ on uniformly bounded, i.e. $\bigcup_{k\geq k_0} M(f_k, K)$ is bounded*

(c) $\overline{\lim}_k M(f_k, K) \subset M(f_0, K)$

(d) $\inf f_k(K) \to \inf f(K)$

(e) *from $x_k \in M(f_k, K)$ for $k \in \mathbb{N}$ it follows that $f(x_k) \to \inf f_0(K)$*

(f) *if $M(f_0, K)$ consists of a single point $\{x_0\}$, then $x_k \in M(f_k, K)$ for $k \in \mathbb{N}$ implies the convergence $x_k \to x_0$.*

Proof. We define

$$\begin{aligned} g : K \times \mathbb{R} &\to \mathbb{R} \\ (x,t) &\mapsto g(x,t) = t. \end{aligned}$$

We consider the sequence of minimization problems

$$\inf\{g(x,t) \,|\, (x,t) \in \text{Epi}(f_k)\}.$$

Apparently we have for all $k \in \mathbb{N}_0$

$$(x, \inf(f_k, K)) \in M(g, K \times \mathbb{R}) \Leftrightarrow x \in M(f_k, K).$$

In particular the boundedness of the non-empty set $M(f_0, K)$ carries over to the boundedness of the set $M(g, K \times \mathbb{R})$. The sequence of the functions to be minimized is in this context constant and equal to g and hence obviously pointwise convergent.

Since the f_k are lower semi-continuous, the sets $\text{Epi}(f_k)$ are closed and thus the conditions of the previous theorem are satisfied. □

Kuratowski Convergence of Level Sets

The level sets of pointwise convergent sequences of convex functions turn out to be Kuratowski convergent, if the level set of the limit function satisfies a *Slater condition*:

Theorem 5.3.28. *Let U be an open subset of a Banach space and $(h_n)_{n\in\mathbb{N}}$ with $h_n : U \to \mathbb{R}$ be a sequence of continuous convex functions, which converge pointwise to the convex function $h : U \to \mathbb{R}$. Let there exist a $\bar{x} \in U$ with $h(\bar{x}) < 0$, then the sequence $(S_n)_{n\in\mathbb{N}}$ of the level sets $S_n := \{x \in U \mid h_n(x) \le 0\}$ converges in the sense of Kuratowski to the level set $S := \{x \in U \mid h(x) \le 0\}$.*

Proof. Let $x_{n_k} \in S_{n_k}$ with $x_{n_k} \to x$, then the continuous convergence implies: $h_{n_k}(x_{n_k}) \to h(x) \le 0$, hence $x \in S$. Therefore $\overline{\lim}\, S_n \subset S$.

Conversely let now $x \in S$. If $h(x) < 0$, then due to the pointwise convergence we have $h_n(x) < 0$ for $n > N$, i.e. $x \in S_n$ for $n > N$. We obtain

$$\mathrm{Int}(S) \subset \underline{\lim}\, S_n.$$

To complete the proof we show that the boundary points of S are also in $\underline{\lim}\, S_n$: let $h(\bar{x}) < 0$ and $h(x_0) = 0$. Consider the sequence $(h_n(x_0))$: if $h_n(x_0) \le 0$, put $x_n := x_0$, if $h_n(x_0) > 0$, then due to the intermediate value theorem there is a $x_n \in [\bar{x}, x_0)$ with $h_n(x_n) = 0$ for $n > N$, where N is chosen such that $h_n(\bar{x}) < 0$ for $n > N$. In both cases $x_n \in S_n$ for $n > N$. Suppose there is a subsequence x_{n_k}, which converges to a $\tilde{x} \in [\bar{x}, x_0)$, then due to continuous convergence $h_{n_k}(x_{n_k}) \to h(\tilde{x}) < 0$. But by construction we then have $x_{n_k} = x_0$ for k large enough, a contradiction. Therefore $x_n \to x_0$ and hence $x_0 \in \underline{\lim}\, S_n$. □

We need the following lemma for the stability assertion of the subsequent theorem:

Lemma 5.3.29. *Let U be an arbitrary set, T a compact metric space and $g : T \times U \to \mathbb{R}$ with $g(\cdot, z)$ continuous on T for all $z \in U$. Let further $(T_n)_{n\in\mathbb{N}}$ be a sequence of compact subsets of T, converging in the sense of Kuratowski to T_0. Let now $h_n : U \to \mathbb{R}$ be defined by $h_n(z) := \max_{t\in T_n} g(t, z)$ for all $z \in U$ and correspondingly $h_0 : U \to \mathbb{R}$ by $h_0(z) := \max_{t\in T_0} g(t, z)$ for all $z \in U$. Then we obtain*

$$\lim_{n\to\infty} h_n(z) = h_0(z) \quad \textit{for all } z \in U,$$

i.e. the sequence $(h_n)_{n\in\mathbb{N}}$ converges pointwise on U to h_0.

Proof. Let $\tau_0 \in T_0$ be chosen, such that $g(\tau_0, z) = h_0(z)$. The Kuratowski convergence implies the existence of a sequence $(t_n)_{n\in\mathbb{N}}$ with $t_n \in T_n$ for all $n \in \mathbb{N}$, which converges to τ_0. From the continuity of $g(\cdot, z)$ and $t_n \in T_n \subset T$ we obtain: $g(t_n, z) \to_{n\to\infty} g(\tau_0, z) = h_0(z)$ i.e.

$$\underline{\lim}\, h_n(z) \ge h_0(z).$$

On the other hand let $\tau_n \in T_n$ with $g(\tau_n, z) = h_n(z)$ and $s := \overline{\lim}\, g(\tau_n, z)$. Then there is a subsequence $(\tau_{n_k})_{k\in\mathbb{N}}$ converging to a $\tau \in T$, such that $\lim_{k\to\infty} g(\tau_{n_k}, z) = s$. The continuity of $g(\cdot, z)$ implies $\lim_{k\to\infty} g(\tau_{n_k}, z) = g(\tau, z)$. But the Kuratowski convergence yields $\tau \in T_0$, i.e.

$$g(\tau_{n_k}, z) = h_{n_k}(z) \xrightarrow{k\to\infty} s = g(\tau, z) \leq h_0(z). \qquad \square$$

Theorem 5.3.30. *Let U be an open subset of a Banach space, T a compact metric space and g a real-valued mapping on $T \times U$. Let $g(\cdot, z)$ be continuous on T for all $z \in U$ and $g(t, \cdot)$ convex for all $t \in T$. Let further $(T_n)_{n\in\mathbb{N}}$ be a sequence of compact subsets of T, which converges in the sense of Kuratowski to T_0, and let $(S_n)_{n\in\mathbb{N}_0}$ be a sequence of subsets of U defined by*

$$S_n := \{z \in U \mid g(t, z) \leq 0 \text{ for all } t \in T_n\}.$$

Let now $(f_n : U \to \mathbb{R})_{n\in\mathbb{N}_0}$ be a sequence of convex functions, which converges pointwise to f_0 on all of U. Then the following stability statement holds:

$$\overline{\lim_{n\to\infty}}\, M(f_n, S_n) \subset M(f_0, S_0).$$

The above theorem can be applied to semi-infinite optimization (see corresponding section in Chapter 2).

5.4 Convex Operators

Definition 5.4.1. Let X be a real vector space. A subset P of X is called a *convex cone* in X, if P has the following properties:

(a) $0 \in P$

(b) $\forall \alpha \in \mathbb{R} \forall x \in P : \alpha \geq 0 \Rightarrow \alpha x \in P$

(c) $\forall x_1, x_2 \in P : x_1 + x_2 \in P$.

A relation $\leq$ is called an *order* on X, if $\leq$ has the following properties:

(a) $\leq$ is reflexive, i.e. $\forall x \in X : x \leq x$

(b) $\leq$ is transitive, i.e. $\forall x, y, z \in X : x \leq y$ and $y \leq z \Rightarrow x \leq z$

(c) $\leq$ is compatible with vector addition, i.e. $\forall x, y, z \in X : x \leq y \Rightarrow x+z \leq y+z$

(d) $\leq$ is compatible with scalar multiplication, i.e. $\forall \alpha \in \mathbb{R} \forall x, y \in X : 0 \leq \alpha$ and $x \leq y \Rightarrow \alpha x \leq \alpha y$.

If P is a convex cone in X resp. $\leq$ an order on X, then we denote the pair (X, P) resp. $(X, \leq)$ as an *ordered vector space*.

Remark 5.4.2. Directly from the definition follows:

(a) If P is a convex cone in X, then the relation $\leq_P$, defined by

$$\forall x, y \in X : x \leq_P y :\Leftrightarrow y - x \in P$$

is an order on X.

(b) If $\leq$ is an order on X, then the set

$$P := \{x \in X \,|\, 0 \leq x\}$$

is a convex cone.

Hence there is a one-to-one correspondence between an order on X and a convex cone in X.

Example 5.4.3. Let (T, Σ, μ) be a measure space and $L^\Phi(\mu)$ the corresponding Orlicz space, then the cone

$$P := \{x \in L^\Phi(\mu) \,|\, x(t) \geq 0 \ \mu\text{-almost everywhere}\}$$

is called the natural cone.

In problems, where not only the order but also topological properties play a role, the notion of normal cones becomes important. The natural cones of the function spaces, that are particulary relevant for applications, are normal but do not have an interior.

Definition 5.4.4. Let A be a subset of a vector space Y ordered by a convex cone C. By the full hull $[A]_C$ of A we denote

$$[A]_C := \{z \in Y \,|\, x \leq_C z \leq_C y \text{ for } x \in A, y \in A\}.$$

Hence $[A]_C = (A + C) \cap (A - C)$. A is called *full*, if $A = [A]_C$.

A convex cone C is called *normal*, if the full hull $[B]_C$ of the unit ball B is bounded. A family F of convex cones is called *uniformly normal*, if the union

$$\bigcup_{C \in F} [B]_C$$

is bounded.

A criterion for this is

Theorem 5.4.5. *Let* $R := \{\|z\| \,|\, \exists C \in F, y \in B \text{ such that } 0 \leq_C z \leq_C y\}$. *If* R *is bounded, then* F *is uniformly normal.*

Proof. Let $x \in \bigcup_{C\in F}[B]_C$. Then there are $y_1, y_2 \in B$ and a $C \in F$, such that $y_1 \leq_C x \leq_C y_2$ or $0 \leq_C x - y_1 \leq_C y_2 - y_1$. Let r be an upper bound of R. From $y_2 - y_1 \in 2B$ it follows that

$$\frac{\|x - y_1\|}{2} \leq r \quad \text{and hence } \|x\| \leq 2r + 1. \qquad \square$$

Examples for normal cones C in normed spaces are the natural cones in

(a) $\mathbb{R}^n$

(b) $C(T)$, where T is a compact metric space

(c) $L^\Phi(\mu)$.

Proof. Ad (a): For the closed unit ball B w.r.t. the maximum norm in $\mathbb{R}^n$ we even have $B = [B]_C$.

Ad (b) and (c): Let $y \in B$, then $0 \leq_C z \leq_C y$ implies due to the monotonicity of the norm $\|z\| \leq \|y\| \leq 1$. The previous theorem yields the assertion. $\square$

Convex Mappings

Let X and Y be vector spaces and C a cone in Y. The mapping $A : X \to Y$ is called C-convex, if for all $0 \leq \alpha \leq 1$ and all $u, v \in X$

$$A(\alpha u + (1-\alpha)v) \leq_C \alpha A(u) + (1-\alpha)A(v).$$

Example 5.4.6. $Y = \mathbb{R}^m$, C the natural cone in $\mathbb{R}^m$ and for $i = 1, \ldots, m$ $f_i : X \to \mathbb{R}$ convex. Then $A = (f_1, \ldots, f_m)$ is a C-convex mapping from X to $\mathbb{R}^m$.

Uniform Boundedness

The following theorem is a generalization of the theorem of Banach on uniform boundedness to families of convex operators.

Theorem 5.4.7. *Let Q be a convex and open subset of a Banach space X and Y a normed space. Let further $\{C_i\}_{i\in I}$ a family of uniformly normal cones in Y and $A_i : Q \to Y$ a C_i-convex continuous mapping for all $i \in I$. If the family $\{A_i\}_{i\in I}$ is pointwise norm-bounded, then $\{A_i\}_{i\in I}$ locally uniformly Lipschitz continuous, i.e. for each $x \in Q$ there is a neighborhood U of x and a number $L > 0$, such that for all $u, v \in U$ and all $i \in I$ we have*

$$\|A_i(u) - A_i(v)\| \leq L\|u - v\|.$$

Proof. The family $\{A_i\}_{i\in I}$ is pointwise norm-bounded, i.e.

$$s(x) := \sup_{i\in I} \|A_i(x)\| < \infty \quad \text{for all } x \in Q.$$

At first we show: the function $s : Q \to \mathbb{R}$ is norm-bounded on an open ball Q_1: otherwise for each $k \in \mathbb{N}$ the set

$$D_k := \{x \in Q \mid s(x) > k\}$$

would be dense in Q. Being the supremum of continuous functions s is lower semi-continuous and hence D_k open for all $k \in \mathbb{N}$. Every Banach space is of second Baire category (see [113], p. 27), and hence

$$\bigcap_{k=1}^{\infty} D_k \neq \emptyset.$$

But $y_0 \in \bigcap_{k=1}^{\infty} D_k$ contradicts $s(y_0) < \infty$.

In the next step we show that every point $x \in Q$ has a neighborhood, on which s is bounded. Let w.l.o.g. 0 be the center of Q_1. Since Q is open, there exists a $0 < \alpha < 1$, such that $(1+\alpha)x \in Q$ and $U := \frac{\alpha}{1+\alpha} Q_1 + x \subset Q$. Let $x' \in U$, i.e. $x' = x + \frac{\alpha}{1+\alpha} z$ with $z \in Q_1$. Then we obtain

$$\begin{aligned} A_i(x') &= A_i\left(\frac{1+\alpha}{1+\alpha} x + \frac{\alpha}{1+\alpha} z\right) \\ &\leq_{C_i} \frac{1}{1+\alpha} A_i((1+\alpha)x) + \frac{\alpha}{1+\alpha} A_i(z) =: \beta_i(z). \end{aligned}$$

On the other hand

$$\begin{aligned} A_i(x') &= (1+\alpha)\left(\frac{1}{1+\alpha} A_i(x') + \frac{\alpha}{1+\alpha} A_i\left(-\frac{z}{1+\alpha}\right)\right) - \alpha A_i\left(-\frac{z}{1+\alpha}\right) \\ &\geq_{C_i} (1+\alpha)\left(A_i\left(\frac{x'}{1+\alpha} - \frac{\alpha}{(1+\alpha)^2} z\right)\right) - \alpha A_i\left(-\frac{z}{1+\alpha}\right) \\ &= (1+\alpha) A_i\left(\frac{x}{1+\alpha}\right) - \alpha A_i\left(-\frac{z}{1+\alpha}\right) =: \alpha_i(z). \end{aligned}$$

Since $\{A_i\}$ is on Q pointwise norm-bounded and on Q_1 uniformly norm-bounded, there exists a number $r > 0$, such that for all $z \in Q_1$ and all $i \in I$ we have

$$\alpha_i(z), \beta_i(z) \in K(0, r).$$

The family $\{C_i\}_{i \in I}$ is uniformly normal. Therefore there is a ball $K(0, R)$ with $[K(0, r)]_{C_i} \subset K(0, R)$ and hence also

$$A_i(x') \in [\alpha_i(z), \beta_i(z)]_{C_i} \subset K(0, R),$$

i.e. $\|A_i(x')\| \leq R$ for all $i \in I$ and all $x' \in U$.

In the final step we turn our attention to the uniform Lipschitz continuity.

Let B be the unit ball in X and let $\delta > 0$ be chosen such that s is bounded on $x + \delta B + \delta B \subset Q$, i.e. there is a $l > 0$ with

$$s(x + \delta B + \delta B) \subset [0, l]. \tag{5.5}$$

For $y_1, y_2 \in x + \delta B$ and $y_1 \neq y_2$ we have

$$z := y_1 + \frac{\delta(y_1 - y_2)}{\|y_1 - y_2\|} \in x + \delta B + \delta B.$$

Let $\lambda := \frac{\|y_1 - y_2\|}{\delta + \|y_1 - y_2\|}$, then, because of $y_1 = (1 - \lambda)y_2 + \lambda z$

$$A_i(y_1) \leq_{C_i} (1 - \lambda)A_i(y_2) + \lambda A_i(z) = A_i(y_2) + \lambda(A_i(z) - A_i(y_2)),$$

i.e.

$$A_i(y_1) - A_i(y_2) \leq_{C_i} \lambda(A_i(z) - A_i(y_2)).$$

Correspondingly for $v := y_2 + \frac{\delta(y_1 - y_2)}{\|y_1 - y_2\|} \in x + \delta B + \delta B$ we obtain

$$A_i(y_2) - A_i(y_1) \leq_{C_i} \lambda(A_i(v) - A_i(y_1)),$$

i.e.

$$A_i(y_1) - A_i(y_2) \in \lambda[A_i(y_1) - A_i(v), A_i(z) - A_i(y_2)]_{C_i}.$$

By (5.5) we have for all $y_1, y_2 \in x + \delta B$ and all $i \in I$ both

$$A_i(y_1) - A_i(v) \quad \text{and} \quad A_i(z) - A_i(y_2) \in K(0, 2l).$$

Since $\{C_i\}$ is uniformly normal, there exists a ball $K(0, l_1)$, such that $[K(0, 2l)]_{C_i} \subset K(0, l_1)$ for all $i \in I$. Thus

$$A_i(y_2) - A_i(y_1) \in \lambda K(0, l_1),$$

i.e.

$$\|A_i(y_2) - A_i(y_1)\| \leq \lambda \cdot l_1 \leq \frac{\|y_1 - y_2\|}{\delta} \cdot l_1 = L\|y_1 - y_2\|$$

with $L = \frac{l_1}{\delta}$. □

Corollary 5.4.8. *Let $\{A_i\}_{i \in I}$ as in the above theorem, then $\{A_i\}_{i \in I}$ is equicontinuous.*

Component-wise Convex Mappings

The theorem proved in 5.4.7 can be extended to component-wise convex mappings.

Theorem 5.4.9. *Let for each $j \in \{1, \dots, n\}$ X_j be a Banach space and U_j an open and convex subset of X_j. Let Y be a normed space, containing the uniformly normal family $\{C_{ij} |\, i \in I, j = \{1, \dots, n\}\}$ of convex cones.*

Furthermore let a family of mappings $F = \{A_i : U_1 \times \cdots \times U_n \to Y\}_{i \in I}$ be given, which is pointwise bounded and has the property that for all $i \in I$ and $j \in \{1, \dots, n\}$ the component $A_{ij} : U_j \to Y$ is continuous and C_{ij}-convex.

Then every point in $U_1 \times \cdots \times U_n$ has a neighborhood U, on which F is uniformly bounded and uniformly Lipschitz continuous, i.e. there is a $L > 0$ such that for all $u, v \in U$ and all $i \in I$

$$\|A_i(u) - A_i(v)\| \le L\|u - v\|$$

holds.

Proof. The proof is performed by complete induction over n. Theorem 5.4.7 yields the start of the induction for $n = 1$.

We assume, the assertion holds for $n - 1$.

For the uniform boundedness it apparently suffices to show the following property: for all sequences $x_k = (x_{k,1}, \dots, x_{k,n})$, which converge to a $(\bar{x}_1, \dots, \bar{x}_n)$, and all sequences $(A_k)_{k \in \mathbb{N}}$ in F the sequence $(\|A_k(x_k)\|)_{k \in \mathbb{N}}$ is bounded.

Let the norm in $X_1 \times \cdots \times X_n$ be given by $\|\cdot\|_{X_1} + \cdots + \|\cdot\|_{X_n}$. For all $z \in \tilde{U} = U_1 \times \cdots \times U_{n-1}$ the sequence $\{A_k(z, \cdot) \,|\, k \in \mathbb{N}\}$ is pointwise bounded and by Theorem 5.4.7 equicontinuous at $\bar{x}_n$. Since $x_{k,n} \to_{k \to \infty} \bar{x}_n$ we have for all $z \in \tilde{U}$

$$\{A_k(z, x_{k,n})\}_{k \in \mathbb{N}} \text{ is bounded,} \tag{5.6}$$

i.e. the family $\{A_k(\cdot, x_{k,n})\}_{k \in \mathbb{N}}$ is pointwise bounded and by induction hypothesis equicontinuous at $(\bar{x}_1, \dots, \bar{x}_{n-1})$.

Since $(x_{k,1}, \dots, x_{k,n-1}) \to (\bar{x}_1, \dots, \bar{x}_{n-1})$ we have

$$A_k(x_{k,1}, \dots, x_{k,n}) - A_k(\bar{x}_1, \dots, \bar{x}_{n-1}, x_{k,n}) \to 0.$$

Using Statement (5.6) the uniform boundedness follows.

Therefore there exist open neighborhoods V_j in U_j, $j \in \{1, \dots, n\}$ and $\alpha \in \mathbb{R}$, such that for all $v_j \in V_j$ and all $A \in F$

$$\|A(v_1, \dots, v_n)\| \le \alpha.$$

Let $Q := V_1 \times \cdots \times V_{n-1}$. We consider the family $\{A(\cdot, v) : Q \to Y \,|\, A \in F, v \in V_n\}$. It is pointwise bounded. Let $x_0 \in Q$ and $x_n \in V_n$. By the induction hypothesis there exists a neighborhood $W \subset Q$ of x_0 and a $L_1 > 0$, such that for all $u_1, u_2 \in W$ and all $v \in V_n$

$$\|A(u_1, v) - A(u_2, v)\| \le L_1\|u_1 - u_2\|$$

holds. In analogy to the above the family $\{A(w, \cdot) : V_n \to Y \mid A \in F, w \in V_1 \times \cdots \times V_{n-1}\}$ is pointwise bounded, and by Theorem 5.4.7 there exists a neighborhood $\tilde{V}_n$ of x_n with $\tilde{V} \subset V_n$ and $L_2 > 0$, such that for all $v_1, v_2 \in \tilde{V}$ and $u \in Q$

$$\|A(u, v_1) - A(u, v_2)\| \le L_2\|v_1 - v_2\|.$$

For all $(x, y), (u, v) \in W \times \tilde{V}$ and all $A \in F$ we then obtain with $L := \max\{L_1, L_2\}$:

$$\begin{aligned}\|A(u, v) - A(x, y)\| &\le \|A(u, v) - A(x, v)\| + \|A(x, v) - A(x, y)\|\\ &\le L(\|u - x\| + \|v - y\|) = L\|(u, v) - (x, y)\|. \qquad \square\end{aligned}$$

5.5 Quantitative Stability Considerations in $\mathbb{R}^n$

In order to better appreciate the subsequent theorem, we will draw a connection to the numerical realization of the Polya algorithm: when solving the approximating problems, one can use the norm of the derivative of the function to be minimized as a stop criterion. It will turn out that stability is essentially preserved if the approximating problems are only solved approximatively by a $\tilde{x}_k$, if the sequence $(\|\nabla f_k(\tilde{x}_k)\|)$ tends to 0.

Theorem 5.5.1. *Let $(f_k : \mathbb{R}^n \to \mathbb{R})_{k\in\mathbb{N}}$ be a sequence of differentiable convex functions, which converge pointwise to a function $f : \mathbb{R}^n \to \mathbb{R}$. Let further the set of minimal solutions of f on $\mathbb{R}^n$ be non-empty and bounded and let $(\tilde{x}_k)_{k\in\mathbb{N}}$ be a sequence in $\mathbb{R}^n$ with the property*

$$\lim_{k\to\infty} \|\nabla f_k(\tilde{x}_k)\| = 0.$$

Then

(a) *The set of points of accumulation of the sequence $(\tilde{x}_k)$ is non-empty and contained in $M(f, \mathbb{R}^n)$.*
(b) $f_k(\tilde{x}_k) \to \inf f(\mathbb{R}^n)$.
(c) $f(\tilde{x}_k) \to \inf f(\mathbb{R}^n)$.
(d) *Let Q be an open bounded superset of $M(f, \mathbb{R}^n)$ and*

$$\varepsilon_k := \sup_{x\in Q} |f(x) - f_k(x)|,$$

then for k sufficiently large $\tilde{x}_k \in Q$ and

$$|\inf f(\mathbb{R}^n) - f(\tilde{x}_k)| = O(\varepsilon_k) + O(\|\nabla f_k(\tilde{x}_k)\|).$$

Proof. Let $\varepsilon > 0$ be given, $C := \{x \in \mathbb{R}^n \mid d(x, M(f, \mathbb{R}^n)) \geq \varepsilon\}$ and $S := \{x \in \mathbb{R}^n \mid d(x, M(f, \mathbb{R}^n)) = \varepsilon\}$. Furthermore let x_0 be a minimal solution of f. Suppose there is a subsequence $(\tilde{x}_i)$ of $(\tilde{x}_k)$, which is contained in C. Then there are numbers $\lambda_i \in (0, 1]$ with $\lambda_i \tilde{x}_i + (1 - \lambda_i)x_0 \in S$. Since the derivative of a convex function is a monotone mapping, we have

$$\langle \nabla f_i(\tilde{x}_i) - \nabla f_i(\lambda_i \tilde{x}_i + (1 - \lambda_i)x_0), (1 - \lambda_i)(\tilde{x}_i - x_0)\rangle \geq 0.$$

Therefore

$$\begin{aligned}\left\langle \nabla f_i(\tilde{x}_i), \frac{\tilde{x}_i - x_0}{\|\tilde{x}_i - x_0\|} \right\rangle &\geq \left\langle \nabla f_i(\lambda_i \tilde{x}_i + (1 - \lambda_i)x_0), \frac{\lambda_i(\tilde{x}_i - x_0)}{\|\lambda_i(\tilde{x}_i - x_0)\|} \right\rangle \\ &\geq \frac{f_i(\lambda_i \tilde{x}_i + (1 - \lambda_i)x_0) - f_i(x_0)}{\|\lambda_i(\tilde{x}_i - x_0)\|},\end{aligned}$$

where the last inequality follows from the subgradient inequality. Let now be $d := \frac{1}{3}(\inf f(S) - \inf f(\mathbb{R}^n))$. As the sequence (f_i) converges uniformly on S to f (see Theorem 5.3.6) we have for large i

$$f_i(\lambda_i \tilde{x}_i + (1 - \lambda_i)x_0) \geq \inf f(S) - d,$$

and

$$f_i(x_0) \leq \inf f(\mathbb{R}^n) + d.$$

Let now δ be the diameter of $M(f, \mathbb{R}^n)$, then apparently

$$\|\lambda_i(\tilde{x}_i - x_0)\| \leq \varepsilon + \delta.$$

Hence we obtain

$$\left\langle \nabla f_i(\tilde{x}_i), \frac{\tilde{x}_i - x_0}{\|\tilde{x}_i - x_0\|} \right\rangle \geq \frac{d}{\varepsilon + \delta} > 0,$$

i.e. $\|\nabla f_i(\tilde{x}_i)\| \geq \frac{d}{\varepsilon+\delta} > 0$ in contradiction to our assumption.

We now prove the remaining parts of the assertion. By Theorem 5.3.25 there is a $K \in \mathbb{N}$, such that $M(f_i, \mathbb{R}^n) \neq \emptyset$ for $i \geq K$. Let now $x_i \in M(f_i, \mathbb{R}^n)$ for $i \geq K$, then

$$f_i(\tilde{x}_i) - f_i(x_i) \leq \langle \nabla f_i(\tilde{x}_i), \tilde{x}_i - x_i\rangle \leq \|\nabla f_i(\tilde{x}_i)\| \|\tilde{x}_i - x_i\|.$$

According to what has been proved above there is $N \in \mathbb{N}$ such that $\tilde{x}_i, x_i \in Q$ for $i \geq N$. Then using the subsequent theorem

$$\begin{aligned}f(\tilde{x}_i) - \inf f(\mathbb{R}^n) &\leq f_i(\tilde{x}_i) + O(\varepsilon_i) - \inf f(\mathbb{R}^n) \\ &\leq (f_i(\tilde{x}_i) - f_i(x_i)) + (f_i(x_i) - f(x_i)) + O(\varepsilon_i) \\ &\quad + (f(x_i) - \inf f(\mathbb{R}^n)) \\ &\leq \|\nabla f_i(\tilde{x}_i)\| \|\tilde{x}_i - x_i\| + O(\varepsilon_i).\end{aligned}$$

Due to the boundedness of the sequences $(\tilde{x}_i)$ and (x_i) the assertion follows. □

Remark. If the limit function f has a unique minimal solution, then we obtain for the estimates in (d)

$$|\inf f(\mathbb{R}^n) - f(\tilde{x}_k)| = O(\varepsilon_k) + o(\|\nabla f_k(\tilde{x}_k)\|).$$

In the previous theorem we have derived quantitative assertions about the speed of convergence of approximative values. For his purpose we have made use of the following theorem, which extends investigations of Peetre [90] on norms to sequences of convex functions.

Theorem 5.5.2. *Let* $(f_k : \mathbb{R}^n \to \mathbb{R})_{k\in\mathbb{N}}$ *be a sequence of convex functions with bounded level sets, which converges pointwise to the function* $f : \mathbb{R}^n \to \mathbb{R}$*. Let further* K *be a closed convex subset of* $\mathbb{R}^n$ *and* $x_k \in M(f_k, K)$*.*

(a) *Let* Q *be an open bounded superset of* $M(f, K)$ *and*

$$\varepsilon_k := \sup_{x\in Q} |f(x) - f_k(x)|,$$

then

i. $f(x_k) - \inf f(K) = O(\varepsilon_k)$

ii. $|f_k(x_k) - \inf f(K)| = O(\varepsilon_k)$.

(b) *If there is a sequence* $(\varepsilon_k)_{k\in\mathbb{N}}$ *tending to zero, such that for all* $x \in \mathbb{R}^n$ *and* $k \in \mathbb{N}$ *the inequality* $|f(x) - f_k(x)| \le \varepsilon_k |f(x)|$ *holds, then*

i. $f(x_k) - \inf f(K) \le 2\varepsilon_k |f(x_k)| \le \frac{2\varepsilon_k}{1-\varepsilon_k} |f_k(x_k)|$

ii. $|f_k(x_k) - \inf f(K)| \le 3\varepsilon_k |f(x_k)| \le \frac{3\varepsilon_k}{1-\varepsilon_k} |f_k(x_k)|$.

Proof. By Theorem 5.3.25 there is a $N \in \mathbb{N}$, such that $x_k \in Q$ for $k \ge N$. Due to the uniform convergence of f_k to f on $\overline{Q}$ the sequence (ε_k) tends to zero. Thus for $k \ge N$ and $x_0 \in M(f, K)$

$$f(x_k) - f(x_0) \le (f(x_k) - f_k(x_k)) + (f_k(x_0) - f(x_0)) \le 2\varepsilon_k,$$

and

$$|f_k(x_k) - f(x_0)| \le |f_k(x_k) - f(x_k)| + |f(x_k) - f(x_0)| \le 3\varepsilon_k.$$

Ad (b): In a similar way as above one obtains for $k \in \mathbb{N}$

$$f(x_k) - f(x_0) \le \varepsilon_k (f(x_k) + f(x_0)) \le 2\varepsilon_k f(x_k),$$

and

$$|f_k(x_k) - f(x_0)| \le 3\varepsilon_k f(x_k).$$

Due to $f_k(x_k) \ge (1 - \varepsilon_k) f(x_k)$ the remaining part of the assertion follows. □

The following theorem gives a framework for obtaining estimates for the sequence of distances $(d(x_k, M(f, \mathbb{R}^n)))$, from the convergence of the sequence of function values $(f(x_k))$.

Theorem 5.5.3. *Let* $(f : \mathbb{R}^n \to \mathbb{R})$ *be a convex function with bounded level sets. Then there is a strictly monotonically increasing function* $\gamma : \mathbb{R}_+ \to \mathbb{R}$ *with* $\gamma(0) = 0$*, such that for all* $x \in \mathbb{R}^n$

$$f(x) \geq \inf f(\mathbb{R}^n) + \gamma(d(x, M(f, \mathbb{R}^n)))$$

holds.

Proof. Let $S_r := \{x \in \mathbb{R}^n \,|\, d(x, M(f, \mathbb{R}^n)) = r\}$ for $r > 0$ and

$$K_r := \{x \in \mathbb{R}^n \,|\, d(x, M(f, \mathbb{R}^n)) \geq r\},$$

K_r is apparently closed. Let $c_r := \inf f(K_r)$ and let (y_n) be a sequence in K_r with $f(y_n) \to c_r$. Let $\varepsilon > 0$ then for n large enough we have $y_n \in S_f(c_r + \varepsilon)$. Since the level sets of f are bounded, there is a convergent subsequence $y_{n_k} \to x_r \in K_r$ yielding $f(x_r) = c_r$. Hence f attains its minimum x_r on K_r. Apparently $x_r \in S_r$, because otherwise x_r would be an interior point of K_r and hence a local, thus a global minimum of f on $\mathbb{R}^n$. Let $c := \inf f(\mathbb{R}^n)$, then we define

$$\gamma(r) := f(x_r) - c = \inf f(K_r) - c > 0.$$

Apparently $\gamma(r_1) \geq \gamma(r_2)$ for $r_1 > r_2$. Let now $\bar{x}_{r_1} \in M(f, \mathbb{R}^n)$ such that $\|x_{r_1} - \bar{x}_{r_1}\| = r_1$. Then there is a $\lambda_0 \in (0, 1)$ with $x_{\lambda_0} := \lambda_0 \bar{x}_{r_1} + (1 - \lambda_0) x_{r_1} \in S_{r_2}$.

Let $h(\lambda) := f(\lambda \bar{x}_{r_1} + (1 - \lambda) x_{r_1})$ for $\lambda \in [0, 1]$. Apparently h is convex on $[0, 1]$ and we have: $h(\lambda_0) \leq (1 - \lambda_0) h(0) + \lambda_0 h(1)$. Since $h(1) = c$ and $h(\lambda_0) = f(x_{\lambda_0}) \geq \gamma(r_2) + c$, as well as $h(0) = f(x_{r_1}) = \gamma(r_1) + c$ it follows that $h(\lambda_0) \leq (1 - \lambda_0)\gamma(r_1) + c$. Putting everything together we obtain $\gamma(r_2) < \gamma(r_1)$. □

Corollary. *Let* $f : \mathbb{R}^n \to \mathbb{R}$ *a convex function with bounded level sets and let* $(x_k)_{k \in \mathbb{N}}$ *be a minimizing sequence* $f(x_k) \to \inf f(\mathbb{R}^n)$*. Let* $\gamma : \mathbb{R}_+ \to \mathbb{R}$ *be the strictly monotonically increasing function of the previous theorem, then*

$$d(x_k, M(f, \mathbb{R}^n)) \leq \gamma^{-1}(|f(x_k) - \inf f(\mathbb{R}^n)|).$$

5.6 Two-stage Optimization

For a suitable choice of a parametric family $\{f_k\}$ of functions, convergence to a particular solution of the original problem can occur even in the case of non-uniqueness of the original problem. This limit can often be characterized as the solution of a two-stage optimization problem. The corresponding framework will be developed in the sequel.

In the treatment of optimization problems the original problem is frequently replaced by a sequence of approximating problems. If the approximating sequence is determined, only certain solutions of the original problem will be available as limits. Frequently they will turn out to be solutions of a second stage, where the second stage is described by a function, which implicitly depends on the choice of the sequence of the approximating problems.

Definition 5.6.1. Let C be a set and g_1, g_2 be two functions on C with values in $\mathbb{R}$. The following problem is called the *two-stage minimization problem* (g_1, g_2, C). Among the minimal solutions of g_1 on C those are selected, which are minimal w.r.t. g_2.

The solutions, i.e. the minimal solutions of g_2 on $M(g_1, C)$, we call two-stage solutions of the problem (g_1, g_2, C). The set of solutions is denoted by $M(g_1, g_2, C)$.

In the following theorem we assume the unique solvability of the approximating problems. Furthermore we assume that every point of accumulation of the sequence of these minimal solutions is an element of the set of solutions of the original problem ("closedness of the algorithm").

In this context we are interested in a characterization of these points of accumulation. Under favorable circumstances this characterization will enable us to enforce convergence of the sequence and to describe the limit as a solution of a two-stage optimization problem.

Theorem 5.6.2. *Let X be a metric space, $f : X \to \mathbb{R}$, and let for the sequence of functions $(f_n : X \to \mathbb{R})_{n\in\mathbb{N}}$ hold: for every $n \in \mathbb{N}$ the minimization problem (f_n, X) has a unique solution $x_n \in X$ and let every point of accumulation of the sequence $(x_n)_{n\in\mathbb{N}}$ be in $M(f, X)$. Let now $(a_n)_{n\in\mathbb{N}}$ be a sequence of non-negative numbers such that the sequence of the functions $(a_n(f_n - f))_{n\in\mathbb{N}}$ converges lower semi-continuously to a function $g : X \to \mathbb{R}$. Then every point of accumulation of the sequence $(x_n)_{n\in\mathbb{N}}$ is in $M(f, g, X)$.*

Proof. Let $y \in M(f, X)$ and $\bar{x} = \lim x_{n_i}$ with $x_{n_i} \in M(f_{n_i}, X)$. We have

$$(f_{n_i}(x_{n_i}) - f_{n_i}(y)) + (f(y) - f(x_{n_i})) \le 0,$$

hence

$$a_{n_i}(f_{n_i}(x_{n_i}) - f(x_{n_i})) \le a_{n_i}(f_{n_i}(y) - f(y)).$$

The lower semi-continuous convergence implies: $g(\bar{x}) \le g(y)$. □

Remark 5.6.3. The sequence $(a_n(f_n - f))_{n\in\mathbb{N}}$ converges lower semi-continuously to g, if it converges pointwise and the following condition is satisfied: there exists a sequence $(\alpha_n)_{n\in\mathbb{N}}$ tending to zero, such that $a_n(f_n - f) + \alpha_n \geq g$ and g is lower semi-continuous.

This means uniform convergence from below on the whole space.

We will now discuss a special case of this approach: the *regularization method of Tikhonov*:

Here $f_n = f + \alpha_n g$ where (α_n) is a sequence of positive numbers tending to zero and g an explicitly given lower semi-continuous function. Then for $a_n = \frac{1}{\alpha_n}$ for all $n \in \mathbb{N}$ we have $a_n(f_n - f) = g$.

As an example we consider the function sequence for the approximate treatment of best approximation in the mean, which was already mentioned in the introduction to Chapter 1. Let the set M be a convex and closed subset of $\mathbb{R}^m$. Let $x \in \mathbb{R}^m$ and $z \in M$, then put $f_n(z) = \sum_{i=1}^m \Phi_n(x_i - z_i)$ where $\Phi_n(s) = |s| + \frac{1}{n}s^2$ and $f(z) = \sum_{i=1}^m |x_i - z_i|$. In order to specify the second stage, choose $a_n = n$ and obtain

$$g(z) = \sum_{i=1}^{m} (x_i - z_i)^2.$$

These specifications have a regularizing effect in the following sense: every approximating problem is uniquely solvable and the sequence of the corresponding solutions converges to the particular best approximation in the mean, which has minimal Euclidean distance to x. Here we have also applied the stability theorem in $\mathbb{R}^m$ (see Theorem 5.3.25).

In the section on applications in Chapter 8 we will revisit the regularization method of Tikhonov under the aspect of local uniform and level uniform convexity in connection to strong solvability in reflexive spaces.

For the Tikhonov regularization the function g is explicitly chosen and in this way the second stage of the optimization is determined. Frequently a second stage is implicitly contained in the choice of the approximating function sequence (f_n).

We will prove the above theorem in a slightly generalized form, in order to simplify the determination of the second stage.

Theorem 5.6.4. *Let X be a metric space, $f : X \to \mathbb{R}$, and for the sequence of functions $(f_n : X \to \mathbb{R})_{n\in\mathbb{N}}$ we assume: for every $n \in \mathbb{N}$ the problem (f_n, X) has a unique solution $x_n \in X$ and every point of accumulation of the sequence $(x_n)_{n\in\mathbb{N}}$ is in $M(f, X)$. Let now $(\gamma_n)_{n\in\mathbb{N}}$ be a sequence of non-decreasing functions, such that the sequence of functions $(\gamma_n(f_n - f))_{n\in\mathbb{N}}$ converges lower semi-continuously to a function $g : X \to \mathbb{R}$. Then every point of accumulation of the sequence $(x_n)_{n\in\mathbb{N}}$ is in $M(f, g, X)$.*

Proof. Let $y \in M(f, X)$ and $\bar{x} = \lim x_{n_i}$ with x_{n_i} minimal solution of f_{n_i}. Then

$$\begin{aligned} f_{n_i}(x_{n_i}) - f_{n_i}(y) + f(y) - f(x_{n_i}) \leq 0 \\ f_{n_i}(x_{n_i}) - f(x_{n_i}) \leq f_{n_i}(y) - f(y) \\ \gamma_{n_i}(f_{n_i}(x_{n_i}) - f(x_{n_i})) \leq \gamma_{n_i}(f_{n_i}(y) - f(y)). \end{aligned}$$

The lower semi-continuous convergence implies $g(\bar{x}) \leq g(y)$. □

Corollary. *If one replaces in the above theorem non-decreasing by non-increasing and lower semi-continuous by upper semi-continuous, then one obtains: every point of accumulation of the sequence is a maximal solution of g on $M(f, X)$.*

The condition of lower semi-continuous convergence is in many cases naturally satisfied, because according to Theorem 5.2.1 a monotone sequence of lower semi-continuous functions, which converges pointwise to a lower semi-continuous function, is already lower semi-continuously convergent.

Example 5.6.5. As an example we consider the treatment of best approximation in the mean in $\mathbb{R}^m$, where the approximating problems are given by the modulars for the Young functions $\Phi_n(s) = |s| - \frac{1}{n}\log(1 + n|s|)$. If now

$$f_n(x) = \sum_{i=1}^{m} \left(|x_i| - \frac{1}{n} \log(1 + n|x_i|) \right),$$

and $f(x) = \sum_{i=1}^{m} |x_i|$, then one can deduce $g(x) := \prod_{i=1}^{m}(|x_i|)$ as a second stage, because

$$\begin{aligned} f_n(x) - f(x) &= \sum_{i=1}^{m} \left(|x_i| - \frac{1}{n} \log(1 + n|x_i|) \right) - \sum_{i=1}^{m} |x_i| \\ &= -\sum_{i=1}^{m} \frac{1}{n} \log(1 + n|x_i|) = -\frac{1}{n} \log \prod_{i=1}^{m} (1 + n|x_i|) \\ &= -\frac{1}{n} \log \left(n^m \prod_{i=1}^{m} \left(\frac{1}{n} + |x_i| \right) \right). \end{aligned}$$

Let now γ_n be the monotonically decreasing function $s \mapsto \gamma_n(s) = \frac{1}{n^m} \exp(-ns)$. Then

$$\gamma_n(f_n(x) - f(x)) = \prod_{i=1}^{m} \left(\frac{1}{n} + |x_i| \right),$$

and we obtain

$$\gamma_n(f_n(x) - f(x)) \xrightarrow{n \to \infty} \prod_{i=1}^{m} (|x_i|),$$

and the convergence is lower semi-continuous, since the sequence is monotonically dencreasing and pointwise convergent.

Second-stage by Differentiation w.r.t. a Parameter

When determining a second stage of an optimization problem one can frequently use differential calculus. We assume a real parameter $\alpha > 0$ and put $F(x,\alpha) := f_\alpha(x)$ as well as $F(x,0) = f(x)$. For many functions F the partial derivative w.r.t. α is a candidate for the second stage. Only the right-sided derivative is needed in this context, which exists for a broad class functions of a real variable.

$$g(x) := \frac{\partial}{\partial\alpha_+} F(x,0) = \lim_{\alpha\to 0} \frac{F(x,\alpha) - F(x,0)}{\alpha}.$$

If for fixed x the mapping $\alpha \mapsto F(x,\alpha)$ is convex, then – due to the monotonicity of the difference quotient – lower semi-continuous convergence is already guaranteed by pointwise convergence w.r.t. x.

Theorem 5.6.6. *Let X be a metric space and let the function $F : X \times [0,a] \to \mathbb{R}$ satisfy the following conditions:*

(a) *$F(x,\cdot) : [0,a] \to \mathbb{R}$ is for all $x \in X$ twice continuously differentiable*

(b) *$\frac{\partial}{\partial\alpha}F(\cdot,0)$ is lower semi-continuous. There exists a $\beta \in \mathbb{R}$, such that $\frac{\partial^2}{\partial\alpha^2}F(\cdot,\alpha) \ge \beta$ for all $x \in X$ and all $\alpha \in [0,a]$.*

Then the first stage is given by the function $f := F(\cdot,0)$ and the second stage by $g := \frac{\partial}{\partial\alpha}F(\cdot,0)$. Let $(\alpha_n)_{n\in\mathbb{N}}$ be a sequence tending to zero, $f_n := F(\cdot,\alpha_n)$ and $x_n \in M(f_n,X)$, then every point of accumulation of the sequence $(x_n)_{n\in\mathbb{N}}$ is in $M(f,g,X)$.

Proof. By the expansion theorem of Taylor there is a $\overline{\alpha} \in (0,\alpha_n)$, such that

$$F(x,\alpha_n) = F(x,0) + \alpha_n \frac{\partial}{\partial\alpha}F(x,0) + \frac{\alpha_n^2}{2}\frac{\partial^2}{\partial\alpha^2}F(x,\overline{\alpha}).$$

Therefore

$$\frac{F(x,\alpha_n) - F(x,0)}{\alpha_n} \ge \frac{\partial}{\partial\alpha}F(x,0) + \frac{\alpha_n}{2}\beta.$$

Due to Remark 5.6.3 the convergence is lower semi-continuous and by Theorem 5.6.2 the assertion follows. □

Testing of the boundedness of the second partial derivative can be omitted, if F is a convex function of the parameter α for fixed x. We then have

Theorem 5.6.7. *Let X be a metric space and let the function $F : X \times [0,a] \to \mathbb{R}$ satisfy the following conditions:*

(a) *$F(x,\cdot) : [0,a] \to \mathbb{R}$ is convex for all $x \in X$.*

(b) *The right-sided partial derivative w.r.t. the parameter $\frac{\partial}{\partial\alpha +}F(\cdot,0) : X \to \overline{\mathbb{R}}$ is lower semi-continuous.*

Then the second stage is given by the function $g := \frac{\partial}{\partial\alpha +}F(\cdot,0)$. Let $(\alpha_n)_{n\in\mathbb{N}}$ be a decreasing sequence of positive numbers tending to zero, $f_n := F(\cdot,\alpha_n)$ and $x_n \in M(f_n,X)$, then every point of accumulation of the sequence $(x_n)_{n\in\mathbb{N}}$ is in $M(f,g,X)$, where $f := F(\cdot,0)$.

Proof. The monotonicity of the difference quotient of convex functions yields:

$$\frac{F(x,\alpha_n) - F(x,0)}{\alpha_n} \to \frac{\partial}{\partial\alpha +}F(x,0)$$

monotonically decreasing. Therefore the convergence is lower semi-continuous. □

Remark. If in the above theorem the function g is strictly convex, then one has even convergence of the sequence (x_n) to the corresponding two-stage solution.

As a first application we describe the limits one obtains if the problem of best approximation in the mean (L^1-approximation) is replaced by a sequence of modular approximation problems. A particularly interesting special case is that of best L^p-approximations ($p > 1$) for $p \to 1$. In order to compute the second stage we put

$$F(x,\alpha) := \int_T |x(t)|^{1+\alpha} d\mu.$$

For each fixed $t \in T$ the function

$$\alpha \mapsto e^{(1+\alpha)\log|x(t)|}$$

is convex. Due to the monotonicity of the integral the corresponding pointwise convexity inequality remains valid. Hence also

$$\alpha \mapsto F(x,\alpha)$$

is convex for every fixed x. Due to the monotonicity of the difference-quotient one can interchange differentiation w.r.t. the parameter and integration (see theorem on monotone convergence of integration theory) and one obtains

$$g(x) := \frac{\partial}{\partial \alpha} F(x, 0) = \int_T |x(t)| \log |x(t)| d\mu. \tag{5.7}$$

The function $-g$ is called *entropy function*.

Let now (T, Σ, μ) be a finite measure space, V a finite-dimensional subspace of $L^{p_0}(\mu)$ for $p_0 > 1$ (resp. a closed convex subset thereof) and $\alpha \in [0, p_0 - 1]$. Then for these α all the functions $F(\cdot, \alpha)$ are finite on V. Also the function g is finite on V, because of the growth properties of the logarithm (and because of the boundedness of $u \mapsto u \log u$ from below for $u \geq 0$). Apparently g is strictly convex and hence in particular continuous on V. Using the stability theorem we obtain the following assertion.

Theorem 5.6.8. *Let (T, Σ, μ) be a finite measure space, V a finite-dimensional subspace of $L^{p_0}(\mu)$ for $p_0 > 1$ (resp. a closed convex subset thereof) and $(p_n)_{n \in \mathbb{N}}$ a sequence in $[1, p_0]$ with $p_n \to 1$. Then the sequence of best L^{p_n}-approximations of an element $x \in L^{p_0}(\mu)$ w.r.t. V converges to the best $L^1(\mu)$-approximation of largest entropy.*

The following theorem provides a topological version of Theorem 5.6.6, which will come into play in the context of level uniform convexity and the weak topology.

Theorem 5.6.9. *Let C be a compact topological space and let the function $F : C \times [0, a] \to \mathbb{R}$ satisfy the following conditions:*

(a) *$F(x, \cdot) : [0, a] \to \mathbb{R}$ is twice continuously differentiable for all $x \in C$.*

(b) *$\frac{\partial}{\partial \alpha} F(\cdot, 0)$ and $F(\cdot, 0)$ are lower semi-continuous.*

(c) *There exists a $\beta \in \mathbb{R}$, such that $\frac{\partial^2}{\partial \alpha^2} F(\cdot, \alpha) \geq \beta$ for all $x \in C$ and all $\alpha \in [0, a]$.*

Then the first stage is given by the function $f := F(\cdot, 0)$ and the second stage by $g := \frac{\partial}{\partial \alpha} F(\cdot, 0)$. Let $(\alpha_n)_{n \in \mathbb{N}}$ be a positive sequence tending to zero, define $f_n := F(\cdot, \alpha_n)$, then

(a) *$M(f_n, C)$ and $S := M(f, g, C))$ are non-empty.*

(b) *Let $x_n \in M(f_n, C)$, then the set of points of accumulation of this sequence is non-empty and contained in $M(f, g, X)$.*

(c) *Let $x_0 \in M(f, g, X)$ then*

   i. $f(x_n) \to f(x_0)$

   ii. $g(x_n) \to g(x_0)$

   iii. $\frac{f(x_n) - f(x_0)}{\alpha_n} \to 0$.

Proof. Being lower semi-continuous on a compact set, g is bounded from below by a number β_1. The function $\frac{\partial^2}{\partial\alpha^2}F(x_0,\cdot)$ is bounded from above by a number β_2. Let $x \in C$ and $\alpha \in [0,a]$. By the expansion theorem of Taylor there are a $\alpha', \alpha'' \in (0,\alpha)$, such that

$$\begin{aligned}F(x,\alpha) - F(x_0,\alpha) &= F(x,0) - F(x_0,0) + \alpha\left\langle \frac{\partial}{\partial\alpha}F(x,0) - \frac{\partial}{\partial\alpha}F(x_0,0)\right\rangle \\ &\quad + \frac{\alpha^2}{2}\left\langle \frac{\partial^2}{\partial\alpha^2}F(x,\alpha') - \frac{\partial^2}{\partial\alpha^2}F(x_0,\alpha'')\right\rangle \\ &\geq f(x) - f(x_0) + \alpha(\beta_1 - g(x_0)) + \frac{\alpha^2}{2}(\beta - \beta_2)\end{aligned}$$

for β_1 and β_2 independent of x and α. Let now $x = x_n$ and $\alpha = \alpha_n$ then

$$0 \geq f_n(x_n) - f_n(x_0) \geq \underbrace{f(x_n) - f(x_0)}_{\geq 0} + \alpha_n(\beta_1 - g(x_0)) + \frac{\alpha_n^2}{2}(\beta - \beta_2),$$

and hence $f(x_n) \to f(x_0)$.

Let further (x_{n_i}) be a convergent subsequence such that $x_{n_i} \to x$. Since $\lim f(x_{n_i}) = f(x_0) = \inf f(C)$, the lower semi-continuity of f yields $f(x) \leq f(x_{n_i}) + \varepsilon$ for large n_i, hence

$$f(x) \leq f(x_0),$$

and therefore $x \in M(f,C)$.

Suppose ii. does not hold, then there is $r > 0$ and a convergent subsequence (x_{n_j}) of (x_n) with $x_{n_j} \to x_1$ and $|g(x_{n_j}) - g(x_0)| > r$.

This will lead to a contradiction, since for large n_j and $g(x_{n_j}) - g(x_0) > r$

$$0 \geq f_{n_j}(x_{n_j}) - f_{n_j}(x_0) \geq f(x_{n_j}) - f(x_0) + \alpha_{n_j}\left(r + \frac{\alpha_{n_j}}{2}(\beta - \beta_2)\right) > 0.$$

Suppose now $g(x_0) - g(x_{n_j}) > r$ then due to lower semi-continuity of g we obtain $g(x_0) \geq g(x_1) + r$ contradicting $x_1 \in M(f,C)$ and $x_0 \in M(f,g,C)$. Using the lower semi-continuity of g we obtain (b) from ii.

$$g(x_0) = \underline{\lim}\, g(x_{n_j}) \geq g(x_1).$$

Since $x_1 \in M(f,C)$ it follows that $x_1 \in M(f,g,C)$. It remains to be shown that iii. holds

$$0 \geq f(x_n) - f(x_0) + \alpha_n(g(x_n) - g(x_0)) + \frac{\alpha_n^2}{2}(\beta - \beta_2).$$

Division by α_n together with ii. yields iii. □

5.6.1 Second Stages and Stability for Epsilon-solutions

Theorem 5.6.10. *Let $(f_k : \mathbb{R}^n \to \mathbb{R})_{k\in\mathbb{N}}$ be a sequence of differentiable convex functions, which converges pointwise to a function $f : \mathbb{R}^n \to \mathbb{R}$. Let further the set of minimal solutions of f on $\mathbb{R}^n$ be non-empty and bounded. Let now $(\alpha_k)_{k\in\mathbb{N}}$ be a sequence of non-negative numbers tending to zero such that the sequence of functions $(\frac{f_k-f}{\alpha_k})_{k\in\mathbb{N}}$ converges lower semi-continuously to a function $g : Q \to \mathbb{R}$ on an open bounded superset Q of $M(f,\mathbb{R}^n)$.*

Let further $(\delta_k)_{k\in\mathbb{N}}$ be a sequence of non-negative numbers tending to zero with $\lim_{k\to\infty} \frac{\delta_k}{\alpha_k} = 0$ and $(\tilde{x}_k)_{k\in\mathbb{N}}$ a sequence in $\mathbb{R}^n$ with the property $\|\nabla f_k(\tilde{x}_k)\| \le \delta_k$.

Then the set of points of accumulation of the sequence $(\tilde{x}_k)$ is non-empty and contained in $M(f,g,\mathbb{R}^n)$.

Proof. Let $x_k \in M(f_k,\mathbb{R}^n)$ for $k \in \mathbb{N}$. By Theorem 5.5.1 for $k > K_0$ the sequences x_k and $\tilde{x}_k$ are contained in Q. If $y \in M(f,\mathbb{R}^n)$, then one has

$$f_k(x_k) - f_k(y) + f(y) - f(\tilde{x}_k) \le 0.$$

Adding to both sides of the inequality the expression $f_k(\tilde{x}_k) - f_k(x_k)$, one obtains

$$\begin{aligned} f_k(\tilde{x}_k) - f_k(y) + f(y) - f(\tilde{x}_k) \le f_k(\tilde{x}_k) - f_k(x_k) &\le \|\nabla f_k(\tilde{x}_k)\|\|\tilde{x}_k - x_k\| \\ &\le \delta_k\|\tilde{x}_k - x_k\|. \end{aligned}$$

Division of both sides by α_k yields

$$\frac{f_k(\tilde{x}_k) - f(\tilde{x}_k)}{\alpha_k} \le \frac{f_k(y) - f(y)}{\alpha_k} + \frac{\delta_k}{\alpha_k}\|\tilde{x}_k - x_k\|.$$

The sequences (x_k) and $(\tilde{x}_k)$ are bounded and hence also the last factor on the right-hand side. Let $\bar{x}$ be a point of accumulation of $(\tilde{x}_k)$, then we obtain by the lower semi-continuous convergence

$$g(\bar{x}) \le g(y).$$ □

Remark. If in the above theorem the function g is strictly convex, then the sequence $(\tilde{x}_k)$ converges to the corresponding two-stage solution.

Example 5.6.11. We consider once more the best L^p-approximations ($p > 1$) for $p \to 1$. Let (α_k) be a sequence of positive numbers tending to zero. If we put

$$f_k(x) := \int_T |x(t)|^{1+\alpha_k} d\mu,$$

and

$$f(x) := \int_T |x(t)| d\mu,$$

then the sequence $(\frac{f_k - f}{\alpha_k})$ converges, as seen above, lower semi-continuously to the strictly convex function

$$g(x) = \int_T |x(t)| \log |x(t)| d\mu.$$

If one determines the best L^{p_k}-solutions with $p_k = 1 + \alpha_k$ only in an approximate sense, but with increasing accuracy, such that the stop criteria (for the norms of the gradients) converges more rapidly to zero than the sequence (α_k), then the sequence of the thus determined approximate solutions converges to the best $L^1(\mu)$-approximation of largest entropy.

As a consequence of the above Theorem 5.6.10 we obtain

Theorem 5.6.12. *Let $(f_k : \mathbb{R}^n \to \mathbb{R})_{k\in\mathbb{N}}$ be a sequence of differentiable convex functions, which converge pointwise to a function $f : \mathbb{R}^n \to \mathbb{R}$. Let further the set of minimal solutions of f on $\mathbb{R}^n$ be non-empty and bounded. Let now Q be an open and bounded superset of $M(f, \mathbb{R}^n)$ and*

$$\varepsilon_k := \sup_{x\in Q} |f(x) - f_k(x)|.$$

Let further $(\alpha_k)_{k\in\mathbb{N}}$ be a sequence of non-negative numbers tending to zero, such that $\lim_{k\to\infty} \frac{\varepsilon_k}{\alpha_k} = 0$ holds.

Let now the function $g : \mathbb{R}^n \to \mathbb{R}$ be differentiable and convex, and let the sequence of functions $(h_k : \mathbb{R}^n \to \mathbb{R})_{k\in\mathbb{N}}$ be defined by

$$h_k := \alpha_k g + f_k.$$

Let further $(\delta_k)_{k\in\mathbb{N}}$ be a sequence of positive numbers tending to zero, such that $\lim_{k\to\infty} \frac{\delta_k}{\alpha_k} = 0$ holds, and let $(\tilde{x}_k)_{k\in\mathbb{N}}$ be a sequence in $\mathbb{R}^n$ with the property $\|\nabla h_k(\tilde{x}_k)\| \le \delta_k$.

Then every point of accumulation of the sequence $(\tilde{x}_k)$ is a solution of the two-stage optimization problem $(f, g, \mathbb{R}^n)$.

If in addition g is strictly convex, then the sequence $(\tilde{x}_k)$ converges to the uniquely determined second-stage solution.

Proof. Apparently the sequence (h_k) converges pointwise to the limit function f. Due to the previous theorem it suffices to show the lower semi-continuous convergence of the sequence $(h_k - f)/\alpha_k$ to g on Q. But this follows using Remark 5.6.3 from

$$\frac{h_k - f}{\alpha_k} = g + \frac{f_k - f}{\alpha_k} \ge g - \frac{\varepsilon_k}{\alpha_k}. \qquad \square$$

Together with the convergence estimates in Chapter 2 we are now in the position to present a robust regularization of the Polya algorithm for the Chebyshev approximation in the spirit of Tikhonov, and hence obtain convergence to the second-stage solution w.r.t. an arbitrarily chosen strictly convex function g.

Theorem 5.6.13 (Regularization of Polya Algorithms). *Let T be a compact metric space and V a finite dimensional subspace of $C(T)$ (resp. a closed convex subset thereof) and let $x \in C(T)$.*

Let $(\Phi_k)_{k\in\mathbb{N}}$ be a sequence of differentiable Young functions converging pointwise to Φ_∞.

Let further $(\varepsilon_k)_{k\in\mathbb{N}}$ be a sequence tending to zero such that

$$\big|\|y\|_{(\Phi_k)} - \|y\|_\infty\big| \leq \varepsilon_k \|y\|_\infty$$

for all $y \in x + V$.

We now choose a sequence $(\alpha_k)_{k\in\mathbb{N}}$ of non-negative numbers tending to zero such that $\lim_{k\to\infty} \frac{\varepsilon_k}{\alpha_k} = 0$ holds, and a strictly convex function $g : x + V \to \mathbb{R}$. As an outer regularization of the Luxemburg norms we then define the function sequence $(h_k : x + V \to \mathbb{R})_{k\in\mathbb{N}}$ with $h_k := \alpha_k g + \|\cdot\|_{(\Phi_k)}$.

If the minimal solutions $\tilde{x}_k$ of h_k are determined approximately but with growing accuracy such that the norms of the gradients converge to zero more rapidly than the sequence (α_k), then the sequence $(\tilde{x}_k)$ converges to the best Chebyshev approximation, which is minimal w.r.t. g.

5.7 Stability for Families of Non-linear Equations

For sequences of convex functions, equicontinuity and hence continuous convergence already follows from pointwise convergence (see Theorem 5.3.13 and Theorem 5.3.8). Subsequently we will show that a similar statement holds for sequences of monotone operators. Continuous convergence in turn implies stability of solutions under certain conditions on the limit problem.

Both, solutions of equations and optimization problems, can be treated in the framework of variational inequalities. In fact, sequences of variational inequalities show a similar stability behavior, where again continuous convergence is of central significance (see [59], Satz 1, p. 245). We employ this scheme in the context of two-stage solutions.

We have treated stability questions for minimal solutions of pointwise convergent sequences of convex functions in Section 5.3. It turns out that stability can be guaranteed if the set of minimal solutions of the limit problem is bounded (see Theorem 5.3.25, see also [59]). The question arises, whether a corresponding statement holds on the equation level for certain classes of mappings that are not necessarily potential operators. Questions of this type arise e.g. in the context of smooth projection methods for semi-infinite optimization (see [62]).

A general framework for treating stability questions involving non-linear equations for sequences of continuous operators on $\mathbb{R}^n$ is given by the following scheme (see [62]):

Theorem 5.7.1. *Let $U \subset \mathbb{R}^n$ and $(A_k : U \to \mathbb{R}^n)_{k \in \mathbb{N}}$ be a sequence of continuous operators that converges continuously on U to a continuous operator $A : U \to \mathbb{R}^n$ with the property: there exists a ball $\overline{K}(x_0, r) \subset U, r > 0$ such that*

$$\langle Ax, x - x_0 \rangle > 0 \tag{5.8}$$

for all x in the sphere $S(x_0, r)$.

Then there exists $k_0 \in \mathbb{N}$ such that for all $k \geq k_0$ each equation $A_k x = 0$ has a solution x_k in $K(x_0, r)$.

Furthermore every point of accumulation of $(x_k)_{k \in \mathbb{N}}$ is a solution of $Ax = 0$.

The above theorem is a consequence of the following well-known

Lemma 5.7.2. *Let $A : U \to \mathbb{R}^n$ be continuous. If there is $r > 0$ and a ball $\overline{K}(x_0, r) \subset U$ such that $\langle Ax, x - x_0 \rangle \geq 0$ for all $x \in S(x_0, r)$, then the non-linear equation $Ax = 0$ has a solution in $\overline{K}(x_0, r)$.*

Proof. Otherwise Brouwer's fixed point theorem applied to the mapping

$$x \mapsto g(x) := -r \left(\frac{Ax}{\|Ax\|} \right) + x_0$$

would lead to a contradiction, because then $\|Ax\| \neq 0$ on $\overline{K}(x_0, r)$, hence g continuous there and thus has a fixed point $x \neq x_0$ on $\overline{K}(x_0, r)$. Hence

$$\left\langle Ax, r \frac{x - x_0}{\|x - x_0\|} \right\rangle = -\|x - x_0\| \|Ax\| < 0. \qquad \square$$

5.7.1 Stability for Monotone Operators

A large class of operators can be treated using the above stability principle, where pointwise convergence already implies uniform convergence on compact subsets. Among them are the monotone operators according to

Definition 5.7.3. Let X be a normed space and let U be a subset of X. A mapping $A : U \to X^*$ is called monotone on U if for all $x, y \in U$ the following inequality holds:

$$\langle Ax - Ay, x - y \rangle \geq 0.$$

Lemma 5.7.4 (see [110]). *Let U be an open subset of $\mathbb{R}^n$ and let $(A_k : U \to \mathbb{R}^n)_{k\in\mathbb{N}}$ be a sequence of continuous monotone operators that converges pointwise on U to an operator $A : U \to \mathbb{R}^n$. Then for every sequence $(x_k) \subset U$ that converges in U it follows that the sequence $(A_k x_k)_{k\in\mathbb{N}}$ is bounded.*

Proof. Assume that there is a sequence (x_k) in U with $\lim x_k = x_0 \in U$ such that $(A_k x_k)$ is unbounded. Then there is a subsequence $(A_{k_i} x_{k_i})$ with the property $\|A_{k_i} x_{k_i}\| \geq i$ for all $i \in \mathbb{N}$. As A_{k_i} is monotone we obtain for all $z \in U$

$$\langle A_{k_i} x_{k_i} - A_{k_i} z, x_{k_i} - z\rangle \geq 0.$$

Let now $y_{k_i} := \frac{A_{k_i} x_{k_i}}{\|A_{k_i} x_{k_i}\|}$ then we can w.l.o.g. assume that the sequence (y_{k_i}) converges to some y in the unit sphere. If the above inequality is divided by $\|A_{k_i} x_{k_i}\|$ we obtain for all $z \in U$

$$\left\langle y_{k_i} - \frac{A_{k_i} z}{\|A_{k_i} x_{k_i}\|}, x_{k_i} - z \right\rangle \geq 0.$$

pointwise convergence implies $A_{k_i} z \to Az$ and hence

$$\lim_{i\to\infty} \left\langle y_{k_i} - \frac{A_{k_i} z}{\|A_{k_i} x_{k_i}\|}, x_{k_i} - z \right\rangle = \langle y, x_0 - z\rangle \geq 0 \quad \forall z \in U.$$

As U is open it follows that $y = 0$, a contradiction. □

The following theorem states that pointwise convergence of continuous monotone operators already implies continuous convergence:

Theorem 5.7.5 (see [110]). *Let U be an open subset of $\mathbb{R}^n$ and let $(A_k : U \to \mathbb{R}^n)_{k\in\mathbb{N}}$ be a sequence of continuous monotone operators that converges pointwise on U to a continuous operator $A : U \to \mathbb{R}^n$ then (A_k) is equicontinuous on U.*

Proof. According to Theorem 5.3.6 it is sufficient to show the following: convergence of a sequence (x_k) in U to an element $x_0 \in U$ implies $\lim A_k x_k = Ax_0$.

Assume that there is a sequence (x_k) in U convergent to $x_0 \in U$ such that $(A_k x_k)$ does not converge to Ax_0, i.e. there is $\varepsilon > 0$ and a subsequence $(A_{k_i} x_{k_i})$ with the property

$$\|A_{k_i} x_{k_i} - Ax_0\| \geq \varepsilon$$

for all $i \in \mathbb{N}$. By Lemma 5.7.4 $(A_{k_i} x_{k_i})$ is bounded and w.l.o.g. we can assume that it converges to some $g \in \mathbb{R}^n$. Because of the previous inequality we have $\|g - Ax_0\| \geq \varepsilon$. On the other hand we obtain by the monotonicity of A_{k_i} for all $u \in U$

$$\langle A_{k_i} x_{k_i} - A_{k_i} u, x_{k_i} - u\rangle \geq 0,$$

and hence using pointwise convergence

$$\langle g - Au, x_0 - u\rangle \geq 0$$

for all $u \in U$. By Theorem 5.7.6 below it follows that $g = Ax_0$, a contradiction. □

Theorem 5.7.6 (Browder and Minty (see [109])). *Let E be a Banach space and U an open subset of E. Let $A : U \to E^*$ be a hemi-continuous operator. If for a pair $u_0 \in U$ and $v_0 \in E^*$ and for all $u \in U$ the inequality*

$$\langle Au - v_0, u - u_0 \rangle \geq 0$$

holds, then $v_0 = Au_0$.

An immediate consequence of the theorem of Browder and Minty is the following characterization theorem for solutions of the equation $Ax = 0$, if A is a monotone operator:

Theorem 5.7.7 (see [109]). *Let E be a Banach space and U an open subset of E. Let $A : U \to E^*$ a continuous and monotone operator. Then the following characterization holds: $Au_0 = 0$ for $u_0 \in U$ if and only if for all $u \in U$ the following inequality holds:*

$$\langle Au, u - u_0 \rangle \geq 0.$$

Proof. The "if"-part follows from Theorem 5.7.6 for $v_0 = 0$. Let now $Au_0 = 0$ then, from the monotonicity of A, we obtain

$$0 \leq \langle Au - Au_0, u - u_0 \rangle = \langle Au, u - u_0 \rangle. \qquad \square$$

Remark 5.7.8. If U is convex then the above theorem directly implies that the set

$$S_A := \{x \in U \mid Ax = 0\}$$

is convex.

For monotone operators we obtain the following existence theorem which is in a way a stronger version of Lemma 5.7.2:

Theorem 5.7.9. *Let $U \subset \mathbb{R}^n$ be convex and $A : U \to \mathbb{R}^n$ be a continuous monotone operator. If there exists a ball $\overline{K}(x_0, r) \subset U$ and $r > 0$ such that $\langle Ax, x - x_0 \rangle > 0$ for all $x \in S(x_0, r)$. Then for the set of solutions S_A of the non-linear equation $Ax = 0$ the following statement holds:*

$$\emptyset \neq S_A \subset K(x_0, r).$$

Proof. The first part follows from Lemma 5.7.2. For the second part let $\lambda, \mu \in \mathbb{R}$ with $\lambda > \mu$ and let $x \in S(x_0, r)$. Then monotonicity of A yields

$$\langle A(\lambda(x - x_0) + x_0) - A(\mu(x - x_0) + x_0), (\lambda - \mu)(x - x_0) \rangle \geq 0.$$

Let I be the intersection of U with the straight line passing through x and x_0. From the above inequality it follows that $g : I \to \mathbb{R}$ with $g(\lambda) := \langle A(\lambda(x - x_0) + x_0), x - x_0 \rangle$ is an increasing function. In particular $g(1) = \langle Ax, x - x_0 \rangle > 0$. Suppose there is a $1 < \lambda_* \in I$ such that $A(\lambda_*(x - x_0) + x_0) = 0$ then $g(\lambda_*) = 0$, a contradiction. $\square$

We are now in the position to present a stronger version of Theorem 5.7.1 for sequences of monotone operators:

Theorem 5.7.10. *Let $U \subset \mathbb{R}^n$ be open and convex and $(A_k : U \to \mathbb{R}^n)_{k\in\mathbb{N}}$ be a sequence of continuous monotone operators that converges pointwise on U to a continuous operator $A : U \to \mathbb{R}^n$ with the Property* (5.8)*: there exists a ball $\overline{K}(x_0, r) \subset U, r > 0$ such that $\langle Ax, x - x_0\rangle > 0$ for all x on the sphere $S(x_0, r)$.*

Then there exists $k_0 \in \mathbb{N}$ such that the set of the solutions of the equation $A_k x = 0$ is non-empty for all $k \geq k_0$ and contained in $K(x_0, r)$.

Furthermore, let $x_k \in \{x \in U \mid A_k x = 0\}, k \geq k_0$ then every point of accumulation of $(x_k)_{k\in\mathbb{N}}$ is a solution of $Ax = 0$.

Property (5.8) is satisfied by various classes of operators, among them are derivatives of convex functions.

Lemma 5.7.11. *If a monotone operator A defined on $\mathbb{R}^n$ has a convex potential f with a bounded set of minimal solutions $M(f, \mathbb{R}^n)$ (which, of course, coincides with the set of solutions of $Ax = 0$), then A satisfies Property* (5.8).

Proof. Apparently for each $x_0 \in M(f, \mathbb{R}^n)$ there is a sphere $S(x_0, r)$ such that $f(x) - f(x_0) > 0$ for all $x \in S(x_0, r)$. As $A = f'$ the subgradient inequality yields

$$0 < f(x) - f(x_0) \leq \langle Ax, x - x_0\rangle$$

on that sphere. □

For monotone operators, in general, such a statement is not available, i.e. Property (5.8) does not follow from the boundedness of the solutions of $Ax = 0$, as the following example shows:

Example 5.7.12. Let $A : \mathbb{R}^2 \to \mathbb{R}^2$ be a linear operator that represents a $\frac{\pi}{2}$-rotation. A is monotone as

$$\langle Ax - Ay, x - y\rangle = \langle A(x - y), x - y\rangle = 0,$$

but on any sphere around the origin we have $\langle Ax, x\rangle = 0$. Obviously, $\{x \mid Ax = 0\} = \{0\}$.

An important class of operators in this context are the Fejér contractions according to

Definition 5.7.13. An operator $P : \mathbb{R}^n \to \mathbb{R}^n$ is called *Fejér contraction* w.r.t. x_0 (see [11]) or strictly quasi-non-expansive (see [26]) if x_0 is a fixed point of P and there is an $r > 0$ such that $\|P(x) - x_0\| < \|x - x_0\|$ for all $x \notin K(x_0, r)$.

Remark 5.7.14. The above definition of a Fejér contraction differs somewhat from that given in [11].

Remark 5.7.15. It follows immediately from the definition that the set of fixed points of a Fejér contraction w.r.t. x_0 is bounded.

If P is a Fejér contraction w.r.t. x_0 then the operator $A := I - P$ has Property (5.8) as the following lemma shows:

Lemma 5.7.16. *Let $P : \mathbb{R}^n \to \mathbb{R}^n$ be a Fejér contraction w.r.t. x_0 then for $A := I - P$ we obtain*

$$\langle Ax, x - x_0 \rangle > 0$$

for all $x \notin K(x_0, r)$.

Proof. Let $x \notin K(x_0, r)$, then we obtain

$$\begin{aligned} \langle Ax, x - x_0 \rangle &= \langle x - x_0 - (P(x) - x_0), x - x_0 \rangle \\ &= \|x - x_0\|^2 - \langle P(x) - x_0, x - x_0 \rangle \\ &\geq \|x - x_0\|^2 - \|P(x) - x_0\| \|x - x_0\| > 0. \end{aligned}$$

□

Remark 5.7.17. If P is also non-expansive on $\mathbb{R}^n$ then $A = I - P$ is apparently monotone and continuous.

It is easily seen that a projection P onto a bounded convex set is a non-expansive Fejér contraction. It can be shown that the same is true for certain compositions of projections (see [62]).

5.7.2 Stability for Wider Classes of Operators

A large class of operators can be treated using the above stability principle, where pointwise convergence already implies continuous convergence. To illustrate this, consider a continuous operator $A : \mathbb{R}^n \to \mathbb{R}^n$ satisfying Property (5.8). Then it is easily seen that in the following situations continuous convergence, and hence stability of the solutions follows from Theorem 5.7.1:

(a) Let (α_k) be a sequence in $\mathbb{R}_+$ tending to 0, let $P : \mathbb{R}^n \to \mathbb{R}^n$ be continuous, and $A_k := \alpha_k P + A$.

Proof. Let $x_k \to x_0$ then the sequence $(P(x_k))$ is bounded, hence $\alpha_k P(x_k) \to 0$ and $A_k(x_k) \to A(x_0)$ because of the continuity of A. □

(b) Let $A_k : \mathbb{R}^n \to \mathbb{R}^n$ be continuous and let $A_k \to A$ component-wise monotone, i.e. for $A_k(x) = (f_k^{(i)}(x))_{i=1}^n$ and $A(x) = (f^{(i)}(x))_{i=1}^n$ one has pointwise monotone convergence of $f_k^{(i)} \to f^{(i)}$ for $i = 1, \dots, n$ on $\mathbb{R}^n$.

Proof. This follows from the theorem of Dini (see 5.2.2) applied to the components of A_k and A respectively. □

(c) Let $A_k : \mathbb{R}^n \to \mathbb{R}^n$ be continuous and $A_k \to A$ pointwise on $\mathbb{R}^n$ and let $A_k - A$ be monotone for all $k \in \mathbb{N}$.

Proof. We have $A_k - A \to 0$ pointwise on $\mathbb{R}^n$ and from Theorem 5.7.10 continuous convergence follows. □

(d) Compositions of continuously convergent sequences of functions preserves continuous convergence, i.e. let $g_k : \mathbb{R}^n \to \mathbb{R}^n$ be continuously convergent to $g : \mathbb{R}^n \to \mathbb{R}^n$ and let $f_k : \mathbb{R}^n \to \mathbb{R}^n$ be continuously convergent to $f : \mathbb{R}^n \to \mathbb{R}^n$ then $f_k \circ g_k$ converges continuously to $f \circ g$.

A special case is obtained if either (f_k) or (g_k) is a constant sequence of a continuous function, e.g. let $B : \mathbb{R}^n \to \mathbb{R}^n$ linear and let $A_k : \mathbb{R}^n \to \mathbb{R}^n$ be continuous and monotone, and let $A_k \to A$ pointwise on $\mathbb{R}^n$ then $B \circ A_k$ converges continuously to $B \circ A$.

5.7.3 Two-stage Solutions

Two-stage solutions we have studied in Section 5.6 (see also [59], [65], [66]), in particular for sequences of convex functions. The following theorem (see [59], p. 246) gives a framework for sequences of non-linear equations, where the second stage is described in terms of a variational inequality:

Theorem 5.7.18. *Let X, Y be a normed spaces, $(A : X \to Y)$ continuous and for the sequence of continuous operators $(A_k : X \to Y)_{k\in\mathbb{N}}$ let $L = \overline{\lim}_{k\to\infty}\{x \mid A_k x = 0\} \subset \{x \mid Ax = 0\} =: S_A$. Let further $(a_k)_{k\in\mathbb{N}}$ be a sequence of positive numbers and $B : X \to X^*$ a continuous mapping with $B(0) = 0$ such that*

(a) *$B \circ A : X \to X^*$ is monotone*

(b) *$a_k(B \circ A_k - B \circ A)$ converges continuously to a mapping $D : X \to X^*$.*

Let $\bar{x} \in L$ then for all $x \in S_A$ the inequality $\langle D\bar{x}, x - \bar{x}\rangle \geq 0$ holds.

Proof. Let $x_k \in \{x \mid A_k x = 0\}$ such that (x_k) converges to an $\bar{x} \in X$, i.e. $\bar{x} \in L$. Let $x \in S_A$, i.e. $Ax = 0$. Since $B \circ A$ monotone and continuous it follows that

$$a_k\langle (B \circ A_k - B \circ A)x_k, x - x_k\rangle = a_k\langle B \circ Ax_k, x_k - x\rangle \geq 0.$$

Since $a_k(B\circ A_k - B\circ A)$ converges continuously to D it follows that $a_k(B\circ A_k - B\circ A)x_k$ converges to $D\bar{x}$ in the norm, and hence inequality $\langle D\bar{x}, x-\bar{x}\rangle \geq 0$ follows. □

Example 5.7.19. If $A_k := \alpha_k P + A$ (compare class (a) of previous section) where A is monotone, P a positive definite linear operator, and (α_k) a sequence of positive numbers tending to 0 then, choosing $a_k := \frac{1}{\alpha_k}$, $D = P$ and the inequality $\langle P\bar{x}, x - \bar{x}\rangle \geq 0$ for all $x \in S_A$ is the characterization of a minimal solution of the strictly convex functional $x \mapsto \langle Px, x\rangle$ on the convex set S_A (compare Remark 5.7.8). In this case, convergence of (x_k) to $\bar{x}$ follows.

Example 5.7.20. Let $A, C : \mathbb{R}^n \to \mathbb{R}^m$ be linear operators, $b \in \mathbb{R}^m$ and let $S_A := \{x \in \mathbb{R}^n \mid Ax = b\}$ be non-empty. Let further (α_k) be a sequence in $\mathbb{R}_+$ tending to zero and let $A_k := \alpha_k C + A$. Then for $a_k := \frac{1}{\alpha_k}$ and $B := A^T$ it follows that

$$a_k(BA_k - BA) = \frac{1}{\alpha_k}(\alpha_k A^T C + A^T A - A^T A) = A^T C =: D,$$

and hence

$$\langle A^T C\bar{x}, x - \bar{x}\rangle \geq 0$$

for all x in the affine subspace S_A, in other words: $A^T C\bar{x}$ is orthogonal to kernel of A.

Example 5.7.21 (see [62]). LP-problem: let $c, x \in \mathbb{R}^n$, $A \in L(\mathbb{R}^n, \mathbb{R}^m), b \in \mathbb{R}^m$, then we consider the following problem:

$$\min\{\langle c, x\rangle \mid Ax = b, x \geq 0\}$$

with bounded and non-empty set of solutions. We choose the mapping $P = P_m \circ P_{m-1} \circ \cdots \circ P_1$ of the successive projections P_i onto the hyperplanes

$$H_i = \{s \in \mathbb{R}^n \mid \langle a_i, s\rangle = b_i\} \text{ for } i \in \{1, \ldots, m\},$$

where a_i denotes the i-th row of A, and – in addition to that – the projection P_K onto the positive cone $\mathbb{R}^n_{\geq 0}$, given by

$$P_K(x) := ((x_1)_+, \ldots, (x_n)_+).$$

As P_K is non-differentiable a smoothing of the projection P_K is obtained by replacing the function $s \mapsto (s)_+$ by a smooth function $\varphi_\alpha : \mathbb{R} \to \mathbb{R}$ ($\alpha > 0$) that approximates the $(\cdot)_+$-function.

The projection P_K is then replaced by $P_\alpha = (\varphi_\alpha(x_1), \ldots, \varphi_\alpha(x_n))$. By use of the Newton method the non-linear equation

$$F_\alpha(x) := x - P_\alpha \circ P(x) + \alpha c = 0.$$

can be solved very efficiently.

It can be shown that $P_K \circ P$ is a non-expansive Fejér contraction w.r.t. any $\hat{x} \in S := \{x \in \mathbb{R}^n \mid Ax = b, x \geq 0\}$ and that $P_\alpha \circ P$ is non-expansive. Let (α_k) be a positive sequence tending to 0. Stability then follows from Lemma 5.7.16 together with Theorem 5.7.10 for the sequence of monotone operators $A_k = F_{\alpha_k}$ converging pointwise to the monotone operator $A = I - P_K \circ P$ satisfying Property (5.8).

Application of Theorem 5.7.18 yields a condition for φ_{α_k} that enforces continuous convergence. We have for $a_k = \frac{1}{\alpha_k}$:

$$a_k(A_k - A) = a_k(-P_{\alpha_k} \circ P + \alpha_k c + P_K \circ P) = \frac{1}{\alpha_k}(P_K \circ P - P_{\alpha_k} \circ P) + c.$$

It follows that, if $\frac{1}{\alpha_k}(\varphi_{\alpha_k} - (\cdot)_+) \to 0$ uniformly on compact subsets of $\mathbb{R}$, then $a_k(A_k - A)$ converges continuously to c. If $\bar{x}$ is any limit point of the sequence of solutions of the equations $(A_k x = 0)$ then for all $x \in S$ we obtain $\langle c, x - \bar{x}\rangle \geq 0$.

Remark 5.7.22. Convex optimization problems with linear constraints can be treated in the same manner: let $f : \mathbb{R}^n \to \mathbb{R}^n$ be convex and differentiable, then consider

$$\min\{f(x) \mid Ax = b, x \geq 0\}.$$

The (monotone) operator F_α becomes

$$F_\alpha(x) := x - P_\alpha \circ P(x) + \alpha f'(x),$$

and one obtains $\langle f'(\bar{x}), x - \bar{x}\rangle \geq 0$ for every point of accumulation $\bar{x}$ and all $x \in S$, i.e. $\bar{x}$ is a minimal solution of f on S, according to the Characterization Theorem of Convex Optimization 3.4.3.

Chapter 6
Orlicz Spaces

In this chapter we will develop the theory of Orlicz spaces equipped with the Luxemburg norm for arbitrary measure spaces and arbitrary Young functions. In the next chapter we will then discuss dual spaces and the Orlicz norm that arises naturally in this context.

At first we will investigate the properties of (one-dimensional) Young functions and their conjugates.

6.1 Young Functions

Definition 6.1.1. Let $\Phi : \mathbb{R} \to \overline{\mathbb{R}}_{\geq 0}$ a lower semi-continuous, symmetric, convex function with $\Phi(0) = 0$, where 0 is an interior point of $\mathrm{Dom}(\Phi)$. Then Φ is called a *Young function*.

Remark 6.1.2. A Young function is monotone on $\mathbb{R}_{\geq 0}$, because let $0 \leq s_1 < s_2$, then we have

$$\Phi(s_1) \leq \frac{s_2 - s_1}{s_2}\Phi(0) + \frac{s_1}{s_2}\Phi(s_2) \leq \Phi(s_2).$$

Let Φ be a finite Young function and let $s_0 := \sup\{s \in \mathbb{R} \mid \Phi(s) = 0\}$, then Φ is strictly monotonically increasing on $[s_0, \infty) \cap \mathrm{Dom}(\Phi)$, because let $s_0 \leq s_1 < s_2$, then $s_1 = \lambda s_2 + (1 - \lambda)s_0$ with $0 \leq \lambda < 1$ and we obtain

$$\Phi(s_1) \leq \lambda\Phi(s_2) + (1 - \lambda)\Phi(s_0) < \Phi(s_2).$$

One of the goals of our consideration is, to anticipate many phenomena of the (typically infinite dimensional) function spaces from the properties of the one-dimensional Young functions. In particular we will see that the duality relations for norms and modulars are connected to the duality (conjugate) of the corresponding Young functions.

Subdifferential and One-sided Derivatives

Essentially the subdifferential of a Young function consists in the interval between left-sided and right-sided derivative. Without any restriction this holds for the interior points of the domain where Φ is finite, denoted by $\mathrm{Dom}(\Phi)$. At the boundary of $\mathrm{Dom}(\Phi)$ the subdifferential can be empty, even though Φ_- and Φ_+ exist in $\overline{\mathbb{R}}$.

Example 6.1.3. Let

$$\Phi(s) = \begin{cases} 1 - \sqrt{1 - s^2} & \text{for } |s| \le 1 \\ \infty & \text{otherwise} \end{cases}$$

then: $1 \in \text{Dom}(\Phi)$, $\Phi'_-(1) = \Phi'_+(1) = \infty$ and $\partial\Phi(1) = \emptyset$ holds.

Example 6.1.4. Let

$$\Phi(s) = \begin{cases} |s| & \text{for } |s| \le 1 \\ \infty & \text{otherwise} \end{cases}$$

then: $1 \in \text{Dom}(\Phi)$, $\Phi'_-(1) = 1$, $\Phi'_+(1) = \infty$ and $\partial\Phi(1) = [1, \infty)$ holds.

Example 6.1.5. Let

$$\Phi(s) = \begin{cases} \tan|s| & \text{for } |s| < \frac{\pi}{2} \\ \infty & \text{otherwise} \end{cases}$$

then: $\frac{\pi}{2} \notin \text{Dom}(\Phi)$ holds.

In the next section on the conjugate of a Young function we will prove the fundamental relation between the generalized inverses in connection with Young's equality. The 'complications' mentioned above require a careful treatment in order to substantiate the simple geometrical view described below.

The following theorem yields an extension of the theorem of Moreau–Pschenitschnii 3.10.4 for Young functions:

Theorem 6.1.6. *Let $s_0 \ge 0$ and let $s_0 \in \text{Dom}(\Phi)$, then $\Phi'_-(s_0)$ and $\Phi'_+(s_0)$ exist in $\overline{\mathbb{R}}$ and we have $\Phi'_-(s_0) \le \Phi'_+(s_0)$. Moreover:*

(a) *If beyond that $\partial\Phi(s_0) \neq \emptyset$, then $\Phi'_-(s_0)$ is finite and the following inclusion holds:*

$$[\Phi'_-(s_0), \Phi'_+(s_0)) \subset \partial\Phi(s_0) \subset [\Phi'_-(s_0), \Phi'_+(s_0)].$$

(b) *If, however, $\partial\Phi(s_0) = \emptyset$, then $\Phi'_-(s_0) = \Phi'_+(s_0) = \infty$.*

(c) *If, in particular $s_0 \in \text{Int}(\text{Dom}(\Phi))$, then $\Phi'_+(s_0)$ is finite and we obtain for the subdifferential at s_0*

$$[\Phi'_-(s_0), \Phi'_+(s_0)] = \partial\Phi(s_0).$$

Proof. Let $t \ge 0$ and $\alpha \in \{1, -1\}$, then we define

$$h_\alpha(t) := \Phi(s_0 + \alpha t) - \Phi(s_0).$$

Apparently h_α is convex and for $0 < s \le t$

$$h_\alpha(s) = h_\alpha\left(\frac{s}{t}t + \frac{t-s}{t}0\right) \le \frac{s}{t}h_\alpha(t) + \frac{t-s}{t}h_\alpha(0) = \frac{s}{t}h_\alpha(t)$$

holds. Therefore

$$\frac{\Phi(s_0 + \alpha s) - \Phi(s_0)}{s} \leq \frac{\Phi(s_0 + \alpha t) - \Phi(s_0)}{t}.$$

In particular the difference quotients $\frac{\Phi(s_0+t)-\Phi(s_0)}{t}$ resp. $\frac{\Phi(s_0-t)-\Phi(s_0)}{-t}$ are monotonically decreasing resp. increasing for $t \downarrow 0$.

Hence there exist $\Phi'_-(s_0) = \lim_{t\downarrow 0} \frac{\Phi(s_0-t)-\Phi(s_0)}{-t}$ and $\Phi'_+(s_0) = \lim_{t\downarrow 0} \frac{\Phi(s_0+t)-\Phi(s_0)}{t}$ in $\overline{\mathbb{R}}$.

Due to

$$\Phi(s_0) = \Phi\left(\frac{1}{2}(s_0 - t) + \frac{1}{2}(s_0 + t)\right) \leq \frac{1}{2}\Phi(s_0 - t) + \frac{1}{2}\Phi(s_0 + t),$$

we have for $t > 0$

$$\frac{\Phi(s_0 - t) - \Phi(s_0)}{-t} \leq \frac{\Phi(s_0 + t) - \Phi(s_0)}{t},$$

and thus $\Phi'_-(s_0) \leq \Phi'_+(s_0)$.

(a) Let now $\partial\Phi(s_0) \neq \emptyset$ and $\varphi \in \partial\Phi(s_0)$, then

$$\varphi \cdot (-t) \leq \Phi(s_0 - t) - \Phi(s_0),$$

thus for $t > 0$ and t small enough (0 is an interior point of $\mathrm{Dom}(\Phi)$)

$$\varphi \geq \frac{\Phi(s_0 - t) - \Phi(s_0)}{-t} \in \mathbb{R}.$$

As the above difference quotient for $t \downarrow 0$ is monotonically increasing, it follows that $\varphi \geq \Phi'_-(s_0) \in \mathbb{R}$.

Furthermore

$$\varphi \cdot t \leq \Phi(s_0 + t) - \Phi(s_0),$$

and hence

$$\varphi \leq \frac{\Phi(s_0 + t) - \Phi(s_0)}{t},$$

thus $\varphi \leq \Phi'_+(s_0)$.

Conversely let $\varphi \in \mathbb{R}$ and $\Phi'_-(s_0) \leq \varphi \leq \Phi'_+(s_0)$, i.e. because of the monotonicity of the difference quotient for $t > 0$

$$\frac{\Phi(s_0 - t) - \Phi(s_0)}{-t} \leq \varphi \leq \frac{\Phi(s_0 + t) - \Phi(s_0)}{t},$$

hence $\varphi \cdot t \leq \Phi(s_0 + t) - \Phi(s_0)$ and $\varphi \cdot (-t) \leq \Phi(s_0 - t) - \Phi(s_0)$, and therefore $\varphi \in \partial\Phi(s_0)$.

(b) If $\partial\Phi(s_0) = \emptyset$, then $\Phi'_-(s_0) = \infty$ must hold, because suppose $\Phi'_-(s_0) < \infty$, then, due to the result obtained above $\Phi'_-(s_0) \in \partial\Phi(s_0)$.

(c) Let now $s_0 \in \mathrm{Int}(\mathrm{Dom}(\Phi))$, then by Theorem 3.10.2 $\partial\Phi(s_0) \neq \emptyset$. In particular by (a). $\Phi'_-(s_0)$ finite. Furthermore $\frac{\Phi(s_0+t)-\Phi(s_0)}{t} \in \mathbb{R}$ for $t > 0$ and sufficiently small, and hence due to the monotonicity of the difference quotient $\Phi'_+(s_0) \in \mathbb{R}$. We obtain

$$\frac{\Phi(s_0 - t) - \Phi(s_0)}{-t} \leq \Phi'_-(s_0) \leq \Phi'_+(s_0) \leq \frac{\Phi(s_0 + t) - \Phi(s_0)}{t},$$

and thus $\Phi'_+(s_0) \in \partial\Phi(s_0)$. □

Remark 6.1.7. For $s_0 < 0$ analogous assertions hold. If in particular $s_0 \in \mathrm{Int}(\mathrm{Dom}(\Phi))$, then also $\Phi'_+(s_0)$ is finite and

$$[\Phi'_-(s_0), \Phi'_+(s_0)] = \partial\Phi(s_0)$$

holds.

Theorem 6.1.8. *Let* $0 \leq u < v \in \mathrm{Dom}(\Phi)$, *then:* $\Phi'_+(u) \leq \Phi'_-(v)$ *holds.*

Proof. We have $u \in \mathrm{Int}(\mathrm{Dom}(\Phi))$ and hence $\Phi'_+(u)$ is finite by the previous theorem. If $\Phi'_-(v)$ is finite, then we also have by the previous theorem $\Phi'_-(v) \in \partial\Phi(v)$. Using the monotonicity of the subdifferential (see Theorem 3.10.3) we then obtain

$$(\Phi_-(v) - \Phi'_+(u)) \cdot (v - u) \geq 0,$$

and since $v - u > 0$ the assertion. If $\Phi'_-(v) = \infty$, then there is nothing left to show. □

The Conjugate of a Young Function

For the conjugate Ψ of a Young function Φ one obtains according to Definition 3.11.1

$$\Psi(s) = \sup\{s \cdot t - \Phi(t) \mid t \in \mathbb{R}\}.$$

Apparently Ψ is again a Young function, because Ψ is, being the supremum of affine functions (see Remark 3.5.2) lower semi-continuous. From the definition immediately (as in the general case) Young's inequality

$$\Phi(t) + \Psi(s) \geq t \cdot s$$

follows.

From the theorem of Fenchel–Moreau 3.11.7 we obtain for $\Psi = \Phi^*$

$$\Phi^{**} = \Psi^* = \Phi.$$

Thus Φ and Ψ are mutual conjugates. For Young's equality we obtain the following equivalences (see Theorem 3.11.9):

Theorem 6.1.9. *The following assertions are equivalent:*

(a) $s \in \partial\Phi(t)$

(b) $t \in \partial\Psi(s)$

(c) $\Phi(t) + \Psi(s) = s \cdot t$.

As a consequence we obtain the following

Lemma 6.1.10. *Let Φ be a finite Young function. Then: $\Psi(\Phi'_+(t)) < \infty$ holds for all $t \in \mathbb{R}$. In particular $\Psi(\Phi'_+(0)) = 0$ holds. A corresponding statement holds for the left-sided derivative Φ'_-.*

Proof. The fact that Φ is finite implies by Theorem 6.1.6 and Remark 6.1.7 that $\Phi'_+(t) \in \partial\Phi(t)$ and hence

$$\Phi(t) + \Psi(\Phi'_+(t)) = \Phi'_+(t) \cdot t.$$

The same proof holds for the left-sided derivative. □

We will now demonstrate a geometrical interpretation of the right-sided derivative of Ψ as a generalized inverse of the right-sided derivative of Φ using the relationship of the subdifferential with the one-sided derivatives shown in the previous section (compare Krasnosielskii [72]) for N-functions).

Lemma 6.1.11. *Let $t \in (\Psi'_-(s), \Psi'_+(s))$, then $\partial\Phi(t)$ consists only of a single point.*

Proof. By Theorem 6.1.9 $s \in \partial\Phi(t)$. Suppose there is $s_1 \in \partial\Phi(t)$ with $s_1 \neq s$, then, again by Theorem 6.1.9 $t \in \partial\Psi(s)$. Therefore according to Theorem 6.1.8

$$s_1 > s \Rightarrow \Psi'_-(s_1) \geq \Psi'_+(s) > t$$
$$s_1 < s \Rightarrow \Psi'_+(s_1) \leq \Psi'_-(s) < t$$

i.e. $t \notin \partial\Psi(s_1)$ a contradiction. □

Notation 6.1.12.

$$\phi(t) := \begin{cases} \Phi'_+(t) & \text{for } t \in \operatorname{Dom}(\Phi) \\ \infty & \text{otherwise} \end{cases}$$

$$\psi(s) := \begin{cases} \Psi'_+(s) & \text{for } s \in \operatorname{Dom}(\Psi) \\ \infty & \text{otherwise.} \end{cases}$$

Lemma 6.1.13. *ϕ is increasing and right-sided continuous on $D := \mathbb{R}_+ \cap \operatorname{Dom} \phi$.*

Proof. Let $0 \le t_1 < t_2 \in \operatorname{Dom}\phi$. Then

$$\phi(t_1) = \Phi'_+(t_1) \le \Phi'_-(t_2) \le \Phi'_+(t_2) = \phi(t_2).$$

Let $t_0 \in D$ and (t_n) a sequence in D such that $t_n \downarrow t_0$, then $\Phi'_+(t_n) = \phi(t_n) \downarrow \bar{s}$. Furthermore $\Phi'_+(t_n) \ge \Phi'_+(t_0)$ for all $n \in \mathbb{N}$ and hence $\bar{s} \ge \Phi'_+(t_0)$. On the other hand we have according to Theorem 6.1.9

$$\Phi(t_n) + \Psi(\phi(t_n)) = t_n \cdot \phi(t_n)$$

and hence

$$\Phi(t_0) + \Psi(\bar{s})) = t_0 \cdot \bar{s}.$$

Again by Theorem 6.1.9 we have $\bar{s} \in \partial\Phi(t_0)$ and hence due to Theorem 6.1.6 $\bar{s} \le \Phi'_+(t_0)$. □

Remark 6.1.14. A corresponding statement holds for the left-sided derivative Φ'_-: Φ'_- is increasing and left-sided continuous on $D := \mathbb{R}_+ \cap \operatorname{Dom}\Phi'_-$.

The duality relation between a Young function and its conjugate can now be described in the following way: the right-sided derivative of the conjugate is the generalized inverse of the right-sided derivative of the Young function (for N-functions see Krasnosielskii, [72]). This is substantiated in the following theorem:

Theorem 6.1.15. *Let $s \ge 0$, then*

$$\psi(s) = \sup\{\tau \mid \phi(\tau) \le s\}$$

holds.

Proof. Case 1: Let the interior of $\partial\Psi(s)$ non-empty and let t be in the interior, then by Lemma 6.1.11: $s = \Phi'(t)$. At first we show the following inclusion:

$$(\Psi'_-(s), \Psi'_+(s)) \subset \{\tau \mid \Phi'(\tau) = s\} \subset \partial\Psi(s).$$

Let $u \in \{\tau \mid \Phi'(\tau) = s\}$, then apparently $s \in \partial\Phi(u)$ and by Theorem 6.1.9 $u \in \partial\Psi(s)$. Let on the other hand $u \in (\Psi'_-(s), \Psi'_+(s))$, then $s = \Phi'(u)$ by Lemma 6.1.11 and hence $u \in \{\tau \mid \Phi'(\tau) = s\}$. According to Theorem 6.1.6

$$\Psi'_+(s) = \sup\{\partial\Psi(s)\} = \sup\{\tau \mid \Phi'(\tau) = s\}$$

holds.

Case 2: Let $\partial\Psi(s)$ consist only of a single point and let $t = \Psi'(s)$ then $s \in \partial\Phi(t)$ by Theorem 6.1.9. Furthermore $t \ge 0$, because if $s = 0$, then $t = \Psi'(s) = 0$, if $s > 0$, then $t = \Psi'(s) \ge \frac{\Psi(s)}{s} \ge 0$.

If $\partial\Phi(t)$ also consists of a single point, then $t = \Psi'(\Phi'(t))$.

If the interior of $\partial\Phi(t)$ is non-empty let $s \in (\Phi'_-(t), \Phi'_+(t))$, then we obtain for $\bar{u} > t$ by Theorem 6.1.8: $\Phi'_-(\bar{u}) \geq \Phi'_+(t)$ provided that $\partial\Phi(\bar{u}) \neq \emptyset$, i.e. $\bar{u} \notin \{u \,|\, \Phi'_+(u) \leq s\}$. For $u < t$ we obtain

$$\Phi'_+(t) > s > \Phi'_-(t) \geq \Phi'_+(u),$$

i.e. $u \in \{\tau \,|\, \Phi'_+(\tau) \leq s\}$ for all $u < t$. Therefore

$$\Psi'(s) = t = \sup\{u \,|\, \Phi'_+(u) \leq s\}.$$

Let now $s = \Phi'_+(t)$, then $\Phi'_+(t) \in \partial\Phi(t)$, i.e. $\Psi'(\Phi'_+(t)) = t$ by Theorem 6.1.9.

Let finally $s = \Phi'_-(t)$, then as above $\Phi'_-(t) \in \partial\Phi(t)$,

$$\Psi'(s) = \Psi'(\Phi'_-(t)) = t = \sup\{u \,|\, \Phi'_+(u) \leq s\},$$

since $s = \Phi'_-(t) \geq \Phi'_+(u)$ for all $u < t$ by Theorem 6.1.8.

Case 3: $\partial\Psi(s) = \emptyset$: i.e. $s > 0$. Then $\Phi'_+(u) \leq s$ for all $u \in \mathbb{R}_{\geq 0}$. For suppose there is $u_1 \in \mathbb{R}_{\geq 0}$ with $\Phi'_+(u_1) := s_1 > s$ then $u_1 \in \partial\Psi(s_1)$ and $u_1 \geq \Psi'_-(s_1) \geq \Psi'_+(s)$ hence $\Psi'_+(s)$ finite, a contradiction. Therefore

$$\sup\{u \,|\, \Phi'_+(u) \leq s\} = \infty. \qquad \square$$

Corollary 6.1.16. *Let ϕ be continuous, strictly monotone, and let* $\lim_{s\to\infty} \phi(s) = \infty$, *then for $s \geq 0$*

$$\psi(s) = \phi^{-1}(s)$$

holds.

Proof. According to our assumption there is a unique $\tau \geq 0$ with $\phi(\tau) = s$, hence by Theorem 6.1.15 $\psi(s) = \tau = \phi^{-1}(s)$. $\square$

Theorem 6.1.17. *If* $\lim_{t\to\infty} \phi(t) = \infty$, *then ψ is finite. If beyond that ϕ is finite, then* $\lim_{s\to\infty} \psi(s) = \infty$ *holds.*

Proof. Let $s \geq 0$ be arbitrary, then there is $t_s \in \mathbb{R}$ with $\phi(t_s) > s$ and hence by definition: $\psi(s) < t_s$.

Let now ϕ be finite and $t_n \to \infty$, hence $\phi(t_n) \to \infty$. Let (s_n) be chosen, such that $s_n \geq \phi(t_n)$, then by definition $t_n \leq \psi(s_n) \to \infty$. $\square$

The relationship between the right-sided derivatives of Φ and Ψ established above we will now employ to substantiate the duality of differentiability and strict convexity.

Definition 6.1.18. Φ is called *continuous in the extended sense*, if Φ is continuous for $t < t_0 = \sup\{r \,|\, \Phi(r) < \infty\}$ and $\lim_{t\uparrow t_0} \Phi(t) = \infty$.

Φ is called (continuously) *differentiable in the extended sense*, if ϕ continuous in the extended sense, i.e. if ϕ is continuous for $t < t_0$ and $\lim_{t\uparrow t_0} \phi(t) = \infty$.

Remark 6.1.19. For every Young function Φ we have $\lim_{t\to\infty}\Phi(t) = \infty$, thus all continuous (i.e. finite) Young functions are also continuous in the extended sense. The continuity of ϕ in the extended sense is, however, a stronger requirement, in particular ϕ is continuous in the extended sense, if $\mathbb{R} \subset \phi(\mathbb{R})$.

Theorem 6.1.20. *Let ϕ be finite. ϕ is strictly monotone, if and only if ψ is continuous in the extended sense.*

Proof. Let at first ϕ be strictly monotone and for $s \geq 0$

$$\psi(s) = \sup\{t \in \mathbb{R} \mid \phi(t) \leq s\}.$$

Let $s \in \phi(\mathbb{R})$, then there is a unique $t_s \in \mathbb{R}_+$ with $\phi(t_s) = s$. Then apparently $\psi(s) = t_s$.

Let now $U := \{u \in \mathbb{R} \mid \phi(u) \leq s\}$ and $V := \{v \in \mathbb{R} \mid \phi(v) > s\}$.

(a) $V \neq \emptyset$: apparently: $\psi(s) = \sup U = \inf V =: t_s$ holds. Let (v_n) be a sequence in V with $v_n \downarrow t_s$, then due to the right-sided continuity of ϕ (see Lemma 6.1.13)

$$r_n := \phi(v_n) \downarrow \phi(t_s) =: \bar{s} \geq s,$$

hence in particular $\psi(\bar{s}) = t_s$ and furthermore

$$v_n = \psi(r_n) \downarrow t_s = \psi(\bar{s}).$$

Let on the other hand (u_n) be a sequence in U with $u_n \uparrow t_s$, then we obtain

$$s_n := \phi(u_n) \uparrow \phi(t_s-) =: \underline{s} \leq s,$$

and therefore

$$u_n = \psi(s_n) \uparrow t_s = \psi(\bar{s}) \leq \psi(\underline{s}).$$

The monotonicity of ψ implies

$$\psi(\bar{s}) \leq \psi(\underline{s}) \leq \psi(s) \leq \psi(\bar{s}).$$

If now $\phi(t_s-) = \phi(t_s)$, i.e. $\underline{s} = \bar{s}$, then ψ is continuous at s, if $\phi(t_s-) < \phi(t_s)$, then ψ is constant on $[\underline{s}, \bar{s}]$ and – due to the results of our previous discussion – continuous at the boundary points of the interval.

(b) $V = \emptyset$: apparently then ϕ is bounded. Let $s_\infty := \sup\{\phi(t) \mid t \in \mathbb{R}\}$ and let (u_n) be a strictly monotone sequence with $s_n := \phi(u_n) \uparrow s_\infty$. Suppose (u_n) is bounded, then $u_n \to u_0 \in \mathbb{R}$. Since $\phi(u_n) < \phi(u_0)$ we then obtain $s_\infty \leq \phi(u_0) \leq s_\infty$, hence ϕ constant on $[u_0, \infty]$, a contradiction to the strict monotonicity of ϕ. Therefore $\psi(s_n) = u_n \to \infty$.

If ϕ is unbounded, then Theorem 6.1.17 implies $\lim_{s\to\infty}\psi(s) = \infty$.

Conversely let ψ be continuous in the extended sense. Let $t \geq 0$, then $\{s \mid \psi(s) = t\} \neq \emptyset$ and bounded. We have $\phi(t) = \sup\{s \mid \psi(s) \leq t\}$ and hence $\phi(t) = s_t = \sup\{s \mid \psi(s) = t\}$. Due to the continuity of ψ we then have $\psi(s_t) = t$. Let now $0 \leq t_1 < t_2$, then $\psi(s_{t_1}) = t_1 < t_2 = \psi(s_{t_2})$. Since ψ is monotonically increasing, we obtain $\phi(t_1) = s_{t_1} < s_{t_2} = \phi(t_2)$. □

Theorem 6.1.21. *Let* $s \in \mathrm{Int}(\mathrm{Dom}(\Phi))$ *and* $s \geq 0$, *then*

$$\Phi(s) = \int_0^s \Phi'_+(t)dt = \int_0^s \Phi'_-(t)dt$$

holds.

Proof. By Theorem 6.1.13 and 6.1.8 the right-sided derivative Φ'_+ is monotonically increasing and bounded on $[0, s]$ and hence Riemann integrable. Let now $0 = t_0 < t_1 < \cdots < t_n = s$ be a partition of the interval $[0, s]$, then by Theorem 6.1.6 and the subgradient inequality

$$\Phi'_-(t_{k-1}) \leq \Phi'_+(t_{k-1}) \leq \frac{\Phi(t_k) - \Phi(t_{k-1})}{t_k - t_{k-1}} \leq \Phi'_-(t_k) \leq \Phi'_+(t_k).$$

Furthermore

$$\Phi(s) - \Phi(0) = \sum_{k=1}^{n} (\Phi(t_k) - \Phi(t_{k-1})),$$

and hence

$$\sum_{k=1}^{n} \Phi'_+(t_{k-1})(t_k - t_{k-1}) \leq \Phi(s) - \Phi(0) \leq \sum_{k=1}^{n} \Phi'_+(t_k)(t_k - t_{k-1}).$$

Since the two Riemann sums converge to the integral, the assertion follows with $\Phi(0) = 0$ (and in an analogous manner for the second integral). □

If we admit (if necessary) the value ∞, then we obtain the representation of a Young function Φ and its conjugate Ψ as integrals of the generalized inverses

$$\Phi(s) = \int_0^s \phi(t)dt$$

resp.

$$\Psi(s) = \int_0^s \psi(t)dt.$$

We are now in the position to state the duality relation announced earlier:

Theorem 6.1.22. *Let* Φ *and* Ψ *be finite, then* Φ *is strictly convex, if and only if* Ψ *is (continuously) differentiable.*

Theorem 6.1.23. *Ψ is finite, if and only if $\frac{\Phi(t)}{t} \to \infty$.*

Proof. Let $s \geq 0$.

If $\frac{\Phi(t)}{t} \leq M$ for all t, then $\Psi(s) = \sup\{t \cdot (s - \frac{\Phi(t)}{t})\} = \infty$ for $s > M$ holds. Conversely let $\frac{\Phi(t)}{t} \to \infty$, then there is a t_s, such that $s - \frac{\Phi(t)}{t}$ is negative for $t > t_s$, i.e.

$$\Psi(s) = \sup\{ts - \Phi(t) \,|\, 0 \leq t \leq t_s\} \leq s \cdot t_s. \qquad \square$$

Theorem 6.1.24. *Ψ is finite, if and only if $\phi(t) \to \infty$.*

Proof. Due to the subgradient inequality we have $\phi(t)(0 - t) \leq \Phi(0) - \Phi(t)$, we obtain for $t > 0$: $\frac{\Phi(t)}{t} \leq \phi(t)$. If Ψ finite, then by Theorem 6.1.23 $\phi(t) \to \infty$. Conversely let $\phi(t) \to \infty$, then by Theorem 6.1.17 ψ is finite and using the above integral representation the assertion follows. $\square$

Remark 6.1.25. Let $t_0 := \sup\{t \,|\, \Phi(t) = 0\}$, then Φ is strictly monotonically increasing on $[t_0, \infty)$, because let $t_0 \leq t_1 < t_2$, then there is a $\lambda \in (0, 1]$ with $t_1 = \lambda t_0 + (1 - \lambda)t_2$ and hence $\Phi(t_1) \leq \lambda\Phi(t_0) + (1 - \lambda)\Phi(t_2) < \Phi(t_2)$.

Lemma 6.1.26. *Let Φ be finite, then the function $s \mapsto s\Phi^{-1}(\frac{1}{s})$ is monotonically increasing on $(0, \infty)$. If in addition Ψ is finite, then: $\frac{\Phi^{-1}(n)}{n} \to_{n\to\infty} 0$.*

Proof. The first part can be seen as follows: for $t > t_0 \geq 0$ let $\Phi(t) > 0$ and $\Phi(t_0) = 0$. Apparently, for each $s > 0$ there is a unique $t > 0$ such that $s = \frac{1}{\Phi(t)}$. Put $s(t) := \frac{1}{\Phi(t)}$ and let $t_0 < t_1 \leq t_2$, then $s(t_1) \geq s(t_2)$ and, due to the monotonicity of the difference quotient $\frac{\Phi(t)}{t}$ at 0

$$s(t_1)\Phi^{-1}\left(\frac{1}{s(t_1)}\right) = \frac{t_1}{\Phi(t_1)} \geq \frac{t_2}{\Phi(t_2)} = s(t_2)\Phi^{-1}\left(\frac{1}{s(t_2)}\right).$$

We now discuss part 2: let $s_n := \Phi^{-1}(n)$, then due to the finiteness of Ψ we have $\frac{s_n}{\Phi(s_n)} \to 0$. $\square$

Stability of the Conjugates of Young Functions

Theorem 6.1.27. *Let $(\Phi_n)_{n\in\mathbb{N}}$ be a sequence of Young functions, converging pointwise and monotonically to the Young function Φ. In the case of a monotonically increasing sequence let in addition Ψ be finite. Then the sequence of the Young functions Ψ_n converges pointwise and monotonically to Ψ, increasing if (Φ_n) is decreasing, decreasing, if (Φ_n) is increasing.*

Proof. By definition we have for the conjugates

$$\Psi_n(t) = \sup\{st - \Phi_n(s) \mid s \geq 0\} = -\inf\{\Phi_n(s) - st \mid s \geq 0\}.$$

Let $t \geq 0$ be chosen. By Theorem 5.2.1 the sequence (φ_n) with $\varphi_n(s) := \Phi_n(s) - st$ converges lower semi-continuously to φ with $\varphi(s) := \Phi(s) - st$. If the sequence (Φ_n) is decreasing, then also the sequence (φ_n) and we obtain according to the Stability Theorem of Monotone Convergence 5.2.3

$$-\Psi_n(t) = \inf\{\varphi_n(s) \mid s \geq 0\} \to \inf\{\varphi(s) \mid s \geq 0\} = -\Psi(t).$$

If the sequence (Φ_n) is monotonically increasing, then from lower semi-continuous convergence (see Theorem 5.1.6) the Kuratowski convergence of the epigraphs $\text{Epi}(\varphi_n) \to \text{Epi}(\varphi)$ follows. If now Ψ is finite, then for $t > 0$ we obtain by Young's equality $M(\varphi, \mathbb{R}) = \partial\Psi(t)$. By Theorem 6.1.6 we have $\partial\Psi(t) = [\Psi'_-(t), \Psi'_+(t)]$. By Theorem 6.1.6 we have $\Psi'_+(t) < \infty$, also $\Psi'_-(t) \geq 0$ holds, since $\Psi'_-(t) \geq \frac{\Psi(t)}{t} \geq 0$.

If $t = 0$, then $M(\varphi, \mathbb{R}) = \{s \in \mathbb{R} \mid \Phi(s) = 0\}$. Thus in any case $M(\varphi, \mathbb{R})$ is non-empty and bounded. From the Stability Theorem on the Epigraph Convergence 5.3.27 we obtain by definition of the conjugates

$$-\Psi_n(t) = \inf\{\varphi_n(s) \mid s \geq 0\} \to \inf\{\varphi(s) \mid s \geq 0\} = -\Psi(t). \qquad \square$$

6.2 Modular and Luxemburg Norm

Let (T, Σ, μ) be an arbitrary measure space. On the vector space of μ-measurable real-valued functions on T we introduce the following relation: two functions are called equivalent if they differ only on a set of μ-measure zero. Let E be the vector space of the equivalence classes (quotient space). Let further $\Phi : \mathbb{R} \to \overline{\mathbb{R}}$ be a Young function. The set

$$L^\Phi(\mu) := \left\{ x \in E \,\middle|\, \text{there exists } \alpha > 0 \text{ with } \int_T \Phi(\alpha x)d\mu < \infty \right\}$$

is called Orlicz space. Apparently $L^\Phi(\mu)$ is a subspace of E, because – due to the convexity of Φ – the arithmetic mean of two elements is still in $L^\Phi(\mu)$. This is also true for an arbitrary multiple of an element.

Frequently we will consider the special case of the sequence spaces: let $T = \mathbb{N}$, Σ the power set of $\mathbb{N}$ and $\mu(t_i) = 1$ for $t_i \in T$ arbitrary, then we denote $L^\Phi(\mu)$ by ℓ^Φ.

Theorem 6.2.1. *The Minkowski functional*

$$x \mapsto \inf\left\{ c > 0 \,\middle|\, \int_T \Phi\left(\frac{x}{c}\right) d\mu \leq 1 \right\}$$

defines a norm on $L^\Phi(\mu)$. *It is called* Luxemburg norm, *denoted by* $\|\cdot\|_{(\Phi)}$.

Proof. Let

$$K := \left\{ x \in L^\Phi(\mu) \,\middle|\, \int_T \Phi(x) d\mu \le 1 \right\}.$$

According to Theorem 3.2.7 we have to show that K is a convex, symmetric linearly bounded set with the origin as algebraically interior point. It is straightforward to establish that the set K is convex and symmetric. Furthermore K has 0 as an algebraically interior point. In order to see this, let $x \in L^\Phi$ and $\alpha > 0$ such that $\int_T \Phi(\alpha x) d\mu < \infty$, then we define the function $h : [0,1] \to \mathbb{R}$ by $h(\lambda) := \int_T \Phi(\lambda\alpha x) d\mu$. Since h is convex and $h(0) = 0$ we obtain

$$h(\lambda) \le \lambda \cdot h(1) < \infty.$$

Hence there exists a $\lambda_0 > 0$ with $h(\lambda) \le 1$ for $\lambda \le \lambda_0$.

We now discuss the linear boundedness: let $y \in L^\Phi \setminus \{0\}$, then there is $M \in \Sigma$, such that $\mu(M) > 0$ and $\varepsilon > 0$ with $|y(t)| \ge \varepsilon$ for all $t \in M$. Since $\lim_{s\to\infty} \Phi(s) = \infty$, we obtain

$$\int_T \Phi(\alpha y) d\mu \ge \mu(M)\Phi(\alpha\varepsilon) \xrightarrow{\alpha\to\infty} \infty.$$ □

Remark 6.2.2. The Luxemburg norm is apparently monotone, i.e. for $x, y \in L^\Phi$ with $0 \le x \le y$ it follows that $\|x\|_{(\Phi)} \le \|y\|_{(\Phi)}$, because for $\varepsilon > 0$ we have

$$1 \ge \int_T \Phi\left(\frac{y}{\|y\|_{(\Phi)} + \varepsilon}\right) d\mu \ge \int_T \Phi\left(\frac{x}{\|y\|_{(\Phi)} + \varepsilon}\right) d\mu,$$

hence $\|x\|_{(\Phi)} \le \|y\|_{(\Phi)} + \varepsilon$.

In order to establish the completeness of Orlicz spaces the following lemma turns out to be helpful.

Lemma 6.2.3. *Let $0 \le x_n \uparrow x$ almost everywhere and $x_n \in L^\Phi(\mu)$ for all $n \in \mathbb{N}$. Then either $x \in L^\Phi(\mu)$ and $\|x_n\|_{(\Phi)} \to \|x\|_{(\Phi)}$ or $\|x_n\|_{(\Phi)} \to \infty$.*

Proof. Let $\varepsilon > 0$ and $\beta := \sup\{\|x_n\|_{(\Phi)} \in \mathbb{N}\} < \infty$. Then for all $n \in \mathbb{N}$ we have

$$\int_T \Phi\left(\frac{x_n}{\beta + \varepsilon}\right) d\mu \le 1.$$

Apparently the sequence $\Phi(\frac{x_n}{\beta+\varepsilon})$ converges pointwise a.e. to $\Phi(\frac{x}{\beta+\varepsilon})$. This is obvious, if Φ is continuous in the extended sense (see Definition 6.1.18 and Remark 6.1.19). If $\Phi(s)$ is finite for $|s| \le a$ and infinite for $|s| > a$ and if $\Phi(\frac{x(t)}{\beta+\varepsilon}) = \infty$ then $\frac{x_n(t)}{\beta+\varepsilon} \uparrow \frac{x(t)}{\beta+\varepsilon} > a$.

By the theorem of monotone convergence of integration theory we obtain

$$1 \geq \sup_n \int_T \Phi\left(\frac{x_n}{\beta+\varepsilon}\right) d\mu = \int_T \Phi\left(\frac{x}{\beta+\varepsilon}\right) d\mu,$$

i.e. $x \in L^\Phi(\mu)$ and $\beta + \varepsilon \geq \|x\|_{(\Phi)}$ and thus $\beta \geq \|x\|_{(\Phi)}$.

On the other hand let $0 < \beta_1 < \beta$. Then – due to the monotonicity of the sequence $(\|x_n\|_{(\Phi)})$ – there is $n_0 \in \mathbb{N}$, such that for $n > n_0$ we have: $\|x_n\|_{(\Phi)} > \beta_1$, i.e.

$$\int_T \Phi\left(\frac{x_n}{\beta_1}\right) d\mu > 1,$$

and by the monotonicity of the integral

$$\int_T \Phi\left(\frac{x}{\beta_1}\right) d\mu > 1,$$

i.e. $\|x\|_{(\Phi)} \geq \beta_1$. □

Theorem 6.2.4. *The Orlicz space $L^\Phi(\mu)$ is a Banach space.*

Proof. Let $(x_n)_n$ be a Cauchy sequence in $L^\Phi(\mu)$, i.e. $\lim_{m,n\to\infty} \|x_n - x_m\|_{(\Phi)} = 0$. Then there exists a subsequence $(y_k)_k$ of $(x_n)_n$ such that

$$\sum_{k=1}^{\infty} \|y_{k+1} - y_k\|_{(\Phi)} < \infty.$$

Let $z_n := |y_1| + \sum_{k=1}^{n} |y_{k+1} - y_k|$, then apparently $z_n \in L^\Phi(\mu)$. By the above lemma the sequence $(z_n)_n$ converges a.e. to a function $z \in L^\Phi(\mu)$. Hence also $\sum_{k=1}^{\infty} |y_{k+1} - y_k|$ is convergent a.e. and thus also $\sum_{k=1}^{\infty}(y_{k+1} - y_k)$. Let now $y := y_1 + \sum_{k=1}^{\infty}(y_{k+1} - y_k)$, then we have

$$\begin{aligned} y - y_n = \sum_{k=1}^{\infty}(y_{k+1} - y_k) + y_1 - y_n &= \sum_{k=1}^{\infty}(y_{k+1} - y_k) - \sum_{k=1}^{n-1}(y_{k+1} - y_k) \\ &= \sum_{k=n}^{\infty}(y_{k+1} - y_k), \end{aligned}$$

and thus

$$\|y - y_n\|_{(\Phi)} \leq \sum_{k=n}^{\infty} \|y_{k+1} - y_k\|_{(\Phi)} \xrightarrow{n\to\infty} 0.$$

Since $(x_n)_n$ is a Cauchy sequence, the whole sequence converges to y w.r.t. the Luxemburg norm. □

Closely related to the Luxemburg norm is the modular f^Φ, which we will employ frequently below.

Definition 6.2.5. The functional $f^\Phi : L^\Phi(\mu) \to \overline{\mathbb{R}}$ with

$$f^\Phi(x) = \int_T \Phi(x) d\mu$$

we call *modular*.

6.2.1 Examples of Orlicz Spaces

In particular we obtain by the choice $\Phi_\infty : \mathbb{R} \to \overline{\mathbb{R}}$ defined by

$$\Phi_\infty(s) := \begin{cases} 0 & \text{for } |s| \leq 1 \\ \infty & \text{otherwise} \end{cases}$$

and the corresponding Luxemburg norm

$$\|x\|_\infty = \inf\left\{c > 0 \,\middle|\, \int_T \Phi_\infty\left(\frac{x}{c}\right) d\mu \leq 1\right\}$$

the space $L^\infty(\mu)$.

For the Young functions $\Phi_p : \mathbb{R} \to \mathbb{R}$ with $\Phi_p(s) := |s|^p$ one obtains the $L^p(\mu)$-spaces with the well-known notation $\|\cdot\|_p$ for the corresponding norms.

In the sequel we need a direct consequence of the closed graph theorem, which is known as the two-norm theorem.

Theorem 6.2.6 (Two-norm theorem). *Let X be a Banach space w.r.t. the norms $\|\cdot\|_a$ and $\|\cdot\|_b$ and let*

$$\|\cdot\|_a \leq c \cdot \|\cdot\|_b.$$

Then the two norms are equivalent.

Proof. We consider the identical mapping

$$\mathrm{id} : (X, \|\cdot\|_a) \to (X, \|\cdot\|_b).$$

Let now (x_n) be a sequence in X with $x_n \to x$ w.r.t. $\|\cdot\|_a$ and $x_n \to y$ w.r.t. $\|\cdot\|_b$. Then $x_n \to y$ w.r.t. $\|\cdot\|_a$ immediately follows and hence $y = x = \mathrm{id}(x)$. By the closed graph theorem the identity is continuous and hence bounded on the unit ball, i.e.

$$\left\|\frac{x}{\|x\|_a}\right\|_b \leq D,$$

therefore $\|\cdot\|_b \leq D \cdot \|\cdot\|_a$. □

Lemma 6.2.7. *Let Φ_1 and Φ_2 be Young functions with $\Phi_2 \leq \Phi_1$, then*

(a) $L^{\Phi_1}(\mu) \subset L^{\Phi_2}(\mu)$

(b) $\|\cdot\|_{(\Phi_2)} \leq \|\cdot\|_{(\Phi_1)}$ *on* $L^{\Phi_1}(\mu)$

(c) $\Psi_1 \geq \Psi_2$

(d) $L^{\Psi_2}(\mu) \subset L^{\Phi_1}(\mu)$

(e) $\|\cdot\|_{(\Psi_2)} \geq \|\cdot\|_{(\Psi_1)}$ *on* $L^{\Psi_2}(\mu)$.

Proof. Let $x \in L^{\Phi_1}(\mu)$, then for $c > \|x\|_{(\Phi_1)}$ we obtain

$$1 \geq \int_T \Phi_1\left(\frac{x}{c}\right) d\mu \geq \int_T \Phi_2\left(\frac{x}{c}\right) d\mu.$$

The remaining part follows from

$$\Psi_1(r) = \sup\{rs - \Phi_1(s)\} \geq \sup\{rs - \Phi_2(s)\} = \Psi_2(r). \qquad \square$$

Definition 6.2.8. The Young function Φ is called *definite*, if $\Phi(s) > 0$ for $s > 0$, otherwise *indefinite*.

Lemma 6.2.9. *Let Ψ not be finite and let $a > 0$, such that Ψ is ∞ on (a, ∞) and finite on $[0, a)$. Then we have for the Young function Φ_1 with $\Phi_1(s) = a|s|$ the relation $\Phi \leq \Phi_1$, where Φ is the conjugate function of Ψ.*

Let now

(a) *Ψ be definite then there are positive numbers β, s_0 and a Young function Φ_2 with*

$$\Phi_2(s) = \begin{cases} 0 & \textit{for } 0 \leq |s| \leq s_0 \\ \beta(|s| - s_0) & \textit{for } |s| > s_0 \end{cases}$$

such that $\Phi_2 \leq \Phi$.

(b) *Ψ indefinite, then there is a $b > 0$ and a Young function Φ_2 with $\Phi_2(s) = b|s|$ and $\Phi_2 \leq \Phi$.*

Proof. For $s \geq 0$ we have

$$\Phi(s) = \sup_{u \geq 0}\{su - \Psi(u)\} = \sup_{0 \leq u \leq a}\{su - \Psi(u)\} \leq \sup_{0 \leq u \leq a}\{su\} = as = \Phi_1(s).$$

(a) Let Ψ be definite. Then $\Psi'_+ > 0$ on $(0, a)$. Let $u_0 \in (0, a)$, then $0 < \Psi'_+(u_0) < \infty$ by Theorem 6.1.6. Choose $s_0 = \Psi'_+(u_0)$, then by Theorem 6.1.15

$$\Phi'_+(s_0) = \sup\{u \,|\, \Psi'_+(u) \leq s_0\} = u_0.$$

Concerning the construction of Φ_2: Let $\beta := \Phi'_+(s_0)$. Then for

$$\Phi_2(s) := \begin{cases} 0 & \text{for } 0 \leq |s| \leq s_0 \\ \beta(|s| - s_0) & \text{for } |s| > s_0 \end{cases}$$

with the right-sided derivative

$$\phi_2(s) = \begin{cases} 0 & \text{for } 0 \leq s < s_0 \\ \beta & \text{for } s \geq s_0 \end{cases}$$

and hence for $s \geq 0$: $\phi_2(s) \leq \Phi'_+(s)$, therefore $\Phi_2(s) \leq \Phi(s)$.

(b) Let Ψ be indefinite, i.e. there is a $s_1 > 0$ with $\Psi(s) = 0$ for all $s \in [0, s_1]$. Then for $s \geq 0$

$$\Phi(s) = \sup_u \{su - \Psi(u)\} \geq \sup_{0 \leq u \leq s_1} \{su - \Psi(u)\} = s_1 \cdot s.$$

If we put $b := s_1$, we obtain the assertion. □

Theorem 6.2.10. *Let Ψ be not finite and for infinite measure let in addition Ψ be indefinite, then we have*

(a) $L^\Phi(\mu) = L^1(\mu)$

(b) $\|\cdot\|_{(\Phi)}$ *is equivalent to* $\|\cdot\|_1$

(c) $L^\Psi(\mu) = L^\infty(\mu)$

(d) $\|\cdot\|_{(\Psi)}$ *is equivalent to* $\|\cdot\|_\infty$.

Proof. Let $a > 0$, such that Ψ is equal to ∞ on (a, ∞) and finite on $[0, a)$. By Lemma 6.2.9 we have for the Young function Φ_1 with $\Phi_1(s) = a|s|$ the relation $\Phi \leq \Phi_1$. Apparently $L^{\Phi_1}(\mu) = L^1(\mu)$ and $\|\cdot\|_{(\Phi_1)} = a\|\cdot\|_1$. By Lemma 6.2.7 we obtain $L^1(\mu) \subset L^\Phi(\mu)$ and $\|\cdot\|_{(\Phi)} \leq a\|\cdot\|_1$.

Let $x \in L^\Psi(\mu)$ and $c > 0$, such that $\int_T \Psi(\frac{x}{c})d\mu \leq 1$. Suppose $x \notin L^\infty(\mu)$, then there is a number $r > a$ and a set $A \in \Sigma$ with $\mu(A) > 0$ and $|\frac{x(t)}{c}| \geq r$ for all $t \in A$. Then $\Psi(\frac{x(t)}{c}) = \infty$ on all of A, a contradiction. Hence we have at first: $L^\Psi \subset L^\infty$.

Let now:

i) Ψ be indefinite, i.e. there is a $s_0 > 0$ with $\Psi(s) = 0$ for all $s \in [0, s_0]$. Let now $x \in L^\infty(\mu)$, then we have $|\frac{x}{\|x\|_\infty} s_0| \leq s_0$ almost everywhere and hence

$$\int_T \Psi\left(\frac{x}{\|x\|_\infty} s_0\right) d\mu = 0.$$

Therefore: $\|x\|_{(\Psi)} \leq \frac{1}{s_0}\|x\|_\infty$ and thus $L^\infty(\mu)$ and $L^\Psi(\mu)$ are equal as sets. The equivalence of the norms follows by the two-norm theorem.

By Lemma 6.2.9 there is a $b > 0$ with $b\|\cdot\|_1 \leq \|\cdot\|_{(\Phi)}$. Then $L^1(\mu) = L^\Phi(\mu)$ and the norms are equivalent.

ii) Ψ definite and $\mu(T) < \infty$. Then by Lemma 6.2.9 there is a Young function Φ_2 with

$$\Phi_2(s) = \begin{cases} 0 & \text{for } 0 \leq |s| \leq s_0 \\ \beta(|s| - s_0) & \text{for } |s| > s_0 \end{cases}$$

and $\Phi_2 \leq \Phi$, hence $L^\Phi \subset L^{\Phi_2}$. Let now $x \in L^{\Phi_2}(\mu)$. We have to show: $x \in L^1(\mu)$:

Let $s_0 > 0$ and let $T_0 := \{t \in T \mid \frac{|x(t)|}{\|x\|_{(\Phi_2)}+\varepsilon} \leq s_0\}$. Then we have

$$\begin{aligned} 1 &\geq \int_T \Phi_2\left(\frac{x}{\|x\|_{(\Phi_2)} + \varepsilon}\right) d\mu = \int_{T\setminus T_0} \beta\left(\left|\frac{x}{\|x\|_{(\Phi_2)} + \varepsilon}\right| - s_0\right) d\mu \\ &= \frac{\beta}{\|x\|_{(\Phi_2)} + \varepsilon} \int_{T\setminus T_0} |x| d\mu - \frac{\beta s_0}{\|x\|_{(\Phi_2)} + \varepsilon} \mu(T \setminus T_0). \end{aligned}$$

Therefore

$$\frac{\|x\|_{(\Phi_2)} + \varepsilon}{\beta} + s_0\mu(T \setminus T_0) \geq \int_{T\setminus T_0} |x| d\mu,$$

put $c := \|x\|_{\Phi_2} + \varepsilon$ and due to $\int_{T_0} |x| d\mu \leq \mu(T_0) \cdot s_0 \cdot c$ we obtain

$$\|x\|_1 \leq \frac{c}{\beta} + s_0\mu(T)(1 + c).$$

Hence $x \in L^1(\mu)$, i.e. in all $L^1(\mu) = L^\Phi(\mu)$. The equivalence of the norms $\|\cdot\|_1$ and $\|\cdot\|_{(\Phi)}$ follows by the two-norm theorem.

Let $\Psi(s)$ be finite for all $s \in [0, s_1]$ and let $x \in L^\infty(\mu)$. Then $|\frac{x}{\|x\|_\infty}\lambda| \leq \lambda$ a.e. for arbitrary positive λ. Ψ has a (finite) inverse Ψ^{-1} on $[0, \Psi(s_1)]$ due to the definiteness of Ψ. Let now either

i. $\frac{1}{\mu(T)} \leq \Psi(s_1)$, then there exists $\lambda := \Psi^{-1}(\frac{1}{\mu(T)})$

ii. or $\frac{1}{\mu(T)} > \Psi(s_1)$ then put $\lambda := s_1$.

Thus we obtain

$$\int_T \Psi\left(\frac{x}{\|x\|_\infty}\lambda\right) d\mu \leq \int_T \Psi(\lambda) d\mu \leq \int_T \frac{1}{\mu(T)} d\mu = 1.$$

Therefore: $L^\infty(\mu) = L^\Psi(\mu)$ and $\|x\|_{(\Psi)} \leq \frac{1}{\lambda}\|x\|_\infty$.

The equivalence of the norms follows again by the two-norm theorem. □

Corollary 6.2.11. *Let Φ be not finite, let $T_0 \in \Sigma$ be a subset with finite measure, and let (T_0, Σ_0, μ_0) be an induced measure space on T_0, then $L^\Phi(\mu)$ contains a closed subspace, isomorphic to $L^\infty(\mu_0)$.*

Proof. By Theorem 6.2.10 items (c) and (d) we have $L^{\Phi}(\mu_0) = L^{\infty}(\mu_0)$ as sets and the corresponding norms are equivalent. □

If Ψ is definite, then the assertion of Theorem 6.2.10 for infinite measure does not hold in general, as can be seen by the following example of sequence spaces.

Example 6.2.12. Let

$$\Phi(s) = \begin{cases} 0 & \text{for } 0 \le |s| \le 1 \\ (|s| - 1) & \text{for } |s| > 1 \end{cases}$$

then for $\Psi = \Phi^*$ we obtain

$$\Psi(s) = \begin{cases} |s| & \text{for } 0 \le |s| \le 1 \\ \infty & \text{for } |s| > 1. \end{cases}$$

First of all we find: ℓ^{Φ} and ℓ^{∞} are equal as sets: if x is bounded, then $f^{\Phi}(\frac{x}{\|x\|_\infty}) = 0 < \infty$, i.e. $x \in \ell^{\Phi}$. If on the other hand y unbounded, $\frac{y}{c}$ is also unbounded for arbitrary $c > 0$ and there is a subsequence of components $(\frac{y_{i_k}}{c})$, whose absolute value tends to infinity and hence

$$f^{\Phi}\left(\frac{y}{c}\right) \ge \sum_{k=k_0}^{\infty} \left(\left|\frac{y_{i_k}}{c}\right| - 1\right) = \infty$$

for $c > 0$ arbitrary, i.e. $y \notin \ell^{\Phi}$. Let now $x \in \ell^{\Phi}$, then $f^{\Phi}(\frac{x}{\|x\|_\infty}) = 0 < 1$, i.e. $\|x\|_\infty \ge \|x\|_{(\Phi)}$. By the two-norm theorem both norms are equivalent.

We now consider ℓ^{Ψ} and show at first: ℓ^1 and ℓ^{Ψ} are equal as sets: let $x \in \ell^1$, then x is in particular bounded and

$$f^{\Psi}\left(\frac{x}{\|x\|_\infty}\right) = \sum_{i=1}^{\infty} \left|\frac{x_i}{\|x\|_\infty}\right| = \frac{\|x\|_1}{\|x\|_\infty} < \infty,$$

i.e. $x \in \ell^{\Psi}$.

Conversely, let $x \in \ell^{\Psi}$, then there is a $c > 0$, such that $f^{\Psi}(\frac{x}{c}) < \infty$. Apparently x is bounded and $c \ge \|x\|_\infty$ must hold, hence

$$f^{\Psi}\left(\frac{x}{c}\right) = \sum_{i=1}^{\infty} \left|\frac{x_i}{c}\right| = \frac{1}{c}\sum_{i=1}^{\infty} |x_i| < \infty,$$

i.e. $x \in \ell^1$. Let now $x \in \ell^{\Psi}$. Since $\|x\|_1 \ge \|x\|_\infty$, we have

$$f^{\Psi}\left(\frac{x}{\|x\|_1}\right) = \sum_{i=1}^{\infty} \left|\frac{x_i}{\|x\|_1}\right| = 1$$

and hence $\|x\|_{(\Psi)} \le \|x\|_1$. In order to show equality, we distinguish two cases

(a) $\|x\|_1 = \|x\|_\infty$: in this case only a single component is different from zero. Let $c < \|x\|_1$, then

$$f^\Psi\left(\frac{x}{c}\right) = \Psi\left(\frac{\|x\|_\infty}{c}\right) = \infty$$

and hence (due to the left-sided continuity of Ψ on $\mathbb{R}_+$): $\|x\|_{(\Psi)} > c$.

(b) $\|x\|_1 > \|x\|_\infty$: let $\|x\|_1 > c > \|x\|_\infty$, then

$$f^\Psi\left(\frac{x}{c}\right) = \sum_{i=1}^{\infty}\left|\frac{x_i}{c}\right| = \frac{1}{c}\|x\|_1 > 1$$

and hence $c < \|x\|_{(\Psi)}$.

In both cases $c < \|x\|_1$ implies $c < \|x\|_{(\Psi)}$, altogether $\|x\|_1 = \|x\|_{(\Psi)}$.

Example 6.2.13. Let $1 < p < \infty$ and

$$\Psi(s) = \begin{cases} |s|^p & \text{for } 0 \leq |s| \leq 1 \\ \infty & \text{for } |s| > 1. \end{cases}$$

Then $\ell^\Psi = \ell^p$ (unit ball identical), hence $\ell^\Psi \neq \ell^\infty$. Moreover, with $\frac{1}{p} + \frac{1}{q} = 1$

$$\Phi(s) = \begin{cases} c_p|s|^q & \text{for } 0 \leq |s| \leq p \\ |s| - 1 & \text{for } |s| > 1 \end{cases}$$

with $c_p = (\frac{1}{p})^{\frac{q}{p}} - (\frac{1}{p})^q = (\frac{1}{p})^{\frac{q}{p}} \cdot \frac{1}{q}$. Apparently $\ell^\Phi \neq \ell^1$.

6.2.2 Structure of Orlicz Spaces

Definition 6.2.14. In the sequel we denote by $M^\Phi(\mu)$ the closed subspace of $L^\Phi(\mu)$ spanned by the step functions in $L^\Phi(\mu)$ with finite support, i.e.

$$\left\{\sum_{i=1}^{n} a_i \chi_{A_i} \,\middle|\, \mu(A_i) < \infty, a_i \in \mathbb{R}, i = 1, \ldots, n\right\}.$$

The domain of finite values of the modular f^Φ

$$C^\Phi(\mu) := \{x \in L^\Phi(\mu) \mid f^\Phi(x) < \infty\}$$

is introduced as the *Orlicz class*. Furthermore we define

$$H^\Phi(\mu) := \{x \in L^\Phi(\mu) \mid f^\Phi(k \cdot x) < \infty \text{ for all } k \in \mathbb{N}\}.$$

Since

$$f^\Phi(k(x+y)) = f^\Phi\left(\frac{1}{2}(2kx + 2ky)\right) \leq \frac{1}{2}f^\Phi(2kx) + \frac{1}{2}f^\Phi(2ky) < \infty$$

for $x, y \in H^\Phi(\mu)$ the set $H^\Phi(\mu)$ is a subspace.

For future use we need the following

Lemma 6.2.15. *Let $x \in H^{\Phi}(\mu)$ and $x \neq 0$, then*

$$f^{\Phi}\left(\frac{x}{\|x\|_{(\Phi)}}\right) = 1$$

holds. If on the other hand $f^{\Phi}(x) = 1$, then $\|x\|_{(\Phi)} = 1$.

Proof. The mapping $h : \mathbb{R} \to \mathbb{R}$ with $\lambda \mapsto f^{\Phi}(\lambda x)$ is apparently convex and hence continuous (see Theorem 5.3.11). Let $0 < \varepsilon < \|x\|_{(\Phi)}$ and $\lambda_1 := 1/(\|x\|_{(\Phi)} + \varepsilon)$, then $f^{\Phi}(\lambda_1 x) \leq 1$. Moreover we obtain with $\lambda_2 := 1/(\|x\|_{(\Phi)} - \varepsilon)$ the inequality $f^{\Phi}(\lambda_2 x) > 1$. By the intermediate value theorem there is a $\overline{\lambda} \in [\lambda_1, \lambda_2]$, such that $f^{\Phi}(\overline{\lambda} x) = 1$. Let $\lambda_0 := \sup\{\lambda \mid h(\lambda) = 0\}$. Then h is strictly monotone on $[\lambda_0, \infty)$, because let $\lambda_0 \leq \lambda_1 < \lambda_2$, then $\lambda_1 = s\lambda_0 + (1-s)\lambda_2$ for a $s \in (0,1)$ and hence

$$h(\lambda_1) \leq s \cdot h(\lambda_0) + (1-s)h(\lambda_2) < h(\lambda_2). \qquad \square$$

Remark 6.2.16. (a) Apparently we have: $H^{\Phi}(\mu) \subset C^{\Phi}(\mu)$.

(b) The Orlicz class is a convex set, because for $x_1, x_2 \in C^{\Phi}(\mu)$ we obtain for $0 \leq \lambda \leq 1$

$$f^{\Phi}(\lambda x_1 + (1-\lambda)x_2) \leq \lambda f^{\Phi}(x_1) + (1-\lambda) f^{\Phi}(x_2) < \infty.$$

Theorem 6.2.17. *If Φ is finite, then $M^{\Phi}(\mu) \subset C^{\Phi}(\mu)$ holds.*

Proof. Let $x \in M^{\Phi}(\mu)$ and y a step function with finite support, for which

$$2\|x - y\|_{(\Phi)} + \varepsilon \leq 1$$

holds ($\varepsilon > 0$ appropriately chosen). Then y is in $C^{\Phi}(\mu)$ and hence also $2y$ – being a step function – in $C^{\Phi}(\mu)$. Due to the convexity of f^{Φ} we obtain

$$\frac{1}{2\|x-y\|_{(\Phi)} + \varepsilon} \int_T \Phi(2(x-y))d\mu \leq \int_T \Phi\left(\frac{2(x-y)}{2\|x-y\|_{(\Phi)} + \varepsilon}\right) d\mu \leq 1.$$

Therefore

$$\int_T \Phi(2(x-y))d\mu \leq 2\|x-y\|_{(\Phi)} + \varepsilon,$$

i.e. $2(x-y) \in C^{\Phi}(\mu)$. Since $C^{\Phi}(\mu)$ is convex, we obtain

$$x = \frac{1}{2} 2y + \frac{1}{2}(2(x-y)) \in C^{\Phi}(\mu). \qquad \square$$

The subsequent theorem asserts that for finite measure and finite Young function Φ the bounded functions are always contained in the closure of the step functions.

Theorem 6.2.18. *Let $\mu(T) < \infty$ and Φ finite, then L^∞ is a linear subspace of M^Φ and $\overline{L^\infty} = M^\Phi$ holds, where the closure is understood w.r.t. $\|\cdot\|_{(\Phi)}$.*

Proof. Let $x \in L^\Phi$ bounded, $a \le x \le b$ a.e. for $a, b \in \mathbb{R}$, and $\varepsilon > 0$ be given. Let further $\delta := \varepsilon\Phi^{-1}(\frac{1}{\mu(T)})$ and n_0 the smallest natural number, such that $\frac{b-a}{\delta} \le n_0$ and let

$$T_k := \{t \in T \mid a + (k-1)\delta \le x(t) < a + k\delta\}$$

for $k = 1, \dots, n_0$, then for the step function $y := \sum_{k=1}^{n_0}(a + (k-1)\delta)\chi_{T_k}$ we have

$$f^\Phi\left(\frac{x-y}{\varepsilon}\right) \le f^\Phi\left(\frac{\delta}{\varepsilon}\right) \le \Phi\left(\frac{\delta}{\varepsilon}\right)\mu(T) = 1,$$

and hence $\|x - y\|_{(\Phi)} \le \varepsilon$. On the other hand the step functions are, of course, contained in L^∞. □

By a similar construction as in the above theorem one can substantiate that for finite measure the step functions are dense in $L^\infty(\mu)$ w.r.t. the $\|\cdot\|_\infty$-norm.

Theorem 6.2.19. *Let $\mu(T) < \infty$, then: $L^\infty(\mu) = M^\infty(\mu)$ holds.*

For infinite measure this statement is false in general, as the example of the sequence spaces shows:

Example 6.2.20. The step functions are not dense in ℓ^∞, as can be seen from the sequence $x = (x_i)_{i\in\mathbb{N}}$ with $x_i = 1$ for $i \in \mathbb{N}$. The space m^∞ consists of the sequences tending to zero. In the literature this space is denoted by c_0.

From convergence in the Luxemburg norm always follows convergence in the modular.

Theorem 6.2.21. *From $\|x_n - x_0\|_{(\Phi)} \to 0$ always $f^\Phi(x_n - x_0) \to 0$ follows.*

Proof. Let $0 < \lambda \le 1$, then

$$f^\Phi(x) = f^\Phi\left(\lambda\frac{x}{\lambda} + (1-\lambda)0\right) \le \lambda f^\Phi\left(\frac{x}{\lambda}\right).$$

Let now $1 > \varepsilon > 0$, then $\int_T \Phi(\frac{x_n - x_0}{\|x_n - x_0\|_{(\Phi)} + \varepsilon})d\mu \le 1$ and thus for n sufficiently large and $\lambda := \|x_n - x_0\|_{(\Phi)} + \varepsilon$

$$\begin{aligned} f^\Phi(x_n - x_0) &\le (\|x_n - x_0\|_{(\Phi)} + \varepsilon)\int_T \Phi\left(\frac{x_n - x_0}{\|x_n - x_0\|_{(\Phi)} + \varepsilon}\right)d\mu \\ &\le (\|x_n - x_0\|_{(\Phi)} + \varepsilon). \end{aligned}$$

□

If Φ satisfies the Δ_2-condition (see Definition 6.2.25), then, as we will see, also the converse of this theorem holds.

From the previous theorem the closedness of $H^\Phi(\mu)$ immediately follows

Theorem 6.2.22. *$H^\Phi(\mu)$ is a closed subspace of $L^\Phi(\mu)$.*

Proof. Let $x_0 \in L^\Phi(\mu)$ and let (x_n) be a sequence in $H^\Phi(\mu)$, converging to x_0. From $x_n \to x_0$ it follows for $k \in \mathbb{N}$ arbitrary $2kx_n \to 2kx_0$, and hence $f^\Phi(2kx_n - 2kx_0) \to 0$, i.e. $2kx_n - 2kx_0 \in C^\Phi(\mu)$ for n sufficiently large. We then obtain

$$\begin{aligned} f^\Phi(kx_0) &= f^\Phi\left(\frac{1}{2}2kx_n - \frac{1}{2}(2kx_n - 2kx_0)\right) \\ &\le \frac{1}{2}f^\Phi(2kx_n) + \frac{1}{2}f^\Phi(2kx_n - 2kx_0) < \infty. \end{aligned}$$ □

Remark 6.2.23. Apparently for Φ finite the step functions with finite support are contained in $H^\Phi(\mu)$. Since $H^\Phi(\mu)$ is closed, we conclude $M^\Phi(\mu) \subset H^\Phi(\mu)$. We will now establish that for Φ finite and a σ-finite measure space already $M^\Phi(\mu) = H^\Phi(\mu)$ holds:

Theorem 6.2.24. *Let (T, Σ, μ) be a σ-finite measure space and let Φ be finite, then $M^\Phi(\mu) = H^\Phi(\mu)$ holds.*

Proof. Let $T = \bigcup_{j=1}^\infty B_j$ with $\mu(B_j) < \infty$. Let further $B^n := \bigcup_{j=1}^n B_j$, let $x \in H^\Phi(\mu)$, and let

$$x_n(t) := \begin{cases} x(t) & \text{for } |x(t)| \le n \text{ and } t \in B^n \\ 0 & \text{otherwise.} \end{cases}$$

then $x_n \in L^\Phi(\mu)$, because $|x_n(t)| \le |x(t)|$ i.e. $\|x_n\|_{(\Phi)} \le \|x\|_{(\Phi)}$. We will now approximate x_n by a sequence of step functions. Let $\delta_{nk} := \frac{n}{k}$ for $k \in \mathbb{N}$ and let for $r \in \mathbb{Z}$ with $0 \le r \le k$

$$C_{nkr} := \{t \in B^n \mid r\delta_{nk} \le x_n(t) < (r+1)\delta_{nk}\},$$

and for $-k \le r < 0$:

$$C_{nkr} := \{t \in B^n \mid (r-1)\delta_{nk} < x_n(t) \le r\delta_{nk}\}.$$

Then we define for $r = -k, \dots, k$

$$x_{nk}(t) := \begin{cases} r\delta_{nk} & \text{for } t \in C_{nkr} \\ 0 & \text{otherwise.} \end{cases}$$

Then we have by construction $|x_{nk}(t)| \leq |x_n(t)|$ and $|x_{nk} - x_n| \leq \delta_{nk}$ (hence uniform convergence on B^n) and thus for $\varepsilon > 0$ arbitrary

$$\int_T \Phi\left(\frac{x_{nk} - x_n}{\varepsilon}\right) d\mu \leq \int_{B_n} \Phi\left(\frac{\delta_{nk}}{\varepsilon}\right) d\mu = \mu(B_n)\Phi\left(\frac{\delta_{nk}}{\varepsilon}\right) \xrightarrow{k\to\infty} 0,$$

and therefore for n fixed

$$\lim_{k\to\infty} \|x_{nk} - x_n\|_{(\Phi)} = 0.$$

In particular by Remark 6.2.23 we conclude $x_n \in H^\Phi(\mu)$.

As a next step we show: $\lim_{n\to\infty} \|x - x_n\| = 0$: since $x - x_n \in H^\Phi(\mu)$ we obtain by Lemma 6.2.15

$$1 = \int_T \Phi\left(\frac{x - x_n}{\|x - x_n\|_{(\Phi)}}\right) d\mu.$$

Suppose there exists a subsequence (x_{n_k}) with $\|x - x_{n_k}\|_{(\Phi)} \geq \delta > 0$, then

$$1 \leq \int_T \Phi\left(\frac{x - x_{n_k}}{\delta}\right) d\mu.$$

On the other hand we have by construction $\frac{|x - x_{n_k}|}{\delta} \leq \frac{|x|}{\delta}$ and hence $\Phi(\frac{|x-x_{n_k}|}{\delta}) \leq \Phi(\frac{|x|}{\delta})$. But we also have $\int_T \Phi(\frac{|x|}{\delta}) d\mu < \infty$ and $\Phi(\frac{|x-x_{n_k}|}{\delta}) \to 0$ almost everywhere. Lebesgue's convergence theorem then yields

$$\int_T \Phi\left(\frac{x - x_{n_k}}{\delta}\right) d\mu \to 0,$$

a contradiction. We obtain: $x \in M^\Phi(\mu)$ and together with Remark 6.2.23 the assertion. □

6.2.3 The Δ_2-condition

In this section we will see that with an appropriate growth condition on Φ, the so-called Δ_2-condition, Orlicz class, Orlicz space, and the closure of the step functions coincide. In addition these conditions play an essential role for the description of the dual space (see next chapter).

Definition 6.2.25. The Young function satisfies the

(a) Δ_2-condition, if there is a $\lambda \in \mathbb{R}$, such that for all $s \in \mathbb{R}$

$$\Phi(2s) \leq \lambda\Phi(s),$$

(b) Δ_2^∞-condition, if there is a $\lambda \in \mathbb{R}$ and a $k > 0$, such that for all $s \geq k$

$$\Phi(2s) \leq \lambda\Phi(s),$$

(c) Δ_2^0-condition, if there is a $\lambda \in \mathbb{R}$ and a $k > 0$, such that for all $0 \leq s \leq k$

$$\Phi(2s) \leq \lambda \Phi(s).$$

Remark 6.2.26. (a) If Φ satisfies the Δ_2^∞-condition for a $k > 0$ and if Φ is definite, Φ satisfies the Δ_2^∞-condition for arbitrary $k > 0$.

(b) if Φ is finite and satisfies the Δ_2^0-condition for a $k > 0$, Φ satisfies the Δ_2^0-condition for arbitrary $k > 0$. In particular Φ is definite.

(c) If Φ satisfies the Δ_2^0-condition and the Δ_2^∞-condition, Φ satisfies the Δ_2-condition.

Proof. The reason is found in the fact that the function $s \mapsto \frac{\Phi(2s)}{\Phi(s)}$ is continuous on $\mathbb{R}_{>0}$ and hence bounded on any compact subinterval of $\mathbb{R}_{>0}$. □

Remark 6.2.27. The Δ_2^∞-condition is satisfied, if and only if there is a $\rho > 1$ and a κ, such that: $\Phi(\rho s) \leq \kappa \Phi(s)$ for $s \geq k > 0$, because (see Krasnosielskii [72]) let $2^n \geq \rho$, then we obtain for $s > k$

$$\Phi(\rho s) \leq \Phi(2^n s) \leq \lambda^n \Phi(s) = \kappa \Phi(s).$$

Conversely let $2 \leq \rho^n$, then $\Phi(2s) \leq \Phi(\rho^n s) \leq \kappa^n \Phi(s)$.

If $\Phi(\rho s) \leq \kappa \Phi(s)$ for $s \geq 0$, then the Δ_2-condition holds for Φ.

Theorem 6.2.28. *If Ψ is not finite, then Φ satisfies the Δ_2^∞-condition. If in addition Ψ is indefinite, then Φ satisfies even the Δ_2-condition.*

Proof. Let $a > 0$, such that Ψ is infinite on (a, ∞) and finite on $[0, a)$. Then by Lemma 6.2.9 $\Phi(s) \leq a \cdot s$ for $s \geq 0$. On the other hand there is by Lemma 6.2.9 a Young function $\Phi_2 \leq \Phi$ with

$$\Phi_2(s) := \begin{cases} 0 & \text{for } 0 \leq |s| \leq s_0 \\ \beta(|s| - s_0) & \text{for } |s| > s_0. \end{cases}$$

Hence for $s \geq 2s_0$

$$\begin{aligned} \Phi(2s) \leq 2as &= \frac{2a}{\beta}\beta(s - s_0) + \frac{2a}{\beta}\beta(2s_0 - s_0) \\ &= \frac{2a}{\beta}(\Phi_2(s) + \Phi_2(2s_0)) \leq \frac{2a}{\beta} 2\Phi_2(s) \leq \frac{4a}{\beta}\Phi(s). \end{aligned}$$

If Ψ is indefinite, then by Lemma 6.2.9 there are a, b with $b|\cdot| \leq \Phi \leq a|\cdot|$. We then have

$$\Phi(2s) \leq 2a|s| = 2\frac{a}{b} b|s| \leq 2\frac{a}{b}\Phi(s).$$ □

Theorem 6.2.29. *If Φ satisfies the Δ_2-condition, then*

$$C^\Phi(\mu) = L^\Phi(\mu)$$

holds. This equality also holds, if $\mu(T) < \infty$ and Φ satisfies only Δ_2^∞-condition.

Proof. Let $x \in L^\Phi(\mu)$, then there is an $\alpha \in \mathbb{R}_{>0}$ with $\int_T \Phi(\alpha x)d\mu < \infty$. Let $\frac{1}{2^k} \le \alpha$, then

$$\int_T \Phi(x)d\mu = \int_T \Phi\left(\frac{2^k}{2^k}x\right)d\mu \le \lambda^k \int_T \Phi\left(\frac{1}{2^k}x\right)d\mu \le \lambda^k \int_T \Phi(\alpha x)d\mu < \infty.$$

Let now $\mu(T) < \infty$, let Φ satisfy the Δ_2^∞-condition for a $s_0 > 0$, and let T_0 be defined by

$$T_0 := \{t \in T \,|\, 2^{-k}|x(t)| \ge s_0\},$$

then

$$\int_T \Phi(x)d\mu = \int_{T_0} \Phi(x)d\mu + \int_{T\setminus T_0} \Phi(x)d\mu \le \int_{T_0} \Phi(x)d\mu + \mu(T \setminus T_0)\Phi(2^k s_0).$$

For the integral over T_0 we obtain as above

$$\int_{T_0} \Phi(x)d\mu \le \lambda^k \int_{T_0} \Phi(\alpha x)d\mu < \infty,$$

and thus the assertion. □

Remark. The above theorem immediately implies that the modular is finite on all of $L^\Phi(\mu)$, provided that Φ satisfies the Δ_2-condition (resp. for finite measure the Δ_2^∞-condition).

For not purely atomic measures also the converse of the above theorem holds. In order to prove this we need the following (see [114]):

Lemma 6.2.30 (Zaanen). *Let E be of finite positive measure and contain no atoms, then for every number β with $0 < \beta < \mu(E)$ there is a subset F of E with the property $\mu(F) = \beta$.*

Theorem 6.2.31. *Let (T, Σ, μ) be a not purely atomic measure space with $\mu(T) < \infty$, then the following implication holds*

$$C^\Phi(\mu) = L^\Phi(\mu) \quad \Rightarrow \quad \Phi \text{ satisfies the } \Delta_2^\infty\text{-condition.}$$

Proof. Let A be the set of atoms and let $\lambda := \mu(T \setminus A)$:

Suppose Φ does not satisfy the Δ_2^∞-condition, then there is due to Remark 6.2.27 a monotonically increasing sequence (t_n) with $t_n \to_{n\to\infty} \infty$, where $\Phi(t_n) > 1$ for all $n \in \mathbb{N}$ and

$$\Phi\left(\left(1+\frac{1}{n}\right)t_n\right) > 2^n\Phi(t_n) \quad \text{for } n \in \mathbb{N}.$$

Let further (T_n) be a sequence of disjoint subsets of $T \setminus A$ with

$$\mu(T_n) = \frac{\lambda}{2^n\Phi(t_n)},$$

which we construct in the following way: let $T_1 \subset T \setminus A$ be chosen according to Lemma 6.2.30, such that $\mu(T_1) = \frac{\lambda}{2\Phi(t_1)}$ and let $T_{n+1} \subset (T \setminus A) \setminus \bigcup_{i=1}^n T_i$ with $\mu(T_{n+1}) = \frac{\lambda}{2^{n+1}\Phi(t_{n+1})}$. This choice is possible (again employing Lemma 6.2.30), since

$$\mu\left((T \setminus A) \setminus \bigcup_{i=1}^n T_i\right) = \mu(T \setminus A) - \sum_{i=1}^n \mu(T_i) = \lambda\left(1 - \sum_{i=1}^n \frac{1}{2^i\Phi(t_i)}\right) \geq \frac{\lambda}{2^n}.$$

We now define a function

$$x(t) := \begin{cases} t_n & \text{for } t \in T_n \\ 0 & \text{for } t \in T \setminus \bigcup_{n=1}^\infty T_n. \end{cases}$$

We conclude $x \in C^\Phi(\mu)$, because

$$f^\Phi(x) = \int_T \Phi(x)d\mu = \sum_{n=1}^\infty \Phi(t_n)\mu(T_n) = \lambda \sum_{n=1}^\infty \frac{1}{2^n} < \infty.$$

But we will now show: βx is not in $C^\Phi(\mu)$ for arbitrary $\beta > 1$: we then have $\beta > 1 + \frac{1}{n}$ for $n > n_0$, hence

$$\int_{T_n} \Phi(\beta x)d\mu = \Phi(\beta t_n)\mu(T_n) > \Phi\left(\left(1+\frac{1}{n}\right)t_n\right)\mu(T_n) \geq 2^n\Phi(t_n)\frac{\lambda}{2^n\Phi(t_n)} = \lambda,$$

and therefore

$$f^\Phi(\beta x) = \int_T \Phi(\beta x)d\mu = \sum_{n=1}^\infty \int_{T_n} \Phi(\beta x)d\mu = \infty,$$

a contradiction. □

The connection between Δ_2-condition and the equality of Orlicz class and Orlicz space is not present for every measure space, as the example of the $\mathbb{R}^n$ shows:

Example. Let Φ be finite and let $T = \{t_1, \ldots, t_n\}$ with $\mu(t_i) = 1$ for $i = 1, \ldots, n$. Then $C^\Phi = L^\Phi$, even if Φ does not satisfy a Δ_2-condition.

Theorem 6.2.32. *If Φ satisfies the Δ_2^0-condition, then*

$$c^\Phi = \ell^\Phi.$$

Proof. Let $x \in \ell^\Phi$, then there is an $\alpha \in \mathbb{R}_{>0}$ with $c := f^\Phi(\alpha x) = \sum_{t_i \in T} \Phi(\alpha x(t_i)) < \infty$, hence for all $i \in \mathbb{N}$ $\Phi(\alpha x(t_i)) \leq c$ holds, i.e.

$$|x(t_i)| \leq \frac{1}{\alpha} \Phi^{-1}(c) =: s_0.$$

Let now $\frac{1}{2^k} \leq \alpha$. Using the Δ_2^0-condition for $0 \leq s \leq s_0$ the remaining part of the proof can be performed in analogy to that of Theorem 6.2.29. □

If Φ satisfies the Δ_2- resp. the Δ_2^0-condition, then, as we have seen, the Orlicz class and Orlicz space agree. Thus the Orlicz class is a linear space, and contains all real multiples. An immediate consequence is the

Theorem 6.2.33. *If Φ satisfies the Δ_2-condition, then*

$$H^\Phi(\mu) = C^\Phi(\mu).$$

This equality also holds, if $\mu(T) < \infty$ and Φ satisfies only the Δ_2^∞-condition.

In the same way we obtain

Theorem 6.2.34. *If Φ satisfies the Δ_2^0-condition, then*

$$h^\Phi(\mu) = c^\Phi(\mu).$$

The theorem of Lindenstrauss–Tsafriri, developed in the sequel, asserts in particular that also the converse of the previous Theorem 6.2.34 holds. As a preparation we need the following

Lemma 6.2.35. *Let Φ be a finite Young function. Then h^Φ is a closed subspace of ℓ^Φ, and the unit vectors $\{e_i\}$ form a basis of h^Φ.*

Proof. Let $x \in h^\Phi$ and let $x_n = \sum_{i=1}^n x(t_i)e_i$. By definition $\frac{x}{\varepsilon} \in h^\Phi$ holds for all $\varepsilon > 0$, i.e.

$$f^\Phi\left(\frac{x}{\varepsilon}\right) = \sum_{i=1}^\infty \Phi\left(\frac{x(t_i)}{\varepsilon}\right) < \infty.$$

In particular the above series converges and we obtain for n sufficiently large

$$\sum_{i=n+1}^{\infty}\Phi\left(\frac{x(t_i)}{\varepsilon}\right)\le 1 \quad\Rightarrow\quad f^{\Phi}\left(\frac{x-x_n}{\varepsilon}\right)\le 1 \quad\Rightarrow\quad \|x-x_n\|_{(\Phi)}\le\varepsilon.$$

Therefore $h^\Phi \subseteq \overline{[e_i]}$ and hence $\bar{h}^\Phi \subseteq \overline{[e_i]}$. On the other hand, apparently $[e_i]\subseteq h^\Phi$ holds and hence $\overline{[e_i]}\subseteq \bar{h}^\Phi$. We obtain $\overline{[e_i]}=\bar{h}^\Phi$.

Let $x\in\bar{h}^\Phi$, then there exists a sequence $(x_n)\subset h^\Phi$, $x_n=\sum_{i=1}^{\infty}x_n(t_i)e_i$ with $x_n\to x$.

Let $\tilde{x}_n=\sum_{i=1}^{N(n)}x_n(t_i)e_i$ be chosen such that $\|\tilde{x}_n-x_n\|_{(\Phi)}\le\frac{1}{n}$. Apparently

$$\|x-\tilde{x}_n\|_{(\Phi)}\le\|x-x_n\|_{(\Phi)}+\frac{1}{n},$$

i.e. $\lim_{n\to\infty}\tilde{x}_n=x$. Let now $c_n:=\|x-\tilde{x}_n\|_{(\Phi)}$, then

$$1\ge f^{\Phi}\left(\frac{x-\tilde{x}_n}{c_n}\right)=\sum_{i=N(n)+1}^{\infty}\Phi\left(\frac{x(t_i)}{c_n}\right)+\sum_{i=1}^{N(n)}\Phi\left(\frac{x(t_i)-x_n(t_i)}{c_n}\right).$$

We obtain $\sum_{i=1}^{\infty}\Phi(\frac{x(t_i)}{c_n})<\infty$ for a sequence (c_n) tending to zero and hence $x\in h^\Phi$, because let $k\in\mathbb{N}$ be arbitrary, then $c_n\le\frac{1}{k}$ for $n>N$ and hence

$$\sum_{i=N}^{\infty}\Phi(kx(t_i))\le\sum_{i=N}^{\infty}\Phi\left(\frac{x(t_i)}{c_n}\right)<\infty.$$

We will now establish the uniqueness of the representation:

Suppose, $s_n=\sum_{k=1}^{n}x_ke_k$ and $s'_n=\sum_{k=1}^{n}x'_ke_k$ converge to $x\in h^\Phi$ and suppose there is an $n_0\in\mathbb{N}$ with $x_{n_0}\ne x'_{n_0}$, then $u_n=s_n-s'_n=\sum_{k=1}^{n}(x_k-x'_k)e_k\ne 0$ for $n\ge n_0$. Then due to the monotonicity of the norm for $n\ge n_0$

$$\|u_n\|_{(\Phi)}\ge\|(x_{n_0}-x'_{n_0})e_{n_0}\|_{(\Phi)}=|x_{n_0}-x'_{n_0}|\frac{1}{\Phi^{-1}(1)}>0.$$

Therefore the sequence $(\|u_n\|_{(\Phi)})$ cannot tend to zero, a contradiction. □

Remark. In Theorem 6.2.22 we have established the closedness of h^Φ in a different way.

In the theorem of Lindenstrauss–Tsafriri it will be shown, that ℓ^Φ is separable, if Φ satisfies the Δ_2^0-condition.

Theorem 6.2.36. *ℓ^∞ is not separable.*

Proof. Let $U := \{x \in \ell^\infty \mid x = (x_i), x_i \in \{0,1\} \text{ for all } i \in \mathbb{N}\}$ and let $x, y \in U$ with $x \neq y$, then apparently $\|x - y\|_\infty = 1$. Using Cantor's method one can show that U is uncountable. Thus no countable dense subset of ℓ^∞ can exist. □

In the subsequent consideration we need the following notion

Definition 6.2.37. A basis $\{x_n\}$ of a Banach space is called *boundedly complete*, if for every number sequence (a_n) with $\sup_n \|\sum_{i=1}^n a_i x_i\| < \infty$ the series $\sum_{n=1}^\infty a_n x_n$ converges.

Theorem 6.2.38 (Lindenstrauss–Tsafriri). *Let Φ be a (definite) finite Young function. Then the following statements are equivalent:*

a) *Φ satisfies the Δ_2^0-condition.*

b) $\ell^\Phi = h^\Phi$.

c) *The unit vectors form a boundedly complete basis of ℓ^Φ.*

d) *ℓ^Φ is separable.*

e) *ℓ^Φ contains no subspace isomorphic to ℓ^∞.*

f) $\ell^\Phi = m^\Phi$.

g) $\ell^\Phi = c^\Phi$.

Proof. a) $\Rightarrow$ b): By Theorem 6.2.32 we have $\ell^\Phi = c^\Phi$. Hence c^Φ is a linear space, i.e. $c^\Phi = h^\Phi$.

b) $\Rightarrow$ c): Let $\sup_n \|\sum_{i=1}^n a_i e_i\|_{(\Phi)} = q < \infty$ for a sequence (a_n), then $c_n = \|\sum_{i=1}^n \frac{a_i}{q} e_i\|_{(\Phi)} \leq 1$ holds, hence

$$1 \geq f^\Phi\left(\frac{\sum_{i=1}^n \frac{a_i}{q} e_i}{c_n}\right) \geq f^\Phi\left(\sum_{i=1}^n \frac{a_i}{q} e_i\right).$$

Therefore $1 \geq \sum_{i=1}^\infty \Phi(\frac{a_i}{q})$ and thus $(a_1, a_2, \dots) \in \ell^\Phi = h^\Phi$. By Lemma 6.2.35 the series $\sum_{i=1}^\infty a_i e_i$ converges in h^Φ.

c) $\Rightarrow$ d): The linear combinations of the unit vectors with rational coefficients $\{e_i\}$ form a countable dense subset of ℓ^Φ.

d) $\Rightarrow$ e): Since ℓ^∞ is not separable, ℓ^Φ cannot contain a subspace isomorphic to ℓ^∞.

e) $\Rightarrow$ a): Suppose Φ does not satisfy the Δ_2^0-condition. Then there is a sequence $(t_n)_{n\in\mathbb{N}}$, such that

$$\Phi(2t_n)/\Phi(t_n) > 2^{n+1} \quad \text{and} \quad \Phi(t_n) \leq 2^{-(n+1)}$$

(if necessary choose a subsequence).

Choose a sequence (k_n) of natural numbers in the following way: let k_n be the smallest natural number, such that $2^{-(n+1)} < k_n\Phi(t_n)$, then apparently

$$2^{-(n+1)} < k_n\Phi(t_n) \le 2^{-n},$$

and therefore $\sum_{n=1}^{\infty} k_n\Phi(t_n) \le \sum_{n=1}^{\infty} 2^{-n} = \frac{1}{1-\frac{1}{2}} - 1 = 1$, while on the other hand $k_n\Phi(2t_n) \ge 2^{n+1}\Phi(t_n)\cdot k_n > 1$.

Let $a := (a_n)$ be a bounded sequence of numbers and let $T : \ell^\infty \to \ell^\Phi$ the linear mapping defined by

$$T(a) = x := (\underbrace{a_1t_1,\dots,a_1t_1}_{k_1-\text{times}},\underbrace{a_2t_2,\dots,a_2t_2}_{k_2-\text{times}},\dots,\underbrace{a_nt_n,\dots,a_nt_n}_{k_n-\text{times}},\dots),$$

then for $a \neq 0$

$$f^\Phi\left(\frac{x}{\sup_k(a_k)}\right) = \sum_{n=1}^{\infty} k_n \cdot \Phi\left(\frac{a_nt_n}{\sup_k(a_k)}\right) \le \sum_{n=1}^{\infty} k_n\Phi(t_n) \le 1$$

holds and hence $\|x\|_{(\Phi)} \le \sup_k |a_k|$.

On the other hand let $0 \le \varepsilon < \sup_k |a_k|$

$$f^\Phi\left(\frac{x}{(\sup_k|a_k|-\varepsilon)\cdot 2^{-1}}\right) = \sum_{n=1}^{\infty} k_n\Phi\left(2\frac{a_nt_n}{(\sup_k|a_k|-\varepsilon)}\right) \ge k_{n_0}\Phi(2t_{n_0}) > 1$$

for a suitable n_0, i.e. $\|x\|_{(\Phi)} \ge 2^{-1}\sup_k |a_k|$. In other words

$$2^{-1}\|a\|_\infty \le \|T(a)\|_{(\Phi)} \le \|a\|_\infty.$$

Therefore $T(\ell^\infty)$ is isomorphic to ℓ^∞.

a) $\Rightarrow$ f): follows from b) together with Theorem 6.2.24.

f) $\Rightarrow$ b): This follows immediately from $m^\Phi \subset h^\Phi$.

a) $\Rightarrow$ g): This follows from Theorem 6.2.34.

g) $\Rightarrow$ b): g) implies that c^Φ is a linear space and hence contained in h^Φ. □

In the not purely atomic case similar assertions can be made as in the case of sequence spaces. The following theorem can be found in [107].

Theorem 6.2.39. *Let Φ be a finite Young function and let (T,Σ,μ) be a not purely atomic measure space. Then the following statement holds: if Φ does not satisfy the Δ_2^∞-condition, $L^\Phi(\mu)$ contains a subspace, which is isometrically isomorphic to ℓ^∞.*

Proof. Let Φ not satisfy the Δ_2^∞-condition, then there is a monotonically increasing sequence (t_n) with $t_n \to_{n\to\infty} \infty$, where $\Phi(t_n) > 1$ for all $n \in \mathbb{N}$ and

$$\Phi\left(\left(1+\frac{1}{n}\right)t_n\right) > 2^n\Phi(t_n) \quad \text{for } n \in \mathbb{N}.$$

Let $S \in \Sigma$ non-atomic with $0 < \mu(S) \leq 1$. Let further (S_n) be a sequence of disjoint subsets of S with

$$\mu(S_n) = \frac{\mu(S)}{2^n}.$$

Then we construct a sequence $(x_n)_n$ in $L^\Phi(\mu)$ with $\operatorname{supp} x_n \subset S_n$, such that $f^\Phi(x_n) < \frac{1}{2^n}$.

For the construction of x_n we choose a sequence of measurable, pairwise disjoint sets $S_{n,k} \subset S_n$ with the property $\mu(S_{n,k}) = \frac{\mu(S_n)}{2^k \Phi(t_k)}$. Then we put

$$x_n := \sum_{k=1}^{\infty} t_k \chi_{S_{n,k}}.$$

In this way we obtain

$$f^\Phi(x_n) = \sum_{k=1}^{\infty} \mu(S_{n,k}) \Phi(t_k) = \sum_{k=1}^{\infty} \frac{\mu(S_n)}{2^k} = \frac{\mu(S)}{2^n} \leq \frac{1}{2^n}.$$

Let $a := (a_n)$ be a bounded sequence of numbers and $T : \ell^\infty \to L^\Phi(\mu)$ a linear mapping defined by

$$x := T(a) = \sum_{n=1}^{\infty} a_n x_n,$$

then for $a \neq 0$

$$f^\Phi\left(\frac{x}{\|a\|_\infty}\right) = \sum_{n=1}^{\infty} \int_{S_n} \Phi\left(\frac{a_n x_n}{\|a\|_\infty}\right) d\mu \leq \sum_{n=1}^{\infty} \int_{S_n} \Phi(x_n) d\mu \leq \sum_{n=1}^{\infty} \frac{1}{2^n} = 1$$

holds and hence $\|x\|_{(\Phi)} \leq \|a\|_\infty$. On the other hand let $0 \leq \varepsilon < \|a\|_\infty$ and $n_0 \in \mathbb{N}$ be chosen, such that $\frac{|a_{n_0}|}{\|a\|_\infty - \varepsilon} = 1 + \delta > 1$ then

$$\begin{aligned}
f^\Phi\left(\frac{x}{\|a\|_\infty - \varepsilon}\right) &= \sum_{n=1}^{\infty} \int_{S_n} \Phi\left(\frac{a_n x_n}{\|a\|_\infty - \varepsilon}\right) d\mu \geq \int_{S_{n_0}} \Phi\left(\frac{a_{n_0} x_{n_0}}{\|a\|_\infty - \varepsilon}\right) d\mu \\
&= \int_{S_{n_0}} \Phi((1+\delta) x_{n_0}) d\mu \\
&= \sum_{k=1}^{\infty} \mu(S_{n_0,k}) \Phi((1+\delta) t_k) \geq \sum_{k=k_0}^{\infty} \mu(S_{n_0,k}) \Phi\left(t_k\left(1 + \frac{1}{k}\right)\right) \\
&> \sum_{k=k_0}^{\infty} \mu(S_{n_0,k}) 2^k \Phi(t_k) = \sum_{k=k_0}^{\infty} \mu(S_{n_0}) = \infty,
\end{aligned}$$

i.e. $\|x\|_{(\Phi)} \geq \|a\|_\infty - \varepsilon$. In other words $\|T(a)\|_{(\Phi)} = \|a\|_\infty$. Therefore $T(\ell^\infty)$ is isometrically isomorphic to ℓ^∞. □

For infinite measure we obtain a somewhat weaker assertion:

Theorem 6.2.40. *Let (T, Σ, μ) be a σ-finite, non-atomic measure space with $\mu(T) = \infty$, then the following statement holds: if $L^\Phi(\mu)$ contains no subspace isomorphic to ℓ^∞, then Φ satisfies the Δ_2-condition.*

Proof. The Δ_2^∞-condition we obtain by Theorem 6.2.39.

In order to establish the validity of the Δ_2^0-condition in this situation, we construct a subspace of $L^\Phi(\mu)$, which is isometrically isomorphic to ℓ^Φ:

Let $T = \bigcup_{k=1}^\infty B_k$, where the B_k are pairwise disjoint and of finite measure. Then we choose a sequence of integers k_i with $k_0 = 0$, such that $\mu(\bigcup_{k=k_0+1}^{k_1} B_k) > 1$ and $\mu(\bigcup_{k=k_n+1}^{k_{n+1}} B_k) > 1$. Now choose $S_n \subset \bigcup_{k=k_n+1}^{k_{n+1}} B_k$ with $\mu(S_n) = 1$ (using Lemma 6.2.30). Those $x \in L^\Phi(\mu)$ that are constant on all S_n and are identical to zero on $T \setminus \bigcup_{n=1}^\infty S_n$, then form a subspace of $L^\Phi(\mu)$, which is isometrically isomorphic to ℓ^Φ: let $T : \ell^\Phi \to L^\Phi(\mu)$ be defined by $c \mapsto x := \sum_{i=1}^\infty c_i \chi_{S_i}$, then

$$1 \geq \int_T \Phi\left(\frac{x}{\|x\|_{(\Phi)} + \varepsilon}\right) d\mu = \sum_{i=1}^\infty \Phi\left(\frac{c_i}{\|x\|_{(\Phi)} + \varepsilon}\right)\mu(S_i) = \sum_{i=1}^\infty \Phi\left(\frac{c_i}{\|x\|_{(\Phi)} + \varepsilon}\right),$$

i.e. $\|c\|_{(\Phi)} \leq \|x\|_{(\Phi)} + \varepsilon$, and similarly one obtains $\|c\|_{(\Phi)} \geq \|x\|_{(\Phi)} - \varepsilon$. Then, according to our assumption ℓ^Φ cannot contain a subspace isomorphic to ℓ^∞. Employing the theorem of Lindenstrauss–Tsafriri this yields the Δ_2^0-condition for Φ. □

Corollary 6.2.41. *The above theorem also holds, if (T, Σ, μ) is a σ-finite, not purely atomic measure space with $\mu(T \setminus A) = \infty$, if A denotes the set of atoms.*

As was established above, the Δ_2-condition has the effect that Orlicz space, Orlicz class, and the closure of the step functions coincide. This fact is recorded in the following

Theorem 6.2.42. *Let (T, Σ, μ) be a σ-finite measure space and let Φ satisfy the Δ_2-condition. Then*

$$M^\Phi(\mu) = L^\Phi(\mu) = C^\Phi(\mu) = H^\Phi(\mu).$$

This also holds, if $\mu(T) < \infty$ and Φ satisfies the Δ_2^∞-condition.

Proof. Since Φ satisfies the Δ_2- resp. the Δ_2^∞-condition, Φ is in particular finite and hence $M^\Phi(\mu) = H^\Phi(\mu)$ by Theorem 6.2.24, due to Theorem 6.2.29 we obtain $L^\Phi(\mu) = C^\Phi(\mu)$, and by Theorem 6.2.33 $C^\Phi(\mu) = H^\Phi(\mu)$. □

For not purely atomic measures and sequence spaces we even obtain that the above equality is equivalent to the corresponding Δ_2-conditions.

Theorem 6.2.43. *Let (T, Σ, μ) be a not purely atomic measure space with $\mu(T) < \infty$ and let Φ be finite, then the following statements are equivalent:*

(a) *Φ satisfies the Δ_2^∞-condition*

(b) $H^\Phi = L^\Phi$

(c) $C^\Phi = L^\Phi$

(d) $M^\Phi = L^\Phi$.

Proof. (d) $\Leftrightarrow$ (b) follows from Theorem 6.2.24.

(b) $\Rightarrow$ (c) due to $H^\Phi \subset C^\Phi$.

(c) $\Rightarrow$ (b) since C^Φ is a linear space, hence $C^\Phi \subset H^\Phi$.

(a) $\Rightarrow$ (c) follows from Theorem 6.2.29.

(c) $\Rightarrow$ (a) follows from Theorem 6.2.31. □

If Φ does not satisfy the Δ_2^∞-condition, one can construct a function $x \in C^\Phi(\mu)$ such that the distance of $3x$ to all step functions is at least 1, as the subsequent example shows:

Example 6.2.44. Let Φ not satisfy the Δ_2^∞-condition, then there is a monotonically increasing sequence (t_n) with $t_n \to_{n\to\infty} \infty$, where $\Phi(t_n) \geq 1$ and

$$\Phi(2t_n) > 2^n \Phi(t_n) \quad \text{for } n \in \mathbb{N}.$$

Let further (T_n) be a sequence of disjoint subsets of T, constructed in a similar manner as in Theorem 6.2.31, such that

$$\mu(T_n) = \frac{1}{2^n \Phi(t_n)} \leq \frac{\mu(T)}{2}$$

for n sufficiently large. We now define a function

$$x(t) := \begin{cases} t_n & \text{for } t \in T_n \\ 0 & \text{for } t \in T \setminus \bigcup_{n=1}^\infty T_n. \end{cases}$$

We obtain $x \in C^\Phi(\mu)$, because

$$f^\Phi(x) = \int_T \Phi(x)d\mu = \sum_{n=1}^\infty \Phi(t_n)\mu(T_n) = \sum_{n=1}^\infty \frac{1}{2^n} < \infty.$$

Moreover: $3x \in L^\Phi$ but not in $M^\Phi(\mu)$: in order to see this let y be an arbitrary step function, then y is in particular bounded, i.e. there is a number a with $|y| \leq a$ and hence for n sufficiently large

$$\begin{aligned} f^\Phi(3x - y) &\geq \int_{T_n} \Phi(3x - y)d\mu \geq \Phi(3t_n - a)\mu(T_n) \geq \Phi(2t_n)\mu(T_n) \\ &> 2^n \Phi(t_n) \frac{1}{2^n \Phi(t_n)} = 1, \end{aligned}$$

and therefore $\|3x - y\|_{(\Phi)} \geq 1$.

For infinite and non-atomic measures we have a theorem that corresponds to Theorem 6.2.43:

Theorem 6.2.45. *Let (T, Σ, μ) a σ-finite, non-atomic measure space with $\mu(T) = \infty$ and let Φ be finite, then the following statements are equivalent:*

(a) *Φ satisfies the Δ_2-condition*

(b) $H^\Phi(\mu) = L^\Phi(\mu)$

(c) $C^\Phi(\mu) = L^\Phi(\mu)$

(d) $M^\Phi(\mu) = L^\Phi(\mu)$.

Proof. (d) $\Rightarrow$ (a): Since $M^\Phi(\mu) = H^\Phi(\mu)$ by Theorem 6.2.24 and $H^\Phi(\mu) \subset C^\Phi(\mu)$ the equality $C^\Phi(\mu) = L^\Phi(\mu)$ follows and hence by Theorem 6.2.31 the Δ_2^∞-condition for Φ. Using the same construction for ℓ^Φ as in Theorem 6.2.40, the Orlicz space $L^\Phi(\mu)$ contains a subspace isomorphic to ℓ^Φ. If Φ does not satisfy the Δ_2^0-condition, ℓ^Φ contains a subspace isomorphic to ℓ^∞, in which the step functions are not dense, a contradiction.

(a) $\Rightarrow$ (c) follows from Theorem 6.2.29.

(c) $\Rightarrow$ (b) follows from Theorem 6.2.33.

(b) $\Rightarrow$ (d) follows from Theorem 6.2.24. □

Corollary 6.2.46. *The above theorem also holds, if (T, Σ, μ) is a σ-finite, not purely atomic measure space with $\mu(T \setminus A) = \infty$, if A denotes the set of atoms.*

6.3 Properties of the Modular

6.3.1 Convergence in Modular

We have seen above (see Theorem 6.2.21) that norm convergence always implies convergence in the modular.

The subsequent theorem states that the converse also applies, provided that Φ satisfies the Δ_2-condition.

Theorem 6.3.1. *Let Φ satisfy the Δ_2-condition and let $(x_n)_{n\in\mathbb{N}}$ be a sequence of elements in $L^\Phi(\mu)$. then*

$$f^\Phi(x_n) \xrightarrow{n\to\infty} 0 \quad \Rightarrow \quad \|x_n\|_{(\Phi)} \xrightarrow{n\to\infty} 0$$

holds. This is also true for finite measure, if Φ is definite and only satisfies the Δ_2^∞-condition.

Proof. Let $\varepsilon > 0$ and $2^{-m} < \varepsilon$ for a $m \in \mathbb{N}$, then for a $\lambda > 0$

$$\int_T \Phi(2^m(x_n))d\mu \le \lambda^m \int_T \Phi(x_n)d\mu \xrightarrow{n\to\infty} 0.$$

Hence there is a $N \in \mathbb{N}$, such that for all $n > N$

$$\int_T \Phi\left(\frac{x_n}{2^{-m}}\right)d\mu = \int_T \Phi(2^m(x_n))d\mu \le 1$$

holds. Therefore

$$\|x_n\|_{(\Phi)} \le 2^{-m} < \varepsilon.$$

We now consider the case $\mu(T) < \infty$. Since Φ is definite, the Δ_2^∞-condition holds for arbitrary $k > 0$ (see Remark 6.2.26). Let now $s_0 > 0$ be chosen, such that $\Phi(s_0) \cdot \mu(T) < \frac{1}{2}$. Let T_n be defined by

$$T_n := \{t \in T \mid 2^m|x_n(t)| \ge s_0\},$$

and m chosen as above, then

$$\begin{aligned}\int_T \Phi(2^m(x_n))d\mu &= \int_{T_n} \Phi(2^m(x_n))d\mu + \int_{T\setminus T_n} \Phi(2^m(x_n))d\mu \\ &\le \int_{T_n} \Phi(2^m(x_n))d\mu + \mu(T\setminus T_n)\Phi(s_0)\end{aligned}$$

holds. For the integral over T_n we obtain with $k = \frac{s_0}{2^m}$ and the Δ_2^∞-condition for $s \ge k$ as above

$$\int_{T_n} \Phi(2^m(x_n))d\mu \le \lambda^m \int_{T_n} \Phi(x_n)d\mu \xrightarrow{n\to\infty} 0.$$

Altogether we obtain for n sufficiently large

$$\int_T \Phi(2^m(x_n))d\mu \le 1,$$

and therefore the assertion. □

Example 6.3.2. Let $\Phi(s) = e^{|s|} - 1$. Then Φ does not satisfy the Δ_2^∞-condition. Let further $T = [0, 1]$, $\alpha > 0$, and let

$$x_\alpha(t) = \begin{cases} \ln \frac{\alpha}{\sqrt{t}} & \text{for } 0 < t \le \alpha^2 \\ 0 & \text{for } \alpha^2 < t \le 1. \end{cases}$$

For $c > \frac{1}{2}$ we then obtain

$$f^{\Phi}\left(\frac{x_\alpha}{c}\right) = \int_0^{\alpha^2} \left(\left(\frac{\alpha}{\sqrt{t}}\right)^{\frac{1}{c}} - 1\right) dt = \int_0^{\alpha^2} (\alpha^{\frac{1}{c}} t^{-\frac{1}{2\cdot c}} - 1) dt$$

$$= \left[\alpha^{\frac{1}{c}} \frac{1}{1-\frac{1}{2c}} t^{1-\frac{1}{2c}} - t\right]_0^{\alpha^2} = \left(\alpha^{\frac{1}{c}} \alpha^{2-\frac{1}{c}} \cdot \frac{1}{1-\frac{1}{2c}} - \alpha^2\right)$$

$$= \alpha^2 \left(\frac{1}{1-\frac{1}{2c}} - 1\right) = \alpha^2 \cdot \frac{\frac{1}{2c}}{1-\frac{1}{2c}}.$$

For $c = 1$ we have $f^{\Phi}(x_\alpha) = \alpha^2$ and for $c = \frac{1}{2}(\alpha^2 + 1)$

$$f^{\Phi}\left(\frac{x_\alpha}{c}\right) = \alpha^2 \frac{\frac{1}{\alpha^2+1}}{1 - \frac{1}{\alpha^2+1}} = \frac{\frac{\alpha^2}{1+\alpha^2}}{\frac{\alpha^2}{1+\alpha^2}} = 1,$$

i.e. $\|x_\alpha\|_{(\Phi)} = \frac{1}{2}(\alpha^2 + 1)$.

For $\alpha \to 0$ we obtain $f^{\Phi}(x_\alpha) \to 0$ but $\|x_\alpha\|_{(\Phi)} \to \frac{1}{2}$, showing that convergence in the modular does not imply norm convergence.

Moreover, for $0 < \rho < 2$ we find

$$f^{\Phi}(\rho x_1) = \int_0^1 (t^{-\frac{\rho}{2}} - 1) dt = \frac{\frac{\rho}{2}}{1 - \frac{\rho}{2}} < \infty,$$

but $f^{\Phi}(2 \cdot x_1) = \int_0^1 (e^{2 \ln \frac{1}{\sqrt{t}}} - 1) d\mu = \int_0^1 (\frac{1}{t} - 1) dt = \infty$ i.e. $C^{\Phi} \neq L^{\Phi}$.

For sequence spaces we obtain a correspondence to the previous theorem:

Theorem 6.3.3. *Let Φ satisfy the Δ_2^0-condition and let $(x_n)_{n \in \mathbb{N}}$ be a sequence of elements in ℓ^{Φ}. Then*

$$f^{\Phi}(x_n) \xrightarrow{n\to\infty} 0 \quad \Rightarrow \quad \|x_n\|_{(\Phi)} \xrightarrow{n\to\infty} 0.$$

Proof. We show as a first step the uniform boundedness of the elements of the sequence: from $f^{\Phi}(x_n) \to_{n\to\infty} 0$ we obtain a $c > 0$, such that

$$f^{\Phi}(x_n) = \sum_{t_i \in T} \Phi(x_n(t_i)) \leq c,$$

i.e. $|x_n(t_i)| \leq \Phi^{-1}(c)$ for all $i \in \mathbb{N}$. Let $\varepsilon > 0$ and $2^{-m} < \varepsilon$ for a $m \in \mathbb{N}$. If we put $s_0 := 2^m \Phi^{-1}(c)$, then, using the Δ_2^0-condition for $0 \leq s \leq s_0$ the remaining part of the proof can be performed in analogy to that of Theorem 6.3.1. □

Example 6.3.4. Let

$$\Phi(s) = \begin{cases} 0 & \text{for } s = 0 \\ e^{-\frac{1}{s}} & \text{for } 0 < s \leq \frac{1}{2} \\ \left(s - \frac{1}{2}\right)\left(\frac{2}{e}\right)^2 + e^{-2} & \text{for } s > \frac{1}{2}. \end{cases}$$

For $0 \leq s \leq \frac{1}{4}$ we have

$$\frac{\Phi(2s)}{\Phi(s)} = \frac{e^{-\frac{1}{2s}}}{e^{-\frac{1}{s}}} = e^{-\frac{1}{2s}} \cdot e^{\frac{2}{2s}} = e^{\frac{1}{2s}} \xrightarrow{s \to 0} \infty,$$

i.e. Φ does not satisfy the Δ_2^0-condition. The convexity of Φ for $s \leq \frac{1}{2}$ follows from

$$\Phi'(s) = \frac{1}{s^2} e^{-\frac{1}{s}}$$

$$\Phi''(s) = -\frac{2}{s^3} e^{-\frac{1}{s}} + \frac{1}{s^4} e^{-\frac{1}{s}} = \left(-2 + \frac{1}{s}\right) \frac{1}{s^3} e^{-\frac{1}{s}} \geq 0.$$

We now consider the space l^Φ. Let $n > e^2$, then we define $x_n(t_k) := \frac{1}{\ln n^2}$ for $1 \leq k \leq n$ and equal to zero for $k > n$, then

$$f^\Phi(x_n) = \sum_{k=1}^{n} \Phi\left(\frac{1}{\ln n^2}\right) = \sum_{k=1}^{n} \frac{1}{e^{\ln n^2}} = n \cdot \frac{1}{n^2} = \frac{1}{n} \xrightarrow{n \to \infty} 0,$$

but $\|x_n\|_{(\Phi)} = \frac{1}{2}$, because

$$f^\Phi\left(\frac{x_n}{\frac{1}{2}}\right) = \sum_{k=1}^{n} \Phi\left(\frac{\frac{1}{\ln n^2}}{\frac{1}{2}}\right) = n \cdot \Phi\left(\frac{2}{2 \ln n}\right) = n \cdot \Phi\left(\frac{1}{\ln n}\right) = n \cdot \frac{1}{e^{\ln n}} = 1,$$

i.e. convergence in the modular does not imply norm convergence.

Moreover, $c^\Phi \neq \ell^\Phi$, because let $x(t_n) := \frac{1}{\ln n^2}$ for $n \in \mathbb{N}$ and $n \geq 2$, then $f^\Phi(x) = \sum_{n=2}^{\infty} \frac{1}{n^2} < \infty$ but

$$f^\Phi(2x) = \sum_{n=2}^{\infty} \Phi\left(\frac{2}{2 \ln n}\right) = \sum_{n=2}^{\infty} \frac{1}{n} = \infty.$$

6.3.2 Level Sets and Balls

In the sequel we will establish a relation between level sets of the modular and balls w.r.t. the Luxemburg norm.

Notation 6.3.5. The ball about x with radius r w.r.t. the Luxemburg norm we denote by

$$K_{(\Phi)}(x, r) := \{y \in L^\Phi(\mu) \mid \|y - x\|_{(\Phi)} < r\}.$$

Lemma 6.3.6. *Let* $1 > \varepsilon > 0$ *then* $K_{(\Phi)}(0, 1-\varepsilon) \subset S_{f^\Phi}(1) \subset \overline{K}_{(\Phi)}(0,1)$ *holds.*

Proof. Let $x \in S_{f^\Phi}(1)$ then

$$\int_T \Phi\left(\frac{x}{1}\right) d\mu = \int_T \Phi(x) d\mu \leq 1.$$

By definition this implies $\|x\|_{(\Phi)} \leq 1$. On the other hand let $x \in K_{(\Phi)}(0, 1-\varepsilon)$ and $x \neq 0$, then by definition of the Luxemburg norm there is a sequence (c_n) of positive numbers such that $c_n > \|x\|_{(\Phi)}$ and $\lim_{n\to\infty} c_n = \|x\|_{(\Phi)} \leq 1 - \varepsilon$, i.e. $c_n \leq 1 - \frac{\varepsilon}{2}$ for n sufficiently large. Due to the convexity of Φ we then obtain

$$1 \geq \int_T \Phi\left(\frac{x}{c_n}\right) d\mu \geq \frac{1}{c_n} \int_T \Phi(x) d\mu,$$

and thus $\int_T \Phi(x) d\mu \leq c_n < 1$ i.e. $x \in S_{f^\Phi}(1)$. □

Remark 6.3.7. In Theorem 7.3.5 we will show that $S_{f^\Phi}(1)$ is closed.

Remark 6.3.8. From the above lemma and Theorem 7.3.5 we conclude

$$S_{f^\Phi}(1) = \overline{K}_{(\Phi)}(0,1).$$

By the characterization of the continuity of real-valued convex functions (see Theorem 3.5.4) we know that boundedness on a ball guarantees continuity. But according to the above lemma we have $K_{(\Phi)}(0, 1-\varepsilon) \subset S_{f^\Phi}(1)$ and hence $f^\Phi(x) \leq 1$ for $x \in K_{(\Phi)}(0, 1-\varepsilon)$.

In particular we obtain the following

Theorem 6.3.9. *Let* Φ *be finite, then* $f^\Phi : H^\Phi(\mu) \to \mathbb{R}$ *is a continuous function.*

Proof. By definition f^Φ is finite on H^Φ. □

6.3.3 Boundedness of the Modular

A convex function on a normed space is called bounded, if it is bounded on bounded sets.

Theorem 6.3.10. *If* Φ *satisfies the* Δ_2*-condition (resp. for* T *of finite measure the* Δ_2^∞*-condition), then* $f^\Phi : L^\Phi(\mu) \to \mathbb{R}$ *is a bounded function.*

Proof. Let at first Φ satisfy the Δ_2-condition, let $M > 0$ and $x \in K_{(\Phi)}(0, M)$. Let further $2^n \geq M$, then we obtain

$$f^\Phi(x) \leq f^\Phi\left(\frac{2^n}{M} x\right) \leq \lambda^n f^\Phi\left(\frac{x}{M}\right) \leq \lambda^n.$$

Let now T be of finite measure and let Φ satisfy the Δ_2^∞-condition, then put

$$T_0 := \left\{ t \in T \,\middle|\, \frac{|x(t)|}{M} \geq s_0 \right\},$$

and we obtain

$$\begin{aligned} f^\Phi(x) &= \int_{T_0} \Phi(x)d\mu + \int_{T\setminus T_0} \Phi(x)d\mu \leq \int_{T_0} \Phi\left(\frac{2^n}{M}x\right)d\mu + \mu(T)\Phi(Ms_0) \\ &\leq \lambda^n \int_{T_0} \Phi\left(\frac{x}{M}\right)d\mu + \mu(T)\Phi(Ms_0) \leq \lambda^n + \mu(T)\Phi(Ms_0). \qquad \square \end{aligned}$$

For sequence spaces we obtain

Theorem 6.3.11. *Let Φ be finite and let Φ satisfy the Δ_2^0-condition, then $f^\Phi : \ell^\Phi \to \mathbb{R}$ is a bounded function.*

Proof. Let Φ satisfy the Δ_2^0-condition on every interval $[0, a]$ (see Remark 6.2.26), let $M > 0$ and $x \in K_{(\Phi)}(0, M)$. Then we have: $\sum_{i=1}^\infty \Phi(\frac{x_i}{M}) \leq 1$ and hence $\frac{|x_i|}{M} \leq \Phi^{-1}(1)$. Let further $2^n \geq M$, then we obtain for $a := 2^n\Phi^{-1}(1)$

$$f^\Phi(x) \leq f^\Phi\left(\frac{2^n}{M}x\right) \leq \lambda^n f^\Phi\left(\frac{x}{M}\right) \leq \lambda^n. \qquad \square$$

Chapter 7

Orlicz Norm and Duality

7.1 The Orlicz Norm

The functional N_Φ defined below will turn out to be a norm, whose domain of finite values will emerge to be $L^\Phi(\mu)$. This norm will be called the *Orlicz norm.*

Definition 7.1.1. Let Φ be a Young function and Ψ its conjugate. Let E be the space of equivalence classes of μ-measurable functions, then we define the following functional:

$$N_\Phi : E \to \overline{\mathbb{R}},$$

where

$$N_\Phi(u) := \sup\left\{ \left| \int_T v \cdot u d\mu \right| \,\middle|\, v \in S_{f^\Psi}(1) \right\}.$$

Remark. The functional defined above is monotone in the following sense: from $|u_1| \leq |u_2|$ it follows that $N_\Phi(u_1) \leq N_\Phi(u_2)$, because

$$\begin{aligned}\left| \int_T v u_1 d\mu \right| &= \left| \int_T \operatorname{sign}(u_1) v |u_1| d\mu \right| \leq \int_T |v|\,|u_1| d\mu \leq \int_T |v|\,|u_2| d\mu \\ &= \left| \int_T \operatorname{sign}(u_2) |v| u_2 d\mu \right| \leq N_\Phi(u_2)\end{aligned}$$

holds, and hence apparently $N_\Phi(u_1) \leq N_\Phi(u_2)$.

Theorem 7.1.2. *Let Ψ be the conjugate function of Φ, then*

(a) *N_Φ is finite on $L^\Phi(\mu)$*

(b) *$L^\Phi(\mu) \subset L^\Psi(\mu)^*$*

(c) *$N_\Phi \leq 2\|\cdot\|_{(\Phi)}$*

hold.

Proof. Let $x \in L^\Phi(\mu)$ and $y \in S_{f^\Psi}(1)$. Let further $\lambda > \|x\|_{(\Phi)}$, then due to Young's inequality

$$\left| \int_T x \cdot y d\mu \right| \leq \lambda \int_T \left| \frac{x}{\lambda} \right| |y| d\mu \leq \lambda \left(\int_T \Phi\left(\frac{x}{\lambda}\right) d\mu + \int_T \Psi(y) d\mu \right) \leq \lambda(1+1) = 2\lambda. \tag{7.1}$$

By Lemma 6.3.6 we have $K_{(\Psi)}(0, 1-\varepsilon) \subset S_{f^\Psi}(1)$. Thus the functional $y \mapsto \int_T x \cdot y d\mu$ defined by x is bounded on $K_{(\Psi)}(0, 1-\varepsilon)$ and hence a continuous functional on $L^\Psi(\mu)$. Moreover we obtain $N_\Phi(x) \leq 2\lambda$ and hence (c). □

The functional N_Φ is apparently a norm on $L^\Phi(\mu)$, which we will call the Orlicz norm, and it will turn out to be equivalent to the Luxemburg norm (see Theorem 7.5.4).

Definition 7.1.3. On $L^\Phi(\mu)$ we define the *Orlicz norm* by

$$\|x\|_\Phi := \sup\left\{\left|\int_T x \cdot y d\mu\right| \,\middle|\, y \in S_{f^\Psi}(1)\right\}.$$

Remark. Since

$$S_{f^\Psi}(1) = \overline{K}_{(\Psi)}(0, 1),$$

(see Remark 6.3.8) it follows that

$$\|x\|_\Phi = \sup\left\{\left|\int_T x \cdot y d\mu\right| \,\middle|\, \|y\|_{(\Psi)} \leq 1\right\}.$$

Thus the Orlicz norm corresponds to the canonical norm of the dual space, i.e. $(L^\Phi(\mu), \|\cdot\|_\Phi)$ is isometrically imbedded into the dual space of $L^\Psi(\mu)$.

7.2 Hölder's Inequality

Theorem 7.2.1. *The following inequalities hold:*

(a) $\|x\|_\Phi \leq 2\|x\|_{(\Phi)}$ *for all* $x \in L^\Phi(\mu)$

(b) Hölder's inequality $|\int_T x \cdot y d\mu| \leq \|x\|_\Phi \cdot \|y\|_{(\Psi)}$.

Proof. (a) follows from Theorem 7.1.2.

Let further be $\varepsilon > 0$, then by Lemma 6.3.6 $\frac{y}{\|y\|_{(\Psi)}+\varepsilon} \in S_{f^\Psi}(1)$, i.e.

$$\left|\int_T x \cdot \frac{y}{\|y\|_{(\Psi)} + \varepsilon} d\mu\right| \leq \|x\|_\Phi,$$

whence Hölder's inequality follows. □

Since the conjugate of Ψ is again Φ, we can interchange the roles of Φ and Ψ in the above consideration and we obtain $L^\Psi(\mu)$ as a subspace of $L^\Phi(\mu)^*$.

We will investigate under what conditions these two spaces are equal in the sequel. It will emerge that in this context the Δ_2-condition plays a central role.

7.3 Lower Semi-continuity and Duality of the Modular

Beyond the description of the dual spaces we are also interested in this chapter in the duality relation between modulars: the fact that Young functions are mutual conjugates carries completely over to the corresponding modulars.

Definition 7.3.1. $(f^\Psi)^* : (L^\Psi(\mu))^* \to \overline{\mathbb{R}}$

$$(f^\Psi)^*(x^*) = \sup_{y \in L^\Psi(\mu)} \left\{ \int_T x^* y d\mu - f^\Psi(y) \right\}.$$

Remark. $(f^\Psi)^*$ is thereby also defined on $L^\Phi(\mu)$, because by Theorem 7.1.2 we have $L^\Phi(\mu) \subset (L^\Psi(\mu))^*$.

We will use the following lemma for real-valued functions f on Cartesian products of arbitrary sets as a tool to interchange the order of least upper bounds for the values of f:

Lemma 7.3.2. *Let A, B be sets and $f : A \times B \to \overline{\mathbb{R}}$, then*

$$\sup_{A\times B} f(a,b) = \sup_A \sup_B f(a,b) = \sup_B \sup_A f(a,b)$$

holds.

Proof. Let $(a_n, b_n)_n$ be a sequence in $A \times B$ with

$$f(a_n, b_n) \xrightarrow{n\to\infty} \sup_{A\times B} f(a,b).$$

Let first of all $\sup_{A\times B} f(a,b) < \infty$, then for $n > N$ and given $\varepsilon > 0$

$$f(a_n, b_n) \geq \sup_{A\times B} f(a,b) - \varepsilon$$

holds, moreover

$$f(a_n, b_n) \leq \sup_B f(a_n, b) \leq \sup_A \sup_B f(a,b)$$

$$f(a_n, b_n) \leq \sup_A f(a, b_n) \leq \sup_B \sup_A f(a,b).$$

Since $\sup_B f(a,b) \leq \sup_{A\times B} f(a,b)$ for all $a \in A$ and

$$\sup_A f(a,b) \leq \sup_{A\times B} f(a,b) \quad \text{for all } b \in B,$$

we obtain

$$\sup_{A\times B} f(a,b) - \varepsilon \leq \left\{ \begin{matrix} \sup_A \sup_B f(a,b) \\ \sup_B \sup_A f(a,b) \end{matrix} \right\} \leq \sup_{A\times B} f(a,b).$$

The case $\sup_{A\times B} f(a,b) = \infty$ can be treated in a corresponding manner. □

Lemma 7.3.3. *Let Φ be a non-finite Young function. Then there is a sequence $(\Phi_n)_{n\in\mathbb{N}}$ of finite Young functions converging pointwise monotonically increasing to Φ. The sequence $(\Psi_n)_{n\in\mathbb{N}}$ of the corresponding conjugates then converges pointwise monotonically decreasing to Ψ, the conjugate of Φ.*

Proof. Let Φ be finite for $|s| < a$ and infinite for $|s| > a$, and let (s_n) be a monotonically increasing sequence in $(0, a)$ with $s_n \to a$. Then we define for $s \geq 0$

$$\Phi_n(s) := \begin{cases} \Phi(s) & \text{for } 0 \leq s \leq s_n \\ \Phi(s_n) + \Phi'_+(s_n)(s - s_n) & \text{for } s_n \leq s < a \\ \Phi(s_n) + \Phi'_+(s_n)(a - s_n) + n(s - a)^2 & \text{for } s \geq a. \end{cases}$$

For $s < 0$ we put: $\Phi_n(s) := \Phi_n(-s)$.

We now discuss the monotonicity:

(a) $0 \leq s \leq s_n$: here we have: $\Phi_n(s) = \Phi_{n+1}(s)$.

(b) $s_n \leq s \leq s_{n+1}$: according to the subgradient inequality we have

$$\Phi_n(s) = \Phi(s_n) + \Phi'_+(s_n)(s - s_n) \leq \Phi(s) = \Phi_{n+1}(s). \tag{7.2}$$

(c) $s_{n+1} \leq s \leq a$: and again by the subgradient inequality

$$\Phi_n(s_{n+1}) = \Phi(s_n) + \Phi'_+(s_n)(s_{n+1} - s_n) \leq \Phi(s_{n+1}) = \Phi_{n+1}(s_{n+1}),$$

apparently

$$\Phi_n(s) = \Phi_n(s_{n+1}) + \Phi'_+(s_n)(s - s_{n+1}),$$

and hence, using (7.2) and the monotonicity of the right-sided derivative

$$\Phi_n(s) \leq \Phi_{n+1}(s_{n+1}) + \Phi'_+(s_{n+1})(s - s_{n+1}) = \Phi_{n+1}(s),$$

(d) $s > a$: since, again due to (7.2) and the monotonicity of the right-sided derivative, we have

$$\begin{aligned}\Phi_n(a) &= \Phi_n(s_n) + \Phi'_+(s_n)((a - s_{n+1}) + (s_{n+1} - s_n)) \\ &\leq \Phi(s_{n+1}) + \Phi'_+(s_n)(a - s_{n+1}) \leq \Phi_{n+1}(a),\end{aligned}$$

monotonicity also holds in this case.

As to pointwise convergence we remark: for $0 \leq s < a$ we have for n sufficiently large: $s < s_n$ and hence $\Phi_n(s) = \Phi(s)$.

For $s = a$ we find: $0 \leq \Phi'_+(s_n)(a - s_n) \leq \Phi(a) - \Phi(s_n)$ and hence

$$\Phi(s_n) \leq \Phi(s_n) + \Phi'_+(s_n)(a - s_n) = \Phi_n(a) \leq \Phi(a).$$

Due to the lower semi-continuity of Φ we have: $\Phi(s_n) \to \Phi(a)$ and hence (Φ_n) converges pointwise to Φ.

This also holds, if $\Phi(a)$ is finite and $\Phi'_+(s_n) \to \infty$, but also, if $\Phi(a) = \infty$, because Φ is due to its lower semi-continuity continuous in the extended sense.

Moreover, Ψ is by Lemma 6.2.9 finite. Thus the pointwise convergence of the sequence (Ψ_n) to Ψ follows from Theorem 6.1.27. □

Theorem 7.3.4. *Let Φ be a Young function and Ψ its conjugate, then for arbitrary $x \in L^\Phi(\mu)$*

$$f^\Phi(x) = \sup_{y \in L^\Psi(\mu)} \left\{ \int_T xy d\mu - f^\Psi(y) \right\} = (f^\Psi)^*(x)$$

holds.

Proof. At first let Φ be finite. By Young's inequality we have for $y \in L^\Psi$ a.e.: $\Phi(x(t)) \geq x(t) \cdot y(t) - \Psi(y(t))$ and hence: $f^\Phi(x) \geq \int_T xy d\mu - f^\Psi(y)$. Therefore

$$f^\Phi(x) \geq \sup_{y \in L^\Psi(\mu)} \left\{ \int_T xy d\mu - f^\Psi(y) \right\}.$$

If $0 \leq u \in E$ is bounded with finite support, then the same is true for $\Psi(\Phi'_+(u))$, since according to Lemma 6.1.10 $\Psi(\Phi'_+(0)) = 0$ and $\Psi(\Phi'_+(\cdot))$ is finite and apparently monotonically increasing. Hence we obtain $f^\Psi(y) < \infty$ for $y = \Phi'_+(u)$, in particular also $y \in L^\Psi(\mu)$. Young's equality then implies $\Phi(u) = u\Phi'_+(u) - \Psi(\Phi'_+(u))$ and therefore

$$f^\Phi(u) = \int_T uy d\mu - f^\Psi(y).$$

Let now $x \in L^\Phi(\mu)$ be arbitrary, $T = \bigcup_{j=1}^\infty B_j$, $\mu(B_j) < \infty$ for $j \in \mathbb{N}$, $B^n := \bigcup_{j=1}^n B_j$ and

$$x_n(t) := \begin{cases} x(t) & \text{for } |x(t)| \leq n,\ t \in B^n \\ 0 & \text{otherwise} \end{cases}$$

then all functions x_n are bounded, have a support of finite measure and $|x_n| \uparrow |x|$ a.e. Employing the theorem on monotone convergence of integration theory it follows that $f^\Phi(x_n) \uparrow f^\Phi(x)$ and for $y_n = \Phi'_+(x_n)$ we obtain

$$f^\Phi(x_n) = \int_T x_n y_n d\mu - f^\Psi(y_n).$$

We now distinguish two cases:

(a) $f^\Phi(x) = \infty$: let $c > 0$ be arbitrary, then there is a $N \in \mathbb{N}$, such that for $n > N$

$$f^\Phi(x_n) = \int_T x_n y_n d\mu - f^\Psi(y_n) > c.$$

For n fixed then $x_k y_n \to_{k\to\infty} xy_n$ pointwise a.e., and by construction $|x_k|\,|y_n| \leq |x|\,|y_n|$, and $\int_T |x|\,|y_n| d\mu < \infty$ due to Hölder's inequality. Lebesgue's convergence theorem then implies that

$$\int_T |x_k|\,|y_n| d\mu \to \int_T |x|\,|y_n| d\mu,$$

and hence

$$\int_T xy_n d\mu - f^\Psi(y_n) > c - \varepsilon$$

for n sufficiently large, hence

$$\sup_{y\in L^\Psi(\mu)} \left\{ \int_T xyd\mu - f^\Psi(y) \right\} = \infty.$$

(b) $f^\Phi(x) < \infty$: let $0 < \varepsilon < f^\Phi(x)$, then for n sufficiently large

$$f^\Phi(x_n) = \int_T x_n y_n d\mu - f^\Psi(y_n) > f^\Phi(x) - \varepsilon,$$

and hence in a similar way as above

$$\sup_{y\in L^\Psi(\mu)} \left\{ \int_T xyd\mu - f^\Psi(y) \right\} \geq f^\Phi(x).$$

Let now Φ not be finite, then there is a sequence (Φ_n) of finite Young functions, converging pointwise monotonically increasing to Φ (see Lemma 7.3.3). By Lemma 6.2.7 we have in particular $L^\Phi \subset L^{\Phi_n}$. The sequence (Ψ_n) of the conjugate Young functions converges pointwise monotonically decreasing to Ψ (see Theorem 6.1.27 and Lemma 7.3.3).

Let now $y \in L^\Psi$, then we obtain by Hölder's inequality

$$\left| \int_T xyd\mu \right| \leq \|x\|_{(\Phi)} \cdot \|y\|_\Psi < \infty.$$

According to our above consideration we conclude

$$f^{\Phi_n}(x) = \sup_{y\in L^{\Psi_n}} \left\{ \int_T xyd\mu - f^{\Psi_n}(y) \right\} = \sup_{y\in L^\Psi} \left\{ \int_T xyd\mu - f^{\Psi_n}(y) \right\},$$

where the last equation can be justified as follows: by Lemma 6.2.7 we have $L^{\Psi_n} \subset L^\Psi$, but for $y \in L^\Psi \setminus L^{\Psi_n}$ we have $f^{\Psi_n}(y) = \infty$.

The theorem on monotone convergence of integration theory yields for $x \in L^\Phi$

$$f^\Phi(x) = \sup_{n\in\mathbb{N}} f^{\Phi_n}(x),$$

and for $y \in L^\Psi$

$$f^\Psi(y) = \inf_{n\in\mathbb{N}} f^{\Psi_n}(y) = -\sup_{n\in\mathbb{N}}\{-f^{\Psi_n}(y)\}.$$

Altogether we conclude using the Supsup-Lemma 7.3.2

$$\begin{aligned} f^\Phi(x) &= \sup_n f^{\Phi_n}(x) = \sup_n \sup_{y\in L^\Psi}\left\{\int_T |xy|d\mu - f^{\Psi_n}(y)\right\} \\ &= \sup_{y\in L^\Psi}\left\{\int_T |xy|d\mu - \inf_n f^{\Psi_n}(y)\right\} = \sup_{y\in L^\Psi}\left\{\int_T |xy|d\mu - f^\Psi(y)\right\}. \end{aligned}$$ □

An important consequence of the above theorem is the following:

Theorem 7.3.5. $f^\Phi : L^\Phi(\mu) \to \overline{\mathbb{R}}$ *is lower semi-continuous and thus* $S_{f^\Phi}(1)$ *is closed.*

Proof. Being the supremum of continuous functionals $\int_T(\cdot)y d\mu - f^\Psi(y)$ on $L^\Phi(\mu)$ with $y \in C^\Psi(\mu)$ then according to Remark 3.5.2 f^Φ is lower semi-continuous on $L^\Phi(\mu)$. □

7.4 Jensen's Integral Inequality and the Convergence in Measure

As an application of Hölder's inequality it will turn out that norm convergence always implies convergence in measure. This fact will become important in the next chapter.

Definition 7.4.1. We say, a sequence of measurable functions $(x_n)_{n\in\mathbb{N}}$ *converges in measure* to a measurable function x_0, if there is a sequence $(Q_n)_{n\in\mathbb{N}}$ with $\mu(Q_n) \to_{n\to\infty} 0$, such that

$$\lim_{n\to\infty} \sup_{t\in T\setminus Q_n} |x_n(t) - x_0(t)| = 0.$$

For the proof of the implication we have announced above, we can make use of the following

Theorem 7.4.2 (Jensen's Integral Inequality). *Let A be a measurable set of finite measure and Φ a Young function and let $v \in C^\Phi(\mu)$, then*

$$\Phi\left(\frac{1}{\mu(A)}\int_A v d\mu\right) \le \frac{1}{\mu(A)}\int_A \Phi(v)d\mu$$

holds.

Proof. Let $v \in C^{\Phi}(\mu)$, then, due to Hölder's inequality, $|v|$ is integrable over A, because

$$\int_A v d\mu = \int_T v\chi_A d\mu \leq \|v\|_{(\Phi)}\|\chi_A\|_{\Psi}.$$

Let now

$$v_n(t) := \begin{cases} v(t) & \text{for } |v(t)| \leq n \text{ and } t \in A \\ 0 & \text{otherwise.} \end{cases}$$

Then $v_n \in C^{\Phi}(\mu)$, since $|v_n(t)| \leq |v(t)|$ i.e. $f^{\Phi}(v_n) \leq f^{\Phi}(v)$. We will now approximate v_n by a sequence of step functions: let $\delta_{nk} := \frac{n}{k}$ for $k \in \mathbb{N}$ and let for $r \in \mathbb{Z}$ with $0 \leq r \leq k$

$$C_{nkr} := \{t \in A \mid r\delta_{nk} \leq v_n(t) < (r+1)\delta_{nk}\}$$

and for $-k \leq r < 0$

$$C_{nkr} := \{t \in B^n \mid (r-1)\delta_{nk} < v_n(t) \leq r\delta_{nk}\}.$$

Then we define for $r = -k, \dots, k$

$$v_{nk}(t) := \begin{cases} a_{nkr} := r\delta_{nk} & \text{for } t \in C_{nkr} \\ 0 & \text{otherwise.} \end{cases}$$

Then $|v_{nk}(t)| \leq |v_n(t)|$ and $|v_{nk}(t) - v_n(t)| \leq \delta_{nk}$ for all $t \in A$ and hence

$$\begin{aligned}\frac{1}{\mu(A)}\int_A \Phi(v)d\mu &\geq \frac{1}{\mu(A)}\int_A \Phi(v_n)d\mu \geq \frac{1}{\mu(A)}\int_A \Phi(v_{nk})d\mu \\ &= \frac{1}{\mu(A)}\sum_r \Phi(a_{nkr})\mu(C_{nkr}) \geq \Phi\Big(\frac{1}{\mu(A)}\sum_r a_{nkr}\mu(C_{nkr})\Big) \\ &= \Phi\Big(\frac{1}{\mu(A)}\int_A v_{nk}d\mu\Big),\end{aligned}$$

where in particular the latter sequence is bounded. By construction we have

$$\frac{1}{\mu(A)}\left|\int_A v_{nk}d\mu - \int_A v_n d\mu\right| \leq \delta_{nk},$$

and the lower semi-continuity of Φ yields $\frac{1}{\mu(A)}\int_A v_n d\mu \in \operatorname{Dom}\Phi$. Let now $\varepsilon > 0$, then, due to the continuity of Φ on $\operatorname{Dom}\Phi$, for $k \in \mathbb{N}$ sufficiently large

$$\Phi\Big(\frac{1}{\mu(A)}\int_A v_{nk}d\mu\Big) \geq \Phi\Big(\frac{1}{\mu(A)}\int_A v_n d\mu\Big) - \varepsilon.$$

Hence

$$\frac{1}{\mu(A)}\int_A \Phi(v)d\mu \geq \Phi\Big(\frac{1}{\mu(A)}\int_A v_n d\mu\Big) - \varepsilon.$$

We conclude by using Lebesgue's convergence theorem $\int_A v_n d\mu \to \int_A v d\mu$. □

For the Luxemburg norm of characteristic functions we obtain:

Lemma 7.4.3. *Let Φ be an arbitrary Young function and Q a measurable set with $\mu(Q) < \infty$, then*

(a) $\|\chi_Q\|_{(\Phi)} = \frac{1}{\Phi^{-1}(\frac{1}{\mu(Q)})}$ *holds, provided that $\frac{1}{\mu(Q)}$ is an element of the range of Φ,*

(b) $\|\chi_Q\|_{(\Phi)} = \frac{1}{t_\infty}$ *otherwise, where $\Phi(t) = \infty$ for $t > t_\infty$ and $\Phi(t) < \infty$ on $[0, t_\infty)$.*

Proof. (a) For $c = \frac{1}{\Phi^{-1}(\frac{1}{\mu(Q)})}$ we obtain

$$1 = \mu(Q)\Phi\left(\frac{1}{c}\right) = \int_T \Phi\left(\frac{\chi_Q}{c}\right) d\mu,$$

hence $\|\chi_Q\|_{(\Phi)} = \frac{1}{\Phi^{-1}(\frac{1}{\mu(Q)})}$.

(b) In this case we have: $\inf\{c > 0 \mid \mu(Q)\Phi(\frac{1}{c}) \le 1\} = \frac{1}{t_\infty}$, since for $t \in [0, t_\infty)$ we have $\Phi(t) < \frac{1}{\mu(Q)}$, hence for $\delta > 0$: $\Phi(\frac{t_\infty}{1+\delta}) < \frac{1}{\mu(Q)}$. Therefore $\|\chi_Q\|_{(\Phi)} \le \frac{1+\delta}{t_\infty}$, on the other hand: $\Phi(\frac{t_\infty}{1-\delta}) = \infty$ and thus $\|\chi_Q\|_{(\Phi)} \ge \frac{1-\delta}{t_\infty}$. □

In order to obtain the Orlicz norm of characteristic functions we can employ Jensen's inequality.

Lemma 7.4.4. *Let Q be a measurable set with $\mu(Q) < \infty$ and let Ψ be finite, then $\|\chi_Q\|_\Phi = \mu(Q)\Psi^{-1}(\frac{1}{\mu(Q)})$ holds.*

Proof. By definition: $\|\chi_Q\|_\Phi = \sup\{\int_T \chi_Q v d\mu \mid f^\Psi(v) \le 1\}$. Due to the previous lemma we have: $\|\chi_Q\|_{(\Psi)} = \frac{1}{\Psi^{-1}(\frac{1}{\mu(Q)})}$. If we put $v := \chi_Q \cdot \Psi^{-1}(\frac{1}{\mu(Q)})(1-\varepsilon)$, then we obtain

$$\|\chi_Q\|_\Phi \ge \mu(Q)\Psi^{-1}\left(\frac{1}{\mu(Q)}\right)(1-\varepsilon)$$

for every $0 < \varepsilon < 1$.

On the other hand we obtain using Jensen's integral inequality for $f^\Psi(v) \le 1$

$$\Psi\left(\frac{1}{\mu(Q)}\int_Q v d\mu\right) \le \frac{1}{\mu(Q)}\int_Q \Psi(v) d\mu \le \frac{1}{\mu(Q)} f^\Psi(v) \le \frac{1}{\mu(Q)},$$

therefore $\int_T \chi_Q v d\mu \le \mu(Q)\Psi^{-1}(\frac{1}{\mu(Q)})$ for all $v \in S_{f^\Psi}(1)$, hence

$$\|\chi_Q\|_\Phi \le \mu(Q)\Psi^{-1}\left(\frac{1}{\mu(Q)}\right).$$ □

We are now in the position to present the connection between norm convergence and convergence in measure.

Theorem 7.4.5. *Let (T, Σ, μ) be a σ-finite measure space. Then the following statement holds: $x_n \to x_0$ in $L^\Phi(\mu)$ implies convergence in measure of x_n to x_0.*

Proof. Suppose this is not the case, and let (Q_n) be a sequence of measurable sets with $\mu(Q_n) \to_{n\to\infty} 0$. Then there is a $\varepsilon > 0$, such that for all $N \in \mathbb{N}$ there is $n > N$ with the property

$$\sup_{t \in T \setminus Q_n} |x_n(t) - x_0(t)| \geq \varepsilon,$$

i.e. there is a subsequence $P_{n_k} \subset T \setminus Q_{n_k}$ and a $\rho > 0$ with $\rho \leq \mu(P_{n_k})$ and

$$|x_{n_k}(t) - x_0(t)| \geq \varepsilon \quad \text{for all } t \in P_{n_k}.$$

At first we show: there are subsets $R_{n_k} \subset P_{n_k}$ of finite measure, such that $0 < \rho \leq \mu(R_{n_k})$

(a) $\mu(T) < \infty$: put $R_{n_k} := P_{n_k}$

(b) $\mu(T) = \infty$: since the measure space is σ-finite, there is a representation $T = \bigcup_{i=1}^\infty T_i$ with pairwise disjoint T_i of finite measure. Then in particular $P_{n_k} = \bigcup_{i=1}^\infty T_i \cap P_{n_k}$. If $\mu(P_{n_k}) = \infty$, then choose $k(n)$ such that $R_{n_k} = \bigcup_{i=1}^{k(n)} T_i \cap P_{n_k}$ and $\rho < \mu(R_{n_k}) < \infty$. If $\mu(P_{n_k}) < \infty$, then put $R_{n_k} := P_{n_k}$.

We now distinguish two situations (at least one of the two Young functions, Φ or Ψ, is finite):

(a) Φ is finite, then $|x_{n_k} - x_0| \geq |x_{n_k} - x_0|\chi_{R_{n_k}}$ and – due to the monotonicity of the Luxemburg norm

$$\|x_{n_k} - x_0\|_{(\Phi)} \geq \varepsilon \frac{1}{\Phi^{-1}(\frac{1}{\mu(R_{n_k})})} \geq \varepsilon \frac{1}{\Phi^{-1}(\frac{1}{\rho})} > 0,$$

a contradiction.

(b) Ψ finite: using Hölder's inequality we obtain

$$\int_T |x_{n_k} - x_0|\chi_{R_{n_k}} d\mu \leq \|x_{n_k} - x_0\|_{(\Phi)} \|\chi_{R_{n_k}}\|_\Psi \leq 2\|x_{n_k} - x_0\|_{(\Phi)} \|\chi_{R_{n_k}}\|_{(\Psi)}.$$

On the other hand

$$\int_T |x_{n_k} - x_0|\chi_{R_{n_k}} d\mu \geq \varepsilon\mu(R_{n_k}).$$

Hence we obtain

$$2\|x_{n_k} - x_0\|_{(\Phi)} \geq \varepsilon\mu(R_{n_k})\Psi^{-1}\left(\frac{1}{\mu(R_{n_k})}\right) \geq \varepsilon\rho\Psi^{-1}\left(\frac{1}{\rho}\right) > 0,$$

since the function $s \mapsto s\Psi^{-1}(\frac{1}{s})$ is according to Lemma 6.1.26 monotonically increasing on $(0, \infty)$, again a contradiction. □

Remark. Direct use of the Orlicz norm of the characteristic function in part (b) of the above proof would have yielded the inequality

$$\|x_{n_k} - x_0\|_{(\Phi)} \|\chi_{R_{n_k}}\|_\Psi \geq \varepsilon\mu(R_{n_k}),$$

therefore

$$\|x_{n_k} - x_0\|_{(\Phi)} \geq \varepsilon \frac{1}{\Psi^{-1}(\frac{1}{\mu(R_{n_k})})} \geq \varepsilon \frac{1}{\Psi^{-1}(\frac{1}{\rho})} > 0,$$

since it is easily seen that the function $s \mapsto \frac{1}{\Psi^{-1}(\frac{1}{s})}$ is monotonically increasing on $(0, \infty)$.

Again by use of Hölder's inequality we obtain a corresponding theorem for sequence spaces:

Theorem 7.4.6. *$x_n \to x_0$ in $\ell^\Phi(\mu)$ implies convergence of x_n to x_0 in the supremum norm.*

Proof. Suppose there is a $\varepsilon > 0$, a subsequence (x_m) such that for every m there is a point t_{i_m} with the property: $|x_m(t_{i_m}) - x_0(t_{i_m})| > \varepsilon$, then we obtain

$$\sum_{i=1}^{\infty} |x_m(t_i) - x_0(t_i)| \chi_{t_{i_m}}(t_i) \leq 2\|x_m - x_0\|_{(\Phi)} \|\chi_{t_{i_m}}\|_{(\Psi)},$$

where $\|\chi_{t_{i_m}}\|_{(\Psi)}$ is constant according to Lemma 7.4.3. On the other hand

$$\sum_{i=1}^{\infty} |x_m(t_i) - x_0(t_i)| \chi_{t_{i_m}}(t_i) = |x_m(t_{i_m}) - x_0(t_{i_m})| > \varepsilon,$$

a contradiction. □

7.5 Equivalence of the Norms

In the sequel we will always assume that T is σ-finite.

Lemma 7.5.1. *Let Φ be finite and let Φ'_- denote the left-sided derivative of Φ. Let u be measurable with $N_\Phi(u) \leq 1$. Then $v_0 = \Phi'_-(|u|)$ belongs to $C^\Psi(\mu)$ and we have $f^\Psi(v_0) \leq 1$.*

Proof. First of all we show: let $v \in C^\Psi(\mu)$ then

$$\left| \int_T uv d\mu \right| \leq \begin{cases} N_\Phi(u) & \text{if } f^\Psi(v) \leq 1 \\ N_\Phi(u) \cdot f^\Psi(v) & \text{for } f^\Psi(v) > 1. \end{cases}$$

The first inequality follows by definition.

Let now $f^{\Psi}(v) > 1$, then from $\Psi(0) = 0$ and Ψ convex we conclude

$$\Psi\left(\frac{v}{f^{\Psi}(v)}\right) \leq \frac{\Psi(v)}{f^{\Psi}(v)},$$

and therefore due to $v \in C^{\Psi}(\mu)$

$$f^{\Psi}\left(\frac{v}{f^{\Psi}(v)}\right) = \int_T \Psi\left(\frac{v}{f^{\Psi}(v)}\right) d\mu \leq \frac{1}{f^{\Psi}(v)} \int_T \Psi(v) d\mu = 1.$$

By definition of N_{Φ} we then obtain

$$\left| \int_T u \frac{v}{f^{\Psi}(v)} d\mu \right| \leq N_{\Phi}(u),$$

and hence

$$\left| \int_T uv d\mu \right| \leq N_{\Phi}(u) \cdot f^{\Psi}(v) \quad \text{if } f^{\Psi}(v) > 1. \tag{7.3}$$

Let now $N_{\Phi}(u) \leq 1$ and $u \neq 0$. Let $T = \bigcup_{j=1}^{\infty} B_j$, $\mu(B_j) < \infty$ for $j \in \mathbb{N}$ and $B^n := \bigcup_{j=1}^{n} B_j$. We define

$$u_n(t) := \begin{cases} u(t) & \text{for } |u(t)| \leq n \text{ and } \ t \in B^n \\ 0 & \text{otherwise.} \end{cases}$$

Young's equality yields $\Psi(\Phi'_-)$ finite on all of $\mathbb{R}$ and $\Psi(\Phi'_-(0)) = 0$ (see Lemma 6.1.10), hence

$$\begin{aligned} f^{\Psi}(\Phi'_-(|u_n|)) &= \int_T \Psi(\Phi'_-(|u_n|)) d\mu = \int_{B_n} \Psi(\Phi'_-(|u_n|)) d\mu \leq \int_{B_n} \Psi(\Phi'_-(n)) d\mu \\ &= \mu(B_n) \Psi(\Phi'_-(n)), \end{aligned}$$

therefore $\Phi'_-(|u_n|) \in C^{\Psi}(\mu)$. Since $u \neq 0$ there is $N \in \mathbb{N}$, such that $u_n \neq 0$ for $n \geq N$. We now consider two cases:

(a) Φ definite, then $\Phi(u_n(t)) > 0$ on a set of positive measure.

(b) Φ indefinite, then $s_0 := \max\{s \,|\, \Phi(s) = 0\} > 0$. If $\Phi(u_n(t)) = 0$ a.e. then $|u_n(t)| \leq s_0$ a.e. and hence $\Phi'_-(|u_n(t)|) = 0$ a.e., hence $\Psi(\Phi'_-(|u_n(t)|)) = 0$ a.e. So, if $\Psi(\Phi'_-(|u_n(t)|)) > 0$ on a set T_0 of positive measure, then $|u_n(t)| > s_0$ on T_0, and hence $\Phi(u_n(t)) > 0$ on T_0.

Suppose there is $n_0 \geq N$ with $f^{\Psi}(\Phi'_-(|u_{n_0}|)) = \int_T \Psi(\Phi'_-(|u_{n_0}|)) d\mu > 1$.

In particular $\Phi(u_{n_0}(t)) > 0$ on a set of positive measure. Young's equality then yields

$$\begin{aligned} \Psi(\Phi'_-(|u_{n_0}(t)|)) &< \Phi(u_{n_0}(t)) + \Psi(\Phi'_-(|u_{n_0}(t)|) \\ &= |u_{n_0}(t)| \cdot \Phi'_-(|u_{n_0}(t)|) \end{aligned}$$

on a set of positive measure. Using Inequality (7.3) we obtain by integration of the above inequality

$$f^\Psi(\Phi'_-(|u_{n_0}|)) = \int_T \Psi(\Phi'_-(|u_{n_0}|))d\mu < \int_T |u_{n_0}(t)|\Phi'_-(|u_{n_0}(t)|)d\mu(t)$$
$$\leq N_\Phi(u_{n_0}) \cdot f^\Psi(\Phi'_-(|u_{n_0}|)),$$

and thus $N_\Phi(u_{n_0}) > 1$. But our assumption was $N_\Phi(u) \leq 1$ and by construction $|u_{n_0}| \leq |u|$, contradicting the monotonicity of N_Φ. Hence $f^\Psi(\Phi'_-(|u_n|)) \leq 1$ for all $n \in \mathbb{N}$ with $n \geq N$. The lemma of Fatou then implies $f^\Psi(v_0) \leq 1$ and hence $v_0 \in C^\Psi(\mu)$. □

Lemma 7.5.2. *Let u be measurable and let $N_\Phi(u) \leq 1$, then $f^\Phi(u) \leq N_\Phi(u)$ holds.*

Proof. First of all let Φ be finite and let $v_0 = \Phi'_-(|u|)\,\mathrm{sign}(u)$. By Lemma 7.5.1 we have $f^\Psi(v_0) \leq 1$. Young's equality yields

$$u(t) \cdot v_0(t) = \Phi(u(t)) + \Psi(v_0(t)),$$

and hence

$$f^\Phi(u) = \int_T \Phi(u)d\mu \leq \int_T \Phi(u)d\mu + \int_T \Psi(v_0)d\mu$$
$$= \int_T u \cdot v_0\, d\mu \leq N_\Phi(u).$$

Let now Φ not be finite. Then, as we will see below, we can construct a sequence of finite Young functions (Φ_n), which is monotonically non-decreasing pointwise to Φ. Then by Lemma 7.3.3 $\Psi_n \geq \Psi$, hence $S_{f^{\Psi_n}}(1) \subset S_{f^\Psi}(1)$ and hence $N_{\Phi_n}(u) \leq N_\Phi(u)$ for every measurable function u. Let now $N_\Phi(u) \leq 1$, then also $N_{\Phi_n}(u) \leq 1$ and therefore, according to the first part of the proof of this lemma

$$\int_T \Phi_n(u)d\mu = f^{\Phi_n}(u) \leq N_{\Phi_n}(u) \leq N_\Phi(u).$$

By the theorem on monotone convergence, $\Phi_n(u) \uparrow \Phi(u)$ implies

$$\int_T \Phi(u)d\mu = f^\Phi(u) \leq N_\Phi(u).$$

As to the construction of the sequence (Φ_n): let (α_n) be a monotone sequence of positive numbers tending to zero and $\Psi_n(s) := \alpha_n s^2 + \Psi(s)$ for arbitrary $s \in \mathbb{R}$. Then (Ψ_n) converges monotonically decreasing to Ψ. The sequence (Φ_n) of conjugate functions is finite, since the strongly convex functions $\Psi_n(s) - ts$ have compact level sets for all $t \in \mathbb{R}$. By construction the sequence (Φ_n) is monotonically increasing. The stability theorem on monotone convergence (see Theorem 5.2.3) then implies (by convergence of the values) the pointwise convergence (monotonically increasing) of the sequence (Φ_n) to Φ. □

Theorem 7.5.3. *Let $u \in E$ and $N_\Phi(u) < \infty$, then $u \in L^\Phi(\mu)$ and*

$$N_\Phi(u) \geq \|u\|_{(\Phi)}.$$

Proof. Lemma 7.5.2 implies

$$f^\Phi\left(\frac{u}{N_\Phi(u)}\right) \leq N_\Phi\left(\frac{u}{N_\Phi(u)}\right) = 1,$$

and therefore $N_\Phi(u) \geq \|u\|_{(\Phi)}$. □

Thus Orlicz and Luxemburg norms are equivalent:

Theorem 7.5.4. *For all $u \in L^\Phi(\mu)$*

$$2\|u\|_{(\Phi)} \geq \|u\|_\Phi \geq \|u\|_{(\Phi)}$$

holds.

Proof. Theorem 7.5.3 and Theorem 7.1.2, as well as Definition 7.1.3. □

Remark. Theorem 7.5.4 also provides the following insight: the set

$$M := \{u \in E \mid N_\Phi(u) < \infty\}$$

is equal to $L^\Phi(\mu)$. In order to describe the Orlicz space as a set, the finiteness of the Luxemburg norm is equivalent to the finiteness of the Orlicz norm.

7.6 Duality Theorems

Theorem 7.6.1 (Representation Theorem). *Let (T, Σ, μ) be a σ-finite measure space and let Φ finite, then every continuous linear functional f on $M^\Phi(\mu)$ can be represented by a function $y \in L^\Psi(\mu)$ via the formula*

$$\langle f, x\rangle = \int_T y(t)x(t)d\mu \quad x \in L^\Phi.$$

If $M^\Phi(\mu)$ is equipped with the Luxemburg norm, then $\|f\| = \|y\|_\Psi$.

Proof. Let $T = \bigcup_{j=1}^\infty B_j$ where $\mu(B_j) < \infty$, and let

$$B^n := \bigcup_{j=1}^n B_j.$$

Let Σ_n the σ-algebra induced on B^n and let $B \in \Sigma_n$, n fixed, then $\chi_B \in L^\Phi(\mu)$. We will now show that $\gamma(B) := \langle f, \chi_B \rangle$ is σ-additive and absolutely continuous w.r.t. μ: let $A_i \in \Sigma_n$ for $i \in \mathbb{N}$,

$$A := \bigcup_{i=1}^{\infty} A_i, \; A_i \cap A_j = \emptyset \quad \text{for } i \neq j,$$

then $\sum_{i=1}^\infty \chi_{A_i} = \chi_A$. At first we show

$$\left\| \sum_{i=1}^{m} \chi_{A_i} - \chi_A \right\|_{(\Phi)} \xrightarrow{m \to \infty} 0.$$

Apparently we have $\mu(A) = \sum_{i=1}^\infty \mu(A_i)$ and hence

$$\left\| \sum_{i=1}^{m} \chi_{A_i} - \chi_A \right\|_{(\Phi)} = \frac{1}{\Phi^{-1}(\frac{1}{\mu(\bigcup_{i=m+1}^\infty A_i)})} = \frac{1}{\Phi^{-1}(\frac{1}{\sum_{i=m+1}^\infty \mu(A_i)})} \xrightarrow{m \to \infty} 0.$$

The continuity of the functional f implies

$$\gamma(A) = \langle f, \chi_A \rangle = \sum_{i=1}^{\infty} \langle f, \chi_{A_i} \rangle = \sum_{i=1}^{\infty} \gamma(A_i).$$

Thus γ is σ-additive.

As to the absolute continuity: let $E \in \Sigma_n$, then

$$\gamma(E) = \langle f, \chi_E \rangle \leq \|f\| \|\chi_E\|_{(\Phi)} = \|f\| \frac{1}{\Phi^{-1}(\frac{1}{\mu(E)})}$$

therefore

$$\gamma(E) = \langle f, \chi_E \rangle \xrightarrow{\mu(E) \to 0} 0.$$

Then the theorem of Radon–Nykodym implies: there is a uniquely determined function $y_n(t) \in L^1(B^n)$ with

$$\gamma(B) = \int_{B^n} y_n \cdot \chi_B \, d\mu.$$

By the uniqueness of y_n and $B^n \subset B^m$ for $n \leq m$ we obtain $y_m|_{B^n} = y_n$. Let now $y^{(n)}(t) := y_n(t)$ for $t \in B^n$ and $y^{(n)}(t) = 0$ otherwise. Then $y^{(n)}$ is defined on T, and $\gamma(B) = \int_T y^{(n)} \cdot \chi_B d\mu$ for $B \in \Sigma_n$. Let now y be the pointwise limit of the sequence $y^{(n)}$, then in particular $y_{|B^n} = y_n$ for all $n \in \mathbb{N}$. Then for an arbitrary step function $x = \sum_{i=1}^k c_i \chi_{A_i}$ with $A_i \in \Sigma_n$ for $i = 1, \ldots, k$

$$\langle f, x \rangle = \int_T xy d\mu. \tag{7.4}$$

As a next step we show: (7.4) holds for all $x \in M^\Phi(\mu)$. By definition x is limit of a sequence of step functions. We will repeat here the explicit construction in the proof of Theorem 6.2.24, because we need its monotonicity property for further reasoning. Let now

$$x_n(t) := \begin{cases} x(t) & \text{for } |x(t)| \leq n \text{ and } t \in B^n \\ 0 & \text{otherwise,} \end{cases}$$

then $x_n \in L^\Phi(\mu)$, since $|x_n(t)| \leq |x(t)|$ i.e. $\|x_n\|_{(\Phi)} \leq \|x\|_{(\Phi)}$. We will now approximate x_n by a sequence of step functions: let $\delta_{nk} := \frac{n}{k}$ for $k \in \mathbb{N}$ and let for $r \in \mathbb{Z}$ with $0 \leq r \leq k$

$$C_{nkr} := \{t \in B^n \,|\, r\delta_{nk} \leq x_n(t) < (r+1)\delta_{nk}\},$$

and for $-k \leq r < 0$

$$C_{nkr} := \{t \in B^n \,|\, (r-1)\delta_{nk} < x_n(t) \leq r\delta_{nk}\}.$$

Then we define for $r = -k, \ldots, k$

$$x_{nk}(t) := \begin{cases} r\delta_{nk} & \text{for } t \in C_{nkr} \\ 0 & \text{otherwise.} \end{cases}$$

Then $|x_{nk}(t)| \leq |x_n(t)|$ and $|x_{nk} - x_n| \leq \delta_{nk}$, i.e. (x_{nk}) converges uniformly to x_n, and hence

$$\lim_{k\to\infty} \|x_{nk} - x_n\|_{(\Phi)} = 0.$$

The continuity of the functional f and its representation on the space of step functions (7.4) implies

$$\langle f, x_n \rangle = \lim_{k\to\infty} \langle f, x_{nk} \rangle = \lim_{k\to\infty} \int_T x_{nk} y \, d\mu.$$

By construction we have

$$\lim_{k\to\infty} x_{nk}(t)y(t) = x_n(t)y(t) \quad \text{a.e.}$$

Moreover

$$|x_{nk}(t)y(t)| \leq |x_n(t)y(t)| \leq n|y(t)| \cdot \chi_{B^n}(t).$$

The function $|y|$ is (due to the theorem of Radon–Nikodym) μ-integrable over B^n. Therefore the requirements of Lebesgue's convergence theorem are satisfied and we obtain

$$\langle f, x_n \rangle = \lim_{k\to\infty} \int_T x_{nk} y \, d\mu = \int_T \lim_{k\to\infty} x_{nk} y \, d\mu = \int_T x_n y \, d\mu.$$

By construction we also have $|x_n| \le |x|$ and hence $\|x_n\|_{(\Phi)} \le \|x\|_{(\Phi)}$. Furthermore, we have by construction: $|x_n y| \to_{n\to\infty} |xy|$ a.e. and we obtain using the lemma of Fatou

$$\left|\int_T xy d\mu\right| \le \int_T |xy| d\mu \le \sup_n \int_T |x_n y| d\mu = \sup_n \langle f, |x_n| \operatorname{sign}(y)\rangle \tag{7.5}$$

$$\le \sup_n \|f\| \|x_n\|_{(\Phi)} \le \|f\| \|x\|_{(\Phi)}. \tag{7.6}$$

i.e. $x(t)y(t)$ is integrable. The sequence $(|x_n y|)$ is monotonically increasing and converges a.e. to $|xy|$. Thus Lebesgue's convergence theorem is applicable and we obtain

$$\lim_{n\to\infty} \langle f, x_n\rangle = \lim_{n\to\infty} \int_T x_n y\, d\mu = \int_T xy\, d\mu. \tag{7.7}$$

We will now show: $\lim_{n\to\infty} \|x - x_n\| = 0$: since $x - x_n \in H^\Phi(\mu)$ (see Theorem 6.2.24 and Theorem 6.2.15) we have

$$1 = \int_T \Phi\left(\frac{x - x_n}{\|x - x_n\|_{(\Phi)}}\right) d\mu.$$

Suppose there is a subsequence (x_{n_k}) with $\|x - x_{n_k}\|_{(\Phi)} \ge \delta > 0$, then

$$1 \le \int_T \Phi\left(\frac{x - x_{n_k}}{\delta}\right) d\mu.$$

On the other hand we have by construction $\frac{|x - x_{n_k}|}{\delta} \le \frac{|x|}{\delta}$ and therefore $\Phi(\frac{|x-x_{n_k}|}{\delta}) \le \Phi(\frac{|x|}{\delta})$. But again due to Theorem 6.2.24 we have $\int_T \Phi(\frac{|x|}{\delta}) d\mu < \infty$ and by construction $\Phi(\frac{|x-x_{n_k}|}{\delta}) \to 0$ a.e. Lebesgue's convergence theorem then yields

$$\int_T \Phi\left(\frac{x - x_{n_k}}{\delta}\right) d\mu \to 0,$$

a contradiction.

By the continuity of f and equation (7.7) we obtain

$$\lim_{n\to\infty} \langle f, x_n\rangle = \langle f, x\rangle = \int_T xy\, d\mu,$$

i.e. the representation of the functional f by a measurable function y.

It remains to be shown: $y \in L^\Psi(\mu)$ and $\|y\|_\Psi = \|f\|$. The first part follows, if we can substantiate $N_\Psi(y) < \infty$, this is certainly the case if: $|\int_T zy d\mu| \le \|f\|$ for all $z \in L^\Phi(\mu)$ with $f^\Phi(z) \le 1$: for given $z \in S_{f^\Phi}(1)$ we construct (in analogy to the above sequence (x_n)) a sequence of bounded functions (z_n), which in turn are approximated by step functions. In particular we then have $z_n \in M^\Phi(\mu)$. Then as above the lemma of Fatou implies

$$\left|\int_T zyd\mu\right| \le \int_T |zy|d\mu \le \sup_n \int_T |z_n y|d\mu = \sup_n \langle f, |z_n| \operatorname{sign}(y)\rangle \tag{7.8}$$

$$\le \sup_n \|f\|\|z_n\|_{(\Phi)} \le \|f\|\|z\|_{(\Phi)}. \tag{7.9}$$

By definition we have

$$\|f\| = \sup\{\langle f, x\rangle \mid \|x\|_{(\Phi)} \le 1, x \in M^\Phi(\mu)\},$$

and by virtue of the representation of the functional

$$\|f\| = \sup\left\{\int_T xyd\mu \mid \|x\|_{(\Phi)} \le 1, x \in M^\Phi(\mu)\right\}. \tag{7.10}$$

But Inequality (7.8) even implies together with (7.10)

$$\|f\| = \sup\left\{\int_T zyd\mu \mid \|z\|_{(\Phi)} \le 1, z \in L^\Phi(\mu)\right\} = \|y\|_\Psi$$

where the last equality holds by the definition of the Orlicz norm. □

Remark. Since in the above theorem z is in general not in $H^\Phi(\mu)$, the norm convergence of the sequence (z_n) does not follow, because $\int_T \Phi(\frac{z}{\delta})d\mu < \infty$ does not necessarily hold.

Under additional conditions the roles of Luxemburg norm and Orlicz norm can be interchanged:

Theorem 7.6.2. *Let Φ and Ψ be finite and let $M^\Phi(\mu) = L^\Phi(\mu)$, then*

(a) $(M^\Psi(\mu), \|\cdot\|_\Psi)^* = (L^\Phi(\mu), \|\cdot\|_{(\Phi)})$

(b) $(M^\Psi(\mu), \|\cdot\|_\Psi)^{**} = (L^\Psi(\mu), \|\cdot\|_\Psi)$

hold.

Proof. Let $X := (M^\Phi(\mu), \|\cdot\|_{(\Phi)})$, then by the previous theorem $X^* = (L^\Psi(\mu), \|\cdot\|_\Psi)$. Let now $U := (M^\Psi(\mu), \|\cdot\|_\Psi)$ and let $f \in U^*$. Due to the equivalence of Luxemburg and Orlicz norms there is – according to the representation theorem – a $y \in L^\Phi(\mu) = M^\Phi(\mu)$ with $\langle f, u\rangle = \int_T u \cdot yd\mu$ for all $u \in U$. Thus we obtain

$$\|f\| = \sup\{\langle f, u\rangle \mid u \in M^\Psi, \|u\|_\Psi \le 1\} = \sup\left\{\int_T uy\, d\mu \,\middle|\, u \in M^\Psi, \|u\|_\Psi \le 1\right\}. \tag{7.11}$$

Let now $z \in S_{f^\Psi}(1)$, then we construct in an analogous way as in the previous theorem a sequence

$$z_n(t) := \begin{cases} z(t) & \text{for } |z(t)| \leq n \text{ and } t \in B^n \\ 0 & \text{otherwise,} \end{cases}$$

then as in the previous theorem $z_n \in M^\Psi(\mu)$. Using the monotonicity of the Orlicz norm we obtain, again similar to the previous theorem

$$\begin{aligned} \left| \int_T zyd\mu \right| &\leq \int_T |zy|d\mu \leq \sup_n \int_T |z_n y|d\mu = \sup_n \langle f, |z_n| \operatorname{sign}(y) \rangle \\ &\leq \sup_n \|f\| \|z_n\|_\Psi \leq \|f\| \|z\|_\Psi, \end{aligned}$$

and hence together with (7.11)

$$\|f\| = \sup \left\{ \left| \int_T zyd\mu \right| \,\middle|\, z \in L^\Psi, \|z\|_\Psi \leq 1 \right\}.$$

Defining $\langle f, z \rangle := \int_T zyd\mu$ for $z \in L^\Psi$ we can, due to Hölder's inequality, extend f as a continuous functional to L^Ψ. But by Theorem 7.6.1 we have $L^\Psi = X^*$, i.e. $f \in X^{**}$. According to [39] p. 181, Theorem 41.1 the canonical mapping $I : X \to X^{**}$ with $y \mapsto \int_T (\cdot) yd\mu$ is norm preserving, i.e. in this case: $\|f\| = \|y\|_{(\Phi)}$. Hence equation (a) holds.

Since

$$(L^\Phi(\mu), \|\cdot\|_{(\Phi)})^* = (M^\Phi(\mu), \|\cdot\|_{(\Phi)})^* = (L^\Psi(\mu), \|\cdot\|_\Psi),$$

and (a) also equation (b) follows. □

As a consequence of Theorem 7.6.1 we obtain

Theorem 7.6.3. *Let Φ satisfy the Δ_2-condition. Then every continuous linear functional f on $(L^\Phi(\mu), \|\cdot\|_{(\Phi)})$ is represented by a function $y \in L^\Psi(\mu)$ via the formula*

$$\langle f, x \rangle = \int_T y(t)x(t)d\mu \quad x \in L^\Phi,$$

where $\|f\| = \|y\|_\Psi$. This also holds, if $\mu(T) < \infty$ and Φ satisfies only the Δ_2^∞-condition.

Proof. With Theorem 6.2.42 we have $M^\Phi = L^\Phi$. The remaining part follows from the representation Theorem 7.6.1. □

Remark. In particular $(L^p(\mu))^* = L^q(\mu)$ for $p \geq 1$ and $\frac{1}{p} + \frac{1}{q} = 1$.

For sequence spaces we obtain in analogy to Theorem 7.6.3.

Theorem 7.6.4. *Let Φ satisfy the Δ_2^0-condition. Every continuous linear functional f on $(\ell^\Phi, \|\cdot\|_{(\Phi)})$ is represented by a sequence $y \in \ell^\Psi$ via the formula*

$$\langle f, x\rangle = \sum_{t_i \in T} y(t_i)x(t_i) \quad x \in \ell^\Phi,$$

where $\|f\| = \|y\|_\Psi$.

Proof. This follows from the theorem of Lindenstrauss–Tsafriri 6.2.38 together with Theorem 7.6.1. □

Remark. In particular $(\ell^p)^* = \ell^q$ for $p \geq 1$ and $\frac{1}{p} + \frac{1}{q} = 1$.

If Ψ is not finite, then we obtain the following assertion about the dual space of $L^\Phi(\mu))$:

Theorem 7.6.5. *Let Ψ not be finite and let for infinite measure in addition Ψ be indefinite, then*

$$(L^\Phi(\mu))^* = L^\Psi(\mu).$$

Moreover, $L^\Psi(\mu)$ and $L^\infty(\mu)$ are equal as sets and the Orlicz norm $\|\cdot\|_\Psi$ is equivalent to $\|\cdot\|_\infty$.

Proof. By Theorem 6.2.28 Φ satisfies the Δ_2^∞- (resp. for infinite measure the Δ_2-) condition. Then by Theorem 6.2.42 $M^\Phi(\mu) = L^\Phi(\mu)$. By Theorem 6.2.10 $L^1(\mu)$ and $L^\Phi(\mu)$ are equal as sets and the corresponding norms are equivalent.

By the above remark $(L^1(\mu))^* = L^\infty(\mu)$. Again by Theorem 6.2.10 $L^\Psi(\mu)$ and $L^\infty(\mu)$ are equal as sets and the Luxemburg norm $\|\cdot\|_{(\Psi)}$ is equivalent to $\|\cdot\|_\infty$. By Theorem 7.5.4 the Luxemburg norm $\|\cdot\|_{(\Psi)}$ is equivalent to the Orlicz norm $\|\cdot\|_\Psi$. □

Remark. The above theorem does not hold in general for $\mu(T) = \infty$ and Ψ definite, since for

$$\Psi(s) = \begin{cases} s^2 & \text{for } |s| \leq 1 \\ \infty & \text{otherwise,} \end{cases}$$

we obtain: $\ell^\Psi = \ell^2$ and $\ell^\Phi = \ell^2$.

7.7 Reflexivity

Our insights into dual spaces, provided by the previous section, put us into the position to characterize the reflexivity of Orlicz spaces.

Theorem 7.7.1. *Let Φ and Ψ be mutually conjugate Young functions, then:*

(a) *Let $\mu(T) < \infty$ and let the measure space be not purely atomic, then $L^\Phi(\mu)$ is reflexive, if and only if Φ and Ψ satisfy the Δ_2^∞-condition.*

(b) *Let $\mu(T) = \infty$ and the measure space not purely atomic with $\mu(T \setminus A) = \infty$ (A set of atoms in T), then $L^\Phi(\mu)$ is reflexive, if and only if Φ and Ψ satisfy the Δ_2-condition.*

Proof. At first we show: Φ finite, for suppose Φ not finite, then $L^\Phi(\mu)$ contains a subspace isomorphic to $L^\infty(\mu_0)$: let $T_0 \in \Sigma$ be of finite measure and (Σ_0, μ_0, T_0) the measure space induced on T_0, then $L^\Phi(\mu_0)$ is a closed subspace of $L^\Phi(\mu)$. By Theorem 6.2.10 we have $L^\Phi(\mu_0)$ is isomorphic to $L^\infty(\mu_0)$. According to (see [41]) every closed subspace of a reflexive Banach space is reflexive, hence $L^\Phi(\mu)$ cannot be reflexive if Φ is not finite.

Let now $\mu(T) < \infty$. Suppose the Δ_2^∞-condition is not satisfied, then by Theorem 6.2.39 $L^\Phi(\mu)$ contains a subspace isomorphic to ℓ^∞, contradicting its reflexivity.

Therefore by Theorem 6.2.42 $M^\Phi(\mu) = L^\Phi(\mu)$ and by Theorem 7.6.1 $(L^\Phi(\mu))^* = L^\Psi(\mu)$. Since $(L^\Phi(\mu))^{**} = L^\Phi(\mu)$ then also $L^\Psi(\mu)$ is reflexive, i.e. Ψ also satisfies the Δ_2^∞-condition.

For infinite measure the assertion follows in analogy to the above line of reasoning using Theorem 6.2.40 and Corollary 6.2.41. □

For sequence spaces we obtain a somewhat weaker version of the above theorem:

Theorem 7.7.2. *Let Φ and Ψ be finite, then ℓ^Φ is reflexive, if and only if Φ and Ψ satisfy the Δ_2^0-condition.*

We finally obtain from reflexivity of the closure of the step functions already equality to the Orlicz space.

Theorem 7.7.3. *Let be Φ finite. If $M^\Phi(\mu)$ is reflexive, then $M^\Phi(\mu) = L^\Phi(\mu)$ holds.*

Proof. By definition $M^\Phi(\mu) \subset L^\Phi(\mu)$. By Theorem 7.6.1 we have: $(M^\Phi(\mu))^* = L^\Psi(\mu)$. Reflexivity then implies $M^\Phi(\mu) = (M^\Phi(\mu))^{**} = (L^\Psi(\mu))^*$. But from Hölder's inequality it follows that $L^\Phi(\mu) \subset (L^\Psi(\mu))^*$ and hence

$$L^\Phi(\mu) \subset (L^\Psi(\mu))^* = M^\Phi(\mu) \subset L^\Phi(\mu).$$ □

7.8 Separability and Bases of Orlicz Spaces

7.8.1 Separability

For sequence spaces the theorem of Lindenstrauss–Tsafriri (see Theorem 6.2.38) contains the following assertion:

$$\ell^\Phi \text{ is separable, if and only if } \Phi \text{ satisfies the } \Delta_2^0\text{-condition},$$

provided that Φ is finite.

Furthermore, Theorem 6.2.36 asserts: ℓ^∞ is not separable.

A basis for the investigation of non-atomic measures in this section is the

Theorem 7.8.1 (Lusin). *Let T be a compact Hausdorff space, μ the corresponding Baire measure and let x be a measurable function. Then for every $\varepsilon > 0$ there is a function y continuous on T such that*

$$\mu(\{t \in T \mid x(t) - y(t) \neq 0\}) < \varepsilon.$$

Moreover, if $\|x\|_\infty < \infty$ then y can be chosen such that $\|y\|_\infty \leq \|x\|_\infty$.

Then (compare [72]):

Theorem 7.8.2. *Let T be a compact Hausdorff space, (T, Σ, μ) the corresponding Baire measure space, and let Φ and Ψ be finite. Then the functions continuous on T are dense in $M^\Phi(\mu)$.*

Proof. Let $u \in M^\Phi(\mu)$ be bounded, such that $|u| \leq a$, then by the theorem of Lusin there is a sequence (u_n) of continuous functions with $|u_n| \leq a$, such that $u - u_n$ different from zero only on a set T_n with $\mu(T_n) \leq \frac{1}{n}$. Then by Hölder's inequality and Lemma 7.4.4 and Lemma 6.1.26

$$\begin{aligned}
\|u - u_n\|_\Phi &= \sup_{\|v\|_{(\Psi)} \leq 1} \left| \int_T (u - u_n) v \, d\mu \right| \\
&\leq 2a \sup_{\|v\|_{(\Psi)} \leq 1} \int_{T_n} |v| \, d\mu = 2a \|\chi_{T_n}\|_\Phi \\
&= 2a\mu(T_n) \Psi^{-1}\left(\frac{1}{\mu(T_n)}\right) \leq \frac{2a}{n} \Psi^{-1}(n).
\end{aligned}$$

By Lemma 6.1.26 it follows that: $\lim_{n\to\infty} \|u - u_n\|_\Phi = 0$. Let now $w \in M^\Phi(\mu)$ be arbitrary and $\varepsilon > 0$, then by definition there is a step function u (obviously bounded) with $\|w - u\|_\Phi < \frac{\varepsilon}{2}$, and hence $\|w - u_n\|_\Phi < \varepsilon$ for n sufficiently large. □

Let T be a compact subset of $\mathbb{R}^m$ and μ the Lebesgue measure, then, as a consequence of the theorem of Stone–Weierstraß (see e.g. [59]), for every function x continuous on T there is a sequence of polynomials with rational coefficients, converging uniformly – and hence also in $L^\Phi(\mu)$ – to x. Thus we obtain the following

Theorem 7.8.3. *Let Φ and Ψ be finite. Let T be a compact subset of $\mathbb{R}^m$ and μ the Lebesgue measure, then $M^\Phi(\mu)$ is separable.*

7.8.2 Bases

Definition 7.8.4. A *(Schauder) basis* (φ_i) of a Banach space X is a sequence of elements of X, such that for every $x \in X$ there is a unique sequence (x_i) of real numbers with the property $\lim_n \|\sum_{i=1}^n x_i\varphi_i - x\| = 0$.

Definition 7.8.5. A basis $\{\varphi_i\}$ of a Banach space is called *boundedly complete*, if for every number sequence (x_i) with $\sup_n \|\sum_{i=1}^n x_i\varphi_i\| < \infty$ there is an $x \in X$, such that the series $\sum_{i=1}^\infty x_i\varphi_i$ converges to x.

We will now repeat the theorem of Lindenstrauss–Tsafriri (see 6.2.38) with those characterizations relevant in this context:

Theorem 7.8.6. *Let Φ be a finite Young function. Then the following statements are equivalent:*

a) *Φ satisfies the Δ_2^0-condition.*

b) *The unit vectors form a boundedly complete basis of ℓ^Φ.*

c) *ℓ^Φ is separable.*

Definition 7.8.7. The series $\sum_{i=1}^\infty u_i$ is called *unconditionally convergent* to $x \in X$, if for every permutation p of the natural numbers the series $\sum_{i=1}^\infty u_{p(i)}$ converges to x.

Definition 7.8.8. A basis (φ_i) of a Banach space X is called an *unconditional basis* of X, if for every $x \in X$ the series $\sum_{i=1}^\infty x_i\varphi_i$ converges unconditionally to x.

The following theorem can be found in [102]:

Theorem 7.8.9. *Let μ be the Lebesgue measure on $\mathbb{R}^n$, and let $L^\Phi(\mu)$ be reflexive. Let further $\{\psi_\lambda\}_{\lambda\in\Lambda}$ be an orthonormal wavelet basis with an r-regular multi-scale analysis of $L^2(\mu)$, then $\{\psi_\lambda\}_{\lambda\in\Lambda}$ is an unconditional basis of $L^\Phi(\mu)$.*

For the definition of an r-regular multi-scale analysis of $L^2(\mu)$ see [84].

7.9 Amemiya formula and Orlicz Norm

We will now concern ourselves with a representation of the Orlicz norm involving only the Young function Φ. This representation is based on the fact that conjugate pairs of Young functions carry over to conjugate pairs of the corresponding modulars (see Theorem 7.3.4).

From Young's inequality for the Young functions Φ and Ψ it follows for $x \in L^{\Phi}(\mu)$ and arbitrary $\lambda > 0$ (weak duality)

$$\|x\|_{\Phi} \le \lambda\left(1 + f^{\Phi}\left(\frac{x}{\lambda}\right)\right),$$

since by definition $\|x\|_{\Phi} = \sup\{\int_T xyd\mu \mid f^{\Psi}(y) \le 1\}$ and due to Young's inequality

$$\Phi\left(\frac{x(t)}{\lambda}\right) + \Psi(y(t)) \ge \frac{x(t)}{\lambda}y(t),$$

hence

$$f^{\Phi}\left(\frac{x}{\lambda}\right) + f^{\Psi}(y) \ge \int_T \frac{x}{\lambda}yd\mu,$$

and therefore

$$\lambda f^{\Phi}\left(\frac{x}{\lambda}\right) + \lambda \ge \int_T xyd\mu \quad \text{for } f^{\Psi}(y) \le 1.$$

The theorem of Amemiya states that even equality holds (strong duality), if on the right-hand side the infimum is taken w.r.t. all positive λ

$$\|x\|_{\Phi} = \inf\left\{\lambda\left(1 + f^{\Phi}\left(\frac{x}{\lambda}\right)\right) \,\middle|\, \lambda > 0\right\}.$$

We will now prove this theorem using Lagrange duality (if one considers the definition of the Orlicz norm as an optimization problem under convex restrictions) and the fact that conjugate Young functions correspond to conjugate modulars.

We define, following Musielak (see [88]), the *Amemiya norm* on $L^{\Phi}(\mu)$:

Definition 7.9.1. Let $x \in L^{\Phi}(\mu)$, then put $A_{\Phi}(x) := \inf\{\lambda(1 + f^{\Phi}(\frac{x}{\lambda})) \mid \lambda > 0\}$.

Theorem 7.9.2. *$A_{\Phi} : L^{\Phi}(\mu) \to \mathbb{R}$ is a norm on $L^{\Phi}(\mu)$ with $A_{\Phi}(x) \ge \|x\|_{\Phi}$.*

Proof. Apparently A_{Φ} is finite on $L^{\Phi}(\mu)$, since by definition there is a $\lambda > 0$ with $f^{\Phi}(\frac{x}{\lambda}) < \infty$. First of all we show the triangle inequality: let $x, y \in L^{\Phi}(\mu)$, let $a > 0$ arbitrary, then there are positive numbers u, v, such that

$$u\left(1 + f^{\Phi}\left(\frac{x}{u}\right)\right) < A_{\Phi}(x) + a \quad \text{and} \quad v\left(1 + f^{\Phi}\left(\frac{y}{v}\right)\right) < A_{\Phi}(y) + a.$$

As an immediate consequence we have

$$f^{\Phi}\left(\frac{x}{u}\right) < \frac{A_{\Phi}(x) + a}{u} - 1 \quad \text{and} \quad f^{\Phi}\left(\frac{y}{v}\right) < \frac{A_{\Phi}(y) + a}{v} - 1.$$

By the convexity of f^Φ we then obtain

$$\begin{aligned} f^\Phi\left(\frac{x+y}{u+v}\right) &\le \frac{u}{u+v} f^\Phi\left(\frac{x}{u}\right) + \frac{v}{u+v} f^\Phi\left(\frac{y}{v}\right) \\ &< \frac{u}{u+v}\left(\frac{A_\Phi(x)+a}{u} - 1\right) + \frac{v}{u+v}\left(\frac{A_\Phi(y)+a}{v} - 1\right) \\ &= \frac{A_\Phi(x) + A_\Phi(y) + 2a}{u+v} - 1. \end{aligned}$$

Therefore $(u+v)(1+f^\Phi(\frac{x+y}{u+v})) < A_\Phi(x)+A_\Phi(y)+2a$ and hence for $a > 0$ arbitrary

$$A_\Phi(x+y) < A_\Phi(x) + A_\Phi(y) + 2a.$$

The positive homogeneity can be seen as follows: let $\rho > 0$ and $\alpha := \frac{\lambda}{\rho}$, then

$$A_\Phi(\rho x) = \inf_{\lambda>0} \lambda\left(1 + f^\Phi\left(\frac{\rho x}{\lambda}\right)\right) = \rho \inf_{\alpha>0} \alpha\left(1 + f^\Phi\left(\frac{x}{\alpha}\right)\right).$$

For $x = 0$ apparently $A_\Phi(x) = 0$ holds. Let now $x \neq 0$, then from weak duality apparently $A_\Phi(x) \ge \|x\|_\Phi > 0$. The symmetry of A_Φ follows immediately from the symmetry of f^Φ. □

We are now in the position to show that Orlicz norm and Amemiya norm coincide.

Theorem 7.9.3.

$$\|x\|_\Phi = \inf_{\lambda>0} \lambda\left(1 + f^\Phi\left(\frac{x}{\lambda}\right)\right) = A_\Phi(x).$$

Proof. We have for $x \in L^\Phi(\mu)$ because of $(f^\Psi)^* = f^\Phi$ by Theorem 7.3.4

$$\begin{aligned} A_\Phi(x) = \inf_{\lambda>0} \lambda\left(1 + f^\Phi\left(\frac{x}{\lambda}\right)\right) &= \inf_{\lambda>0} \lambda\left(1 + \sup_{y\in L^\Psi}\left(\left\langle \frac{x}{\lambda}, y\right\rangle - f^\Psi(y)\right)\right) \\ &= \inf_{\lambda>0} \sup_{y\in L^\Psi} (\langle x, y\rangle - \lambda(f^\Psi(y) - 1)) \\ &= -\sup_{\lambda>0} \inf_{y\in L^\Psi} (-\langle x, y\rangle + \lambda(f^\Psi(y) - 1)). \end{aligned}$$

The definition of the Orlicz norm yields

$$\|x\|_\Phi = \sup\{\langle x, y\rangle \mid f^\Psi(y) \le 1\} = -\inf\{-\langle x, y\rangle \mid f^\Psi(y) \le 1\}.$$

Since the restriction set $S_{f^\Psi}(1)$ has the interior point 0 (see Lemma 6.3.6), there is by the theorem on Lagrange multipliers for Convex Optimization (see Theorem 3.14.4) a non-negative λ^*, such that

$$-\inf\{-\langle x, y\rangle \mid f^\Psi(y) \le 1\} = -\inf_{y\in L^\Psi}\{-\langle x, y\rangle + \lambda^*(f^\Psi(y) - 1)\}.$$

The theorem on the saddle point property of Lagrange multipliers 3.14.5 then yields (Lagrange duality)

$$-\inf_{y \in L^\Psi}\{-\langle x, y\rangle + \lambda^*(f^\Psi(y) - 1)\} = -\sup_{\lambda>0}\inf_{y \in L^\Psi}\{-\langle x, y\rangle + \lambda(f^\Psi(y) - 1)\}.$$

Hence

$$\|x\|_\Phi = \inf_{\lambda>0} \lambda\left(1 + (f^\Psi)^*\left(\frac{x}{\lambda}\right)\right) = A_\Phi(x).$$

Altogether we obtain the following representation for the Orlicz norm

$$\|x\|_\Phi = \inf_{\lambda>0} \lambda\left(1 + \int_T \Phi\left(\frac{x}{\lambda}\right) d\mu\right).$$ □

Remark 7.9.4. The mapping $\varphi : \mathbb{R}_{>0} \to \overline{\mathbb{R}}$ with

$$\varphi(\lambda) = \lambda\left(1 + \int_T \Phi\left(\frac{x}{\lambda}\right) d\mu\right)$$

is convex.

Idea of proof. For Φ differentiable one obtains using the monotonicity of the difference quotient and the theorem on monotone convergence

$$\begin{aligned}\varphi'(\lambda) &= 1 + \int_T \left(\Phi\left(\frac{x(t)}{\lambda}\right) - \frac{x(t)}{\lambda}\Phi'\left(\frac{x(t)}{\lambda}\right)\right) d\mu(t)\\ &= 1 - \int_T \Psi\left(\Phi'\left(\frac{x(t)}{\lambda}\right)\right) d\mu(t),\end{aligned}$$

where in the last part we have used Young's equality. The last expression is, however, monotonically increasing w.r.t. λ. If Φ is not differentiable, one can approximate Φ by differentiable Young functions. □

Example 7.9.5. (a)

$$\|x\|_1 = \inf_{\lambda>0} \lambda\left(1 + \int_T \left|\frac{x}{\lambda}\right| d\mu\right) = \min_{\lambda \geq 0} \lambda + \int_T |x| d\mu = \int_T |x| d\mu.$$

(b)

$$\|x\|_\infty = \inf_{\lambda>0} \lambda\left(1 + \int_T \Phi_\infty\left|\frac{x}{\lambda}\right| d\mu\right).$$

For $\lambda \geq \|x\|_\infty$ we have $\varphi_\infty(\lambda) = \lambda$, for $\lambda < \|x\|_\infty$ we have $\varphi_\infty(\lambda) = \infty$.

(c) $1 < p < \infty$:

$$\|x\|_p = \inf_{\lambda>0} \lambda\left(1 + \int_T \left|\frac{x}{\lambda}\right|^p d\mu\right) = \inf_{\lambda>0} \left(\lambda + \lambda^{1-p} \int_T |x|^p d\mu\right).$$

Now we have: $\varphi_p'(\lambda)=1 + (1-p)\lambda^{-p}\int_T |x|^p d\mu$ and hence $\lambda_p=((p-1)\int_T |x|^p d\mu)^{\frac{1}{p}}$ i.e.

$$\|x\|_p = \varphi(\lambda_p) = \omega_p \|x\|_{(p)},$$

where

$$\omega_p = (p-1)^{\frac{1}{p}} + (p-1)^{\frac{1-p}{p}} = \frac{p^{\frac{1}{p}}}{\left(\frac{p-1}{p}\right)^{\frac{p-1}{p}}}.$$

We observe using $\lim_{x\to 0} x \ln x = 1$

(a) $\omega_p > 0$,

(b) $\lim_{p\to 1} \omega_p = 1$,

(c) $\lim_{p\to\infty} \omega_p = 1$.

Chapter 8

Differentiability and Convexity in Orlicz Spaces

8.1 Flat Convexity and Weak Differentiability

For a continuous convex function $f : X \to \mathbb{R}$ the flat convexity of the set $\{x \in X \mid f(x) \leq r\}$ is characterized by the differentiability properties of f.

For a continuous convex function $f : X \to \mathbb{R}$ (X a real normed space) the subdifferential

$$\partial f(x_0) := \{\phi \in X^* \mid \phi(x - x_0) \leq f(x) - f(x_0)\}$$

is a non-empty convex set (see Theorem 3.10.2). For $\phi \in \partial f(x_0)$ the graph of $[f(x_0) + \phi(\cdot - x_0)]$ is a supporting hyperplane of the epigraph $\{(x, s) \in X \times R \mid f(x) \leq s\}$ in $(x_0, f(x_0))$, and each supporting hyperplane can be represented in such a way.

The right-handed derivative $f'_+(x_0, x) := \lim_{t \downarrow 0} \frac{f(x_0+tx)-f(x_0)}{t}$ always exists and is finite (see Corollary 3.3.2) and the equation $f'_-(x_0, x) = -f'_+(x_0, -x)$ holds. Due to the theorem of Moreau–Pschenitschnii (see Theorem 3.10.4) one obtains

$$f'_+(x_0, x) = \max_{\phi \in \partial f(x_0)} \phi(x)$$

$$f'_-(x_0, x) = \min_{\phi \in \partial f(x_0)} \phi(x).$$

A convex set (with non-empty interior) is called flat convex, if every boundary point has a unique supporting hyperplane.

Theorem 8.1.1. *Let X be a real normed space and $f : X \to \mathbb{R}$ a continuous convex function. Then for $r > \inf\{f(x) \mid x \in X\}$ the following statements are equivalent:*

(a) *The convex set $S_f(r) := \{x \in X \mid f(x) \leq r\}$ is flat convex.*

(b) *For all boundary points x_0 of $S_f(r)$ one has*

   (i) $[f'_+(x_0, \cdot) + f'_-(x_0, \cdot)] \in X^*$

   (ii) *there exists a $c > 0$ with*

$$f'_-(x_0, x) = c f'_+(x_0, x)$$

   for all x with $f'_+(x_0, x) \geq 0$.

In particular, $S_f(r)$ is flat convex, if f Gâteaux differentiable in $\{x \in X \mid f(x) = r\}$.

Proof. Let f be not constant. Due to the continuity of f the set $S_f(r)$ has a non-empty interior, and for every boundary point x_0 of $S_f(r)$ one has $f(x_0) = r > \inf f(x)$, hence from the definition we obtain $0 \notin \partial f(x_0)$.

(a) $\Rightarrow$ (b): Let H be the unique supporting hyperplane of $S_f(r)$ in x_0. As for every $\phi \in \partial f(x_0)$ also $x_0 + \operatorname{Ker} \phi$ is a supporting hyperplane we obtain

$$x_0 + \operatorname{Ker} \phi = H.$$

If $\phi_0 \in X^*$ represents the hyperplane H, then $\phi = \lambda\phi_0$ for a $\lambda \in \mathbb{R}$. The theorem of Moreau–Pschenitschnii yields

$$\partial f(x_0) = \{\lambda\phi_0 \mid \lambda_1 \leq \lambda \leq \lambda_2\}$$

and hence

$$\left.\begin{aligned} f'_+(x_0, x) &= \lambda_2\phi_0(x) \\ f'_-(x_0, x) &= \lambda_1\phi_0(x) \end{aligned}\right\} \text{ for } \phi_0(x) \geq 0$$

and

$$\left.\begin{aligned} f'_+(x_0, x) &= \lambda_1\phi_0(x) \\ f'_-(x_0, x) &= \lambda_2\phi_0(x) \end{aligned}\right\} \text{ for } \phi_0(x) < 0.$$

We conclude (i)

$$[f'_+(x_0, \cdot) + f'_-(x_0, \cdot)] = (\lambda_1 + \lambda_2) \cdot \phi_0(\cdot) \in X^*.$$

It remains to verify (ii): as $0 \notin \partial f(x_0)$, the relation $\operatorname{sign} \lambda_1 = \operatorname{sign} \lambda_2 \neq 0$ holds, consequently

$$f'_-(x_0, x) = (\lambda_1\lambda_2^{-1})^{\operatorname{sign} \lambda_1} f'_+(x_0, x)$$

for x with $f'_+(x_0, x) \geq 0$.

(b) $\Rightarrow$ (a): For $\phi \in \partial f(x_0)$ we have

$$f'_+(x_0, x) \geq \phi(x) \geq f'_-(x_0, x) \tag{8.1}$$

for all $x \in X$ and because of (ii)

$$f'_+(x_0, x) \geq \phi(x) \geq cf'_+(x_0, x) = f'_-(x_0, x) \tag{8.2}$$

for those x with $f'_+(x_0, x) \geq 0$. From $\phi(x) = 0$ it follows because of (8.1) $f'_+(x_0, x) \geq 0$ and from (8.2)

$$f'_+(x_0, x) = f'_-(x_0, x) = 0,$$

i.e.

$$\operatorname{Ker} \phi \subset \operatorname{Ker}(f'_+ + f'_-) := H_0.$$

As $f'_+(x_0, \cdot) \geq f'_-(x_0, \cdot)$ and $0 \notin \partial f(x_0)$ it follows from (8.2) $H_0 \neq X$ and hence

$$\operatorname{Ker} \phi = H_0.$$

Let now H be a supporting hyperplane of $S_f(r)$ in x_0. Due to the theorem of Mazur 3.9.6 the affine supporting set $H \times \{f(x_0)\}$ of the epigraph of f in $(x_0, f(x_0))$ can be extended to a supporting hyperplane in $X \times \mathbb{R}$, thus there is $\phi \in \partial f(x_0)$ with

$$H \times \{f(x_0)\} \subset \{(x, f(x_0) + \phi(x - x_0)) \mid x \in X\}.$$

Hence

$$H \subset \{x \mid \phi(x - x_0) = 0\} = x_0 + \operatorname{Ker} \phi = x_0 + H_0,$$

consequently

$$H = x_0 + H_0. \qquad \square$$

Corollary. *Let $S_f(r)$ be flat convex and f not Gâteaux differentiable at x_0. Then f is at x_0 differentiable in direction x if and only if $f'_+(x_0, x) = 0$.*

Proof. If $f'_+(x_0, x) = 0$, then using (ii) it follows that $f'_+(x_0, x) = f'_-(x_0, x)$. Conversely, if $f'_+(x_0, x) = f'_-(x_0, x)$, then, because of (i) we can assume $f'_+(x_0, x) \geq 0$ (exchange x by $-x$ if necessary). As $c \neq 1$, it follows from (ii) that $f'_+(x_0, x) = 0$. $\square$

Theorem 8.1.2. *If f is non-negative and positively homogeneous, then $S_f(r)$ is flat convex if and only if f is Gâteaux differentiable in $\{x \in X \mid f(x) > 0\}$.*

Proof. If $S_f(r)$ is flat convex for an $r > 0$, then, because of the positive homogeneity of f, all level sets $S_f(r)$ are flat convex. Now we have

$$\begin{aligned} f'_+(x_0, x_0) &= \lim_{t \downarrow 0} \frac{f(x_0 + t x_0) - f(x_0)}{t} = f(x_0) \\ &= \lim_{t \downarrow 0} \frac{f(x_0 - t x_0) - f(x_0)}{-t} = f'_-(x_0, x_0). \end{aligned}$$

Hence $c = 1$ in (ii) and thus $f'_+(x_0, x) = f'_-(x_0, x)$ for $f'_+(x_0, x) \geq 0$.

Let now $f'_+(x_0, x) < 0$, then

$$f'_+(x_0, -x) \geq f'_-(x_0, -x) = -f'_+(x_0, x) > 0,$$

and with $c = 1$ in (ii) it follows that

$$f'_+(x_0, -x) = f'_-(x_0, -x).$$

Therefore we obtain using (i)

$$\begin{aligned} 2f'_+(x_0, x) = -2f'_-(x_0, -x) &= -(f'_+(x_0, -x) + f'_-(x_0, -x)) \\ &= f'_+(x_0, x) + f'_-(x_0, x), \end{aligned}$$

hence $f'_+(x_0, x) = f'_-(x_0, x)$. $\square$

If f is a norm, the above consideration yields the

Theorem 8.1.3 (Mazur). *A normed space X is flat convex if and only if the norm is Gâteaux differentiable in $X \setminus \{0\}$.*

Theorem 8.1.4. *Let X be a normed space X whose norm is Gâteaux differentiable in $X \setminus \{0\}$ and let $D : X \setminus \{0\} \to X^*$ denote the Gâteaux derivative. Then $\|D(x)\| = 1$ and $\langle x, D(x)\rangle = \|x\|$ for all $x \in X \setminus \{0\}$.*

Proof. We have, due to the subgradient inequality for all $h \in X$

$$\langle h, D(x)\rangle = \langle x + h - x, D(x)\rangle \leq \|x + h\| - \|x\| \leq \|h\|$$

hence $\|D(x)\| \leq 1$. On the other hand $\langle D(x), 0-x\rangle \leq \|0\| - \|x\|$, hence $\langle D(x), x\rangle \geq \|x\|$, and thus the assertion. □

8.2 Flat Convexity and Gâteaux Differentiability of Orlicz Spaces

Let $\Phi : \mathbb{R} \to \mathbb{R}$ be a finite Young function. Let (T, Σ, μ) be a σ-finite measure space, and $L^\Phi(\mu)$ the Orlicz space determined via Φ and μ, equipped with the Luxemburg norm $\|\cdot\|_{(\Phi)}$.

Let further $M^\Phi(\mu)$ be the closed subspace of $L^\Phi(\mu)$, spanned by the step functions in $L^\Phi(\mu)$.

We consider the unit ball in $M^\Phi(\mu)$. If $x \in M^\Phi(\mu)$, then (see Theorem 6.2.24 and Lemma 6.2.15)

$$\|x\|_{(\Phi)} = 1 \quad \text{if and only if} \quad \int_T \Phi(x)d\mu = 1, \tag{8.3}$$

If Φ satisfies the Δ_2-condition, then (see Theorem 6.2.42) $M^\Phi(\mu) = L^\Phi(\mu)$.

A normed space is called *flat convex*, if the closed unit ball is flat convex.

According to Theorem 8.1.1 the level set $S_f(r)$ is flat convex, if f is Gâteaux differentiable in $\{x \in X \mid f(x) = r\}$.

Lemma 8.2.1. *The function $f^\Phi : M^\Phi(\mu) \to \mathbb{R}$ defined by*

$$f^\Phi(x) = \int_T \Phi(x)d\mu$$

is continuous and convex. Furthermore for $x_0 \in M^\Phi(\mu)$

$$(f^\Phi)'_+(x_0, x) = \int_{\{x>0\}} x\Phi'_+(x_0)d\mu + \int_{\{x<0\}} x\Phi'_-(x_0)d\mu \tag{8.4}$$

$$(f^\Phi)'_-(x_0, x) = \int_{\{x>0\}} x\Phi'_-(x_0)d\mu + \int_{\{x<0\}} x\Phi'_+(x_0)d\mu.$$

If Φ is differentiable, then f^Φ is Gâteaux differentiable and

$$(f^\Phi)'(x_0, x) = \int_T x\Phi'(x_0)d\mu. \tag{8.5}$$

Proof. Convexity of f^Φ follows immediately from the convexity of Φ. As f^Φ is bounded on the unit ball, f^Φ is continuous (see Theorem 6.3.9 and Theorem 6.2.24). For the difference quotient we obtain

$$\begin{aligned}\frac{f^\Phi(x_0+\tau x)-f^\Phi(x_0)}{\tau} &= \int_T \frac{\Phi(x_0(t)+\tau x(t))-\Phi(x_0(t))}{\tau}d\mu \\ &= \int_{x(t)>0} \frac{\Phi(x_0(t)+\tau x(t))-\Phi(x_0(t))}{\tau x(t)}x(t)d\mu \\ &\quad + \int_{x(t)<0} \frac{\Phi(x_0(t)+\tau x(t))-\Phi(x_0(t))}{\tau x(t)}x(t)d\mu.\end{aligned}$$

Through the monotonicity w.r.t. τ of

$$\frac{\Phi(s_0+\tau s)-\Phi(s_0)}{\tau}$$

for all $s_0, s \in \mathbb{R}$, the representation (8.4) of $(f^\Phi)'_+$ and $(f^\Phi)'_-$ follows.

Let Φ be differentiable. According to (8.4) $(f^\Phi)'(x_0, x)$ exists and we have (8.5). □

Lemma 8.2.2. *Let T_0, T_1 and $T_2 \in \Sigma$ be disjoint sets with $0 < \mu(T_i) < \infty$ ($i = 0, 1, 2$). If there exists an $s_0 \geq 0$ with $\Phi'_+(s_0) \neq \Phi'_-(s_0)$ and $\Phi(s_0) \cdot \mu(T_0) < 1$, then $M^\Phi(\mu)$ is not flat convex.*

Proof. The set of discontinuities of a function, monotonically increasing on $[a, b]$, is at most countable (see [89], S. 229). Hence there is $s_1 > 0$ with $\Phi'_+(s_1) = \Phi'_-(s_1)$ and

$$1 - \Phi(s_0)\mu(T_0) - \Phi(s_1)\mu(T_1) > 0.$$

As Φ is continuous, one can choose $s_2 \in \mathbb{R}$ such that

$$\Phi(s_0)\cdot\mu(T_0) + \Phi(s_1)\cdot\mu(T_1) + \Phi(s_2)\cdot\mu(T_2) = 1.$$

For the functions

$$\begin{aligned} x_0 &= s_0\chi_{T_0} + s_1\chi_{T_1} + s_2\chi_{T_2}, \\ x_1 &= \chi_{T_0}, \\ x_2 &= \chi_{T_1} \end{aligned}$$

we have

$$0 < (f^\Phi)'_+(x_0, x_1) = \mu(T_0) \cdot \Phi'_+(s_0) \neq (f^\Phi)'_-(x_0, x_1) = \mu(T_0) \cdot \Phi'_-(s_0),$$
$$0 < (f^\Phi)'_+(x_0, x_2) = \mu(T_1) \cdot \Phi'_+(s_1) = \mu(T_1) \cdot \Phi'_-(s_1) = (f^\Phi)'_-(x_0, x_2).$$

Hence condition (ii) of Theorem 8.1.1 is not satisfied and thus $M^\Phi(\mu)$ not flat convex. □

Theorem 8.2.3. *Let μ be not purely atomic. Then $M^\Phi(\mu)$ is flat convex if and only if Φ is differentiable.*

Proof. Lemma 8.2.2 and Theorem 8.2.1. □

The derivative of the norm $\|\cdot\|_{(\Phi)}$ is now determined as follows. Let $x_0 \in M^\Phi(\mu)$ and $\|x_0\|_{(\Phi)} = 1$. The graph of the function $x \to (f^\Phi)'(x_0, x - x_0) + f^\Phi(x_0)$ is a supporting hyperplane of the epigraph of f^Φ in $(x_0, f^\Phi(x_0)) = (x_0, 1) \in M^\Phi \times \mathbb{R}$. This means that the hyperplane $\{x \in M^\Phi \mid (f^\Phi)'(x_0, x - x_0) = 0\}$ supports the unit ball of M^Φ in x_0. If we denote $\|\cdot\|_{(\Phi)}$ by p_Φ then $p'_\Phi(x_0, x_0) = 1$ (see Theorem 8.1.4) and $p'_\Phi(x_0, \cdot)$ is a multiple of $(f^\Phi)'(x_0, \cdot)$. Taking into account that $\int_T \Phi'(x_0)x_0 d\mu \geq \int_T \Phi(x_0)d\mu = 1$, we obtain that

$$x \to \frac{\int x\Phi'(x_0)d\mu}{\int x_0\Phi'(x_0)d\mu}, \quad \|x_0\|_{(\Phi)} = 1, \tag{8.6}$$

is the derivative of the norm $\|\cdot\|_{(\Phi)}$ (in M^Φ).

If the measure μ is purely atomic, then the differentiability of Φ is not a necessary condition for flat convexity of $M^\Phi(\mu)$.

In the sequel let

$$S = \{s \in \mathbb{R} \mid \Phi \text{ in } s \text{ not differentiable}\}.$$

Then we have

Theorem 8.2.4. *Let (T, Σ, μ) be purely atomic and consist of more than 2 atoms. Then $M^\Phi(\mu)$ is flat convex if and only if for all $s \in S$ and all atoms $A \in \Sigma$*

$$\Phi(s) \cdot \mu(A) \geq 1.$$

Proof. Necessity follows immediately from Lemma 8.2.2.

Let $x_0 \in M^\Phi(\mu)$ and $\|x_0\|_{(\Phi)} = 1$. According to Lemma 8.2.1 and the condition $\Phi(s) \cdot \mu(A) \geq 1$ it is sufficient to consider $x_0 = r\chi_A$ for an atom $A \in \Sigma$ and $r \in S$.

As $\Phi(r) \cdot \mu(A) = 1$ we have $\Phi'_+(r) \cdot r \geq \Phi(r) > 0$ and $\Phi'_-(r) \cdot r \geq \Phi(r) > 0$. In particular $\text{sign}(\Phi'_+(r)) = \text{sign}(\Phi'_+(r)) = \text{sign}(r)$.

Since $0 \notin S$ and $\Phi'(0) = 0$ it follows from Lemma 8.2.1 that

$$(f^{\Phi})'_+(x_0, x) = \int_{\{x>0\}} x\Phi'_+(x_0)d\mu + \int_{\{x<0\}} x\Phi'_-(x_0)d\mu$$
$$= \Phi'_+(r) \int_{\{x>0\}} x \cdot \chi_A d\mu + \Phi'_-(r) \int_{\{x<0\}} x \cdot \chi_A d\mu,$$

and in the same way

$$(f^{\Phi})'_-(x_0, x) = \Phi'_-(r) \int_{\{x>0\}} x\chi_A d\mu + \Phi'_+(r) \int_{\{x<0\}} x\chi_A d\mu.$$

Hence we obtain (i) with $f'_+(x_0, x) + f'_-(x_0, x) = \mu(A)[\Phi'_+(r) + \Phi'_-(r)]x|_A$, where A an atom, i.e. x constant on A. This implies

$$x \mapsto f'_+(x_0, x) + f'_-(x_0, x) = \mu(A)[\Phi'_+(r) + \Phi'_-(r)]x|_A \in (M^{\Phi})^*.$$

Concerning (ii): If $x > 0$ on A then $\int_{\{x<0\}} x \cdot \chi_A d\mu = 0$ and hence

$$f'_+(x_0, x) = \Phi'_+(r) \int_{\{x>0\}} x \cdot \chi_A d\mu.$$

If $\Phi'_+(r) > 0$ then $f'_+(x_0, x) \geq 0$ and we have $f'_-(x_0, x) = \frac{\Phi'_-(r)}{\Phi'_+(r)} f'_+(x_0, x)$. If, however, $x < 0$ on A then

$$f'_+(x_0, x) = \Phi'_-(r) \int_{\{x<0\}} x \cdot \chi_A d\mu.$$

If $\Phi'_-(r) < 0$ then $f'_+(x_0, x) \geq 0$ and $f'_-(x_0, x) = \frac{\Phi'_+(r)}{\Phi'_-(r)} f'_+(x_0, x)$.

From Theorem 8.1.1 it then follows that $M^{\Phi}(\mu)$ is flat convex. □

8.3 A-differentiability and B-convexity

In this section we will consider the duality of A-differentiability and B-convexity of a conjugate pair of convex functions. In this context B is the polar of A. These notions were introduced by Asplund and Rockafellar in [4]. Our main interest is focused on the duality of Fréchet differentiability and $K(0, 1)$-convexity. The results of this section will play an important role for the substantiation of the duality of Fréchet differentiability and local uniform convexity (see Theorem 8.4.10) and – as we will explain in detail in the next section – for the strong solvability of optimization problems.

The exposition of this section is to a large extent based on that in [4].

Let in the sequel X and Y be normed spaces and either $Y = X^*$ or $X = Y^*$ and let $f : X \to \mathbb{R}$ and $g : Y \to \mathbb{R}$ mutually conjugate, continuous, and convex functions.

Lemma 8.3.1. *$S_f(r)$ is bounded for every $r \in \mathbb{R}$.*

Proof. Let $y \in Y$, then

$$g(y) \geq \sup\{\langle x, y\rangle - f(x) \mid x \in S_f(r)\} \geq \sup\{\langle x, y\rangle - r \mid x \in S_f(r)\},$$

hence

$$\sup\{\langle x, y\rangle \mid x \in S_f(r)\} \leq g(y) + r < \infty,$$

and similarly

$$\inf\{\langle x, y\rangle \mid x \in S_f(r)\} \geq -g(-y) - r > -\infty.$$

This means $S_f(r)$ is weakly bounded and therefore, due to Theorem 5.3.15, bounded. □

Lemma 8.3.2. *Let the interior of $S_f(0)$ be non-empty and let $h : Y \to \overline{\mathbb{R}}$ be the support functional of $S_f(0)$, i.e.*

$$h(v) := \sup\{\langle z, v\rangle \mid f(z) \leq 0\}$$

then

$$h(v) = \min\left\{\lambda g\left(\frac{v}{\lambda}\right) \,\middle|\, \lambda > 0\right\}$$

holds.

Proof. By the previous theorem $S_f(0)$ is bounded and hence h finite. Apparently $h(v) = -\inf\{\langle z, -v\rangle \mid f(z) \leq 0\}$ holds. Let now for $\lambda > 0$

$$\phi(\lambda) := \inf\{\langle z, -v\rangle + \lambda f(z) \mid z \in X\} = \inf\{-(\langle z, v\rangle - \lambda f(z)) \mid z \in X\}.$$

Then

$$-\phi(\lambda) = \sup\{\langle z, v\rangle - \lambda f(z) \mid z \in X\} = \lambda \sup\left\{\left\langle z, \frac{v}{\lambda}\right\rangle - f(z) \,\middle|\, z \in X\right\} = \lambda g\left(\frac{v}{\lambda}\right)$$

holds. According to the theorem on Lagrange duality (see Theorem 3.14.5) we then obtain

$$h(v) = -\max\{\phi(\lambda) \mid \lambda > 0\} = \min\{-\phi(\lambda) \mid \lambda > 0\} = \min\left\{\lambda g\left(\frac{v}{\lambda}\right) \,\middle|\, \lambda > 0\right\}.$$

□

Definition 8.3.3. Let $C \subset X$, then we define the *polar* C^0 of C by

$$C^0 := \{v \in Y \mid \langle u, v\rangle \leq 1,\ \forall u \in C\}.$$

The *bipolar* of C is the defined as follows:

$$C^{00} := \{u \in X \mid \langle u, v\rangle \leq 1,\ \forall v \in C^0\}.$$

Then the following theorem holds (see [23], II, p. 51):

Theorem 8.3.4 (Bipolar Theorem). *Let X be a normed space and either $Y = X^*$ or $X = Y^*$ and let $C \subset X$. Then*

$$C^{00} = \overline{\operatorname{conv}(C \cup \{0\})}$$

holds, where the closure has to be taken w.r.t. the weak topology.

Remark 8.3.5. If C is convex, weakly closed, and if C contains the origin, then

$$C^{00} = C$$

holds. By Theorem 3.9.18 a closed convex subset of a normed space is always weakly closed.

The following lemma is of central importance for the subsequent discussion:

Lemma 8.3.6. *Let f and g be continuous mutually conjugate convex functions on X resp. on Y and let for $x \in X$ and $y \in Y$ Young's equality be satisfied*

$$\langle x, y\rangle - f(x) - g(y) = 0,$$

then for every $\delta > 0$

$$\begin{aligned}\{v \,|\, g(y+v) - g(y) - \langle x, v\rangle \le \delta\}^0 &\subset \delta^{-1}\{u \,|\, f(x+u) - f(x) - \langle u, x\rangle \le \delta\}\\ &\subset 2\{v \,|\, g(y+v) - g(y) - \langle x, v\rangle \le \delta\}^0.\end{aligned}$$

Proof. Let $f_\delta(z) := (f(x+\delta z) - f(x) - \langle \delta z, y\rangle - \delta)\delta^{-1}$ and $g_\delta(v) := (g(y+v) - g(y) - \langle x, v\rangle + \delta)\delta^{-1}$. Then f_δ and g_δ are mutual conjugates, because

$$\begin{aligned}f_\delta^*(v) &= \sup\{\langle z, v\rangle - f_\delta(z) \,|\, z \in X\}\\ &= \sup\{\langle z, v\rangle - (f(x+\delta z) - f(x) - \langle \delta z, y\rangle - \delta)\delta^{-1} \,|\, z \in X\}\\ &= \sup\{\langle z, y+v\rangle - (f(x+\delta z) - f(x) - \delta)\delta^{-1} \,|\, z \in X\}\\ &= \sup\{\langle z, y+v\rangle - (f(x+\delta z) + g(y) - \langle x, y\rangle - \delta)\delta^{-1} \,|\, z \in X\}\\ &= \delta^{-1}\sup\{\langle x+\delta z, y+v\rangle - f(x+\delta z) - g(y) - \langle x, v\rangle + \delta \,|\, z \in X\}\\ &= \delta^{-1}(g(y+v) - g(y) - \langle x, v\rangle + \delta) = g_\delta(v).\end{aligned}$$

In the same manner it can be shown that: $f_\delta = g_\delta^*$.

Using the subgradient inequality it follows that: $1 = \inf g_\delta = g_\delta(0)$ and $-1 = f_\delta(0) = \inf f_\delta$.

It remains to be shown

$$S_{g_\delta}(2)^0 \subset C_\delta := S_{f_\delta}(0) \subset 2 \cdot S_{g_\delta}(2)^0.$$

Here C_δ is a convex, closed set, which due to $f_\delta(0) = -1$ contains the origin as an interior point (thus C_δ is also weakly closed by Theorem 3.9.18). The bipolar theorem 8.3.4 together with Remark 8.3.5 then implies $C_\delta^{00} = C_\delta$. The assertion then follows if we can show

$$S_{g_\delta}(2) \supset C_\delta^0 \supset \frac{1}{2} \cdot S_{g_\delta}(2) \tag{$*$}$$

Let now h be the support functional of C_δ, i.e.

$$h(v) = \sup\{\langle z, v\rangle \mid z \in C_\delta\},$$

then by Lemma 8.3.2

$$h(v) = \inf\left\{\lambda g_\delta\left(\frac{v}{\lambda}\right) \,\middle|\, \lambda > 0\right\}$$

holds, since C_δ is bounded due to the finiteness of the conjugate g_δ.

In particular: $h(v) \leq g_\delta(v)$, hence

$$S_{g_\delta}(2) \subset S_h(2) = 2 \cdot S_h(1) = 2C_\delta^0,$$

and thus the right-hand side of $(*)$.

We now prove the left hand side of $(*)$ and show at first

$$\{v | h(v) < 1\} \subset S_{g_\delta}(2).$$

Let $v \in Y$ such that $h(v) < 1$, then there is a $\lambda > 0$ with

$$\lambda g_\delta(\lambda^{-1}v) < 1.$$

On the other hand: $g_\delta(\lambda^{-1}v) \geq 1 = \inf g_\delta$ and hence: $0 < \lambda < 1$. By the convexity of g_δ we then obtain

$$g_\delta(v) \leq (1-\lambda)g_\delta(0) + \lambda g_\delta(\lambda^{-1}v) < 2.$$

As mentioned above, $\{v \mid h(v) \leq 1\} = C_\delta^0$. Let $h(v) = 1$, $\lambda_n \in (0,1)$ with $\lambda_n \to 0$ and $v_n := (1-\lambda_n)v$, then since $0 \in \{v \mid h(v) < 1\}$, since $h(0) = 0$, and the convexity of h also $v_n \in \{v \mid h(v) < 1\}$. Therefore $g_\delta(v_n) < 2$, and, since by construction $v_n \to v$, we obtain $g_\delta(v) \leq 2$, due to the continuity of g_δ. □

Definition 8.3.7. Let A be a non-empty subset of X, $f : X \to \mathbb{R}$, then f is called *A-differentiable* at $x \in X$, if there is a $y \in Y$, such that

$$\limsup_{\lambda \downarrow 0} \sup_{u \in A} \left| \frac{f(x+\lambda u) - f(x)}{\lambda} - \langle u, y\rangle \right| = 0.$$

Then y is called the *A-gradient* of f at x.

Let $g : Y \to \mathbb{R}$ and B subset of Y. We say g is *B-convex* at $y \in Y$ w.r.t. $x \in X$, if for every $\varepsilon > 0$ there is a $\delta > 0$, such that

$$\{v \mid g(y+v) - g(y) - \langle x, v\rangle \leq \delta\} \subset \varepsilon B.$$

Theorem 8.3.8. *Let f and g be finite and mutually conjugate convex functions on X resp. on Y. Let A be non-empty subsets of X and B the polar of A in Y. Let $x \in X$ and $y \in \partial f(x)$, then y is the A-gradient of f at x if and only if g is B-convex at y w.r.t. x.*

Proof. Since $y \in \partial f(x)$ then y is an A-gradient at x, if and only if for every $\varepsilon > 0$ there is a $\mu > 0$, such that

$$\sup_{u\in A}\left\{\frac{f(x+\lambda u) - f(x)}{\lambda} - \langle u, y\rangle\right\} \leq \varepsilon, \quad \forall\, 0 < \lambda \leq \mu.$$

We obtain

$$\begin{aligned}\varepsilon^{-1}A &\subset \varepsilon^{-1}\{u \mid f(x+\mu u) - f(x) - \langle \mu u, y\rangle \leq \mu\varepsilon\}\\ &= \varepsilon^{-1}\mu^{-1}\{z \mid f(x+z) - f(x) - \langle z, y\rangle \leq \mu\varepsilon\} = C_{\mu\varepsilon}.\end{aligned}$$

Lemma 8.3.6 implies

$$C_{\mu\varepsilon} \subset 2\{v \mid g(y+v) - g(y) - \langle x, v\rangle \leq \mu\varepsilon\}^0,$$

hence

$$\frac{1}{2}\varepsilon^{-1}A \subset \{v \mid g(y+v) - g(y) - \langle x, v\rangle \leq \mu\varepsilon\}^0,$$

and therefore

$$2\varepsilon B \supset \{v \mid g(y+v) - g(y) - \langle x, v\rangle \leq \mu\varepsilon\}.$$

If we put $\delta := \mu\varepsilon/2$ then we obtain: g is B-convex at y w.r.t. x.

Conversely, let

$$\varepsilon B \supset \{v \mid g(y+v) - g(y) - \langle x, v\rangle \leq \delta\},$$

then

$$\varepsilon^{-1}A \subset \{v \mid g(y+v) - g(y) - \langle x, v\rangle \leq \delta\}^0,$$

and by Lemma 8.3.6 then also

$$\varepsilon^{-1}A \subset C_\delta = \delta^{-1}\{z \mid f(x+z) - f(x) - \langle z, y\rangle \leq \delta\}.$$

Therefore for all $u \in A$

$$\frac{f(x+\delta\varepsilon^{-1}u) - f(x)}{\delta\varepsilon^{-1}} - \langle u, y\rangle \leq \varepsilon. \qquad \square$$

If we choose for A the unit ball in X, then B is the unit ball in Y and A-differentiability of f coincides with its Fréchet differentiability. Thereby we obtain

Corollary 8.3.9. *y is Fréchet gradient of f at $x \Leftrightarrow g$ is $K(0,1)$-convex at y w.r.t. x, if $K(0,1)$ denotes the unit ball in Y.*

The subsequent theorem will draw a connection between $K(0,1)$-convexity and strong solvability according to

Definition 8.3.10. A function f has a *strong minimum* k_0 on a subset K of a Banach space X, if the set of minimal solutions $M(f,K)$ of f on K only consists of $\{k_0\}$ and if for every sequence $(k_n)_{n\in\mathbb{N}} \subset K$ with

$$\lim_{n\to\infty} f(k_n) = f(k_0) \quad \text{we have} \quad \lim_{n\to\infty} k_n = k_0.$$

The problem of minimizing f on K is then called *strongly solvable*.

Theorem 8.3.11. *g is $K(0,1)$-convex at y w.r.t. x, if and only if y is strong global minimum of $g - \langle x, \cdot\rangle$.*

Proof. Let g be $K(0,1)$-convex at y w.r.t. x. Apparently for all $\varepsilon > 0$ and all $z \in Y$ with $\|z\| > \varepsilon$

$$g(y+z) - g(y) > \langle x, z\rangle,$$

and hence $x \in \partial g(y)$. By the subgradient inequality we obtain for all $z \in Y$

$$g(z) - \langle x, z\rangle \geq g(y) - \langle x, y\rangle,$$

i.e. y is a global minimum of $g - \langle x, \cdot\rangle$. Let now

$$g(y+v_n) - \langle x, y+v_n\rangle \xrightarrow{n\to\infty} g(y) - \langle x, y\rangle,$$

hence

$$g(y+v_n) - \langle x, v_n\rangle - g(y) \to 0.$$

Suppose there is a subsequence (v_{n_k}) and a $\rho > 0$ with $\|v_{n_k}\| \geq \rho$, then for all $\delta > 0$ there is n_k with

$$v_{n_k} \in \{v \mid g(y+v) - g(y) - \langle x, v\rangle \leq \delta\},$$

in contradiction to the $K(0,1)$-convexity of g.

Conversely let y be a strong global minimum of $g - \langle x, \cdot\rangle$, i.e.

$$g(y+v_n) - g(y) - \langle x, v_n\rangle \to 0$$

implies $v_n \to 0$. Since apparently $x \in \partial g(y)$, the $K(0,1)$-convexity of g at y w.r.t. x follows. $\square$

As a corollary of Theorem 8.3.8 we obtain

Corollary 8.3.12. *Let f be Gâteaux differentiable at x and let $y \in \partial f(x)$, then g is H_u-convex at y w.r.t. x, for all half-spaces $H_u := \{v \in Y \mid \langle u, v\rangle \le 1\}$, with $u \in X$ arbitrary.*

Proof. f is A_u-differentiable w.r.t. every one-point subset $A_u := \{u\}$ of X. Then

$$A_u^0 = H_u = \{v \mid \langle u, v\rangle \le 1\}$$

holds. □

We thus obtain a duality relation between Gâteaux differentiability and strict convexity of the conjugate.

Theorem 8.3.13. *Let Y be a normed space and $X = Y^*$ and let f be Gâteaux differentiable, then g is strictly convex.*

Proof. Suppose g is not strictly convex, then there are $y, v \in Y$ with $v \neq 0$ and $x \in \partial g(y)$, such that for all $\lambda \in [-1, 1]$

$$g(y + \lambda v) - g(y) - \langle x, \lambda v\rangle = 0$$

holds and hence for all $\delta > 0$

$$[-v, v] \subset \{z \mid g(y + z) - g(y) - \langle x, z\rangle \le \delta\}.$$

According to the theorem of Hahn–Banach there is a $u \in X$ with $\|u\| = 1$ and $\langle u, v\rangle = \|v\|$, but then $[-v, v]$ is not contained in εH_u for $\varepsilon < \|v\|$, a contradiction to Corollary 8.3.12. □

8.4 Local Uniform Convexity, Strong Solvability and Fréchet Differentiability of the Conjugate

If one wants to generalize the notion of uniform convexity of functions (see [76]) by allowing for different convexity modules for different points of the space, there are different possibilities to do so.

Definition 8.4.1. A monotonically increasing function $\tau : \mathbb{R}_+ \to \mathbb{R}_+$ with $\tau(0) = 0$, $\tau(s) > 0$ for $s > 0$ and $\frac{\tau(s)}{s} \to_{s\to\infty} \infty$ is called a *convexity module*. A continuous function $f : X \to \mathbb{R}$, where X is a normed space, is called *locally uniformly convex* by definition

a) if for all $x \in X$ a convexity module τ_x exists, such that for all $y \in X$ we have
$$\frac{1}{2}(f(y) + f(x)) \geq f\left(\frac{x+y}{2}\right) + \tau_x(\|x-y\|),$$
b) if for all $x \in X$ and all $x^* \in \partial f(x)$ a convexity module τ_{x,x^*} exists, such that for all $y \in X$ we have
$$f(y) - f(x) \geq \langle y - x, x^* \rangle + \tau_{x,x^*}(\|x-y\|),$$
c) if for all $x \in X$ a convexity module τ_x exists, such that for all $y \in X$ we have
$$\frac{1}{2}(f(x+y) + f(x-y)) \geq f(x) + \tau_x(\|y\|).$$

It is easily seen that: a) $\Rightarrow$ b) and b) $\Rightarrow$ c); if the function f satisfies property a), one obtains using the subgradient inequality
$$\begin{aligned}\frac{1}{2}(f(x) + f(y)) &\geq f\left(\frac{x+y}{2}\right) + \tau_x(\|x-y\|)\\ &\geq f(x) + \left\langle \frac{x+y}{2} - x, x^* \right\rangle + \tau_x(\|x-y\|).\end{aligned}$$
Defining $\tau_{x,x^*}(s) := \tau_x(2s)$ then immediately property b) follows. If a function f satisfies property b), then again by the subgradient inequality
$$\begin{aligned}\frac{1}{2}f(x+y) &\geq \frac{1}{2}(\langle y, x^* \rangle + f(x) + \tau_{x,x^*}(\|y\|))\\ &\geq \frac{1}{2}(f(x) - f(x-y) + f(x) + \tau_{x,x^*}(\|y\|)),\end{aligned}$$
and hence property c) for $\tau_x(s) := \frac{1}{2}\tau_{x,x^*}(s)$.

The converse, however is not true, because the function $f(x) = e^x$ satisfies property c) using the convexity module $\tau_x(s) := e^x(\cosh(s) - 1)$, but does not have bounded level sets, and hence cannot satisfy b) (see below).

The strictly convex function $f(x) := (x+1)\log(x+1) - x$ satisfies property b), because for $h > 0$ we obtain due to the strict convexity of f
$$\begin{aligned}0 &< \frac{1}{h}(f(x+h) - f(x) - f'(x)\cdot h)\\ &= \frac{1}{h}\left((x+1)\log\frac{x+h+1}{x+1} + h\left(\log\frac{x+h+1}{x+1} - 1\right)\right) \xrightarrow{h\to\infty} \infty,\end{aligned}$$
but violates a) at the origin, because
$$\lim_{y\to\infty}\frac{1}{y}\left(\frac{1}{2}f(y) - f\left(\frac{y}{2}\right)\right) = \frac{1}{2}\log 2 < \infty.$$
In the sequel we will mean by *locally uniformly convex functions* always those with property b).

Remark. Lovaglia in [78] investigates locally uniformly convex norms. The notion introduced there differs from the one we consider here, the squares of these norms are, however, locally uniformly convex in the sense of b) (see [56]).

Remark 8.4.2. If $f : X \to \mathbb{R}$ is locally uniformly convex then $\frac{f(y)}{\|y\|} \to_{\|y\| \to \infty} \infty$. In particular, the level sets of f are bounded.

Proof. We have for all $x \in X$ and all $x^* \in \partial f(x)$ and all $y \in X$

$$f(y) - f(x) \geq \langle y - x, x^* \rangle + \tau_{x,x^*}(\|x - y\|).$$

Let in particular be $x = 0$ and $x^* \in \partial f(0)$ then

$$\begin{aligned} \frac{f(y)}{\|y\|} &\geq \frac{f(0)}{\|y\|} + \left\langle \frac{y}{\|y\|}, x^* \right\rangle + \frac{\tau_{0,x^*}(\|y\|)}{\|y\|} \\ &\geq \frac{f(0)}{\|y\|} - \|x^*\| + \frac{\tau_{0,x^*}(\|y\|)}{\|y\|} \xrightarrow{\|y\| \to \infty} \infty. \end{aligned}$$ □

Definition 8.4.3. A function f has a *strong minimum* k_0 on a subset K of a Banach space X, if the set of minimal solutions $M(f, K)$ of f on K only consists of $\{k_0\}$ and if for every sequence $(k_n)_{n \in \mathbb{N}} \subset K$ with

$$\lim_{n \to \infty} f(k_n) = f(k_0) \quad \text{we have: } \lim_{n \to \infty} k_n = k_0.$$

The problem of minimizing f on K is then called *strongly solvable*.

Lemma 8.4.4. *Let X be a reflexive Banach space and let $f : X \to \mathbb{R}$ be convex and continuous then the following statements are equivalent:*

(a) *f has a strong minimum on every closed hyperplane and on X*

(b) *f has a strong minimum on every closed half-space and on X*

(c) *f has a strong minimum on every closed convex subset of X.*

Proof. Without limiting generality we can assume that $f(0) = 0$ and $f(x) > 0$ for $x \neq 0$. Otherwise we can consider $g(x) := f(x_0 - x) - f(x_0)$, where $f(x_0) = \min\{f(x) \mid x \in X\}$.

(a) ⇒ (b): Let $G_\alpha := \{x \mid \langle x_0^*, x \rangle \geq \alpha\}$. If $0 \in G_\alpha$, then $g_0 = 0$ is the strong minimum of f on G_α. If $0 \notin G_\alpha$, then it follows that $\alpha \neq 0$. Let x be an interior point of G_α, then $f(x) > 0$ and with $x_\lambda := (1 - \lambda) \cdot 0 + \lambda x$ for $\lambda \in (0, 1)$ it follows that $f(x_\lambda) \leq \lambda f(x) < f(x)$. Hence x cannot be a minimum of f on G_α. Let g_0 be the strong minimum of f on $H_\alpha := \{x \mid \langle x_0^*, x \rangle = \alpha\}$. Then according to the reasoning above $f(g_0) = \inf\{f(g) \mid g \in G_\alpha\}$. Let now $f(g_n) \to f(g_0)$ with $g_n \in G_\alpha$.

Then $\langle x_0^*, g_n\rangle \to \alpha$, because assume there is $\varepsilon > 0$ and a subsequence $\langle x_0^*, g_{n_k}\rangle \geq \alpha + \varepsilon$. Then $g_{n_k} \in G_{\alpha+\varepsilon}$. But we have for every $\varepsilon > 0$

$$\min\{f(x) \,|\, \langle x_0^*, x\rangle \geq \alpha + \varepsilon\} > f(g_0).$$

For let $f(g_\varepsilon) = \inf\{f(g) \,|\, g \in G_{\alpha+\varepsilon}\}$, then we obtain in the same way as above $\langle x_0^*, g_\varepsilon\rangle = \alpha + \varepsilon$. As 0 is strong minimum of f on X, the mapping $\lambda \mapsto f(\lambda g_\varepsilon)$ is strictly monotonically increasing on $[0, 1]$. The mapping $\lambda \mapsto \phi(\lambda) = \langle x_0^*, \lambda g_\varepsilon\rangle$ is continuous there and we have: $\phi(0) = 0$ and $\phi(1) = \alpha + \varepsilon$. Hence there is a $\lambda_\alpha \in (0, 1)$ with $\phi(\lambda_\alpha) = \alpha$, i.e. $g_\alpha := \lambda_\alpha g_\varepsilon \in H_\alpha$. Therefore we obtain

$$f(g_0) \leq f(g_\alpha) < f(g_\varepsilon) = \min\{f(x) \,|\, \langle x_0^*, x\rangle \geq \alpha + \varepsilon\},$$

hence $f(g_{n_k}) \geq f(g_\varepsilon)$, a contradiction, which establishes $\langle x_0^*, g_n\rangle \to \alpha$.

Since $g_n \in G_\alpha$ we have $\frac{\alpha}{\langle x_0^*, g_n\rangle} \leq 1$. This implies

$$f\left(\frac{\alpha}{\langle x_0^*, g_n\rangle} g_n\right) \leq \frac{\alpha}{\langle x_0^*, g_n\rangle} f(g_n) \to f(g_0).$$

Moreover we have

$$\frac{\alpha}{\langle x_0^*, g_n\rangle} g_n \in \{x \,|\, \langle x_0^*, x\rangle = \alpha\},$$

and hence $f(g_0) \leq f(\frac{\alpha}{\langle x_0^*, g_n\rangle} g_n)$. We conclude that $(\frac{\alpha}{\langle x_0^*, g_n\rangle} g_n)$ is a minimizing sequence of f on H_α, and hence

$$\frac{\alpha}{\langle x_0^*, g_n\rangle} g_n \to g_0,$$

thus $g_n \to g_0$.

(b) $\Rightarrow$ (c): Let K be convex and closed, let $0 \notin K$ and let $r := \inf f(K)$, then $r > 0$ and the interior of $S_f(r)$ is non-empty. According to the Separation Theorem of Eidelheit 3.9.14 there is a half space G_α with $K \subset G_\alpha$ and $G_\alpha \cap \operatorname{Int}(S_f(r)) = \emptyset$. Let g_0 be the strong minimum of f on G_α, then $f(g_0) \leq r$, but $f(g_0) < r$ is impossible, because in that case g_0 would belong to the interior of $S_f(r)$. Let (k_n) be a sequence in K with $f(k_n) \to r = \inf(K)$. Then $f(g_0) = r$ since $K \subset G_\alpha$, and (k_n) is a minimizing sequence for f on G_α, hence $k_n \to g_0$. But K is closed, therefore $g_0 \in K$. Thus g_0 is minimal solution of f on K and because of $K \subset G_\alpha$ also strong minimum.

If $0 \in K$ and (k_n) a sequence in K with $f(k_n) \to 0$, we can assume that K is a subset of a half space G. The point 0 is minimum and hence strong minimum of f on G and therefore also strong minimum of f on K. □

Definition 8.4.5. A convex function is called *bounded*, if the image of every bounded set is bounded.

In order to establish a relation between strong solvability and local uniform convexity, we need the subsequent

Lemma 8.4.6. *Let X be a reflexive Banach space and $f : X \to \mathbb{R}$ convex and continuous.*
Then $\frac{f(x)}{\|x\|} \to_{\|x\|\to\infty} \infty$ holds, if and only if the convex conjugate f^* is bounded.

Proof. Let f^* be bounded. Suppose there is a sequence $(x_n)_{n=1}^\infty$ with $\|x_n\| \to_{n\to\infty} \infty$ for which the sequence of expressions $\frac{f(x_n)}{\|x_n\|} < M$ for some $M \in \mathbb{R}$, then there is a sequence

$$(x_n^*)_{n=1}^\infty \subset X^* \quad \text{with } \|x_n^*\| = 1 \text{ and } \langle x_n^*, x_n\rangle = \|x_n\|.$$

Then one obtains

$$\begin{aligned} f^*(2Mx_n^*) &= \sup_{x\in X}\{2M\langle x_n^*, x\rangle - f(x)\} \\ &\geq \|x_n\|\left(2M\left\langle x_n^*, \frac{x_n}{\|x_n\|}\right\rangle - \frac{f(x_n)}{\|x_n\|}\right) \geq M\|x_n\|, \end{aligned}$$

and because of the boundedness of f^* a contradiction.

Conversely let $\|x^*\| \leq r$, then there is $\varrho \in \mathbb{R}$, such that $\frac{f(x)}{\|x\|} \geq r$ for $\|x\| \geq \varrho$. Therefore

$$\begin{aligned} f^*(x^*) &= \sup_{x\in X}\{\langle x, x^*\rangle - f(x)\} \leq \sup_{x\in X}\left\{\left(r - \frac{f(x)}{\|x\|}\right)\|x\|\right\} \\ &\leq \sup_{\|x\|\leq\varrho}\left\{\left(r - \frac{f(x)}{\|x\|}\right)\|x\|\right\} + \sup_{\|x\|\geq\varrho}\left\{\left(r - \frac{f(x)}{\|x\|}\right)\|x\|\right\} \\ &\leq \sup_{\|x\|\leq\varrho}\{r\|x\| - f(x)\} \leq r\varrho - \inf f(\{x \mid \|x\| \leq \varrho\}). \end{aligned}$$

In order to estimate $-\inf f(\{x \mid \|x\| \leq \varrho\})$ from above, let finally $x_0^* \in \partial f(0)$, then $f(x) \geq f(0) - \|x\|\|x_0^*\| \geq f(0) - \rho\|x_0^*\|$ and thus

$$-\inf f(\{x \mid \|x\| \leq \varrho\}) \leq \rho\|x_0^*\| - f(0). \qquad \square$$

Corollary 8.4.7. *Let $f : X \to \mathbb{R}$ be locally uniformly convex then f^* is bounded.*

Proof. Remark 8.4.2 and the previous lemma. $\square$

Lemma 8.4.8. *Let X be a reflexive Banach space, $f : X \to \mathbb{R}$ continuous and convex, and let f^* be bounded, then for every closed convex set K the set of minimal solutions $M(f, K) \neq \emptyset$.*

Proof. As f^* is bounded, then according to Lemma 8.4.6 in particular all level sets of f are bounded and hence due to the theorem of Mazur–Schauder 3.13.5 $M(f, K) \neq \emptyset$ for an arbitrary closed convex set K. □

Lemma 8.4.9. *Let f be bounded, then f is Lipschitz continuous on bounded sets.*

Proof. Let B be the unit ball in X, let $\varepsilon > 0$, and let S be a bounded subset of X. Then also $S + \varepsilon B$ is bounded. Let α_1 and α_2 be lower and upper bounds of f on $S + \varepsilon B$. Let $x, y \in S$ with $x \neq y$ and let $\lambda := \frac{\|y-x\|}{\varepsilon+\|y-x\|}$. Let further $z := y + \frac{\varepsilon}{\|y-x\|}(y - x)$, then $z \in S + \varepsilon B$ and $y = (1 - \lambda)x + \lambda z$. From the convexity of f we obtain

$$f(y) \leq (1 - \lambda)f(x) + \lambda f(z) = f(x) + \lambda(f(z) - f(x)),$$

and hence

$$f(y) - f(x) \leq \lambda(\alpha_2 - \alpha_1) = \frac{\alpha_2 - \alpha_1}{\varepsilon}\frac{\varepsilon}{\varepsilon + \|y - x\|}\|y - x\| < \frac{\alpha_2 - \alpha_1}{\varepsilon}\|y - x\|.$$

If one exchanges the roles of x and y, one obtains the Lipschitz continuity f on S. □

The subsequent theorem provides a connection between strong solvability and Fréchet differentiability.

Theorem 8.4.10. *Let X be a reflexive Banach space and $f : X \to \mathbb{R}$ a strictly convex and bounded function, whose conjugate f^* is also bounded. Then f has a strong minimum on every closed convex set, if and only if f^* is Fréchet differentiable.*

Proof. Let f^* be Fréchet differentiable. According to Lemma 8.4.4 it suffices to show that f has a strong minimum on every closed hyperplane and on X. By Corollary 8.3.9 and Theorem 8.3.11 f^* is Fréchet differentiable at x^*, if and only if the function $f - \langle x^*, \cdot\rangle$ has a strong minimum on X. As f^* is differentiable at 0, f has a strong minimum on X.

Let $H := \{x \mid \langle x_0^*, x\rangle = r\}$ with $x_0^* \neq 0$. By Lemma 8.4.8 we have $M(f, H) \neq \emptyset$. Let $h_0 \in M(f, H)$. According to Theorem 3.10.5 there is a $x_1^* \in \partial f(h_0)$ with $\langle x_1^*, h - h_0\rangle = 0$ for all $h \in H$. Due to the subgradient inequality we obtain $\langle x_1^*, x - h_0\rangle \leq f(x) - f(h_0)$ for all $x \in X$. Setting $f_1 := f - \langle x_1^*, \cdot\rangle$, it follows that $f_1(x) \geq f_1(h_0)$ for all $x \in X$. As f^* is differentiable at x_1^*, h_0 is the strong minimum of f_1 on X. In particular h_0 is also the strong minimum of f on H, due to

$$f|_H = (f_1 + \langle x_1^*, h_0\rangle)|_H,$$

because for all $h \in H$ we have

$$f_1(h) + \langle x_1^*, h_0\rangle = f(h) - \langle x_1^*, h\rangle + \langle x_1^*, h_0\rangle = f(h) - \langle x_1^*, h - h_0\rangle = f(h).$$

In order to prove the converse let $x_0^* \in X^*$. According to Corollary 8.3.9 and Theorem 8.3.11 we have to show that $f_1 := f - \langle x_0^*, \cdot \rangle$ has a strong minimum on X. As $f_1^*(x^*) = f^*(x_0^* + x^*)$ for all $x^* \in X^*$, apparently f_1^* is bounded and hence f_1 has, according to Lemma 8.4.8, a minimum x_0 on X. If $x_0^* = 0$, then $f_1 = f$ and hence x_0 is the strong minimum of f_1 on X.

Let $x_0^* \neq 0$ and let $f_1(x_n) \to f_1(x_0)$.

Let further be $\varepsilon > 0$, let $K_1 := \{x \in X \mid \langle x_0^*, x \rangle \geq \langle x_0^*, x_0 \rangle + \varepsilon\}$ and let $K_2 := \{x \in X \mid \langle x_0^*, x \rangle \leq \langle x_0^*, x_0 \rangle - \varepsilon\}$ then we obtain because of the strict convexity of f_1

$$\min\{\min(f_1, K_1), \min(f_1, K_2)\} > f_1(x_0).$$

It follows $\langle x_0^*, x_n \rangle \to \langle x_0^*, x_0 \rangle$, because otherwise there is a subsequence (x_{n_k}) in K_1 or K_2, contradicting $f_1(x_n) \to f_1(x_0)$.

Then for $H := \{h \mid \langle x_0^*, x_0 \rangle = \langle x_0^*, h \rangle\}$ we have according to the Formula of Ascoli (see Application 3.12.5)

$$\min\{\|x_n - h\| \mid h \in H\} = \frac{|\langle x_0^*, x_n \rangle - \langle x_0^*, x_0 \rangle|}{\|x_0^*\|} \to 0.$$

For $h_n \in M(\|x_n - \cdot\|, H)$ we conclude: $x_n - h_n \to 0$. The level sets of f_1 are bounded due to Lemma 8.4.6, hence also the sequences (x_n) and (h_n).

As f is, according to Lemma 8.4.9, Lipschitz continuous on bounded sets, there is a constant L with

$$|f(x_n) - f(h_n)| \leq L\|x_n - h_n\| \to 0.$$

We obtain

$$\begin{aligned} f_1(h_n) &= f(h_n) - \langle x_0^*, h_n \rangle \\ &= (f(h_n) - f(x_n)) + (f(x_n) - \langle x_0^*, x_n \rangle) - \langle x_0^*, h_n - x_n \rangle \\ &\to f_1(x_0). \end{aligned}$$

On H the functions f and f_1 differ only by a constant. As f has a strong minimum there, this also holds for f_1, hence $h_n \to x_0$ and thus $x_n \to x_0$. □

Theorem 8.4.11. *Let X be a reflexive Banach space and let f be a continuous convex function on X. Let further f be locally uniformly convex. Then f has a strong minimum on every closed convex set.*

Proof. Let f be locally uniformly convex. As f^* is bounded (see Corollary 8.4.7), then due to Lemma 8.4.6 all level sets of f are bounded and hence according to Lemma 8.4.8 $M(f, K) \neq \emptyset$ for an arbitrary closed convex set K. Let now $k_0 \in M(f, K)$, then by Theorems 3.4.3 and 3.10.4 there is

$$x_0^* \in \partial f(k_0) \quad \text{with } \langle k - k_0, x_0^* \rangle \geq 0 \text{ for all } k \in K.$$

If τ is the convexity module of f belonging to k_0 and x_0^*, then for an arbitrary minimizing sequence (k_n) we have

$$f(k_n) - f(k_0) \geq \langle k_n - k_0, x_0^* \rangle + \tau(\|k_n - k_0\|) \geq \tau(\|k_n - k_0\|),$$

thus $\lim_{n\to\infty} k_n = k_0$. □

Strong solvability is inherited, if one adds a continuous convex function to a locally uniformly convex function.

Corollary 8.4.12. *Let X be a reflexive Banach space and let f and g be continuous convex functions on X. Let further f be locally uniformly convex. Then $f + g$ is locally uniformly convex and has a strong minimum on every closed convex set.*

Proof. First we show: $f + g$ is locally uniformly convex: let $x \in X$ and $x_f^* \in \partial f(x)$, then there is a convexity module τ_{x,x_f^*}, such that for all $y \in X$

$$f(y) - f(x) \geq \langle y - x, x_f^* \rangle + \tau_{x,x_f^*}(\|x - y\|).$$

If $x_g^* \in \partial g(x)$, then we obtain using the subgradient inequality

$$f(y) + g(y) - (f(x) + g(x)) \geq \langle y - x, x_f^* + x_g^* \rangle + \tau_{x,x_f^*}(\|x - y\|).$$ □

Remark 8.4.13. Let X be a reflexive Banach space and let f and g be continuous convex functions on X, and let f^* be bounded. Then $(f + g)^*$ is bounded.

Proof. This is according to Theorem 8.4.6 the case if and only if $\frac{f(x)+g(x)}{\|x\|} \to_{\|x\|\to\infty} \infty$. Now we have for $x_g^* \in \partial g(0)$: $g(x) - g(0) \geq \langle x - 0, x_g^* \rangle \geq -\|x\|\|x_g^*\|$, hence $\frac{g(x)}{\|x\|} \geq \frac{g(0)}{\|x\|} - \|x_g^*\| \geq c \in \mathbb{R}$ for $\|x\| \geq r > 0$. Therefore we conclude

$$\frac{f(x) + g(x)}{\|x\|} \geq \frac{f(x)}{\|x\|} + c \xrightarrow{\|x\|\to\infty} \infty.$$ □

In [56] the main content of the subsequent theorem is proved for f^2, without requiring the boundedness of f^*. For the equivalence of strong solvability and local uniform convexity here we need in addition the boundedness of f.

Theorem 8.4.14. *Let X be a reflexive Banach space, and let $f : X \to \mathbb{R}$ be a bounded, strictly convex function.*

Then f^* is bounded and f has a strong minimum on every closed convex set, if and only if f is locally uniformly convex.

Proof. If f is locally uniformly convex, then the boundedness of f^* follows from Corollary 8.4.7 and strong solvability from Theorem 8.4.11.

Conversely, let f have the strong minimum property, then by Theorem 8.4.10 f^* is Fréchet differentiable. If x_0 is the Fréchet gradient of f^* at x_0^*, then $f(\cdot) - \langle \cdot, x_0^* \rangle$ has, according to Corollary 8.3.9 and Theorem 8.3.11 the global strong minimum at x_0. Let

$$\tau(s) := \inf_{\|y\|=s} \{f(x_0 + y) - f(x_0) - \langle y, x_0^* \rangle\}.$$

Then $\tau(0) = 0$ and $\tau(s) > 0$ for $s > 0$. Furthermore τ is monotonically increasing, because let $s_2 > s_1 > 0$ and let $\|z\| = s_1$. Define $y := \frac{s_2}{s_1} z$ then we obtain, due to the monotonicity of the difference quotient

$$\frac{f(x_0 + \frac{s_1}{s_2} y) - f(x_0)}{\frac{s_1}{s_2}} \le f(x_0 + y) - f(x_0),$$

and hence

$$\begin{aligned} f(x_0 + z) - f(x_0) - \langle z, x_0^* \rangle &\le \frac{s_1}{s_2} (f(x_0 + y) - f(x_0) - \langle y, x_0^* \rangle) \\ &\le f(x_0 + y) - f(x_0) - \langle y, x_0^* \rangle. \end{aligned}$$

Finally we conclude using Theorem 8.4.6

$$\begin{aligned} \frac{\tau(s)}{s} &= -\frac{f(x_0)}{s} + \inf_{\|y\|=s} \left\{ \frac{f(x_0 + y)}{\|y\|} - \left\langle \frac{y}{\|y\|}, x_0^* \right\rangle \right\} \\ &\ge -\frac{f(x_0)}{s} - \|x_0^*\| + \inf_{\|y\|=s} \left\{ \frac{f(x_0 + y)}{\|y\|} \right\} \xrightarrow{s \to \infty} \infty. \end{aligned}$$ □

The following example shows, that on a reflexive Banach space the conjugate of a bounded convex function does not have to be bounded.

Example 8.4.15. Let $f : l^2 \to \mathbb{R}$ be defined by $f(x) = \sum_{i=1}^{\infty} \varphi_i(x^{(i)})$, where $\varphi_i : \mathbb{R} \to \mathbb{R}$ is given by

$$\varphi_i(s) := \begin{cases} \frac{s^2}{2} & \text{for } |s| \le 1 \\ \frac{i}{i+1} |s|^{\frac{i+1}{i}} + \frac{1-i}{2(i+1)} & \text{for } |s| > 1. \end{cases}$$

f is bounded, because $f(x) \le \|x\|_2^2$ for all $x \in l^2$. The conjugate function of f is

$$f^*(x) = \sum_{i=1}^{\infty} \psi_i(x^{(i)}) \quad \text{with } \psi_i(s) := \begin{cases} \frac{s^2}{2} & \text{for } |s| \le 1 \\ \frac{|s|^{i+1}}{i+1} + \frac{i-1}{2(i+1)} & \text{for } |s| > 1. \end{cases}$$

Let e_i denote the i-th unit vector in l^2, then we obtain

$$f^*(2e_i) = \frac{2^{i+1}}{i+1} + \frac{i-1}{2(i+1)} \xrightarrow{i \to \infty} \infty.$$

f^* is continuous on l^2 and Gâteaux differentiable.

8.4.1 E-spaces

For a particular class of Banach spaces, the so-called E-spaces, all convex norm-minimization problems are *strongly solvable*.

Definition 8.4.16. Let (Ω, d) be a metric space and Ω_0 a subset of Ω. Ω_0 is called *approximatively compact*, if for each $x \in \Omega$ every minimizing sequence in Ω_0 (i.e. every sequence $(x_n) \subset \Omega_0$, for which $d(x, x_n) \to d(x, \Omega_0)$ holds) has a point of accumulation in Ω_0.

Definition 8.4.17. A Banach space X is an *E-space*, if X is strictly convex and every weakly closed set is approximatively compact.

Such spaces were introduced by Fan and Glicksberg [29].

The following theorem can be found in [42]:

Theorem 8.4.18. *Let X be a Banach space and let $S(X)$ denote its unit sphere. Then X is an E-space, if and only if X is reflexive and strictly convex and from $x_n, x \in S(X)$ with $x_n \rightharpoonup x$ it follows that $\|x_n - x\| \to 0$.*

Proof. Let K be a weakly closed subset of the reflexive Banach space X and let $x \in X \setminus K$. Let further be $(x_n) \subset K$ a minimizing sequence, more precisely

$$\|x - x_n\| \to d(x, K) = d.$$

Then $d > 0$ holds, because suppose $d = 0$, then $x_n \to x$ and hence $\langle x_n, y^* \rangle \to \langle x, y^* \rangle$ for all $y^* \in X^*$, hence $x_n \rightharpoonup x \in K$, a contradiction to K being weakly closed.

Since (x_n) is bounded, there is by the theorem of Eberlein–Shmulian (see [28]) a weakly convergent subsequence (x_{n_k}) with $x_{n_k} \rightharpoonup x_0 \in K$. It follows that

$$\frac{x_{n_k} - x}{\|x_{n_k} - x\|} \rightharpoonup \frac{x_0 - x}{d}.$$

Apparently $\|x_0 - x\| \geq d$, hence $\|x_0 - x\| = d$, for suppose $d < \|x_0 - x\|$, then $\frac{x_0 - x}{d} \notin U(X) := \{y \in X \mid \|y\| \leq 1\}$, in contradiction to $U(X)$ being weakly closed. Hence by assumption $x_{n_k} - x \to x_0 - x$, i.e. $x_{n_k} \to x_0$ which means x_0 is a point of accumulation of the minimizing sequence.

Conversely, let X be an E-space. First of all we show the reflexivity: let $x^* \in S(X^*)$ arbitrary and let $H := \{y \in X \mid \langle y, x^* \rangle = 0\}$. H is weakly closed and hence approximatively compact. Let $x \in X \setminus H$ and $(x_n) \subset H$ with $\|x_n - x\| \to d(x, H)$. By assumption there exists a subsequence (x_{n_k}) with $x_{n_k} \to x_0 \in H$. Therefore $\|x_{n_k} - x\| \to \|x_0 - x\| = d(x, H)$. Hence, because of the strict convexity of X, x_0 is the best approximation of x w.r.t. H. According to the Theorem of Singer 3.12.6 then

$$|\langle x - x_0, x^* \rangle| = \|x - x_0\|$$

holds. Therefore x^* attains the supremum on the unit ball of X. Since x^* was arbitrarily chosen, X is – according to the theorem of James [45] – reflexive.

Let now $(x_n) \subset S(X)$ with $x_n \rightharpoonup x \in S(X)$. We have to show: $x_n \to x$.

We choose $x^* \in S(X^*)$ with $\langle x, x^* \rangle = 1$. Then

$$\langle x_n, x^* \rangle \to \langle x, x^* \rangle = 1, \tag{8.7}$$

and $\langle x_n, x^* \rangle > 0$ for all $n > N$ ($N \in \mathbb{N}$ suitably chosen). Let now $\bar{x}_n := \frac{x_n}{\langle x_n, x^* \rangle}$, then

$$(\bar{x}_n) \subset H_1 := \{z \in X \mid \langle z, x^* \rangle = 1\}.$$

The Formula of Ascoli 3.12.5 yields $d(0, H_1) = |\langle 0, x^* \rangle - 1| = 1$. Furthermore, we obtain

$$\|\bar{x}_n\| = \left\| \frac{x_n}{\langle x_n, x^* \rangle} \right\| = \frac{1}{|\langle x_n, x^* \rangle|} \to 1,$$

i.e. $\|\bar{x}_n - 0\| \to d(0, H_1)$. Hence $(\bar{x}_n)$ is a minimizing sequence for the approximation of 0 w.r.t. H_1. Since H_1 is approximatively compact there exists a subsequence $(\bar{x}_{n_k})$ with $\bar{x}_{n_k} \to \bar{x} \in H_1$. But by assumption and (8.7) $\bar{x}_{n_k} \rightharpoonup x$ and hence $x = \bar{x}$. Again by (8.7): $x_{n_k} \to x$. Since every subsequence of $(\bar{x}_n)$ is a minimizing sequence and hence contains a to x convergent subsequence, we altogether obtain $x_n \to x$. □

The properties of an E-space are closely related to the Kadec–Klee property (see [104]):

Definition 8.4.19. A Banach space X has the *Kadec–Klee property*, if $x_n \rightharpoonup x$ and $\|x_n\| \to \|x\|$ already imply $\|x_n - x\| \to 0$.

Therefore, a strictly convex and reflexive Banach space with Kadec–Klee property is an E-space. Conversely, it is easily seen that an E-space has the Kadec–Klee property.

But the E-space property can also be characterized by Fréchet differentiability of the dual space. In order to substantiate this we need the following

Lemma 8.4.20 (Shmulian). *The norm of a normed space X is Fréchet differentiable at $x \in S(X)$, if and only if every sequence $(x_n^*)_{n\in\mathbb{N}} \subset U(X^*)$ with $\langle x, x_n^* \rangle \to 1$ is convergent.*

Proof. Let $x^* \in S(X^*)$ be the Fréchet derivative of the norm at x. We will show now: each sequence $(x_n^*) \subset U(X^*)$ with $\langle x, x_n^* \rangle \to 1$ converges to x^*.

Since $1 \geq \|x_n^*\| \geq \langle x, x_n^* \rangle \to 1$ it follows that $\|x_n^*\| \to 1$. Let $\bar{x}_n^* := \frac{x_n^*}{\|x_n^*\|}$. Suppose the sequence $(\bar{x}_n^*)$ does not converge to x^*, then there is an $\varepsilon > 0$ and a sequence $(z_n) \subset S(X)$ with $\langle z_n, \bar{x}_n^* - x^* \rangle \geq 2\varepsilon$. Put $x_n := \frac{1}{\varepsilon}(\|x\| - \langle x, \bar{x}_n^* \rangle) z_n$, then

$\langle x, \bar{x}_n^* \rangle = \langle x, x_n^* \rangle \frac{1}{\|x_n^*\|} \to 1 = \|x\|$, hence $x_n \to 0$. By definition the following chain of inequalities holds:

$$\begin{aligned}\frac{\|x + x_n\| - \|x\| - \langle x_n, x^* \rangle}{\|x_n\|} &\geq \frac{\langle x + x_n, \bar{x}_n^* \rangle - 1 - \langle x_n, x^* \rangle}{\|x_n\|} \\ &= \frac{\langle x, \bar{x}_n^* \rangle - 1 + \langle x_n, \bar{x}_n^* - x^* \rangle}{\|x_n\|} \\ &= \frac{\langle x, \bar{x}_n^* \rangle - 1 + \langle z_n, \bar{x}_n^* - x^* \rangle \frac{1 - \langle x, \bar{x}_n^* \rangle}{\varepsilon}}{\frac{1 - \langle x, \bar{x}_n^* \rangle}{\varepsilon}} \\ &= -\varepsilon + \langle z_n, \bar{x}_n^* - x^* \rangle \geq \varepsilon,\end{aligned}$$

a contradiction to x^* being the Fréchet derivative of the norm at x.

In order to prove the converse we remark that the norm at x has a Gâteaux derivative, since otherwise there are two distinct subgradients $x_1^*, x_2^* \in \partial\|x\|$. But then the sequence $x_1^*, x_2^*, x_1^*, x_2^*, \ldots$ violates the assumption since for $i = 1, 2$ we obtain: $\langle h, x_i^* \rangle = \langle x + h - x, x_i^* \rangle \leq \|x + h\| - \|x\| \leq \|h\|$ and hence $\|x_i^*\| \leq 1$. Therefore $\langle x, x_i^* \rangle \leq \|x\| = 1$ and $\langle 0 - x, x_i^* \rangle \leq \|0\| - \|x\| = -1$, hence $\langle x, x_i^* \rangle = 1$. Let then x^* be the Gâteaux derivative of the norm at x. If the norm is not Fréchet differentiable at x, there exists an $\varepsilon > 0$ and a sequence $(x_n) \subset X$ with $x_n \to 0$, such that

$$\frac{\|x + x_n\| - \|x\| - \langle x_n, x^* \rangle}{\|x_n\|} \geq \varepsilon,$$

i.e. $\|x + x_n\| - \langle x + x_n, x^* \rangle \geq \varepsilon \|x_n\|$ and hence

$$-\langle x_n, x^* \rangle \geq \varepsilon \|x_n\| + \langle x, x^* \rangle - \|x + x_n\|. \tag{8.8}$$

Choose a sequence $(x_n^*) \subset S(X^*)$ with $\langle x + x_n, x_n^* \rangle = \|x + x_n\|$, then due to $x_n \to 0$

$$\langle x, x_n^* \rangle = \|x + x_n\| - \langle x_n, x_n^* \rangle \to \|x\| = 1.$$

Therefore by assumption there is $\bar{x}^* \in S(X^*)$ with $x_n^* \to \bar{x}^*$ and we have

$$\langle x, x_n^* \rangle \to \langle x, \bar{x}^* \rangle = \|x\|. \tag{8.9}$$

Since $\langle x + h - x, \bar{x}^* \rangle = \langle x + h, \bar{x}^* \rangle - \langle x, \bar{x}^* \rangle \leq \|x + h\| - \|x\|$, we have $\bar{x}^* \in \partial\|x\|$ hence $x^* = \bar{x}^*$.

On the other hand we obtain by (8.8)

$$\begin{aligned}\|x_n^* - x^*\| \geq \left\langle \frac{x_n}{\|x_n\|}, x_n^* - x^* \right\rangle &= \frac{\langle x_n, x_n^* \rangle - \langle x_n, x^* \rangle}{\|x_n\|} \\ &\geq \frac{\langle x_n, x_n^* \rangle + \varepsilon \|x_n\| + \langle x, x^* \rangle - \|x + x_n\|}{\|x_n\|} \\ &= \frac{\langle x, x^* \rangle - \langle x, x_n^* \rangle}{\|x_n\|} + \varepsilon \geq \varepsilon,\end{aligned}$$

where the last inequality follows from $\langle x, x^* \rangle = \|x\| \geq \langle x, x_n^* \rangle$. Thus the sequence (x_n^*) does not converge to x^*, a contradiction. □

An immediate consequence of the above lemma of Shmulian is the following corollary, which is a special case of the theorem of Phelps 3.8.6:

Corollary 8.4.21. *Let the norm of a normed space X be Fréchet differentiable for all $x \neq 0$. Then the mapping*

$$\begin{aligned} D : X \setminus \{0\} &\to S(X^*) \\ x &\mapsto D(x), \end{aligned}$$

where $D(x)$ denotes the Fréchet gradient at x, is continuous.

Proof. Let $x_n \to x$ then

$$\begin{aligned} |\langle x, D(x_n) - D(x) \rangle| &= |\langle x_n, D(x_n) \rangle + \langle x - x_n, D(x_n) \rangle - \langle x, D(x) \rangle| \\ &\leq |\|x\| - \|x_n\|| + |\langle x - x_n, D(x_n) \rangle| \leq 2\|x - x_n\|, \end{aligned}$$

hence $\langle x, D(x_n) \rangle \to \langle x, D(x) \rangle = 1$, where the last equality is due to Theorem 8.1.4 for $x \in S(X)$. The lemma of Shmulian then implies $D(x_n) \to D(x)$ in X^*. □

By the lemma of Shmulian we are now in the position, to describe the E-space property by the Fréchet differentiability of the dual space:

Theorem 8.4.22 (Anderson). *A Banach space X is an E-space, if and only if X^* is Fréchet differentiable.*

Proof. Let X be an E-space. By Theorem 8.4.18 X is reflexive and, since $X = X^{**}$ is strictly convex, X^* is smooth. Let $x^* \in S(X^*)$ be arbitrary. Now we have to show the validity of the Shmulian criterion at x^*:

Let $(x_n) \subset U(X)$ with $\langle x_n, x^* \rangle \to 1$. We have to show: the sequence (x_n) is convergent.

We have: $1 \geq \|x_n\| \geq \langle x_n, x^* \rangle \to 1$, hence $\|x_n\| \to 1$. Let $\bar{x}_n := \frac{x_n}{\|x_n\|}$ and let $\bar{x}$ be a weak point of accumulation of the sequence $(\bar{x}_n)$ i.e. $\bar{x}_{n_k} \rightharpoonup \bar{x}$ for a subsequence $(\bar{x}_{n_k})$ of $(\bar{x}_n)$, then $\|\bar{x}\| \leq 1$, since the closed unit ball of X is weakly closed.

On the other hand by the theorem of Hahn–Banach there is a $\bar{y}^* \in S(X^*)$ with $\langle \bar{x}, \bar{y}^* \rangle = \|\bar{x}\|$.

Moreover, we have

$$\langle \bar{x}, x^* \rangle = \lim \langle \bar{x}_{n_k}, x^* \rangle = \lim \left\langle \frac{x_{n_k}}{\|x_{n_k}\|}, x^* \right\rangle = 1,$$

hence $\|\bar{x}\| = 1$. Theorem 8.4.18 then implies $\bar{x}_{n_k} \to \bar{x}$.

Since X^* is smooth, $\bar{x}$ is by $\bar{x} \in S(X)$ and $\langle \bar{x}, x^* \rangle = 1$ uniquely determined as the Gâteaux gradient of the norm at x^*, since

$$\langle \bar{x}, y^* - x^* \rangle = \langle \bar{x}, y^* \rangle - \langle \bar{x}, x^* \rangle \leq \|y^*\| - \|x^*\|.$$

Therefore we obtain for the whole sequence $\bar{x}_n \to \bar{x}$ and since $\|x_n\| \to 1$, as seen above, also $x_n \to \bar{x}$, showing that the Shmulian criterion is satisfied and hence X^* Fréchet differentiable.

Conversely, let X^* be Fréchet differentiable. Let $x^* \in S(X^*)$ be arbitrary. By definition there is a sequence $(x_n) \subset U(X)$ with $\langle x_n, x^* \rangle \to 1$. From the lemma of Shmulian we conclude: (x_n) is convergent: $x_n \to \bar{x}$, hence $\langle x_n, x^* \rangle \to \langle \bar{x}, x^* \rangle = 1$ and $\|\bar{x}\| \leq 1$. Therefore x^* attains its supremum at $\bar{x} \in U(X)$. According to the theorem of James X is reflexive, since $x^* \in S(X^*)$ was arbitrarily chosen.

Let now $x_n, x \in S(X)$ with $x_n \rightharpoonup x$, it remains to be shown: $x_n \to x$. Let $x^* \in S(X^*)$ be chosen, such that $\langle x, x^* \rangle = 1$, then $\langle x_n, x^* \rangle \to \langle x, x^* \rangle = 1$. Again by the lemma of Shmulian the sequence (x_n) is convergent: $x_n \to \bar{x}$, hence $\langle x_n, x^* \rangle \to \langle \bar{x}, x^* \rangle = 1$ and $\|\bar{x}\| = 1$. The smoothness of X^* implies $x = \bar{x}$, as seen above. □

The link to strong solvability (see [42]) is provided by the following

Theorem 8.4.23. *A Banach space X is an E-space, if and only if for every closed convex set K the problem $\min(\|\cdot\|, K)$ is strongly solvable.*

Proof. Let X be an E-space and K a closed convex subset, then K is also weakly closed and hence approximatively compact. Let $x \in X \setminus K$ and $(x_n) \subset K$ a minimizing sequence, i.e. $\|x - x_n\| \to d(x, K)$, then there is a subsequence (x_{n_k}) with $x_{n_k} \to x_0 \in K$. Hence $\|x - x_0\| = d(x, K)$. Since X is strictly convex, x_0 is the uniquely determined best approximation of x w.r.t. K. As every subsequence of (x_n) is a minimizing sequence, it again contains a subsequence converging to x_0. Thus we have for the whole sequence $x_n \to x_0$. Therefore the problem $\min(\|x - \cdot\|, K)$ is strongly solvable.

Conversely, let every problem of this kind be strongly solvable. Suppose X is not strictly convex, then there are $x_1, x_2 \in S(X)$ with $x_1 \neq x_2$ and $[x_1, x_2] \subset S(X)$. Then the problem $\min(0, [x_1, x_2])$ is apparently not strongly solvable, a contradiction.

Now we show the reflexivity: let $x^* \in S(X^*)$ arbitrary and let $H := \{y \in X \mid \langle y, x^* \rangle = 0\}$. H is convex and closed. Let $x \in X \setminus H$ and let $(x_n) \subset H$ with $\|x_n - x\| \to d(x, H)$. By our assumption $x_n \to x_0 \in H$. Therefore $\|x_n - x\| \to \|x_0 - x\| = d(x, H)$. Hence x_0 is – due to the strict convexity of X – the best approximation of x w.r.t. H. According to the theorem of Singer 3.12.6 we then obtain

$$\langle x - x_0, x^* \rangle = \|x - x_0\|.$$

This means that x^* attains its supremum on the unit ball of X. Since x^* was arbitrarily chosen, X is, by the theorem of James, reflexive.

That $x_n, x \in S(X)$ with $x_n \rightharpoonup x$ implies that $\|x_n - x\| \to 0$ can be shown in the same way as in the last part of the proof of Theorem 8.4.18. □

8.5 Fréchet differentiability and Local Uniform Convexity in Orlicz Spaces

We obtain as the central result of this paragraph that in a reflexive Orlicz space strong and weak differentiability of Luxemburg and Orlicz norm, as well as strict and local uniform convexity coincide.

8.5.1 Fréchet Differentiability of Modular and Luxemburg Norm

If Φ is differentiable, then f^Φ and $\|\cdot\|_{(\Phi)}$ are Gâteaux differentiable and for the Gâteaux derivatives we have (see Lemma 8.2.1)

$$\langle (f^\Phi)'(x_0), x\rangle = \int_T x\Phi'(x_0)d\mu$$

and (see Equation (8.6)) $\langle \|x_0\|'_{(\Phi)}, x\rangle = \frac{\langle f'_\Phi(y_0), x\rangle}{\langle f'_\Phi(y_0), y_0\rangle}$, where $y_0 := \frac{x_0}{\|x_0\|_{(\Phi)}}$ for $x_0 \neq 0$. If the conjugate function of Φ satisfies the Δ_2-condition we can prove the continuity of the above Gâteaux derivatives.

The following theorem was shown by Krasnosielskii [72] for the Lebesgue measure on a compact subset of the $\mathbb{R}^n$, but in a different way.

Theorem 8.5.1. *Let (T, Σ, μ) be a σ-finite measure space, Φ a differentiable Young function, and let its conjugate function Ψ satisfy the Δ_2-condition. Then the Gâteaux derivatives of f^Φ and $\|\cdot\|_{(\Phi)}$ are continuous mappings from*

$$M^\Phi(\mu) \text{ resp. } M^\Phi(\mu) \setminus \{0\} \text{ to } L^\Psi(\mu).$$

Proof. Let $(x_n)_{n=1}^\infty$ be a sequence in $M^\Phi(\mu)$ with $\lim x_n = x_0$.

First we show that in $L^\Psi(\mu)$ the relation $\lim_{n\to\infty} \Phi'(x_n) = \Phi'(x_0)$ holds. As Ψ satisfies the Δ_2-condition, convergence in the norm is equivalent to convergence w.r.t. the modular f^Ψ (see Theorems 6.2.21, 6.3.1 and 6.3.3), i.e. we have to show that

$$\lim_{n\to\infty} f^\Psi(\Phi'(x_n) - \Phi'(x_0)) = 0.$$

Let now T be represented as a countable union of pairwise disjoint sets T_i, $i = 1, 2, \ldots$, of finite positive measures.

We define

$$S_k := \bigcup_{i=1}^{k} T_i, \quad S'_k := \{t \in T \,|\, |x_0(t)| \le k\},$$
$$D_k := S_k \cap S'_k,$$

and finally

$$R_k := T \setminus D_k.$$

First we show: $\int_{R_k} x_0 \Phi'(x_0) d\mu \to_{k\to\infty} 0$. As $|x_0(t)| > k$ on R_k we obtain

$$\infty > \int_T x_0 \Phi'(x_0) d\mu \ge \int_{R_k} x_0 \Phi'(x_0) d\mu \ge \mu(R_k) k \Phi'(k).$$

As $k\Phi'(k) \ge \Phi(k) \to_{k\to\infty} \infty$, it follows that $\mu(R_k) \to_{k\to\infty} 0$. As $D_k \subset D_{k+1}$ the sequence $(x_0\Phi'(x_0)\chi_{D_k})$ converges monotonically increasing pointwise a. e. to $x_0\Phi'(x_0)$. With $\int_{D_k} x_0\Phi'(x_0)d\mu \to_{k\to\infty} \int_T x_0\Phi'(x_0)d\mu$ it then follows that $\int_{R_k} x_0\Phi'(x_0)d\mu \to_{k\to\infty} 0$. For given $\varepsilon > 0$ we now choose k large enough that

$$\int_{R_k} x_0 \Phi'(x_0) d\mu \le \varepsilon \quad \text{and} \quad \mu(D_k) > 0. \tag{8.10}$$

As Φ' is uniformly continuous on $I := [-k-1, k+1]$, there is a $\delta \in (0,1)$, such that

$$|\Phi'(s) - \Phi'(r)| \le \Psi^{-1}\left(\frac{\varepsilon}{\mu(D_k)}\right)$$

for $|s - r| \le \delta$ and $s, r \in I$.

According to Theorem 7.4.5 the sequence $(x_n)_{n=1}^{\infty}$ converges to x_0 in measure, i.e. there is a sequence of sets $(Q_n)_{n=1}^{\infty}$ with $\lim_{n\to\infty} \mu(Q_n) = 0$, such that

$$\lim_{n\to\infty} \sup_{t \in T\setminus Q_n} |x_0(t) - x_n(t)| = 0.$$

In particular there is a natural number N, such that for $n \ge N$

$$|x_n(t) - x_0(t)| \le \delta \quad \text{for } t \in T \setminus Q_n,$$

and

$$\mu(Q_n) \le \frac{\varepsilon}{k\Phi'(k)}. \tag{8.11}$$

Thus we obtain

$$\int_{T\setminus(Q_n\cup R_k)} \Psi(\Phi'(x_n) - \Phi'(x_0)) d\mu \le \int_{T\setminus(Q_n\cup R_k)} \Psi\left(\Psi^{-1}\left(\frac{\varepsilon}{\mu(D_k)}\right)\right) d\mu \tag{8.12}$$

$$= \frac{\mu(T \setminus (Q_n \cup R_k))}{\mu(D_k)} \varepsilon \le \varepsilon. \tag{8.13}$$

According to Theorem 3.8.5 the derivative of a continuous convex function is demicontinuous, i.e. here: if the sequence $(w_n)_{n=1}^{\infty}$ converges to w_0 in $M^{\Phi}(\mu)$, then $(\Phi'(w_n))_{n=1}^{\infty}$ converges $*$-weak to $\Phi'(w_0)$. Since according to the Uniform Boundedness theorem of Banach (see Theorem 5.3.14) the sequence $(\Phi'(w_n))_{n=1}^{\infty}$ is bounded in $L^{\Psi}(\mu)$, we obtain because of

$$\begin{aligned}&\int_T w_n\Phi'(w_n)d\mu - \int_T w_0\Phi'(w_0)d\mu \\ &\quad = \int_T (w_n - w_0)\Phi'(w_n)d\mu + \int_T w_0(\Phi'(w_n) - \Phi'(w_0))d\mu\end{aligned}$$

using Hölder's inequality

$$\begin{aligned}&\left|\int_T w_n\Phi'(w_n)d\mu - \int_T w_0\Phi'(w_0)d\mu\right| \\ &\quad \le \|w_n - w_0\|_{(\Phi)}\|\Phi'(w_n)\|_{\Psi} + \left|\int_T w_0(\Phi'(w_n) - \Phi'(w_0))d\mu\right|,\end{aligned}$$

where the last expression on the right-hand side converges to zero because of the *-weak convergence of $\Phi'(w_n)$ to $\Phi'(w_0)$. In this way we obtain the relation

$$\lim_{n\to\infty}\int_T w_n\Phi'(w_n)d\mu = \int_T w_0\Phi'(w_0)d\mu. \tag{8.14}$$

Let now: $w_n := \chi_{Q_n\cup R_k}\cdot x_n$, let further be $w_0 := \chi_{R_k}\cdot x_0$, and let $v_n := \chi_{Q_n\cup R_k}\cdot x_0$, then we obtain

$$w_n - w_0 = w_n - v_n + v_n - w_0 = (x_n - x_0)\chi_{Q_n\cup R_k} + x_0(\chi_{Q_n\cup R_k} - \chi_{R_k}).$$

Now we have: $|x_n - x_0| \ge |x_n - x_0|\chi_{Q_n\cup R_k}$ and, using the monotonicity of the Luxemburg norm, we obtain $|x_n - x_0|\chi_{Q_n\cup R_k} \to_{n\to\infty} 0$, moreover using (8.11)

$$\|v_n - w_0\|_{(\Phi)} = \|\chi_{Q_n\setminus R_k}|x_0|\|_{(\Phi)} \le \|\chi_{Q_n}\cdot k\|_{(\Phi)} \le \frac{\varepsilon}{\Phi'(k)} \le \frac{\varepsilon}{\frac{\Phi(k)}{k}} \le \varepsilon,$$

since Ψ is in particular finite (and hence $\frac{\Phi(k)}{k} \to \infty$). We conclude $w_n \to_{n\to\infty} w_0$. Taking

$$\begin{aligned}&\int_{Q_n\cup R_k} x_n\Phi'(x_n)d\mu - \int_{R_k} x_0\Phi'(x_0)d\mu + \int_{R_k} x_0\Phi'(x_0)d\mu - \int_{Q_n\cup R_k} x_0\Phi'(x_0)d\mu \\ &\quad = \left(\int_T w_n\Phi'(w_n)d\mu - \int_T w_0\Phi'(w_0)d\mu\right) \\ &\qquad + \left(\int_T w_0\Phi'(w_0)d\mu - \int_T v_n\Phi'(v_n)d\mu\right)\end{aligned}$$

into account it follows using (8.14)

$$\lim_{n\to\infty}\left(\int_{Q_n\cup R_k} x_n\Phi'(x_n)d\mu - \int_{Q_n\cup R_k} x_0\Phi'(x_0)d\mu\right) = 0.$$

The Δ_2-condition for Ψ together with Young's equality yields a. e.

$$\begin{aligned}\Psi(\Phi'(x_n) - \Phi'(x_0)) &\le \frac{1}{2}(\Psi(2\Phi'(x_n)) + \Psi(2\Phi'(x_0))\\ &\le \frac{\lambda}{2}(\Psi(\Phi'(x_n)) + \Psi(\Phi'(x_0))\\ &\le \frac{\lambda}{2}(x_n\Phi'(x_n) + x_0\Phi'(x_0)).\end{aligned}$$

Therefore

$$\begin{aligned}\int_{Q_n\cup R_k}\Psi(\Phi'(x_n) - \Phi'(x_0))d\mu &\le \frac{\lambda}{2}\left(\int_{Q_n\cup R_k} x_n\Phi'(x_n)d\mu + \int_{Q_n\cup R_k} x_0\Phi'(x_0)d\mu\right)\\ &\le \lambda\left(\int_{Q_n\cup R_k} x_0\Phi'(x_0)d\mu + \varepsilon\right)\end{aligned}$$

for n sufficiently large and, using (8.10) and (8.11), we obtain

$$\int_{Q_n\cup R_k} x_0\Phi'(x_0)d\mu = \int_{R_k} x_0\Phi'(x_0)d\mu + \int_{Q_n\setminus R_k} x_0\Phi'(x_0)d\mu \le 2\varepsilon.$$

Together with (8.12) the first part of the claim follows.

Let now $x_0 \in M^\Phi(\mu)\setminus\{0\}$ and $(x_n)_{n=1}^\infty$ be a sequence that converges in $M^\Phi(\mu)$ to x_0.

Let $y_n := \frac{x_n}{\|x_n\|_{(\Phi)}}$ for $n \in \mathbb{N}\cup\{0\}$, then also $(y_n)_{n=1}^\infty$ converges to y_0, i.e. $\lim_{n\to\infty}\Phi'(y_n) = \Phi'(y_0)$ and, because of (8.14) also

$$\lim_{n\to\infty}\int_T y_n\Phi'(y_n)d\mu = \int_T y_0\Phi'(y_0)d\mu$$

holds, which completes the proof for the derivatives of the Luxemburg norm. □

Remark 8.5.2. To prove the above Theorem 8.5.1 for the sequence space l^Φ, only the Δ_2^0-condition for Ψ is required. The proof can be performed in a similar way.

Remark 8.5.3. If T in Theorem 8.5.1 has finite measure, only the Δ_2^∞-condition for Ψ is needed.

Proof. Φ is differentiable, hence Ψ strict convex according to Theorem 6.1.22, thus in particular definite.

Let now $\Psi(2s) \le \lambda\Psi(s)$ for all $s \ge s_0 \ge 0$, let, different from Theorem 8.5.1, $D_k := S'_k$ and k large enough that

$$\mu(R_k) \le \frac{\varepsilon}{2(\Psi(2s_0)+1)},$$

furthermore $\int_{R_k} x_0\Phi'(x_0)d\mu \le \varepsilon$ and $\frac{\Psi(2s_0)}{k\Phi'(k)} \le 1$ are satisfied.

If we set

$$\begin{aligned} P_n &:= \{t \in Q_n \cup R_k \mid |\Phi'(x_n(t))| \le s_0\} \\ P'_n &:= \{t \in Q_n \cup R_k \mid |\Phi'(x_0(t))| \le s_0\}, \end{aligned}$$

then we obtain in similar fashion as in Theorem 8.5.1

$$\begin{aligned} &\int_{Q_n\cup R_k} \Psi(\Phi'(x_n) - \Phi'(x_0))d\mu \\ &\quad \le \frac{1}{2}\bigg(\int_{Q_n\cup R_k} \Psi(2\Phi'(x_n))d\mu + \int_{Q_n\cup R_k} \Psi(2\Phi'(x_0))d\mu\bigg) \\ &\quad = \frac{1}{2}\bigg(\int_{P_n} \Psi(2\Phi'(x_n))d\mu + \int_{(Q_n\cup R_k)\setminus P_n} \Psi(2\Phi'(x_n))d\mu \\ &\qquad + \int_{P'_n} \Psi(2\Phi'(x_0))d\mu + \int_{(Q_n\cup R_k)\setminus P'_n} \Psi(2\Phi'(x_0))d\mu\bigg) \\ &\quad \le \frac{1}{2}\bigg(\Psi(2s_0)(\mu(P_n) + \mu(P'_n)) \\ &\qquad + \lambda\bigg(\int_{(Q_n\cup R_k)\setminus P_n} x_n\Phi'(x_n)d\mu + \int_{(Q_n\cup R_k)\setminus P'_n} x_0\Phi'(x_0)d\mu\bigg)\bigg) \\ &\quad \le \frac{1}{2}(\mu(R_k) + \mu(Q_n))\Psi(2s_0) + \frac{\lambda}{2}\bigg(2\int_{Q_n\cup R_k} x_0\Phi'(x_0)d\mu + \varepsilon\bigg) \\ &\quad \le \varepsilon(1+3\lambda) \end{aligned}$$

for n sufficiently large. □

We will now demonstrate that for a finite, not purely atomic measure space the following statement holds: if $(L^\Phi(\mu), \|\cdot\|_{(\Phi)})$ is Fréchet differentiable, then $(L^\Phi(\mu), \|\cdot\|_{(\Phi)})$ is already reflexive. For this purpose we need the following

Lemma 8.5.4. *Let (T,Σ,μ) be a σ-finite, not purely atomic measure space. If $(L^\Phi(\mu), \|\cdot\|_{(\Phi)})$ is Fréchet differentiable, then Ψ is finite.*

Proof. For let Ψ not be finite, then Φ' is bounded on $\mathbb{R}$, i.e. there are positive numbers a, b, c such that $a(s-c) \le \Phi(s) \le b\cdot s$ for $s \ge c$.

If T has finite measure, Theorem 6.2.10 implies

(a) $L^1(\mu) = L^\Phi(\mu)$, and the norms $\|\cdot\|_1$ and $\|\cdot\|_{(\Phi)}$ are equivalent.

(b) $L^\infty(\mu) = L^\Psi(\mu)$, and the norms $\|\cdot\|_\infty$ and $\|\cdot\|_\Psi$ are equivalent (here we have also used the equivalence of Orlicz and Luxemburg norms).

Clearly Φ satisfies the Δ_2^∞-condition, hence $M^\Phi(\mu)=L^\Phi(\mu)$ (see Theorem 6.2.42) Therefore we obtain

$$(L^\Psi(\mu), \|\cdot\|_\Psi) = (M^\Phi(\mu), \|\cdot\|_{(\Phi)})^*.$$

Let now A be the set of atoms in T and let $\lambda := \mu(T \setminus A)$, then we choose disjoint sets G_k in $T \setminus A$ with $\mu(G_k) = 2^{-k-1}\lambda$ for $k = 1, 2, \dots$ (see Theorem 6.2.30). Furthermore let $s_0 \in \mathbb{R}_+$ be large enough that

$$\Phi'\left(\frac{s_0}{e^2\Phi^{-1}(\frac{1}{\frac{\lambda}{2}+\mu(A)})}\right) > 0.$$

If we define the functions x_n on T by

$$x_n(t) := \begin{cases} (1-\frac{1}{k})^n s_0 & \text{for } t \in G_k,\ k = 1, 2, \dots \\ 1 & \text{otherwise} \end{cases}$$

for $n = 1, 2, \dots$ and x_0 with

$$x_0(t) := \begin{cases} 0 & \text{for } t \in G_k,\ k = 1, 2, \dots \\ 1 & \text{otherwise,} \end{cases}$$

then the sequence (x_n) converges to x_0 in the L^1-norm. To see this we observe

$$\int_T |x_0 - x_n| d\mu = \sum_{k=1}^\infty \int_{G_k} |x_0 - x_n| d\mu = \lambda s_0 \sum_{k=1}^\infty 2^{-k-1}\left(1 - \frac{1}{k}\right)^n.$$

Let now $\varepsilon > 0$ be given, then we choose a natural number r such that $\sum_{k=r+1}^\infty 2^{-k-1} < \frac{\varepsilon}{2}$. Then

$$\int_T |x_0 - x_n| d\mu \le \lambda s_0 \left(\left(1 - \frac{1}{r}\right)^n \sum_{k=1}^r 2^{-k-1} + \frac{\varepsilon}{2}\right) \le s_0 \lambda \varepsilon$$

for sufficiently large n. Hence also $\|x_n - x_0\|_{(\Phi)} \to 0$, due to the equivalence of the norms.

Thus the number-sequence $(\|x_n\|_{(\Phi)})$ converges because of $\sum_k \mu(G_k) = \frac{\lambda}{2}$ to

$$\|x_0\|_{(\Phi)} = \frac{1}{\Phi^{-1}((\frac{\lambda}{2} + \mu(A))^{-1})}$$

(see Lemma 7.4.3). We now set $y_n := \frac{x_n}{\|x_n\|_{(\Phi)}}$ for $n = 0, 1, 2, \ldots$. The Gâteaux gradient of $\|\cdot\|_{(\Phi)}$ at x_n is

$$\frac{y_n}{\int_T y_n \Phi'(y_n) d\mu}.$$

We first consider the sequence $(\Phi'(y_n))_{n=1}^{\infty}$ in $L^\infty(\mu)$.

For n sufficiently large we obtain because of the monotonicity of Φ' and the choice of s_0

$$\begin{aligned}\|\Phi'(y_n) - \Phi'(y_0)\|_\infty &\geq \operatorname*{ess\,sup}_{t \in G_n} |\Phi'(y_n(t)) - \Phi'(y_0(t))| \\ &= \Phi'\left(\frac{(1-\frac{1}{n})^n s_0}{\|x_n\|_{(\Phi)}}\right) \geq \Phi'\left(\frac{e^{-2} s_0}{\|x_0\|_{(\Phi)}}\right) > 0.\end{aligned}$$

But the number-sequence

$$\left(\int_T y_n \Phi'(y_n) d\mu\right)_{n=1}^{\infty} \text{ converges to } \int_T y_0 \Phi'(y_0) d\mu,$$

because apparently $\{\Phi'(y_n) \,|\, n \in \mathbb{N}\}$ is uniformly bounded in L^∞ and we have

$$\begin{aligned}\int_T y_0(\Phi(y_n) - \Phi'(y_0)) d\mu &= \int_{T \setminus \bigcup_{k=1}^{\infty} G_k} y_0(\Phi(y_n) - \Phi'(y_0)) d\mu \\ &= \frac{1}{\|x_0\|_{(\Phi)}} \left(\Phi'\left(\frac{1}{\|x_n\|_{(\Phi)}}\right) - \Phi'\left(\frac{1}{\|x_0\|_{(\Phi)}}\right)\right) \\ &\quad \cdot \left(\frac{\lambda}{2} + \mu(A)\right).\end{aligned}$$

Thus the Gâteaux derivative of $\|\cdot\|_{(\Phi)}$ is not continuous in y_0 and therefore $\|\cdot\|_{(\Phi)}$ according to Theorem 3.8.6 not Fréchet differentiable.

If T has infinite measure, the we choose a not purely atomic subset T_0 of T with finite measure.

If Σ_0 is the subalgebra of Σ, that consists in the elements of Σ contained in T_0, then we denote the restriction of μ to Σ_0 by μ_0. In the same way as above we can construct a sequence (y_n) of elements of the unit sphere of $L^\Phi(\mu_0)$ converging to y_0 in $L^\Phi(\mu_0)$, for which the sequence $(\Phi'(y_n))$ does not converge to $\Phi'(y_0)$ in $L^\Psi(\mu_0)$. If one considers $L^\Phi(\mu_0)$ and $L^\Psi(\mu_0)$ as subspaces of $L^\Phi(\mu)$ and $L^\Psi(\mu)$ resp., then it is easy to see that the restriction of the Luxemburg norm of $L^\Phi(\mu)$ to $L^\Phi(\mu_0)$ and of the Orlicz norm of $L^\Psi(\mu)$ to $L^\Psi(\mu_0)$ is equal to the Luxemburg norm on $L^\Phi(\mu_0)$ and equal to the Orlicz norm on $L^\Psi(\mu_0)$ resp. Thus for Ψ not finite and $\mu(T) = \infty$ the Luxemburg norm $\|\cdot\|_{(\Phi)}$ is also not Fréchet differentiable on $L^\Phi(\mu) \setminus \{0\}$. □

The Fréchet differentiability of $(L^\Phi, \|\cdot\|_{(\Phi)})$ implies already the Δ_2^∞-condition for Φ:

Theorem 8.5.5. *Let (T, Σ, μ) be a σ-finite, not purely atomic measure space. If $(L^\Phi(\mu), \|\cdot\|_{(\Phi)})$ is Fréchet differentiable, then Φ satisfies the Δ_2^∞-condition.*

Proof. Assume that Φ does not satisfy the Δ_2^∞-condition, then L^Φ contains according to Theorem 6.2.39 a subspace, which is isometrically isomorphic to ℓ^∞. But ℓ^∞ is not Fréchet differentiable, because otherwise ℓ^1 (according to the theorem of Anderson (see Theorem 8.4.22)) is an E-space, hence in particular reflexive, a contradiction. □

Therefore we obtain the following

Theorem 8.5.6. *Let (T, Σ, μ) be a finite, not purely atomic measure space. then the following statement holds: if $(L^\Phi(\mu), \|\cdot\|_{(\Phi)})$ is Fréchet differentiable, then $(L^\Phi(\mu), \|\cdot\|_{(\Phi)})$ is reflexive.*

Proof. According to Theorem 8.5.5 Φ satisfies the Δ_2^∞-condition. Then, due to Theorem 6.2.42 we have $L^\Phi(\mu) = M^\Phi(\mu)$. Therefore, if $\|\cdot\|_{(\Phi)}$ is Fréchet differentiable on $L^\Phi(\mu) \setminus \{0\}$, then Ψ is finite according to Lemma 8.5.4, hence by Theorem 7.6.2 $(M^\Psi(\mu), \|\cdot\|_\Psi)^* = (L^\Phi(\mu), \|\cdot\|_{(\Phi)})$ and thus, according to the theorem of Anderson (see 8.4.22), $(M^\Psi(\mu), \|\cdot\|_\Psi)$ is an E-space, hence in particular reflexive. Using Hölder's inequality we obtain

$$(L^\Psi(\mu), \|\cdot\|_\Psi) \subset (L^\Phi(\mu), \|\cdot\|_{(\Phi)})^* = (M^\Psi(\mu), \|\cdot\|_\Psi)$$

and so we conclude $L^\Psi(\mu) = M^\Psi(\mu)$. □

We are now in the position to characterize the Fréchet differentiable Luxemburg norms. In order to simplify the wording of our results we introduce the following notation:

Definition 8.5.7. We call the measure space (T, Σ, μ) *essentially not purely atomic*, if $\mu(T \setminus A) > 0$, and for $\mu(T) = \infty$ even $\mu(T \setminus A) = \infty$ hold, where A is the set of atoms.

Theorem 8.5.8. *Let (T, Σ, μ) be a σ-finite, essentially not purely atomic measure space, and let $(L^\Phi(\mu), \|\cdot\|_{(\Phi)})$ be reflexive. Then the following statements are equivalent:*

a) *$(L^\Phi(\mu), \|\cdot\|_{(\Phi)})$ is flat convex,*

b) *f^Φ is continuously Fréchet differentiable on $L^\Phi(\mu)$,*

c) *$\|\cdot\|_{(\Phi)}$ is continuously Fréchet differentiable on $L^\Phi(\mu) \setminus \{0\}$,*

d) *Φ is differentiable.*

Proof. a) $\Rightarrow$ b) and c): According to Theorem 8.2.3 flat convexity for a not purely atomic measure space implies the differentiability of Φ and by the theorem of Mazur 8.1.3 the Gâteaux differentiability of $\|\cdot\|_{(\Phi)}$.

Using Theorem 7.7.1, reflexivity of $(L^{\Phi}(\mu), \|\cdot\|_{(\Phi)})$ implies the Δ_2 or Δ_2^{∞}-condition for Φ and Ψ, if T has infinite or finite measure respectively, because otherwise L^{Φ} has a (closed) subspace, isomorphic to ℓ^{∞}, which according to [41] is also reflexive, a contradiction. We conclude, using Theorem 6.2.42 $M^{\Phi}(\mu) = L^{\Phi}(\mu)$.

Due to the equivalence of Luxemburg and Orlicz norm, $(L^{\Phi}(\mu), \|\cdot\|_{\Phi})$ is also reflexive. Moreover, Ψ has to be finite and hence, according to Theorem 7.6.2

$$(L^{\Phi}(\mu), \|\cdot\|_{\Phi})^* = (L^{\Psi}(\mu), \|\cdot\|_{(\Psi)}).$$

Being the dual space of $(L^{\Phi}(\mu), \|\cdot\|_{\Phi})$, the space $(L^{\Psi}(\mu), \|\cdot\|_{(\Psi)})$ is also reflexive. By the Δ_2 or Δ_2^{∞}-condition (depending on infinite or finite measure of T respectively) for Ψ and Theorems 8.5.1 and 8.5.3 it follows that the Gâteaux derivative of f^{Φ} is continuous on $M^{\Phi}(\mu)$ and hence, as in Theorem 8.5.1, the corresponding property for $\|\cdot\|_{(\Phi)}$. By Theorem 3.8.6 continuous Gâteaux differentiability and continuous Fréchet differentiability are equivalent.

a) $\Leftrightarrow$ d): According to Theorem 8.2.3 flat convexity for a not purely atomic measure is equivalent to differentiability of Φ.

b) $\Rightarrow$ a): Due to Theorem 8.1.1 the level set $S_{f^{\Phi}}(1)$ is flat convex. On the other hand $S_{f^{\Phi}}(1)$ is identical to the unit sphere of $M^{\Phi} = L^{\Phi}$. Hence $(L^{\Phi}, \|\cdot\|_{(\Phi)})$ is flat convex.

c) $\Rightarrow$ a): If $\|\cdot\|_{(\Phi)}$ is Fréchet differentiable on $L^{\Phi}(\mu) \setminus \{0\}$, then $(L^{\Phi}(\mu), \|\cdot\|_{(\Phi)})$ is in particular flat convex due to the theorem of Mazur 8.1.3. □

For finite measure a somewhat stronger version is available.

Theorem 8.5.9. *Let (T, Σ, μ) be a finite, not purely atomic measure space. Then the following statements are equivalent:*

a) $(L^{\Phi}(\mu), \|\cdot\|_{(\Phi)})$ *is flat convex and reflexive,*

b) $\|\cdot\|_{(\Phi)}$ *is Fréchet differentiable on* $L^{\Phi}(\mu) \setminus \{0\}$,

c) Φ *is differentiable and* $(L^{\Phi}(\mu), \|\cdot\|_{(\Phi)})$ *is reflexive.*

Proof. Theorems 8.5.8 and 8.5.6. □

For the sequence space l^{Φ} the above theorem can be proved in somewhat weaker form.

Theorem 8.5.10. *Let the Young function Φ be finite and let l^{Φ} be reflexive. then the following statements are equivalent:*

a) $(l^{\Phi}, \|\cdot\|_{(\Phi)})$ *is flat convex,*

b) $\|\cdot\|_{(\Phi)}$ *is Fréchet differentiable on* $l^\Phi \setminus \{0\}$,

c) Φ *is differentiable for all* s *with* $|s| < \Phi^{-1}(1)$.

Proof. The equivalence of a) and c) follows with Theorems 8.2.4 and 7.7.2. Let now $(l^\Phi, \|\cdot\|_{(\Phi)})$ be flat convex. The left-sided derivative Φ'_- of Φ in $\Phi^{-1}(1)$ is finite. If we continue Φ'_- in a continuous, linear, and strictly increasing way beyond $\Phi^{-1}(1)$, and denote that primitive function which is zero at the origin by $\tilde{\Phi}$, then $\tilde{\Phi}$ is a differentiable Young function. Apparently $l^\Phi = l^{\tilde{\Phi}}$ and $\|\cdot\|_{(\tilde{\Phi})} = \|\cdot\|_{(\Phi)}$. If $\tilde{\Psi}$ is the convex conjugate of $\tilde{\Phi}$ then by construction $\tilde{\Psi}$ is finite. Φ and hence $\tilde{\Phi}$ satisfies the Δ_2^0-condition, because otherwise $(l^\Phi, \|\cdot\|_{(\Phi)})$ contains, according to the theorem of Lindenstrauss–Tsafriri a subspace isomorphic to ℓ^∞, a contradiction to the reflexivity of $(l^\Phi, \|\cdot\|_{(\Phi)})$. It follows that $m^{\tilde{\Phi}} = \ell^{\tilde{\Phi}}$. Due to Theorem 7.6.2 we have $(l^{\tilde{\Phi}}, \|\cdot\|_{\tilde{\Phi}})^* = (l^{\tilde{\Psi}}, \|\cdot\|_{(\tilde{\Psi})})$. The equivalence of the norms implies that $(l^{\tilde{\Phi}}, \|\cdot\|_{\tilde{\Phi}})$ is also reflexive and hence $(l^{\tilde{\Psi}}, \|\cdot\|_{(\tilde{\Psi})})$, being the dual space of a reflexive space, is also reflexive. As demonstrated above the Δ_2^0-condition for $\tilde{\Psi}$ follows. Using Remark 8.5.2 we obtain b). By Theorem 8.1.2 a) follows from b). □

8.5.2 Fréchet Differentiability and Local Uniform Convexity

In the sequel we need a characterization of strict convexity of the Luxemburg norm on $H^\Phi(\mu)$.

Theorem 8.5.11. *Let* (T, Σ, μ) *be a not purely atomic measure space and let* Φ *a finite Young function.*

Φ *is strictly convex, if and only if* $(H^\Phi(\mu), \|\cdot\|_{(\Phi)})$ *is strictly convex.*

Proof. Let A be the set of atoms of T. If Φ is not strictly convex, then there are different positive numbers s and t, such that

$$\Phi\left(\frac{s+t}{2}\right) = \frac{1}{2}(\Phi(s) + \Phi(t)).$$

We choose pairwise disjoint sets $T_1, T_2, T_3 \in \Sigma$ of finite measure with $0 < \mu(T_1) = \mu(T_2) \le \min(\frac{\mu(T\setminus A)}{3}, \frac{1}{2(\Phi(s)+\Phi(t))})$ and $0 < \mu(T_3) \le \mu(T \setminus A) - 2\mu(T_1)$.

Furthermore, let $u := \Phi^{-1}(\frac{1-\mu(T_1)(\Phi(s)+\Phi(t))}{\mu(T_3)})$.

The functions

$$x := s\chi_{T_1} + t\chi_{T_2} + u\chi_{T_3},$$

and

$$y := t\chi_{T_1} + s\chi_{T_2} + u\chi_{T_3}$$

are then elements of the unit sphere, because we have

$$\int_T \Phi(x)d\mu = \mu(T_1)\Phi(s) + \mu(T_2)\Phi(t) + \mu(T_3)\Phi(u)$$
$$= \mu(T_2)\Phi(s) + \mu(T_1)\Phi(t) + \mu(T_3)\Phi(u) = \int_T \Phi(y)d\mu = 1.$$

Now, the properties of s, t, T_1 and T_2 imply

$$\int_T \Phi\left(\frac{x+y}{2}\right)d\mu = \mu(T_1)\Phi\left(\frac{s+t}{2}\right) + \mu(T_2)\Phi\left(\frac{s+t}{2}\right) + \mu(T_3)\Phi(u) = 1.$$

Conversely, let Φ be strictly convex, and let x_1, x_2 be elements of the unit sphere with $\|\frac{x_1+x_2}{2}\|_{(\Phi)} = 1$, then Lemma 6.2.15 implies

$$1 = \int_T \Phi(x_1)d\mu = \int_T \Phi(x_2)d\mu = \int_T \Phi\left(\frac{x_1+x_2}{2}\right)d\mu,$$

hence, because of the convexity of Φ, we obtain $\Phi(\frac{x_1+x_2}{2}) - \Phi(x_1) - \Phi(x_2) = 0$ almost everywhere and due to the strict convexity $x_1 = x_2$ almost everywhere, i.e. the Luxemburg norm is strictly convex on $H^\Phi(\mu)$. □

Corollary 8.5.12. *Let (T, Σ, μ) be a not purely atomic measure space and let Φ be a finite Young function.*

If f^Φ strictly convex, then Φ is strictly convex, because then also H^Φ is strictly convex.

For our further reasoning we need the subsequent theorem (see e.g. [49], p. 350), which describes the duality between strict and flat convexity:

Theorem 8.5.13. *If the dual space X^* of a normed space X is strict or flat convex, then X itself is flat or strict convex respectively.*

We are now in the position, to characterize the reflexive and locally uniformly convex Orlicz spaces w.r.t. the Orlicz norm.

Theorem 8.5.14. *Let (T,Σ,μ) a σ-finite, essentially not purely atomic measure space and let $L^\Phi(\mu)$ be reflexive. Then the following statements are equivalent:*

a) *Φ is strictly convex,*

b) *$(L^\Phi(\mu), \|\cdot\|_\Phi)$ is strictly convex,*

c) *$\|\cdot\|_\Phi^2$ is locally uniformly convex,*

d) *f^Φ is locally uniformly convex.*

Proof. The reflexivity implies in particular that Φ and Ψ are finite.

a) $\Rightarrow$ b): If Φ is strictly convex, then its conjugate Ψ is differentiable (see Theorem 6.1.22). Then due to Theorem 8.2.3 $(L^\Psi(\mu), \|\cdot\|_{(\Psi)})$ is flat convex, hence $(L^\Phi(\mu), \|\cdot\|_\Phi)$ strictly convex (see Theorem 8.5.13).

b) $\Leftrightarrow$ c): If $(L^\Phi(\mu), \|\cdot\|_\Phi)$ is strictly convex, then $(L^\Psi(\mu), \|\cdot\|_{(\Psi)})$ is flat convex (see Theorem 8.5.13) and hence due to Theorem 8.5.8 $\|\cdot\|^2_{(\Psi)}/2$ Fréchet differentiable. As $\|\cdot\|^2_\Phi/2$ is the conjugate function of $\|\cdot\|^2_{(\Psi)}/2$ (see Example 3.11.10), then, according to Theorem 8.4.10 and Theorem 8.4.14 $\|\cdot\|^2_\Phi$ is locally uniformly convex.

Apparently b) follows immediately from c).

b) $\Rightarrow$ d): From flat convexity of $(L^\Psi(\mu), \|\cdot\|_{(\Psi)})$ and reflexivity it follows with Theorem 8.5.8 that f^Ψ is Fréchet differentiable. Φ and Ψ satisfy the Δ_2- or the Δ_2^∞-condition respectively, depending on $\mu(T)$ being infinite or finite (see Theorem 7.7.1). Thus f^Ψ and its conjugate f^Φ are bounded (see Theorem 7.3.4 and Theorem 6.3.10). Due to Theorem 8.4.10 and Theorem 8.4.14 this implies the local uniform convexity of f^Φ.

d) $\Rightarrow$ a): This follows immediately from Corollary 8.5.12. □

Remark 8.5.15. In Example 8.6.12 we present a reflexive and strictly convex Orlicz space w.r.t. the Orlicz norm, which is not uniformly convex (compare Milnes in [85], p. 1482).

For the sequence space l^Φ the above theorem can be proved in a somewhat weaker version.

Theorem 8.5.16. *Let Φ and Ψ be finite and let l^Φ be reflexive, then the following statements are equivalent:*

a) $(l^\Phi, \|\cdot\|_\Phi)$ *is strictly convex,*

b) $\|\cdot\|^2_\Phi$ *is locally uniformly convex.*

Proof. As in Theorem 8.5.14 by use of Theorem 8.5.10. □

8.5.3 Fréchet Differentiability of the Orlicz Norm and Local Uniform Convexity of the Luxemburg Norm

Using the relationships between Fréchet differentiability, local uniform convexity and strong solvability presented in the previous paragraphs, we are now in the position to describe the Fréchet differentiability of the Orlicz norm.

Theorem 8.5.17. *Let (T, Σ, μ) be an essentially not purely atomic, σ-finite measure space, and let $L^\Phi(\mu)$ be reflexive. Then the following statements are equivalent:*

a) $(L^\Phi(\mu), \|\cdot\|_\Phi)$ *is flat convex,*

b) Φ *is differentiable,*

c) $\|\cdot\|_\Phi$ *is Fréchet differentiable on* $L^\Phi(\mu) \setminus \{0\}$.

Proof. a) $\Rightarrow$ b): Let Ψ be the conjugate of Φ.

If $(L^\Phi(\mu), \|\cdot\|_\Phi)$ is flat convex, then $(L^\Psi(\mu), \|\cdot\|_{(\Psi)})$ is strictly convex. Due to Theorem 8.5.11 Ψ is strictly convex and hence Φ differentiable according to Theorem 6.1.22.

b) $\Rightarrow$ c): From the differentiability of Φ it follows by Theorem 8.5.8 that f^Φ is Fréchet differentiable. As in the proof of Theorem 8.5.14 the local uniform convexity of f^Ψ follows. Strong and weak sequential convergence agree on the set $S := \{x \mid f^\Psi(x) = 1\}$ because from $x_n \rightharpoonup x$ for $x_n, x \in S$ it follows for $x^* \in \partial f^\Psi(x)$

$$0 = f^\Psi(x_n) - f^\Psi(x) \geq \langle x_n - x, x^* \rangle + \tau(\|x_n - x\|_{(\Psi)}),$$

where τ is the convexity module of f^Ψ belonging to x and x^*, and thus $x_n \to x$. As S is the unit sphere of $L^\Psi(\mu)$ w.r.t. the Luxemburg norm, $(L^\Psi(\mu), \|\cdot\|_{(\Psi)})$ is an E-space according to Theorem 8.4.18, hence $\|\cdot\|_{(\Psi)}$ has a strong minimum on every closed convex set due to Theorem 8.4.23, Apparently, this also holds for $\|\cdot\|_{(\Psi)}^2/2$. Theorem 8.4.10 and Theorem 8.4.14 then imply the Fréchet differentiability of $\|\cdot\|_\Phi^2/2$ and hence of $\|\cdot\|_\Phi$ in $L^\Phi(\mu) \setminus \{0\}$.

c) $\Rightarrow$ a): This follows from the theorem of Mazur. □

It is now a simple task to characterize the locally uniformly convex, reflexive Orlicz spaces w.r.t. the Luxemburg norm.

Theorem 8.5.18. *Let* (T, Σ, μ) *be* σ*-finite, essentially not purely atomic measure space and let* $L^\Phi(\mu)$ *be reflexive. Then the following statements are equivalent:*

a) Φ *is strictly convex,*

b) $(L^\Phi(\mu), \|\cdot\|_{(\Phi)})$ *is strictly convex,*

c) $\|\cdot\|_{(\Phi)}^2$ *is locally uniformly convex.*

Proof. Because of $H^\Phi(\mu) = L^\Phi(\mu)$ the equivalence of a) and b) follows from Theorem 8.5.11. If $(L^\Phi(\mu), \|\cdot\|_{(\Phi)})$ is strictly convex, then $(L^\Psi(\mu), \|\cdot\|_\Psi)$ is flat convex and therefore due to Theorem 8.5.17 $\|\cdot\|_\Psi$ is Fréchet differentiable. Using Theorems 8.4.10 and 8.4.14 c) follows.

c) $\Rightarrow$ b) is obvious. □

The theorems corresponding to Theorems 8.5.17 and 8.5.18 for l^Φ can be stated in the subsequent weaker form.

Theorem 8.5.19. *Let l^Φ be reflexive, Φ differentiable and Ψ finite, then $\|\cdot\|_\Phi$ is Fréchet differentiable on $l^\Phi \setminus \{0\}$.*

Proof. Because of Theorem 7.7.2 Ψ satisfies the Δ_2^0-condition. Hence f^Φ is, due to Remark 8.5.2, Fréchet differentiable. The remaining reasoning follows the lines of Theorem 8.5.17 for b) $\Rightarrow$ c). □

Remark 8.5.20. If the conditions of Theorem 8.5.19 are satisfied then strong and weak differentiability of the Orlicz norm on l^Φ agree.

Theorem 8.5.21. *Let l^Φ be reflexive, let Φ be strictly convex and let Ψ be finite, then $\|\cdot\|^2_{(\Phi)}$ is locally uniformly convex.*

Proof. Ψ is due to Theorem 6.1.22 differentiable and hence $\|\cdot\|_\Psi$ according to Theorem 8.5.19 Fréchet differentiable. Using Theorems 8.4.10 and 8.4.14 the claim of the theorem follows. □

8.5.4 Summary

Based on the theorems proved above we are now able to describe Fréchet differentiability and local uniform convexity by a list of equivalent statements.

Theorem 8.5.22. *Let (T, Σ, μ) be a σ-finite essentially not purely atomic measure space, let Φ be a Young function and Ψ its conjugate, and let $L^\Phi(\mu)$ be reflexive. Then the following statements are equivalent:*

a) *Φ is differentiable,*

b) *$(L^\Phi(\mu), \|\cdot\|_\Phi)$ is flat convex,*

c) *$(L^\Phi(\mu), \|\cdot\|_{(\Phi)})$ is flat convex,*

d) *$\|\cdot\|_\Phi$ is continuously Fréchet differentiable on $L^\Phi(\mu) \setminus \{0\}$,*

e) *$\|\cdot\|_{(\Phi)}$ is continuously Fréchet differentiable on $L^\Phi(\mu) \setminus \{0\}$,*

f) *f_Φ is continuously Fréchet differentiable on $L^\Phi(\mu)$,*

g) *Ψ is strictly convex,*

h) *$L^\Psi(\mu), \|\cdot\|_\Psi)$ is strictly convex,*

i) *$L^\Psi(\mu), \|\cdot\|_{(\Psi)})$ is strictly convex,*

j) *$\|\cdot\|^2_\Psi$ is locally uniformly convex,*

k) *$\|\cdot\|^2_{(\Psi)}$ is locally uniformly convex,*

l) *f^Ψ is locally uniformly convex,*

m) *$\|\cdot\|_\Psi$ has a strong minimum on K,*

n) *$\|\cdot\|_{(\Psi)}$ has a strong minimum on K,*

o) f^Ψ *has a strong minimum on* K,

p) $L^\Psi(\mu), \|\cdot\|_\Psi)$ *is an E-space,*

q) $L^\Psi(\mu), \|\cdot\|_{(\Psi)})$ *is an E-space,*

for every closed convex subset K *of* $L^\Psi(\mu)$.

Proof. a), c), e), f) are equivalent according to Theorem 8.5.8; a), b), d) according to Theorem 8.5.17; g), h), j), l) according to Theorem 8.5.14; g), i), k) according to Theorem 8.5.18. The equivalence of a) and g) is well known (6.1.22), the equivalence of j) and m), k) and n) as well as of l) and o) follow from Theorem 8.4.14. Equivalence of m) and q) ad of n) and p) follows from Theorem 8.4.23. □

For the sequence space l^Φ we obtain the following

Theorem 8.5.23. *Let* Φ *be differentiable,* Ψ *finite and let* l^Φ *be reflexive. Then the following statements hold:*

a) $\|\cdot\|_\Phi$ *is continuously Fréchet differentiable on* $l^\Phi \setminus \{0\}$,

b) $\|\cdot\|_{(\Phi)}$ *is continuously Fréchet differentiable on* $l^\Phi \setminus \{0\}$,

c) f^Φ *is continuously Fréchet differentiable,*

d) $\|\cdot\|^2_\Psi$ *is locally uniformly convex,*

e) $\|\cdot\|^2_{(\Psi)}$ *is locally uniformly convex,*

f) f^Ψ *is locally uniformly convex,*

g) $\|\cdot\|_\Psi$ *has a strong minimum on* K,

h) $\|\cdot\|_{(\Psi)}$ *has a strong minimum on* K,

i) f^Ψ *has a strong minimum on* K

for every closed convex subset K *of* l^Ψ.

Proof. b) follows from Theorem 8.5.10, d) with Theorem 8.5.21, a) with Theorem 8.5.19. From reflexivity we obtain using Remark 8.5.2 statement c) and using Theorem 8.4.10 and Theorem 8.4.14 thereby f). Finally e) follows from a). Statements g), h) and i) follow using Theorems 8.4.10 and 8.4.14. □

8.6 Uniform Convexity and Uniform Differentiability

Whereas in the previous section we have studied the duality between Fréchet differentiability and local uniform convexity (and the related strong solvability) in particular in Orlicz spaces, we will now turn our attention to the duality between uniform convexity and uniform differentiability.

Definition 8.6.1. Let X be a Banach space, then we define the *module of smoothness* by

$$\rho_X(\tau) := \sup_{\|x\|=\|y\|=1} \left(\frac{1}{2}(\|x+\tau y\| + \|x-\tau y\|) - 1 \right)$$

X is called *uniformly differentiable*, if

$$\lim_{\tau \to 0} \frac{\rho_X(\tau)}{\tau} = 0.$$

Köthe (see [49], p. 366 f.) shows that uniform differentiability implies in particular Fréchet differentiability of the norm.

In the context of our treatment of *greedy* algorithms we need the following

Lemma 8.6.2. *Let X be a Banach space of dimension at least 2 and let ρ_X be the corresponding module of smoothness, then $\rho_X(u) > 0$ for $u > 0$. ρ_X is convex (and hence continuous) on $\mathbb{R}_+$. If X is beyond that uniformly differentiable, then $u \mapsto \frac{\rho_X(u)}{u}$ is strictly monotonically increasing.*

Proof. In order to simplify notation put $\rho := \rho_X$. The first part follows from $\rho(\tau) \geq (1+\tau)^{1/2} - 1$ (see Lindenstrauss [77]). ρ is convex, being the supremum of convex functions. Since $\rho(0) = 0$ the function $u \mapsto \frac{\rho(u)}{u}$ being the difference quotient at 0 is monotonically increasing. Suppose there are $0 < t < u$ with $\frac{\rho(u)}{u} = \frac{\rho(t)}{t}$, then $\frac{\rho(\tau)}{\tau}$ is constant on $[u,t]$, i.e. there is a $c > 0$ with $\frac{\rho(\tau)}{\tau} = c$ there, hence $\rho(\tau) = c\tau$ on $[u,t]$. Therefore $\rho'_+(u) = c$ and due to the subgradient inequality

$$c(v-u) = \rho'_+(u)(v-u) \leq \rho(v) - c \cdot u$$

and hence $c \leq \frac{\rho(v)}{v}$ for all $v > 0$. But then $c \leq \lim_{v \to 0} \frac{\rho(v)}{v} = 0$, a contradiction. □

Definition 8.6.3. Let X be a Banach space, then we define the *convexity module* by

$$\delta_X(\varepsilon) := \inf_{\|x\|=\|y\|=1, \|x-y\|=\varepsilon} (2 - \|x+y\|)$$

X is then called *uniformly convex*, if $\delta_X(\varepsilon) > 0$ for all $\varepsilon > 0$.

The following theorem can be found in Lindenstrauss (compare [77]):

Theorem 8.6.4 (Lindenstrauss). *X is uniformly convex, if X^* is uniformly differentiable. Moreover for arbitrary Banach spaces*

$$\rho_{X^*}(\tau) = \sup_{0 \leq \varepsilon \leq 2} (\tau\varepsilon/2 - \delta_X(\varepsilon))$$

$$\rho_X(\tau) = \sup_{0 \leq \varepsilon \leq 2} (\tau\varepsilon/2 - \delta_{X^*}(\varepsilon))$$

for $\tau > 0$.

Remark 8.6.5. That uniform differentiability of X^* follows from uniform convexity of X, can be understood by using the above duality relation between the modules of smoothness and convexity as follows: at first the parallelogram equality in a Hilbert space for $\|x\| = \|y\| = 1$ and $\|x - y\| = \varepsilon$ implies

$$\frac{1}{2}(2 - \|x + y\|) = \frac{1}{2}(2 - \sqrt{4 - \|x - y\|^2}) = 1 - \sqrt{1 - \frac{\varepsilon^2}{4}} = \delta_H(\varepsilon).$$

Apparently then $\delta_H(\varepsilon) = \lambda\varepsilon^2 + o(\varepsilon^2)$. According to Day (see [24]) Hilbert spaces are those Banach spaces with largest convexity module: for an arbitrary Banach space X: $\delta_X(\varepsilon) \leq \delta_H(\varepsilon)$ for all $\varepsilon > 0$. Then in particular: $\lim_{\varepsilon \to 0} \frac{\delta_X(\varepsilon)}{\varepsilon} = 0$.

Let now (τ_n) be a sequence of positive numbers tending to zero, then the equality

$$\frac{\tau_n}{2} = \frac{\delta(\varepsilon)}{\varepsilon}$$

has a solution ε_n for each $n \in \mathbb{N}$ large enough. Then (ε_n), due to $\delta_X(\varepsilon) > 0$ for all $\varepsilon > 0$, also tends to zero. Even though δ_X is in general not convex, the mapping $\varepsilon \mapsto \frac{\delta_X(\varepsilon)}{\varepsilon}$ is monotonically increasing (see [30], Proposition 3). Hence, because of $\frac{\tau_n}{2} - \frac{\delta_X(\varepsilon)}{\varepsilon} \leq 0$ for $\varepsilon \geq \varepsilon_n$

$$\frac{\rho_{X^*}(\tau_n)}{\tau_n} = \sup_{0 \leq \varepsilon \leq \varepsilon_n} \left(\frac{\varepsilon}{\tau_n} \left(\frac{\tau_n}{2} - \frac{\delta_X(\varepsilon)}{\varepsilon} \right) \right) \leq \frac{\varepsilon_n}{2}.$$

Conversely uniform convexity of X can be obtained from uniform differentiability of X^* in the following way: let

$$\begin{aligned}\tilde{\delta}_X(\varepsilon) &:= \sup_{0 \leq \tau \leq 2} (\tau\varepsilon/2 - \rho_{X^*}(\tau)) \\ &= \sup_{0 \leq \tau \leq 2} \left(\tau \left(\varepsilon/2 - \frac{\rho_{X^*}(\tau)}{\tau} \right) \right).\end{aligned}$$

If $\varepsilon > 0$, then for τ small enough the expression $\varepsilon/2 - \frac{\rho_{X^*}(\tau)}{\tau}$ is positive and hence $\tilde{\delta}_X(\varepsilon) > 0$. On the other hand apparently $\delta_X(\varepsilon) \geq \frac{\tau\varepsilon}{2} - \rho_{X^*}(\tau)$ holds for all $\tau > 0$ and $0 \leq \varepsilon \leq 2$ and hence $\delta_X(\varepsilon) \geq \tilde{\delta}_X(\varepsilon)$.

8.6.1 Uniform Convexity of the Orlicz Norm

For non-atomic measures Milne (see [85]) gives the following characterization of uniform convexity of the Orlicz norm:

Theorem 8.6.6. *$(L^\Phi(\mu), \|\cdot\|_\Phi)$ is uniformly convex, if and only if Ψ is differentiable and for each $0 < \varepsilon < 1/4$ there is a constant R_ε with $1 < R_\varepsilon \leq N < \infty$, such that*

(a) *for infinite measure*

i. $\Phi(2u) \le N\Phi(u)$

ii. $\Phi'_+((1-\varepsilon)u) < \frac{1}{R_\varepsilon}\Phi'_+(u)$

for all $u > 0$

(b) *for finite measure*

i. $\limsup_{u\to\infty} \Phi(2u)/\Phi(u) \le N$

ii. $\limsup_{u\to\infty} \frac{\Phi'_+(u)}{\Phi'_+((1-\varepsilon)u)} > R_\varepsilon$

holds.

A remark of Trojanski and Maleev (see [83]) indicates that the Δ_2-condition for Ψ can be expressed in terms of Φ, in fact:

Lemma 8.6.7. *Let* Φ *and* Ψ *be finite and* $\ell < 1$, *then*

$$\Phi(\ell t) \le \frac{\ell}{2}\Phi(t) \Leftrightarrow \Psi(2s) \le \frac{2}{\ell}\Psi(s)$$

holds.

Proof. By the Young's equality

$$\Psi(2s) + \Phi(\Psi'_+(2s)) = 2s\Psi'_+(2s)$$

holds and hence

$$\frac{\ell}{2}\Psi(2s) = \ell s\Psi'_+(2s) - \frac{\ell}{2}\Phi(\Psi'_+(2s)) \le \ell s\Psi'_+(2s) - \Phi(\ell\Psi'_+(2s)).$$

Young's inequality $\Psi(s) - ts \ge -\Phi(t)$ then yields for $t = \ell\Psi'_+(s)$

$$\frac{\ell}{2}\Psi(2s) \le \ell s\Psi'_+(2s) + \Psi(s) - \ell\Psi'_+(s)s = \Psi(s).$$

Interchanging the roles of Φ and Ψ, one obtains the converse:

$$\frac{2}{\ell}\Phi(\ell t) = \frac{2}{\ell}(\ell t\Phi'_+(\ell t)) - \Psi(\Phi'_+(\ell t)) \le 2t\Phi'_+(\ell t) - \Psi(2\Phi'_+(\ell t)) \le \Phi(t),$$

where the last inequality follows from Young's inequality $\Phi(t) - st \ge -\Psi(s)$ for $s = 2\Phi'_+(\ell t)$. □

Remark 8.6.8. In a similar manner one can show for $0 < \ell < 1 < R < \infty$

$$\Phi(\ell t) \le \frac{\ell}{R}\Phi(t) \Leftrightarrow \Psi(Rs) \le \frac{R}{\ell}\Psi(s).$$

By Remark 6.2.27 this is equivalent to the Δ_2-condition for Ψ.

Hence from condition a(ii) resp. b(ii) in the theorem of Milne 8.6.6 the Δ_2-condition for Ψ follows:

Remark 8.6.9. From $\Phi'_+((1-\varepsilon)u) < \frac{1}{R_\varepsilon}\Phi'_+(u)$ we obtain via integration

$$\frac{1}{1-\varepsilon}\Phi((1-\varepsilon)t) = \int_0^t \Phi'_+((1-\varepsilon)u)du < \frac{1}{R_\varepsilon}\int_0^t \Phi'_+(u)du = \frac{1}{R_\varepsilon}\Phi(t).$$

From the theorems of Lindenstrauss and Milne we thus obtain a description of the uniform differentiability of the Luxemburg norm in the non-atomic case.

Theorem 8.6.10. $(L^\Phi(\mu), \|\cdot\|_{(\Phi)})$ *is uniformly differentiable, if and only if* Φ *is differentiable and for each* $0 < \varepsilon < 1/4$ *there is a constant* $1 < R_\varepsilon \leq N < \infty$, *such that*

(a) *for infinite measure*

  i. $\Psi(2u) \leq N\Psi(u)$

  ii. $\Psi'_+((1-\varepsilon)u) < \frac{1}{R_\varepsilon}\Psi'_+(u)$

for all $u > 0$

(b) *for finite measure*

  i. $\limsup_{u\to\infty} \Psi(2u)/\Psi(u) \leq N$

  ii. $\limsup_{u\to\infty} \frac{\Psi'_+(u)}{\Psi'_+((1-\varepsilon)u)} > R_\varepsilon$.

Remark 8.6.11. Let Φ be a differentiable Young function with $\phi := \Phi'$ and $\phi(2s) \leq \lambda\phi(s)$ for all $s \geq 0$, then Φ satisfies the Δ_2-condition with $\Phi(2s) \leq 2\lambda\Phi(s)$ for all $s \geq 0$.

Proof. We obtain by change of variable $\tau = 2t$

$$\Phi(2s) = \int_0^{2s} \phi(\tau)d\tau = 2\int_0^s \phi(2t)dt \leq 2\lambda\int_0^s \phi(t)dt = 2\lambda\Phi(s). \qquad \square$$

We now give an example of a reflexive, w.r.t. the Orlicz norm strictly convex Orlicz space, which is not uniformly convex (compare Milnes in [85], p. 1482).

Example 8.6.12. Let $u_0 = v_0 = 0$ and $0 < \varepsilon < \frac{1}{4}$. Let further $u_n := 2^{n-1}$ and $u'_n := (1+\varepsilon)u_n$, furthermore $v_n := 2^{n-1}$ and $v'_n := (2^{n-1} + \frac{1}{2})$ for $n \in \mathbb{N}$. Let further ϕ be the linear interpolant of these values, more precisely: $\phi(s) = s$ for $0 \leq s \leq u_1$ and for $n \in \mathbb{N}$

$$\phi(s) = \begin{cases} v_n + r'_n(s-u_n) & \text{for } u_n \leq s \leq u'_n \\ v'_n + r''_n(s-u'_n) & \text{for } u'_n \leq s \leq u_{n+1}, \end{cases}$$

where $r'_n = \frac{v'_n - v_n}{u'_n - u_n} = \frac{1}{\varepsilon 2^n}$ and $r''_n = \frac{v_{n+1}-v'_n}{u_{n+1}-u'_n} = \frac{1-\frac{1}{2^n}}{1-\varepsilon}$.

In particular ϕ is strictly monotonically increasing and for $n \in \mathbb{N}$

$$\phi(u_{n+1}) = 2^n = 2 \cdot \phi(u_n)$$

holds and hence for $u_n \leq s \leq u_{n+1}$

$$\phi(u_{n+1}) = \phi(2u_n) \leq \phi(2s) \leq \phi(u_{n+2}) \leq 4 \cdot \phi(u_n) \leq 4\phi(s).$$

The interval $(0, u_1)$ requires a special treatment: let at first $0 < s \leq \frac{1}{2}$, then $\phi(2s) = 2s = 2\phi(s)$. For $\frac{1}{2} < s < 1$ we obtain

$$\phi(2s) = \begin{cases} v_1 + r_1'(2s - u_1) & \text{for } u_1 \leq 2s \leq u_1' \\ v_1' + r_1''(2s - u_1') & \text{for } u_1' \leq 2s \leq u_2. \end{cases}$$

We observe

$$\phi(2s) \leq \begin{cases} (\frac{1}{\varepsilon} - 1)s & \text{for } u_1 \leq 2s \leq u_1' \\ \frac{2-\varepsilon}{1-\varepsilon} s & \text{for } u_1' \leq 2s \leq u_2. \end{cases}$$

Since $\frac{2-\varepsilon}{1-\varepsilon} \leq \frac{8}{3}$ we obtain altogether with $\lambda := \max\{4, \frac{1}{\varepsilon} - 1\}$

$$\phi(2s) \leq \lambda\phi(s) \quad \text{for } s \geq 0,$$

and hence by Remark 8.6.11

$$\Phi(2s) \leq 2\lambda\Phi(s) \quad \text{for } s \geq 0.$$

We now consider the inverse ψ of ϕ. By Corollary 6.1.16 ψ is the derivative of the conjugate Ψ of Φ. We obtain: $\psi(s) = s$ for $0 \leq s \leq v_1$ and for $n \in \mathbb{N}$

$$\psi(s) = \begin{cases} u_n + \tau_n'(s - v_n) & \text{for } v_n \leq s \leq v_n' \\ u_n' + \tau_n''(s - v_n') & \text{for } v_n' \leq s \leq v_{n+1}, \end{cases}$$

where $\tau_n' = \frac{u_n' - u_n}{v_n' - v_n} = \varepsilon 2^n$ and $\tau_n'' = \frac{u_{n+1} - u_n'}{v_{n+1} - v_n'} = \frac{1-\varepsilon}{1-\frac{1}{2^n}}$. As above we obtain

$$\psi(v_{n+1}) = 2^n = 2 \cdot \psi(v_n),$$

and hence for $v_n \leq s \leq v_{n+1}$

$$\psi(v_{n+1}) = \psi(2v_n) \leq \psi(2s) \leq \psi(v_{n+2}) \leq 4 \cdot \psi(v_n) \leq 4\psi(s).$$

The interval $(0, v_1)$ again requires a special treatment: let at first $0 < s \leq \frac{1}{2}$, then $\psi(2s) = 2s = 2\psi(s)$. For $\frac{1}{2} < s < 1$ we obtain

$$\psi(2s) = \begin{cases} u_1 + \tau_1'(2s - v_1) & \text{for } v_1 \leq 2s \leq v_1' \\ u_1' + \tau_1''(2s - v_1') & \text{for } v_1' \leq 2s \leq v_2. \end{cases}$$

We observe

$$\psi(2s) \leq \begin{cases} 3s & \text{for } v_1 \leq 2s \leq v_1' \\ 4s & \text{for } v_1' \leq 2s \leq v_2, \end{cases}$$

and hence in a similar way as above

$$\Psi(2s) \leq 8\Psi(s) \quad \text{for } s \geq 0.$$

Thus Φ and Ψ satisfy the Δ_2-condition. However

$$\frac{\phi((1+\varepsilon)u_n)}{\phi(u_n)} = \frac{\phi(u_n')}{\phi(u_n)} = \frac{v_n'}{v_n} = \frac{2^{n-1}+\frac{1}{2}}{2^{n-1}} \to 1.$$

Therefore condition b(ii) of Theorem 8.6.6 is violated. Let now (T, Σ, μ) be a non-atomic measure space with $\mu(T) < \infty$, then by the Characterization Theorem of Milne 8.6.6 $(L^\Phi(\mu), \|\cdot\|_\Phi)$ is not uniformly convex. But according to the Theorem 8.6.10 $(L^\Psi(\mu), \|\cdot\|_{(\Psi)})$ is not uniformly differentiable. On the other hand apparently Φ and Ψ are strictly convex and differentiable, by Theorem 7.7.1 $L^\Phi(\mu)$ reflexive, and hence due to Theorem 8.5.22 $\|\cdot\|_\Phi^2$ and $\|\cdot\|_{(\Psi)}^2$ Fréchet differentiable and locally uniformly convex and have a strong minimum on every closed convex subset K of L^Φ resp. L^Ψ.

Remark 8.6.13. For the spaces $L^p(\mu)$ for arbitrary measure space (see [77])

$$\rho_X(\tau) = \begin{cases} ((1+\tau)^p + |1-\tau|^p)^{\frac{1}{p}} = (p-1)\tau^2/2 + O(\tau^4) & \text{for } 2 \leq p < \infty \\ (1+\tau)^p - 1 = \tau^p/p + O(\tau^{2p}) & \text{for } 1 \leq p \leq 2 \end{cases}$$

holds.

Maleev and Troyanski (see [83]) construct for a given Young function Φ, satisfying together with its conjugate a Δ_2-condition, an equivalent Young function Λ in the following way: $\Phi_1(t) := \int_0^t \frac{\Phi(u)}{u} du$ and $\Lambda(t) := \int_0^t \frac{\Phi_1(u)}{u} du$. Λ then also satisfies the Δ_2-condition and $(L^\Lambda(\mu), \|\cdot\|_{(\Lambda)})$ is isomorphic to $(L^\Phi(\mu), \|\cdot\|_{(\Phi)})$. They then show

Theorem 8.6.14. *Let $\mu(T) = \infty$, where T contains a subset of infinite measure free of atoms, and let $L^\Phi(\mu)$ be reflexive. Then one obtains for convexity and smoothness modules of $X = (L^\Lambda(\mu), \|\cdot\|_{(\Lambda)})$ the following estimates: $\delta_X(\varepsilon) \geq AF_\Phi(\varepsilon)$, $\varepsilon \in [0,1]$ and $\rho_X(\tau) \leq BG_\Phi(\tau)$, $\tau \in [0,1]$.*

Here F_Φ and G_Φ are defined in the following way:

$$F_\Phi(\varepsilon) = \varepsilon^2 \inf\{\Phi(uv)/u^2\Phi(v) \mid u \in [\varepsilon, 1], v \in \mathbb{R}_+\}$$

$$G_\Phi(\tau) = \tau^2 \sup\{\Phi(uv)/u^2\Phi(v) \mid u \in [\tau, 1], v \in \mathbb{R}_+\}.$$

For finite resp. purely atomic measures similar conditions are given.
It is easily seen: for $\Phi(t) = |t|^p$ one obtains $\Lambda(t) = \frac{1}{p^2}|t|^p$ and

$$G_\Phi(\tau) = \begin{cases} \tau^2 & \text{for } 2 \le p < \infty \\ \tau^p & \text{for } 1 \le p \le 2 \end{cases}$$

$$F_\Phi(\varepsilon) = \begin{cases} \varepsilon^p & \text{for} 2 \le p < \infty \\ \varepsilon^2 & \text{for} 1 \le p \le 2. \end{cases}$$

8.6.2 Uniform Convexity of the Luxemburg Norm

Definition 8.6.15. A Young function we call δ-*convex* if for all $a \in (0,1)$ there is a function $\delta : (0,1) \to (0,1)$, such that for all $s > 0$, and all $b \in [0, a]$

$$\Phi\left(\frac{1+b}{2}s\right) \le (1 - \delta(a))\frac{\Phi(s) + \Phi(b \cdot s)}{2}$$

holds. Φ is called δ^∞-*convex*, if the above inequality holds in some interval (c, ∞) for $c > 0$, and δ^0-*convex*, if the above inequality holds in some interval $(0, c)$.

Remark 8.6.16. According to [3] the δ-convexity is equivalent to the following condition: for all $a \in (0,1)$ there is a function $\delta : (0,1) \to (0,1)$, such that for all $s > 0$

$$\Phi\left(\frac{1+a}{2}s\right) \le (1 - \delta(a))\frac{\Phi(s) + \Phi(a \cdot s)}{2}$$

holds. A corresponding statement holds for the δ^∞-convexity and δ^0-convexity.

Remark 8.6.17. If Φ is strictly convex and δ^∞-convex for $s \ge d > 0$, then Φ is δ^∞-convex for all $c > 0$.

Let $0 < c < d$, then because of the strict convexity of Φ

$$0 < \Phi\left(\frac{1+a}{2}s\right) < \frac{\Phi(s) + \Phi(a \cdot s)}{2}$$

for all $s \in [c, d]$. Hence

$$0 < h(s) := \frac{\Phi(\frac{1+a}{2}s)}{\frac{\Phi(s)+\Phi(a\cdot s)}{2}} < 1.$$

h is continuous on $[c, d]$ and attains its maximum there. Hence

$$\delta'(a) := \max(\delta(a), 1 - \max\{h(s) \mid s \in [c, d]\}).$$

A corresponding statement is obtained for the δ^0-convexity.

Theorem 8.6.18. *Let (T,Σ,μ) be an arbitrary measure space. Then $(L^\Phi, \|\cdot\|_{(\Phi)})$ is uniformly convex, if Φ is δ-convex and satisfies the Δ_2-condition. If $\mu(T) < \infty$, then $(L^\Phi, \|\cdot\|_{(\Phi)})$ is uniformly convex, if Φ is strictly convex and is δ^∞-convex and Φ satisfies the Δ_2^∞-condition.*

Proof. δ-convexity of Φ is *sufficient*.

Case 1: let $\mu(T) = \infty$. We consider at first the modular and show

$$\forall \varepsilon > 0\, \exists q(\varepsilon) : \forall x, y \quad \text{with } f^\Phi(x) = f^\Phi(y) = 1 \text{ and } f^\Phi(x-y) > \varepsilon \tag{8.15}$$

$$\Rightarrow\ f^\Phi\left(\frac{x+y}{2}\right) < 1 - q(\varepsilon). \tag{8.16}$$

W.l.g. we can assume $0 < \varepsilon < 1$. Let $U = \{t \in T \mid |x(t) - y(t)| \geq \frac{\varepsilon}{4}\max\{|x(t)|, |y(t)|\}\}$.

Let $|s - r| \geq \frac{\varepsilon}{4}\max\{s, r\}$, then for $s \geq r \geq 0$ apparently $s \geq r + \frac{\varepsilon}{4}s$, hence $s(1 - \frac{\varepsilon}{4}) \geq r$, thus $r = b \cdot s$ with $0 \leq b \leq 1 - \frac{\varepsilon}{4} =: a$.

If, however $r \geq s \geq 0$, then correspondingly $s = b \cdot r$. We then obtain for $|s - r| \geq \frac{\varepsilon}{4}\max\{s, r\}$

$$\Phi\left(\frac{s+r}{2}\right) \leq \left(1 - \delta\left(1 - \frac{\varepsilon}{4}\right)\right)\frac{\Phi(s) + \Phi(r)}{2},$$

for the modular this means

$$\begin{aligned}
1 - f^\Phi\left(\frac{x+y}{2}\right) &= \frac{f^\Phi(x) + f^\Phi(y)}{2} - f^\Phi\left(\frac{x+y}{2}\right)\\
&= \frac{1}{2}\bigg(\int_U \Phi(x)d\mu + \int_U \Phi(y)d\mu\\
&\qquad + \int_{T\setminus U} \Phi(x)d\mu + \int_{T\setminus U} \Phi(y)d\mu\bigg)\\
&\quad - \int_U \Phi\left(\frac{x+y}{2}\right)d\mu - \int_{T\setminus U} \Phi\left(\frac{x+y}{2}\right)d\mu\\
&\geq \frac{1}{2}\int_U \Phi(x)d\mu + \frac{1}{2}\int_U \Phi(y)d\mu - \int_U \Phi\left(\frac{x+y}{2}\right)d\mu\\
&\geq \frac{1}{2}\int_U \Phi(x)d\mu + \frac{1}{2}\int_U \Phi(y)d\mu\\
&\quad - \frac{1 - \delta(1 - \frac{\varepsilon}{4})}{2}\left(\int_U \Phi(x)d\mu + \int_U \Phi(y)d\mu\right)
\end{aligned}$$

$$= \frac{\delta}{2}\left(1 - \frac{\varepsilon}{4}\right)\left(\int_U \Phi(x)d\mu + \int_U \Phi(y)d\mu\right)$$
$$= \frac{\delta}{2}\left(1 - \frac{\varepsilon}{4}\right)(f^\Phi(x \cdot \chi_U) + f^\Phi(y \cdot \chi_U)).$$

Therefore

$$1 - f^\Phi\left(\frac{x+y}{2}\right) \geq \frac{\delta}{2}\left(1 - \frac{\varepsilon}{4}\right)(f^\Phi(x \cdot \chi_U) + f^\Phi(y \cdot \chi_U)). \tag{8.17}$$

Let now $t \in T \setminus U$, then $|x(t) - y(t)| < \frac{\varepsilon}{4}\max(|x(t)|, |y(t)|)$, hence due to the monotonicity and convexity of the Young function: $\Phi(x(t) - y(t)) \leq \frac{\varepsilon}{2}\Phi(\frac{1}{2}(|x(t)| + |y(t)|))$. We conclude

$$\begin{aligned} f^\Phi((x-y)\chi_{T\setminus U}) &\leq \frac{\varepsilon}{2} f^\Phi\left(\frac{1}{2}(|x| + |y|)\chi_{T\setminus U}\right) \\ &\leq \frac{\varepsilon}{4}(f^\Phi(x\chi_{T\setminus U}) + f^\Phi(y\chi_{T\setminus U})) \\ &\leq \frac{\varepsilon}{4}(f^\Phi(x) + f^\Phi(y)) \leq \frac{\varepsilon}{2}. \end{aligned}$$

If, as assumed, $f^\Phi(x - y) > \varepsilon$, we obtain

$$f^\Phi((x-y)\chi_U) = f^\Phi(x-y) - f^\Phi((x-y)\chi_{T\setminus U}) > \frac{\varepsilon}{2}. \tag{8.18}$$

Using the Δ_2-condition for Φ and Inequality (8.17) we conclude

$$\begin{aligned} \frac{\varepsilon}{2} &< f^\Phi((x-y)\chi_U) \leq f^\Phi((|x| + |y|)\chi_U) \leq \frac{1}{2}(f^\Phi(2x\chi_U) + f^\Phi(2y\chi_U)) \\ &\leq \frac{1}{2}(\lambda(f^\Phi(x\chi_U) + f^\Phi(y\chi_U))) \leq \frac{\lambda}{2}\frac{2}{\delta(1-\frac{\varepsilon}{4})}\left(1 - f^\Phi\left(\frac{x+y}{2}\right)\right). \end{aligned}$$

Solving this inequality for $f^\Phi(\frac{x+y}{2})$ we obtain

$$f^\Phi\left(\frac{x+y}{2}\right) \leq 1 - \frac{\varepsilon}{2\lambda}\delta\left(1 - \frac{\varepsilon}{4}\right). \tag{8.19}$$

Putting $q(\varepsilon) = \frac{\varepsilon}{2\lambda}\delta(1 - \frac{\varepsilon}{4})$ we arrive at Assertion (8.15).

Let now $\|x\|_{(\Phi)} = \|y\|_{(\Phi)} = 1$ and $\|x - y\|_{(\Phi)} > \varepsilon$, then there is a $\eta > 0$ with $f^\Phi(x-y) > \eta$, for suppose there is a sequence (z_n) with $\|z_n\|_{(\Phi)} \geq \varepsilon$ and $f^\Phi(z_n) \to 0$, then by Theorem 6.3.1 $z_n \to 0$, a contradiction. By Inequality (8.19) we then have

$$f^\Phi\left(\frac{x+y}{2}\right) \leq 1 - q(\eta).$$

As a next step we show: there is a $\delta > 0$ with $\|\frac{x+y}{2}\|_{(\Phi)} \leq 1 - \delta$: for suppose there are sequences x_n and y_n with $\|x_n\|_{(\Phi)} = \|y_n\|_{(\Phi)} = 1$ and $f^{\Phi}(\frac{x_n+y_n}{2}) \leq 1 - q(\eta)$ as well as $\|\frac{x_n+y_n}{2}\|_{(\Phi)} \to 1$. If we put $u_n := \frac{x_n+y_n}{2}$, then $\|u_n\|_{(\Phi)} \leq 1$ and $\frac{1}{\|u_n\|_{(\Phi)}} < 2$ for n sufficiently large. Putting $z_n := \frac{u_n}{\|u_n\|_{(\Phi)}}$ it follows for such n

$$\begin{aligned}
1 = f^{\Phi}(z_n) &= f^{\Phi}\left(\left(\frac{1}{\|u_n\|_{(\Phi)}} - 1\right)2u_n + \left(2 - \frac{1}{\|u_n\|_{(\Phi)}}\right)u_n\right) \\
&\leq \left(\frac{1}{\|u_n\|_{(\Phi)}} - 1\right)f^{\Phi}(2u_n) + \left(2 - \frac{1}{\|u_n\|_{(\Phi)}}\right)f^{\Phi}(u_n) \\
&\leq \lambda\left(\frac{1}{\|u_n\|_{(\Phi)}} - 1\right)f^{\Phi}(u_n) + \left(2 - \frac{1}{\|u_n\|_{(\Phi)}}\right)f^{\Phi}(u_n) \\
&\leq \left(\lambda\left(\frac{1}{\|u_n\|_{(\Phi)}} - 1\right) + \left(2 - \frac{1}{\|u_n\|_{(\Phi)}}\right)\right)(1 - q(\eta)) \\
&= \left(1 + (\lambda - 1)\left(\frac{1}{\|u_n\|_{(\Phi)}} - 1\right)\right)(1 - q(\eta)) \\
&\leq 1 - \frac{q(\eta)}{2},
\end{aligned}$$

a contradiction.

Case 2: Let now $\mu(T) < \infty$ and let $c > 0$ be chosen, such that $\Phi(2c)\mu(T) < \frac{\varepsilon}{8}$. Let

$$U = \left\{ t \in T \,\middle|\, |x(t) - y(t)| \geq \frac{\varepsilon}{4}\max(|x(t)|, |y(t)|) \wedge \max(|x(t)|, |y(t)|) \geq c \right\}.$$

Then as above for all $t \in U$

$$\Phi\left(\frac{x(t) + y(t)}{2}\right) \leq \frac{1 - \delta(1 - \frac{\varepsilon}{4})}{2}(\Phi(x(t)) + \Phi(y(t))),$$

and hence as above the Inequality (8.17) holds.

Let now $t \in T \setminus U$, then by definition $|x(t) - y(t)| < \frac{\varepsilon}{4}\max(|x(t)|, |y(t)|)$ or $\max(|x(t)|, |y(t)|) < c$. Then the convexity and monotonicity of Φ on $\mathbb{R}_+$ imply

$$\Phi(x(t) - y(t)) < \Phi\left(\frac{\varepsilon}{4}\max(|x(t)|, |y(t)|)\right) \leq \frac{\varepsilon}{4}(\Phi(x(t)) + \Phi(y(t))),$$

or

$$\Phi(x(t) - y(t)) < \Phi(2\max(|x(t)|, |y(t)|)) \leq \Phi(2c).$$

Therefore

$$\begin{aligned}
\int_{T \setminus U} \Phi(x(t) - y(t))d\mu &\leq \frac{\varepsilon}{4}\left(\int_T \Phi(x(t))d\mu + \int_T \Phi(y(t))d\mu\right) + \Phi(2c)\mu(T) \\
&= \frac{\varepsilon}{2} + \Phi(2c)\mu(T) \leq \frac{5}{8}\varepsilon.
\end{aligned}$$

As above (see (8.18)) we then obtain

$$f^{\Phi}((x-y)\chi_U) > \frac{3}{8}\varepsilon.$$

According to Remark 6.2.26 the Δ_2^∞-condition for Φ holds for $s \geq \frac{\varepsilon}{8}c$ and hence

$$\begin{aligned}\frac{3}{8}\varepsilon &\leq f^{\Phi}((x-y)\chi_U) = f^{\Phi}\Big(2\frac{x-y}{2}\chi_U\Big) = \int_U \Phi\Big(2\frac{x(t)-y(t)}{2}\Big)d\mu \\ &\leq \lambda\int_U \Phi\Big(\frac{x(t)-y(t)}{2}\Big)d\mu \leq \frac{\lambda}{2}(f^{\Phi}(x\chi_U)+f^{\Phi}(y\chi_U)) \\ &\leq \frac{\lambda}{2}\frac{2}{\delta(1-\frac{\varepsilon}{4})}\Big(1-f^{\Phi}\Big(\frac{x+y}{2}\Big)\Big).\end{aligned}$$

Finally we obtain as above

$$f^{\Phi}\Big(\frac{x+y}{2}\Big) \leq 1 - \frac{3}{8}\frac{\varepsilon}{\lambda}\delta\Big(1-\frac{\varepsilon}{4}\Big).$$

□

For non-atomic measures also the converse of the above theorem holds.

Theorem 8.6.19. *Let (T,Σ,μ) be a non-atomic measure space. If then $\mu(T) < \infty$ and if $(L^\Phi, \|\cdot\|_{(\Phi)})$ is uniformly convex, then Φ is δ^∞-convex and satisfies the Δ_2^∞-condition.*

If $\mu(T) = \infty$ and if $(L^\Phi, \|\cdot\|_{(\Phi)})$ is uniformly convex, then Φ is strictly convex and δ-convex and Φ satisfies the Δ_2-condition.

Proof. Suppose $\forall d > 0\ \exists a \in (0,1)\ \forall \delta > 0\ \exists\, u_\delta \geq d$

$$\Phi\Big(\frac{u_\delta + au_\delta}{2}\Big) > (1-\delta)\frac{\Phi(u_\delta)+\Phi(au_\delta)}{2}.$$

Let now $d > 0$ be chosen, such that $\Phi(d)\mu(T) \geq 2$. Then there are $A_\delta, B_\delta \in \Sigma$ with

(a) $A_\delta \cap B_\delta = \emptyset$

(b) $\mu(A_\delta) = \mu(B_\delta)$

(c) $(\Phi(u_\delta)+\Phi(au_\delta))\mu(A_\delta) = 1$.

If we put $x_\delta := u_\delta\chi_{A_\delta} + au_\delta\chi_{B_\delta}$ and $y_\delta := u_\delta\chi_{B_\delta} + au_\delta\chi_{A_\delta}$, we obtain

$$\begin{aligned}f^{\Phi}\Big(\frac{x_\delta - y_\delta}{1-a}\Big) &= f^{\Phi}(u_\delta(\chi_{A_\delta}-\chi_{B_\delta})) = \Phi(u_\delta)(\mu(A_\delta)+\mu(B_\delta)) \\ &\geq (\Phi(u_\delta)+\Phi(au_\delta))\mu(A_\delta) = 1,\end{aligned}$$

and hence $\|x_\delta - y_\delta\|_{(\Phi)} \geq 1 - a$. On the other hand

$$\begin{aligned} f^\Phi\left(\frac{x_\delta + y_\delta}{2(1-\delta)}\right) &\geq \frac{1}{1-\delta} f^\Phi\left(\frac{x_\delta + y_\delta}{2}\right) \\ &= \frac{1}{1-\delta} f^\Phi\left(\frac{u_\delta(1+a)}{2}(\chi_{A_\delta} + \chi_{B_\delta})\right) \\ &= \frac{1}{1-\delta} \Phi\left(\frac{u_\delta(1+a)}{2}\right)(\mu(A_\delta) + \mu(B_\delta)) \\ &\geq \frac{\Phi(u_\delta) + \Phi(a u_\delta)}{2} 2\mu(A_\delta) = (\Phi(u_\delta) + \Phi(a u_\delta))\mu(A_\delta) = 1, \end{aligned}$$

and hence $\|\frac{x_\delta + y_\delta}{2}\|_{(\Phi)} \geq 1 - \delta$. If we put $\varepsilon := 1 - a$ then for all $\delta \in (0,1)$ there are functions x_δ, y_δ with $\|x_\delta - y_\delta\|_{(\Phi)} \geq \varepsilon$ but $\|\frac{x_\delta + y_\delta}{2}\|_{(\Phi)} \geq 1 - \delta$, where $\|x_\delta\|_{(\Phi)} = \|y_\delta\|_{(\Phi)} = 1$, since $f^\Phi(x_\delta) = \Phi(u_\delta)\mu(A_\delta) + \Phi(a u_\delta))\mu(B_\delta) = 1 = f^\Phi(y_\delta)$ because of properties (b) and (c).

The proof for $\mu(T) = \infty$ is performed in a similar way (see [46]). □

For sequence spaces (see [46]) the following theorem holds:

Theorem 8.6.20. $(\ell^\Phi, \|\cdot\|_{(\Phi)})$ *is uniformly convex, if and only if* Φ *is* δ^0*-convex and strictly convex on* $(0, s_0]$ *with* $\Phi(s_0) = \frac{1}{2}$, *and satisfies the* Δ_2^0*-condition.*

8.7 Applications

As applications we discuss

- Tikhonov regularization: this method was introduced for the treatment of ill-posed problems, of which there are a whole lot. The convergence of the method was proved by Levitin and Polyak for uniformly convex regularizing functionals. We show here that locally uniformly convex regularizations are sufficient for that purpose. As we have given a complete description of local uniform convexity in Orlicz spaces we can state such regularizing functionals explicitly.

  In this context we also discuss level uniformly convex regularizations as another generalization of Levitin and Polyak's approach.

- Ritz method: the Ritz method plays an important role in many applications (e.g. in FEM-methods). It is well known that the Ritz procedure generates a minimizing sequence. Actual convergence of the minimal solutions on each subspace is only achieved if the original problem is strongly solvable.

- Greedy algorithms have drawn a growing attention and experienced a rapid development in recent years (see e.g. Temlyakov [104]). The aim is to arrive at a 'compressed' representation of a function in terms of its dominating 'frequencies'. The convergence proof makes use of the Kadec–Klee property of an E-space.

It may be interesting to note that in the convergence proof of the Tikhonov regularization method we make explicit use of local uniform convexity and the convergence of the Ritz method follows from strong solvability, whereas the convergence proof of the greedy algorithm follows from the Kadec–Klee property, so, in a way, three different aspects of E-spaces come into play.

8.7.1 Regularization of Tikhonov Type

Locally Uniformly Convex Regularization

In this section we investigate the treatment of ill-posed problems by Tikhonov's method, using locally uniform convex regularizing functions on reflexive Banach spaces. At first we observe that local uniform convexity of a convex function carries over to the monotonicity of the subdifferential.

Lemma 8.7.1. *Let X be a Banach space, $f : X \to \mathbb{R}$ a continuous locally uniformly convex function, then for all $x, y \in X$ and all $x^* \in \partial f(x)$ and all $y^* \in \partial f(y)$*

$$\tau_{x,x^*}(\|x - y\|) \leq \langle x - y, x^* - y^* \rangle,$$

if τ_{x,x^*} denotes the convexity module belonging to f at x, x^*.

Proof. As f is locally uniformly convex we have

$$\tau_{x,x^*}(\|x - y\|) + \langle y - x, x^* \rangle \leq f(y) - f(x).$$

On the other hand the subgradient inequality yields: $\langle x - y, y^* \rangle \leq f(x) - f(y)$, in total

$$\tau_{x,x^*}(\|x - y\|) + \langle y - x, x^* \rangle \leq \langle y - x, y^* \rangle,$$

as claimed. □

Theorem 8.7.2. *Let X be a reflexive Banach space and let f and g continuous, Gâteaux differentiable convex functions on X. Let further f be locally uniformly convex, let K be a closed, convex subset of X and let $S := M(g, K) \neq \emptyset$.*

Let now $(\alpha_n)_{n\in\mathbb{N}}$ be a positive sequence tending to zero and $f_n := \alpha_n f + g$. Let finally x_n be the (uniquely determined) minimal solution of f_n on K, then the sequence $(x_n)_{n\in\mathbb{N}}$ converges to the (uniquely determined) minimal solutions of f on S.

Proof. According to Theorem 8.4.12 f_n is locally uniformly convex and due to Remark 8.4.2 f_n^* is bounded, hence $M(f_n, K)$ consists of the unique element x_n.

Let now $x \in S$, then because of the monotonicity of the derivative of g

$$\underbrace{\langle x_n - x, g'(x_n) - g'(x)\rangle}_{\geq 0} + \alpha_n \langle x_n - x, f'(x_n)\rangle$$
$$= \underbrace{\langle x_n - x, \alpha_n f'(x_n) + g'(x_n)\rangle}_{\leq 0} - \underbrace{\langle x_n - x, g'(x)\rangle}_{\geq 0} \leq 0.$$

It follows

$$\langle x_n - x, f'(x_n)\rangle \leq 0. \tag{8.20}$$

Let now $\bar{x} \in S$ arbitrary, then

$$0 \geq f_n(x_n) - f_n(\bar{x}) = g(x_n) - g(\bar{x}) + \alpha_n(f(x_n) - f(\bar{x})) \geq \alpha_n(f(x_n) - f(\bar{x})).$$

This implies $f(x_n) \leq f(\bar{x})$, hence $x_n \in S_f(f(\bar{x}))$ for all $n \in \mathbb{N}$. As f^* is bounded, it follows according to Theorem 8.4.6 that the sequence (x_n) is bounded. Let now (x_k) be a subsequence converging weakly to x_0. Since K is weakly closed (see Theorem 3.9.18) we have $x_0 \in K$.

First we show: $\langle y - x_0, g'(x_0)\rangle \geq 0$ for $y \in K$ arbitrary, i.e. $x_0 \in S$. We observe

$$\underbrace{\langle y - x_k, g'(y) - g'(x_k)\rangle}_{\geq 0} + \alpha_k \langle x_k - y, f'(x_k)\rangle$$
$$= \langle y - x_k, g'(y)\rangle + \underbrace{\langle y - x_k, -f_k'(x_k)\rangle}_{\leq 0} \leq \langle y - x_k, g'(y)\rangle.$$

The expression $\langle x_k - y, f'(x_k)\rangle$ is for fixed y bounded from below because

$$\langle x_k - y, f'(x_k)\rangle = \underbrace{\langle x_k - y, f'(x_k) - f'(y)\rangle}_{\geq 0} + \langle x_k - y, f'(y)\rangle$$
$$\geq -\|f'(y)\|\|x_k - y\| \geq -C.$$

Hence we obtain

$$-C \cdot \alpha_k \leq \alpha_k \langle x_k - y, f'(x_k)\rangle \leq \langle y - x_k, g'(y)\rangle.$$

On the other hand the weak convergence of $x_k \rightharpoonup x_0$ implies

$$\langle y - x_k, g'(y)\rangle \xrightarrow{k\to\infty} \langle y - x_0, g'(y)\rangle,$$

and thus

$$\langle y - x_0, g'(y)\rangle \geq 0$$

for all $y \in K$. Let now $1 \geq t > 0$, $z = y - x_0$ and $y = x_0 + tz = ty + (1-t)x_0 \in K$, then the continuity of $t \mapsto \langle z, g'(x_0 + tz)\rangle = \frac{d}{dt} g(x_0 + tz)$ (see Lemma 3.8.2) implies: $0 \leq \langle z, g'(x_0 + tz)\rangle \to_{t\to 0} \langle z, g'(x_0)\rangle$, hence $\langle y - x_0, g'(x_0)\rangle \geq 0$, i.e. $x_0 \in S$.

Now we show the strong convergence of (x_k) to x_0: due to Lemma 8.7.1 the weak convergence and Inequality (8.20) yield

$$\begin{aligned}\tau_{x_0,f'(x_0)}(\|x_0 - x_k\|) &\le \langle x_0 - x_k, f'(x_0) - f'(x_k)\rangle \\ &= \langle x_0 - x_k, f'(x_0)\rangle + \underbrace{\langle x_k - x_0, f'(x_k)\rangle}_{\le 0} \\ &\le \langle x_0 - x_k, f'(x_0)\rangle \to 0,\end{aligned}$$

hence $x_k \to x_0$.

It remains to be shown: x_0 is the minimal solution of f on S: because of the demicontinuity (see Theorem 3.8.5) of f' it follows with Theorem 5.3.15 for $x \in S$

$$0 \le \langle x - x_k, f'(x_k)\rangle \to \langle x - x_0, f'(x_0)\rangle,$$

and by the Characterization Theorem of Convex Optimization 3.4.3 the assertion of the theorem since apparently $x_n \to x_0$ because of the uniqueness of x_0. □

Remark. The proof of above theorem easily carries over to hemi-continuous monotone operators (compare [55]).

As an application of the above theorem in Orlicz spaces we obtain

Theorem 8.7.3. *Let (T, Σ, μ) be a σ-finite, essentially not purely atomic measure space, let $L^\Phi(\mu)$ be reflexive, and let Φ be strictly convex and differentiable. Let either $f := f^\Phi$ or $f := \frac{1}{2}\|\cdot\|^2_{(\Phi)}$ or $f := \frac{1}{2}\|\cdot\|^2_\Phi$ and let $g : L^\Phi(\mu) \to \mathbb{R}$ convex, continuous, and Gâteaux differentiable. Let K be a closed convex subset of $L^\Phi(\mu)$ and let $S := M(g, K) \neq \emptyset$.*

Let now $(\alpha_n)_{n\in\mathbb{N}}$ be a positive sequence tending to zero and $f_n := \alpha_n f + g$. Let finally x_n be the (uniquely determined) minimal solution of f_n on K, then the sequence $(x_n)_{n\in\mathbb{N}}$ converges to the (uniquely determined) minimal solution of f on S.

Proof. Theorem 8.7.2 together with Theorem 8.5.22. □

Level-uniform Convex Regularization

Another generalization of the Levitin–Polyak theorem is the regularization by level-uniformly convex functionals.

Definition 8.7.4. Let X be a Banach space, then the function $f : X \to \mathbb{R}$ is called *level-uniformly convex*, if for every $r \in \mathbb{R}$ there is a function $\tau_r : \mathbb{R}_{\ge 0} \to \mathbb{R}_{\ge 0}$ such that $\tau(0) = 0$, $\tau(s) > 0$ for $s > 0$ such that for all $x, y \in S_f(r)$

$$f\left(\frac{x+y}{2}\right) \le \frac{1}{2}f(x) + \frac{1}{2}f(y) - \tau_r(\|x - y\|).$$

It turns out that boundedness of the level sets is already an inherent property of level-uniform convexity.

Lemma 8.7.5. *The level sets $S_f(r)$ of the level-uniformly convex function f are bounded.*

Proof. The function f is bounded from below by α on the unit ball B, since for $x^* \in \partial f(0)$ we have $-\|x^*\| \le \langle x - 0, x^* \rangle \le f(x) - f(0)$ for all $x \in B$. Suppose $S_f(r_0)$ is not bounded for some $r_0 > f(0)$. Then there exists a sequence (x_n) where $\|x_n\| = 1$ and $nx_n \in S_f(r_0)$. Hence

$$f(nx_n) \ge 2f((n-1)x_n) - f((n-2)x_n) + 2\tau_{r_0}(2).$$

Let $\beta := 2\tau_{r_0}(2)$. Then we obtain by induction for $2 \le k \le n$

$$f(nx_n) \ge kf((n-k+1)x_n) - (k-1)f((n-k)x_n) + \frac{k(k-1)}{2}\beta.$$

For $k = n$ this results in the contradiction

$$\begin{aligned} r_0 \ge f(nx_n) &\ge nf(x_n) - (n-1)f(0) + \frac{n(n-1)}{2}\beta \\ &\ge n\left(\alpha - f(0) + \frac{n-1}{2}\beta\right) + f(0) \xrightarrow{n\to\infty} \infty. \end{aligned}$$ □

Theorem 8.7.6. *Let X be a reflexive Banach space and let f and g continuous convex functions on X. Let further f be level uniformly convex, let K be a closed, convex subset of X and let $S := M(g, K) \ne \emptyset$.*

Let now $(\alpha_n)_{n\in\mathbb{N}}$ be a positive sequence tending to zero and $f_n := \alpha_n f + g$. Let finally x_n be the (uniquely determined) minimal solution of f_n on K, then the sequence $(x_n)_{n\in\mathbb{N}}$ converges to the (uniquely determined) minimal solution of f on S.

Proof. Let now $\bar{x} \in S$ be arbitrary, then

$$0 \ge f_n(x_n) - f_n(\bar{x}) = g(x_n) - g(\bar{x}) + \alpha_n(f(x_n) - f(\bar{x})) \ge \alpha_n(f(x_n) - f(\bar{x})).$$

This implies $f(x_n) \le f(\bar{x}) =: r$, hence $x_n \in S_f(f(\bar{x}))$ for all $n \in \mathbb{N}$. Due to the above lemma the sequence (x_n) is bounded. Hence by Theorem 5.6.9 (the compact topological space C now being $S_f(f(\bar{x})) \cap K$, equipped with the weak topology) (x_n) converges weakly to the minimal solution x_0 of f on $M(g, K)$ and $f(x_n) \to f(x_0)$, since according to Remark 3.11.8 f is weakly lower semi-continuous. Due to the level-uniform convexity

$$\tau_r(\|x_n - x_0\|) \le \frac{1}{2}f(x_n) + \frac{1}{2}f(x_0) - f\left(\frac{x_n + x_0}{2}\right).$$

But $(\frac{x_n+x_0}{2})$ converges weakly to x_0. Since f is weakly lower semi-continuous we have $f(\frac{x_n+x_0}{2}) \geq f(x_0) - \varepsilon$ for $\varepsilon > 0$ and n large enough, hence

$$\tau_r(\|x_n - x_0\|) \leq \frac{1}{2}(f(x_n) + f(x_0)) - f(x_0) + \varepsilon,$$

and thus the strong convergence of (x_n) to x_0. □

8.7.2 Ritz's Method

The following method of minimizing a functional on an increasing sequence of subspaces of a separable space is well known and due to Ritz (compare e.g. [109]). Ritz's method generates a minimizing sequence:

Theorem 8.7.7. *Let X be a separable normed space, let $X = \overline{\text{span}\{\varphi_i, i \in \mathbb{N}\}}$, and let $X_n := \text{span}\{\varphi_1, \dots, \varphi_n\}$. Let further $f : X \to \mathbb{R}$ be upper semi-continuous and bounded from below. If $d := \inf f(X)$ and $d_n := \inf f(X_n)$ for $n \in \mathbb{N}$, then $\lim_{n\to\infty} d_n = d$.*

Proof. Apparently $d_n \geq d_{n+1}$ for all $n \in \mathbb{N}$, hence $d_n \to a \in \mathbb{R}$. Suppose $a > d$. Let $\frac{a-d}{2} > \varepsilon > 0$ and $x \in X$ with $f(x) \leq d + \varepsilon$. As f is upper semi-continuous, there is a neighborhood $U(x)$ with $f(y) \leq f(x) + \varepsilon$ for all $y \in U(x)$, in particular there is $y_m \in U(x)$ with $y_m \in X_m$. It follows that

$$a \leq d_m \leq f(y_m) \leq f(x) + \varepsilon \leq d + 2\varepsilon$$

a contradiction. □

Corollary 8.7.8. *Let $d_n := \inf f(X_n)$, and $(\delta_n)_{n\in\mathbb{N}}$ be a sequence of positive numbers tending to zero, and let $x_n \in X_n$ chosen in such a way that $f(x_n) \leq d_n + \delta_n$, then $(x_n)_{n\in\mathbb{N}}$ is a minimizing sequence for the minimization of f on X.*

For locally uniformly convex functions the convergence of the Ritz's method can be established.

Theorem 8.7.9. *Let X be a separable reflexive Banach space with*

$$X = \overline{\text{span}\{\varphi_i, i \in \mathbb{N}\}},$$

and let $X_n := \text{span}\{\varphi_1, \dots, \varphi_n\}$. Let f and g be continuous convex function on X. Let further f be locally uniformly convex. Let $d := \inf(f+g)(X)$ and $d_n := \inf(f+g)(X_n)$ for $n \in \mathbb{N}$, let $(\delta_n)_{n\in\mathbb{N}}$ be a sequence of positive numbers and let $x_n \in X_n$ be chosen such that $f(x_n)+g(x_n) \leq d_n+\delta_n$ then the minimizing sequence $(x_n)_{n\in\mathbb{N}}$ converges to the (uniquely determined) minimal solution of $f+g$ on X.

Proof. Corollary 8.7.8 together with Theorems 8.4.12 and 8.4.11, and Corollary 8.4.7. □

Remark 8.7.10. By the above theorem we are enabled to regularize the minimization problem $\min(g, X)$ by adding a positive multiple αf of a locally uniformly convex function, i.e. one replaces the above problem by $\min(\alpha f + g, X)$ (compare Theorem 8.7.2).

As an application of the above theorem in Orlicz spaces we obtain

Theorem 8.7.11. *Let (T, Σ, μ) be a σ-finite, essentially not purely atomic measure, let $L^\Phi(\mu)$ separable and reflexive, and let Φ be strictly convex. Let $L^\Phi(\mu) = \overline{\text{span}\{\varphi_i, i \in \mathbb{N}\}}$ and let $X_n := \text{span}\{\varphi_1, \ldots, \varphi_n\}$. Let either $f := f^\Phi$ or $f := \|\cdot\|^2_{(\Phi)}$ or $f := \|\cdot\|^2_\Phi$ and let $g : L^\Phi(\mu) \to \mathbb{R}$ convex and continuous.*

Let $d := \inf(f + g)(L^\Phi(\mu))$ and $d_n := \inf(f + g)(X_n)$ for $n \in \mathbb{N}$, let $(\delta_n)_{n\in\mathbb{N}}$ be a sequence of positive numbers tending to zero and let $x_n \in X_n$ be chosen such that $f(x_n) + g(x_n) \le d_n + \delta_n$, then the minimizing sequence $(x_n)_{n\in\mathbb{N}}$ converges to the (uniquely determined) minimal solution of $f + g$ on $L^\Phi(\mu)$.

Proof. Theorem 8.7.9 and Theorem 8.5.22 . □

8.7.3 A Greedy Algorithm in Orlicz Space

For a compressed approximate representation of a given function in $L^2[a, b]$ by harmonic oscillations it is reasonable to take only the those frequencies with dominating Fourier coefficients into account. This leads to non-linear algorithms, whose generalization to Banach spaces was considered by V. N. Temlyakov. We will discuss the situation in Orlicz spaces.

Definition 8.7.12. Let X be a Banach space, then $D \subset X$ is called a *dictionary* if

(a) $\|\varphi\| = 1$ for all $\varphi \in D$

(b) from $\varphi \in D$ it follows that $-\varphi \in D$

(c) $X = \overline{\text{span}(D)}$.

We consider the following algorithm, which belongs to the class of non-linear m-term algorithms (see [105]) and in [104] is denoted as *Weak Chebyshev Greedy Algorithm* (WCGA) by V. N. Temlyakov:

Algorithm 8.7.13 (WGA). Let X strict convex and let $\tau = (t_k)_{k\in\mathbb{N}}$ with $0 < t_k \le 1$ for all $k \in \mathbb{N}$. Let $x \in X$ be arbitrary, and let $F_x \in S(X^*)$ denote a functional with $F_x(x) = \|x\|$.

Let now $x \in X \setminus \{0\}$ be given, then set $r_0 := x$, and for $m \geq 1$:

(a) Choose $\varphi_m \in D$ with

$$F_{r_{m-1}}(\varphi_m) \geq t_m \sup\{F_{r_{m-1}}(\varphi)|\varphi \in D\}.$$

(b) For $U_m := \text{span}\{\varphi_j, j = 1, \dots, m\}$ let x_m be the best approximation of x w.r.t. U_m.

(c) Set $r_m := x - x_m$ and $m \leftarrow m + 1$, goto (a).

If X is flat convex, then F_x is given by the gradient $\nabla\|x\|$ of the norm at x (see theorem of Mazur 8.1.3). Apparently: $\|F_x\| = 1$.

Before we discuss the general situation, we will consider the case $\tau = (t)$ with $0 < t \leq 1$ and denote the corresponding algorithm by GA.

Convergence of GA in Fréchet Differentiable E-Spaces

The subsequent exposition follows the line of thought of Temlyakov in [104]:

Let $\alpha := d(x, \text{span}\{\varphi_i, i \in \mathbb{N}\})$ and $\alpha_n := \|r_n\| = d(x, U_n)$, then apparently $\alpha_n \downarrow \gamma \geq \alpha$. We obtain

Lemma 8.7.14. *Let X be flat convex and $\alpha > 0$, then there is a $\beta > 0$ and $N \in \mathbb{N}$, such that $F_{r_n}(\varphi_{n+1}) > \beta$ for all $n \geq N$.*

Proof. Since x_n is a best approximation of x w.r.t. U_n, we have $F_{r_n}(\varphi_i) = 0$ for $i = 1, \dots, n$ and $F_{r_n}(r_n) = \|r_n\|$ (see Theorem 3.12.6). Let now $D_1 := D \backslash \{\pm\varphi_i, i \in \mathbb{N}\}$. Since $x \in \overline{\text{span}(D)}$, there are $x' \in \text{span}(D_1)$, a $N \in \mathbb{N}$, and a $x'' \in \text{span}\{\varphi_i\}_{i=1}^N = U_N$, such that $\|x - x' - x''\| < \frac{\alpha}{2}$.

Let now $n > N$, then $F_{r_n}(\varphi_i) = 0$ for $1 \leq i \leq N$, i.e. $F_{r_n}(x'') = 0$. Therefore we obtain, due to

$$\alpha \leq F_{r_n}(r_n) = \|r_n\| = F_{r_n}(x - x_n) = F_{r_n}(x),$$

the subsequent inequality

$$\begin{aligned} F_{r_n}(x') &= F_{r_n}(x' + x'') = F_{r_n}(x - (x - x' - x'')) = F_{r_n}(x) - F_{r_n}(x - x' - x'') \\ &\geq F_{r_n}(x) - \|x - x' - x''\| \geq \alpha - \frac{\alpha}{2} = \frac{\alpha}{2}. \end{aligned}$$

For x' there is a representation as a linear combination of elements in D_1: $x' = \sum_{i=1}^m c_i\psi_i$ with $\psi_i \in D_1$ for $i = 1, \dots, m$. We obtain

$$\begin{aligned} \frac{\alpha}{2} &\leq F_{r_n}(x') = \sum_{i=1}^m c_i F_{r_n}(\psi_i) \leq \sum_{i=1}^m |c_i|\,|F_{r_n}(\psi_i)| \\ &\leq m \cdot \max\{|c_i|\,|\,i = 1, \dots, m\} \cdot \max\{|F_{r_n}(\psi_i)|\,|\,i = 1, \dots, m\}. \end{aligned}$$

Therefore

$$\max\{|F_{r_n}(\psi_i)| \mid i = 1, \dots, m\} \geq \frac{\alpha}{2m \cdot \max\{|c_i| \mid i = 1, \dots, m\}}.$$

We finally have

$$F_{r_n}(\varphi_{n+1}) \geq t \cdot \sup\{F_{r_n}(\varphi) \mid \varphi \in D\} \geq t \cdot \max\{|F_{r_n}(\psi_i)| \mid i = 1, \dots, m\} \geq \beta,$$

if we choose $\beta := t \cdot \frac{\alpha}{2m \cdot \max\{|c_i| \mid i=1,\dots,m\}}$. □

Lemma 8.7.15. *Let the norm of X be Fréchet differentiable, then for every convergent subsequence* $(x_{n_k})_{k\in\mathbb{N}}$ *of the sequence* $(x_n)_{n\in\mathbb{N}}$ *generated by GA*

$$\lim_{k\to\infty} x_{n_k} = x$$

holds.

Proof. Suppose, there is a subsequence $x_{n_k} \to y \neq x$. Then the norm is Fréchet differentiable at $x - y$ and therefore by Corollary 8.4.21 the duality mapping $x \mapsto F_x$ is continuous at $x - y$, i.e. $F_{x-x_{n_k}} \to F_{x-y}$. For arbitrary $n \in \mathbb{N}$ we then have

$$F_{x-y}(\varphi_n) = \lim_{k\to\infty} F_{x-x_{n_k}}(\varphi_n) = 0,$$

and hence

$$F_{x-x_{n_k}}(\varphi_{n_k+1}) = (F_{x-x_{n_k}} - F_{x-y})(\varphi_{n_k+1}) \leq \|F_{x-x_{n_k}} - F_{x-y}\| \to 0. \quad (8.21)$$

Since $\alpha_{n_k} \downarrow \gamma \geq \alpha$ it follows that $\gamma = \|y - x\| > 0$.

But then also $\alpha > 0$, because assume $\alpha = 0$, then there is a subsequence $u_{m_k} \in U_{m_k}$ with $u_{m_k} \to x$. However $\alpha_{m_k} = \|x_{m_k} - x\| \leq \|u_{m_k} - x\| \to 0$, hence $\alpha_{m_k} \to 0 = \gamma$, a contradiction. Therefore $\alpha > 0$. By Lemma 8.7.14 we then obtain $F_{x-x_{n_k}}(\varphi_{n_k+1}) > \beta > 0$, a contradiction to (8.21). □

Lemma 8.7.16. *Let X be a strictly convex Banach space with Kadec–Klee property (see Definition* 8.4.19*), whose norm is Fréchet differentiable, then for the sequence generated by GA the following relation holds:*

$$x_n \to x \Leftrightarrow (x_n)_{n\in\mathbb{N}} \text{ has a weakly convergent subsequence}.$$

Proof. Let $x_{n_k} \rightharpoonup y$, then by Theorem 3.9.18 $y \in \overline{\text{span}\{\varphi_k\}_{k=1}^\infty}$. Let $r_n := x - x_n$, then $\|r_n\| \downarrow \gamma \geq 0$. Apparently $\gamma \geq \alpha := d(x, \overline{\text{span}\{\varphi_k\}_{k=1}^\infty})$. We will show now $\gamma = \alpha$: let $u_n \in \text{span}\{\varphi_k\}_{k=1}^\infty$ and let $u_n \to y$, then $\gamma \leq \|u_n - x\| \to \|y - x\|$, hence $\|y - x\| \geq \gamma$.

If $\gamma = 0$, then the assertion $x_n \to x$ already holds. Suppose now that $\gamma > 0$, then we obtain

$$\gamma \leq \|y - x\| = F_{x-y}(x - y) = \lim_{k\to\infty} F_{x-y}(r_{n_k}) \leq \lim_{k\to\infty} \|r_{n_k}\| = \gamma.$$

The Kadec–Klee property then yields: $r_{n_k} \to y - x$, hence $x_{n_k} \to y$. But by Lemma 8.7.15 we then obtain $x_{n_k} \to x$, contradicting $\|y - x\| \geq \gamma > 0$. □

In the reflexive case we then have the following

Theorem 8.7.17 (Convergence for GA). *Let X be an E-space, whose norm is Fréchet differentiable. Then GA converges for every dictionary D and every $x \in X$.*

Proof. Since X is reflexive, the (bounded) sequence (x_n) has a weakly convergent subsequence. Lemma 8.7.16 together with Theorem 8.4.18 yields the assertion. □

In Orlicz spaces the above theorem assumes the following formulation:

Theorem 8.7.18 (Convergence of GA in Orlicz spaces). *Let (T, Σ, μ) be a σ-finite, essentially not purely atomic measure space, let Φ be a differentiable, strictly convex Young function and Ψ its conjugate, and let $L^\Phi(\mu)$ be reflexive. The GA converges for ever dictionary D and every x in $L^\Phi(\mu)$.*

Proof. Ψ is differentiable. Hence by Theorem 8.5.22 also $(L^\Psi(\mu), \|\cdot\|_{(\Psi)})$ and $(L^\Psi(\mu), \|\cdot\|_\Psi)$ resp. are Fréchet differentiable, thus due to the theorem of Anderson 8.4.22 $(L^\Phi(\mu), \|\cdot\|_{(\Phi)})$ and $(L^\Phi(\mu), \|\cdot\|_\Phi)$ resp. are E-spaces, whose norms are by Theorem 8.5.22 Fréchet differentiable. □

Remark 8.7.19. Reflexivity of $L^\Phi(\mu)$ can – depending on the measure – be characterized by appropriate Δ_2-conditions (see Theorem 7.7.1).

Convergence of WGA in Uniformly Differentiable Spaces

We will now investigate the convergence of WGA, if we drop the limitation $t_m = \tau$. It turns out that this is possible in uniformly differentiable spaces under suitable conditions on (t_m) (see [103]):

For further investigations we need the following

Lemma 8.7.20. *Let $x \in X$ be given. Choose for $\varepsilon > 0$ a $x_\varepsilon \in X$, such that $\|x - x_\varepsilon\| < \varepsilon$ and $\frac{x_\varepsilon}{A(\varepsilon)} \in \operatorname{conv}(D)$, then*

$$\sup_{\phi\in D} F_{r_{m-1}}(\phi) \geq \frac{1}{A(\varepsilon)}(\|r_{m-1}\| - \varepsilon).$$

If $x \in \overline{\mathrm{conv}(D)}$ *then*

$$\sup_{\phi \in D} F_{r_{m-1}}(\phi) \geq \|r_{m-1}\|.$$

Proof. $A(\varepsilon)$ is determined as follows: if $x \in \overline{\mathrm{conv}(D)}$, then choose $A(\varepsilon) = 1$, otherwise choose $x_\varepsilon \in \mathrm{span}(D)$, $x_\varepsilon = \sum_{k=1}^{n} c_k \phi_k$, where due to the symmetry of D the coefficients c_k, $k = 1, \dots, n$, can be chosen to be positive, then put $A(\varepsilon) := \sum_{k=1}^{n} c_k$ and hence $x_\varepsilon / A(\varepsilon) \in \mathrm{conv}(D)$. Apparently

$$\sup_{\phi \in D} F_{r_{m-1}}(\phi) = \sup_{\phi \in \mathrm{conv}(D)} F_{r_{m-1}}(\phi) \geq \frac{1}{A(\varepsilon)} F_{r_{m-1}}(x_\varepsilon)$$

holds.

Since x_{m-1} is the best approximation of x w.r.t. U_{m-1}, we obtain

$$\begin{aligned} F_{r_{m-1}}(x_\varepsilon) &= F_{r_{m-1}}(x + x_\varepsilon - x) = F_{r_{m-1}}(x) - F_{r_{m-1}}(x - x_\varepsilon) \geq F_{r_{m-1}}(x) - \varepsilon \\ &= F_{r_{m-1}}(r_{m-1} + x_{m-1}) - \varepsilon = F_{r_{m-1}}(r_{m-1}) - \varepsilon = \|r_{m-1}\| - \varepsilon. \end{aligned}$$

The two previous considerations then imply

$$\sup_{\phi \in D} F_{r_{m-1}}(\phi) \geq \frac{1}{A(\varepsilon)} (\|r_{m-1}\| - \varepsilon).$$

For $x \in \overline{\mathrm{conv}(D)}$ we have $A(\varepsilon) = 1$ and therefore

$$\sup_{\phi \in D} F_{r_{m-1}}(\phi) \geq \|r_{m-1}\| - \varepsilon$$

for each $\varepsilon > 0$. □

As a consequence we obtain

Lemma 8.7.21. *Let X be uniformly differentiable with smoothness module $\rho(u)$. Let $x \in X$ be given. If one chooses for $\varepsilon > 0$ a $x_\varepsilon \in X$, such that $\|x - x_\varepsilon\| < \varepsilon$ and $\frac{x_\varepsilon}{A(\varepsilon)} \in \mathrm{conv}(D)$, then one obtains for $\lambda > 0$ arbitrary*

$$\|r_m\| \leq \|r_{m-1}\| \left(1 + 2\rho\left(\frac{\lambda}{\|r_{m-1}\|}\right) - \lambda t_m \frac{1}{A(\varepsilon)} \left(1 - \frac{\varepsilon}{\|r_{m-1}\|}\right)\right).$$

Proof. From the definition one immediately obtains for $\lambda > 0$ arbitrary and $x, y \in X$ with $x \neq 0$ and $\|y\| = 1$

$$2\left(1 + \rho\left(\frac{\lambda}{\|x\|}\right)\right) \geq \frac{1}{\|x\|}(\|x + \lambda y\| + \|x - \lambda y\|).$$

For $x = r_{m-1}$ and $y = \varphi_m$ we then have

$$2\left(1+\rho\left(\frac{\lambda}{\|r_{m-1}\|}\right)\right)\|r_{m-1}\| - \|r_{m-1}+\lambda\varphi_m\| \ge \|r_{m-1}-\lambda\varphi_m\|.$$

Moreover, by step 1 of WGA using the previous lemma

$$F_{r_{m-1}}(\varphi_m) \ge t_m \sup_{\phi\in D} F_{r_{m-1}}(\phi) \ge t_m \frac{1}{A(\varepsilon)}(\|r_{m-1}\| - \varepsilon),$$

and thus

$$\begin{aligned}\|r_{m-1}+\lambda\varphi_m\| &\ge F_{r_{m-1}}(r_{m-1}+\lambda\varphi_m) = F_{r_{m-1}}(r_{m-1}) + \lambda F_{r_{m-1}}(\varphi_m)\\ &\ge \|r_{m-1}\| + \lambda t_m \frac{1}{A(\varepsilon)}(\|r_{m-1}\| - \varepsilon).\end{aligned}$$

Altogether we conclude

$$\begin{aligned}\|r_m\| &\le \|r_{m-1}-\lambda\varphi_m\|\\ &\le 2\left(1+\rho\left(\frac{\lambda}{\|r_{m-1}\|}\right)\right)\|r_{m-1}\|\\ &\quad - \lambda t_m \frac{1}{A(\varepsilon)}(\|r_{m-1}\| - \varepsilon) - \|r_{m-1}\|\\ &= \|r_{m-1}\|\left(1+2\rho\left(\frac{\lambda}{\|r_{m-1}\|}\right) - \lambda t_m \frac{1}{A(\varepsilon)}\left(1 - \frac{\varepsilon}{\|r_{m-1}\|}\right)\right).\end{aligned}$$ □

For the convergence proof of WGA we finally need the following

Lemma 8.7.22. *Let $0 \le \alpha_k < 1$ for $k \in \mathbb{N}$ and let $\beta_m := (1-\alpha_1)\cdot(1-\alpha_2)\cdots(1-\alpha_m)$. If $\sum_{k=1}^{\infty}\alpha_k = \infty$, then $\lim_{m\to\infty}\beta_m = 0$ holds.*

Proof. From the subgradient inequality for $-\ln$ at 1 we have for $x > 0$: $\ln x \le x - 1$ and therefore $\ln(1-\alpha_k) \le -\alpha_k$ for $k \in \mathbb{N}$, hence

$$\ln(\beta_m) = \sum_{k=1}^{m}\ln(1-\alpha_k) \le -\sum_{k=1}^{m}\alpha_k,$$

and finally

$$\beta_m = e^{-\sum_{k=1}^{m}\alpha_k} \xrightarrow{m\to\infty} 0.$$ □

Remark 8.7.23. If X is uniformly differentiable with smoothness module $\rho(u)$, then for the (strictly monotone and continuous) function $\gamma(u) := \frac{\rho(u)}{u}$ (see Lemma 8.6.2) it follows that the equation

$$\gamma(u) = \theta t_m$$

has for $0 < \theta \leq \frac{1}{2}$ a unique solution in $(0, 2]$, since for $\|x\| = 1$ we have

$$\gamma(2) \geq \frac{1}{2}\left(\frac{1}{2}(\|x + 2x\| + \|x - 2x\|) - 1\right) = \frac{1}{2}.$$

Theorem 8.7.24. *Let X be uniformly differentiable with smoothness module $\rho(u)$ and let for the solutions ξ_m of the equations $\gamma(u) = \theta t_m$ for each $\theta \in (0, \frac{1}{2}]$ the condition*

$$\sum_{m=1}^{\infty} t_m \xi_m = \infty$$

be satisfied. Then WGA converges for arbitrary $x \in X$.

Proof. Suppose there is $\alpha > 0$ with $\|r_m\| \geq \alpha$. W.l.g. α can be chosen, such that $\theta := \frac{\alpha}{8A(\varepsilon)} \in (0, \frac{1}{2}]$. Then by Lemma 8.7.21 we conclude for arbitrary $\lambda > 0$

$$\|r_m\| \leq \|r_{m-1}\|\left(1 + 2\rho\left(\frac{\lambda}{\alpha}\right) - \lambda t_m \frac{1}{A(\varepsilon)}\left(1 - \frac{\varepsilon}{\alpha}\right)\right).$$

For $\lambda = \alpha \cdot \xi_m$ we obtain

$$\begin{aligned}
\|r_m\| &\leq \|r_{m-1}\|\left(1 + 2\theta t_m \xi_m - \alpha t_m \xi_m \frac{1}{A(\varepsilon)}\left(1 - \frac{\varepsilon}{\alpha}\right)\right) \\
&= \|r_{m-1}\|\left(1 + 2\theta t_m \xi_m - 8\theta t_m \xi_m \left(1 - \frac{\varepsilon}{\alpha}\right)\right) \\
&= \|r_{m-1}\|\left(1 - 2\theta t_m \xi_m \left(4\left(1 - \frac{\varepsilon}{\alpha}\right) - 1\right)\right) = \|r_{m-1}\|(1 - 2\theta t_m \xi_m)
\end{aligned}$$

for $\varepsilon = \frac{\alpha}{2}$. Using Lemma 8.7.22 we arrive at a contradiction. □

Remark 8.7.25. Frequently, as in [103], estimates for the smoothness module are available:

If $\rho_1(u) \leq \rho_2(u)$ and if $\xi_m^{(i)}$ is solution of $\gamma_i(u) = \theta t_m$ for $i = 1, 2$, then apparently $\xi_m^{(2)} \leq \xi_m^{(1)}$.

Condition $\sum_{m=1}^{\infty} t_m \xi_m^{(1)} = \infty$ can then be replaced by $\sum_{m=1}^{\infty} t_m \xi_m^{(2)} = \infty$.

The following theorem describes conditions for the convergence of WGA in Orlicz spaces:

Theorem 8.7.26. *Let (T, Σ, μ) be a non-atomic measure space. Let Φ be differentiable and let for each $0 < \varepsilon < 1/4$ exist a constant $1 < R_\varepsilon \leq N < \infty$, such that*

(a) *for infinite measure*

i. $\Psi(2u) \le N\Psi(u)$

ii. $\Psi'_+((1-\varepsilon)u) < \frac{1}{R_\varepsilon}\Psi'_+(u)$

for all $u > 0$

(b) *for finite measure*

i. $\limsup_{u\to\infty} \Psi(2u)/\Psi(u) \le N$

ii. $\limsup_{u\to\infty} \frac{\Psi'_+(u)}{\Psi'_+((1-\varepsilon)u)} > R_\varepsilon$

then $(L^\Phi(\mu), \|\cdot\|_{(\Phi)})$ *is uniformly differentiable. Let* $\rho(u)$ *be the corresponding smoothness module and let for the solutions* ξ_m *of the equations* $\gamma(u) = \theta t_m$ *for each* $\theta \in (0, \frac{1}{2}]$ *be satisfied that*

$$\sum_{m=1}^{\infty} t_m \xi_m = \infty.$$

Then WGA converges for arbitrary $x \in (L^\Phi(\mu), \|\cdot\|_{(\Phi)})$.

Proof. Theorem 8.7.24 together with Theorem 8.6.10. □

A corresponding theorem is available for the Orlicz norm, making use of the description of uniform convexity of the Luxemburg norm in Theorem 8.6.18.

Theorem 8.7.27. *Let* (T, Σ, μ) *be an arbitrary measure space. Then let* Ψ *be* δ*-convex and satisfy the* Δ_2*-condition.*

If $\mu(T) < \infty$*, let* Ψ *be strictly convex and* δ^∞*-convex and let* Ψ *satisfy the* Δ_2^∞*-condition.*

Then $(L^\Phi(\mu), \|\cdot\|_\Phi)$ *is uniformly differentiable. Let* $\rho(u)$ *be the corresponding smoothness module and let for the solutions* ξ_m *of the equations* $\gamma(u) = \theta t_m$ *for each* $\theta \in (0, \frac{1}{2}]$ *be satisfied that*

$$\sum_{m=1}^{\infty} t_m \xi_m = \infty.$$

Then WGA converges for arbitrary $x \in (L^\Phi(\mu), \|\cdot\|_\Phi)$.

Proof. Theorem 8.7.24 together with Theorems 8.6.18 and 8.6.4. □

If $x \in \overline{\mathrm{conv}(D)}$, then by Lemma 8.7.20 we can replace step (a) of WGA by a condition which is algorithmically easier to verify. In this way we obtain a modification of WGA:

Algorithm 8.7.28 (WGAM). Let X be strictly convex and let $\tau = (t_k)_{k\in\mathbb{N}}$ with $0 < t_k \le 1$ for all $k \in \mathbb{N}$.

Let further $x \in \overline{\text{conv}(D)} \setminus \{0\}$ be given, then put $r_0 := x$, and for $m \geq 1$:

(a) Choose $\varphi_m \in D$ such that

$$F_{r_{m-1}}(\varphi_m) \geq t_m \|r_{m-1}\|.$$

(b) For $U_m := \text{span}\{\varphi_j \mid j = 1, \dots, m\}$ let x_m be the best approximation of x w.r.t. U_m.

(c) Put $r_m := x - x_m$ and $m \leftarrow m + 1$ goto (a).

According to Theorem 8.7.24 WGAM converges for arbitrary dictionaries and arbitrary $x \in \overline{\text{conv}(D)} \setminus \{0\}$, since the inequality in Lemma 8.7.21 also holds for WGAM.

If X is separable, D can be chosen to be countable. The condition in step 1 of WGAM is then satisfied after examining a finite number of elements of the dictionary. Examples for separable Orlicz spaces we have mentioned in Section 7.8.1.

The speed of convergence of WGA resp. WGAM can be described by the subsequent recursive inequality:

Lemma 8.7.29. *Let $x \in \overline{\text{conv}(D)}$. Let further $\sigma(v) := v \cdot \gamma^{-1}(v)$, then*

$$\|r_m\| \leq \|r_{m-1}\| \left(1 - 2\sigma\left(\frac{1}{4} t_m \|r_{m-1}\|\right)\right)$$

holds.

Proof. Let $x \in \overline{\text{conv}(D)}$, then $A(\varepsilon) = 1$. Therefore by Lemma 8.7.21

$$\|r_m\| \leq \|r_{m-1}\| \left(1 + 2\rho\left(\frac{\lambda}{\|r_{m-1}\|}\right) - \lambda t_m \left(1 - \frac{\varepsilon}{\|r_{m-1}\|}\right)\right)$$

for every $\varepsilon > 0$ and arbitrary $\lambda > 0$, i.e.

$$\|r_m\| \leq \|r_{m-1}\| \left(1 + 2\rho\left(\frac{\lambda}{\|r_{m-1}\|}\right) - \lambda t_m\right).$$

Let now λ_m be solution of

$$\frac{1}{4} t_m \|r_{m-1}\| = \gamma\left(\frac{\lambda}{\|r_{m-1}\|}\right).$$

Then we obtain

$$\|r_m\| \leq \|r_{m-1}\| \left(1 - \frac{1}{2} \lambda_m t_m\right),$$

and

$$\gamma^{-1}\left(\frac{1}{4} t_m \|r_{m-1}\|\right) = \frac{\lambda_m}{\|r_{m-1}\|},$$

i.e.

$$\frac{1}{2} \lambda_m t_m = \frac{1}{2} t_m \|r_{m-1}\| \gamma^{-1}\left(\frac{1}{4} t_m \|r_{m-1}\|\right) = 2\sigma\left(\frac{1}{4} t_m \|r_{m-1}\|\right).$$

□

Remark 8.7.30. For $\rho_1 \leq \rho_2$ the inequality $\sigma_2 \leq \sigma_1$ holds.

In the case of Banach spaces with $\rho(u) \leq au^p$ for $1 < p \leq 2$ one obtains: $\sigma(v) \geq bv^q$ where $q = \frac{p}{p-1}$. Temlyakov (see [103]) thereby obtains the estimate for WGA

$$\|r_m\| \leq c\left(1 + \sum_{k=1}^{m} t_k^q\right)^{-1/q}.$$

Then for the special case GA apparently

$$\|r_m\| \leq c(1 + mt^q)^{-1/q} \leq \frac{c}{t} m^{-1/q}$$

holds.

Chapter 9
Variational Calculus

In this chapter we describe an approach to variational problems, where the solutions appear as pointwise (finite dimensional) minima for fixed t of the supplemented Lagrangian. The minimization is performed simultaneously w.r.t. to both the state variable x and $\dot{x}$, different from Pontryagin's maximum principle, where optimization is done only w.r.t. the $\dot{x}$ variable. We use the idea of the Equivalent Problems of Carathéodory employing suitable (and simple) supplements to the original minimization problem. Whereas Carathéodory considers equivalent problems by use of solutions of the Hamilton–Jacobi partial differential equations, we shall demonstrate that quadratic supplements can be constructed, such that the supplemented Lagrangian is convex in the vicinity of the solution. In this way, the fundamental theorems of the Calculus of Variations are obtained. In particular, we avoid any usage of field theory.

Our earlier discussion of strong solvability in the previous chapter has some connection to variational calculus: it turns out that – if the Legendre–Riccati differential equation is satisfied – the variational functional is uniformly convex w.r.t. the Sobolev norm on a δ-neighborhood of the solution.

In the next part of this chapter we will employ the stability principles developed in Chapter 5 in order to establish optimality for solutions of the Euler–Lagrange equations of certain non-convex variational problems.

The principle of pointwise minimization is then applied to the detection of a smooth, monotone trend in time series data in a parameter-free manner. In this context we also employ a Tikhonov-type regularization.

The last part of this chapter is devoted to certain problems in optimal control, where, to begin with, we put our focus on stability questions. In the final section we treat a minimal time problem which turns out to be equivalent to a linear approximation in the mean, and thus closing the circle of our journey through function spaces.

9.1 Introduction

If a given function has to be minimized on a subset (restriction set) of a given set, one can try to modify this function outside of the restriction set by adding a supplement in such a way that the global minimal solution of the supplemented function lies in the restriction set. It turns out that this global minimal solution is a solution of the original (restricted) problem. The main task is to determine such a suitable supplement.

Supplement Method 9.1.1. *Let M be an arbitrary set, $f : M \to \mathbb{R}$ a function and T a subset of M. Let $\Lambda : M \to \mathbb{R}$ be a function, that is constant on T. Then we obtain:*

If $x_0 \in T$ is a minimal solution of the function $f + \Lambda$ on all of M, then x_0 is a minimal solution of f on T.

Proof. For $x \in T$ arbitrary we have

$$f(x_0) + \Lambda(x_0) \leq f(x) + \Lambda(x) = f(x) + \Lambda(x_0). \qquad \square$$

Definition 9.1.2 (Piecewise continuously differentiable functions). We consider functions $x \in C[a, b]^n$ for which there exists a partition $\{a = t_0 < t_1 < \cdots < t_m = b\}$, such that: x is continuously differentiable for each $i \in \{1, \ldots, m\}$ on $[t_{i-1}, t_i)$ and the derivative $\dot{x}$ has a left-sided limit in t_i. The value of the derivative of x in b is then defined as the left-sided derivative in b. Such a function we call piecewise continuously differentiable. The set of these functions is denoted by $\mathrm{RCS}^{(1)}[a, b]^n$.

Definition 9.1.3. Let W be a subset of $\mathbb{R}^\nu \times [a, b]$. We call $G : W \to \mathbb{R}^k$ *piecewise continuous*, if there is a partition $Z = \{a = t_0 < t_1 < \cdots < t_m = b\}$ of $[a, b]$ such that for each $i \in \{1, \ldots, m\}$ the function $G : W \cap (\mathbb{R}^\nu \times [t_{i-1}, t_i)) \to \mathbb{R}^k$ has a continuous extension $M_i : W \cap (\mathbb{R}^\nu \times [t_{i-1}, t_i]) \to \mathbb{R}^k$.

In the sequel we shall consider *variational problems* in the following setting: let $U \subset \mathbb{R}^{2n+1}$, be such that $U_t := \{(p, q) \in \mathbb{R}^{2n} \mid (p, q, t) \in U\} \neq \emptyset$ for all $t \in [a, b]$, let $L : U \to \mathbb{R}$ be piecewise continuous. The restriction set S for given $\alpha, \beta \in \mathbb{R}^n$ is a set of functions that is described by

$$S \subset \{x \in \mathrm{RCS}^1[a, b]^n \mid (x(t), \dot{x}(t), t), (x(t), \dot{x}(t-), t) \in U \ \forall t \in [a, b], \\ x(a) = \alpha, x(b) = \beta\}.$$

The variational functional $f : S \to \mathbb{R}$ to be minimized is defined by

$$f(x) = \int_a^b L(x(t), \dot{x}(t), t)dt.$$

The variational problem with fixed endpoints is then given by

Minimize f on S.

The central idea of the subsequent discussion is to introduce a supplement in integral form that is constant on the restriction set. This leads to a new variational problem with a modified Lagrangian. The solutions of the original variational problem can now be found as minimal solutions of the modified variational functional. Because of the monotonicity of the integral, the variational problem is now solved by pointwise minimization of the Lagrangian w.r.t. the x and $\dot{x}$ variables for every fixed t, employing the methods of finite-dimensional optimization.

This leads to sufficient conditions for a solution of the variational problem. This general approach does not even require differentiability of the integrand. Solutions of the pointwise minimization can even lie at the boundary of the restriction set so that the Euler–Lagrange equations do not have to be satisfied. For interior points the Euler–Lagrange equations will naturally appear by setting the partial derivatives to zero, using a linear supplement potential.

9.1.1 Equivalent Variational Problems

We now attempt to describe an approach to variational problems that uses the idea of the Equivalent Problems of Carathéodory (see [17], also compare Krotov [74]) employing suitable supplements to the original minimization problem. Carathéodory constructs equivalent problems by use of solutions of the Hamilton–Jacobi partial differential equations. In the context of Bellman's Dynamic Programming (see [6]) this supplement can be interpreted as the so-called value function. The technique to modify the integrand of the variational problem already appears in the works of Legendre in the context of the second variation (accessory problem).

In this treatise we shall demonstrate that explicitly given quadratic supplements are sufficient to yield the main results.

Definition 9.1.4. Let $F : [a, b] \times \mathbb{R}^n \to \mathbb{R}$ with $(t, x) \mapsto F(t, x)$ be continuous, w.r.t. x continuously partially differentiable and w.r.t. t piecewise partially differentiable in the sense that the partial derivative F_t is piecewise continuous. Moreover, we require that the partial derivative F_{xx} exists and is continuous, and that F_{tx}, F_{xt} exist in the piecewise sense, and are piecewise continuous such that $F_{tx} = F_{xt}$.

Then we call F a *supplement potential*.

Lemma 9.1.5. *Let $F : [a, b] \times \mathbb{R}^n \to \mathbb{R}$ be a supplement potential. Then the integral over the supplement*

$$\int_a^b [\langle F_x(t, x(t)), \dot{x}(t)\rangle + F_t(t, x(t))]dt$$

is constant on S.

Proof. Let Z_1 be a common partition of $[a, b]$ for F and x, i.e. $Z_1 = \{a = \tilde{t}_0 < \tilde{t}_1 < \cdots < \tilde{t}_j = b\}$, such that the requirement of piecewise continuity for $\dot{x}$ and F_t are satisfied w.r.t. Z_1, then

$$\begin{aligned}
&\int_a^b [\langle F_x(t, x(t)), \dot{x}(t)\rangle + F_t(t, x(t))]dt \\
&\quad = \sum_{i=1}^{j} \int_{\tilde{t}_{i-1}}^{\tilde{t}_i} [\langle F_x(t, x(t)), \dot{x}(t)\rangle + F_t(t, x(t))]dt
\end{aligned}$$

$$= \sum_{i=1}^{j} (F(\tilde{t}_i, x(\tilde{t}_i)) - F(\tilde{t}_{i-1}, x(\tilde{t}_{i-1})))$$
$$= F(b, \beta) - F(a, \alpha). \qquad \square$$

An *equivalent problem* is then given through the supplemented Lagrangian $\tilde{L}$

$$\tilde{L} := L - \langle F_x, \dot{x} \rangle - F_t.$$

9.1.2 Principle of Pointwise Minimization

The aim is to develop sufficient criteria for minimal solutions of the variational problem by replacing the minimization of the variational functional on subsets of a function space by finite dimensional minimization. This can be accomplished by pointwise minimization of an explicitly given supplemented integrand for fixed t using the monotonicity of the integral (application of this method to general control problems was treated in [63]). The minimization is done simultaneously, with respect to the x- and the $\dot{x}$-variables in $\mathbb{R}^{2n}$. This is the main difference as compared to Hamilton/Pontryagin where minimization is done solely w.r.t. the $\dot{x}$-variables, methods, which lead to necessary conditions in the first place.

The principle of pointwise minimization is demonstrated in [61], where a complete treatment of the brachistochrone problem is presented using only elementary minimization in $\mathbb{R}$ (compare also [59], p. 120 or [57]). For a treatment of this problem using fields of extremals see [33], p. 367.

Our approach is based on the following obvious

Lemma 9.1.6. *Let A be a set of integrable real functions on $[a, b]$ and let $l^* \in A$. If for all $l \in A$ and all $t \in [a, b]$ we have $l^*(t) \leq l(t)$ then*

$$\int_a^b l^*(t)dt \leq \int_a^b l(t)dt$$

for all $l \in A$.

Theorem 9.1.7 (Principle of Pointwise Minimization). *Let a variational problem with Lagrangian L and restriction set S be given.*

If for an equivalent variational problem

$$\textit{Minimize } g(x) := \int_a^b \tilde{L}(x(t), \dot{x}(t), t)dt,$$

where

$$\tilde{L} = L - \langle F_x, \dot{x} \rangle - F_t,$$

an $x^* \in S$ can be found, such that for all $t \in [a, b]$ the point $(p_t, q_t) := (x^*(t), \dot{x}^*(t))$ is a minimal solution of the function $(p, q) \mapsto \tilde{L}(p, q, t) =: \varphi_t(p, q)$ on $U_t := \{(p, q) \in \mathbb{R}^{2n} \mid (p, q, t) \in U\}$.

Then x^* is a solution of the original variational problem.

Proof. For the application of Lemma 9.1.6, set $A = \{l_x : [a, b] \to \mathbb{R} \mid t \mapsto l_x(t) = \tilde{L}(x(t), \dot{x}(t), t),\ x \in S\}$ and $l^* := l_{x^*}$. According to Lemma 9.1.5 the integral over the supplement is constant. □

It turns out (see below) that the straightforward approach of a linear (w.r.t. x) supplement already leads to the Euler–Lagrange equation by setting the partial derivatives of $\tilde{L}$ (w.r.t. p and q) to zero.

9.1.3 Linear Supplement

In the classical theory a linear supplement is used where the supplement potential F has the structure

$$(t, x) \mapsto F(t, x) = \langle \lambda(t), x \rangle, \tag{9.1}$$

and $\lambda \in \mathrm{RCS}^1[a, b]^n$ is a function that has to be determined in a suitable way.

As $F_x(t, x) = \lambda(t)$ and $F_t(t, x) = \langle \dot{\lambda}(t), x \rangle$ we obtain for the equivalent problem

$$\text{Minimize } g(x) = \int_a^b L(x(t), \dot{x}(t), t) - \langle \lambda(t), \dot{x}(t) \rangle - \langle \dot{\lambda}(t), x(t) \rangle dt \quad \text{on } S. \tag{9.2}$$

We shall now attempt to solve this problem through pointwise minimization of the integrand. This simple approach already leads to a very efficient method for the treatment of variational problems.

For the approach of pointwise minimization we have to minimize

$$(p, q) \mapsto \ell_t(p, q) := L(p, q, t) - \langle \lambda(t), q \rangle - \langle \dot{\lambda}(t), p \rangle$$

on U_t.

Let W be an open superset of U in the relative topology of $\mathbb{R}^n \times \mathbb{R}^n \times [a, b]$, and let $L : W \to \mathbb{R}$ be piecewise continuous and partially differentiable w.r.t. p and q (in the sense that L_p and L_q are piecewise continuous). If for fixed $t \in [a, b]$ a point $(p_t, q_t) \in \operatorname{Int} U_t$ is a corresponding minimal solution, then the partial derivatives of ℓ_t have to be equal to zero at this point. This leads to the equations

$$L_p(p_t, q_t, t) = \dot{\lambda}(t) \tag{9.3}$$

$$L_q(p_t, q_t, t) = \lambda(t). \tag{9.4}$$

The pointwise minimum (p_t, q_t) yields a function $t \mapsto (p_t, q_t)$. It is our aim to show that this pair provides a solution x^* of the variational problem where $x^*(t) := p_t$ and

$\dot{x}^*(t) = q_t$. In the spirit of the supplement method this means that the global minimum is an element of the restriction set S. The freedom of choosing a suitable function λ is exploited to achieve this goal.

Definition 9.1.8. A function $x^* \in \mathrm{RCS}^1[a,b]^n$ is called an *extremaloid* (see Hestenes [38], p. 60), if it satisfies the Euler–Lagrange equation in integral form, i.e. there is a constant c such that:

$$\int_a^t L_x(x(\tau), \dot{x}(\tau), \tau)d\tau + c = L_{\dot{x}}(x(t), \dot{x}(t), t) \quad \forall t \in (a,b]. \tag{9.5}$$

If the extremaloid x^* is a $C^1[a,b]^n$-function then x^* is called an *extremal*. An extremaloid x^* is called *admissible* if $x^* \in S$.

Remark 9.1.9. An extremaloid x^* always satisfies the *Weierstrass–Erdmann condition*, i.e.

$$t \mapsto L_{\dot{x}}(x^*(t), \dot{x}^*(t), t) \text{ is continuous.}$$

For an extremaloid x^* the definition

$$\lambda(t) := \int_a^t L_x(x^*(\tau), \dot{x}^*(\tau), \tau)d\tau + c \tag{9.6}$$

(c a constant) leads to a $\lambda \in \mathrm{RCS}^1$.

A fundamental question of variational calculus is, under what conditions an extremaloid is a minimal solution of the variational problem.

If x^* is an admissible extremaloid then for every $t \in [a,b]$ the pair $(x^*(t), \dot{x}^*(t)) \in \operatorname{Int} U_t$ satisfies the first necessary condition for a pointwise minimum at t. From the perspective of pointwise minimization we can state the following: if setting the partial derivatives of the integrand equal to zero leads to a pointwise (global) minimum on U_t for every $t \in [a,b]$, then indeed x^* is a solution of the variational problem. The principle of pointwise minimization also provides a criterion to decide which of the extremaloids is the global solution.

For convex integrands, an extremaloid already leads to the sufficient conditions for a pointwise minimum.

If for every $t \in [a,b]$ the set U_t is convex and the Lagrangian $L(\cdot,\cdot,t) : U_t \to \mathbb{R}$ is a convex function, then an admissible extremaloid is a solution of the variational problem. This remains true if the extremaloid lies partially on the boundary of U (i.e. if the pointwise minimum lies on the boundary of U_t). As we require continuous differentiability in an open superset of U_t the vector of the partial derivatives (i.e. the gradient) represents the (total) derivative, all directional derivatives are equal to zero at $(x^*(t), \dot{x}^*(t))$. The Characterization Theorem of Convex Optimization (see Theorem 3.4.3) guarantees that this point is indeed a pointwise minimum.

Theorem 9.1.10. *Let $X \subset \mathbb{R}^n \times \mathbb{R}^n$ be open and $\phi : X \to \mathbb{R}$ be differentiable. Let K be a convex subset of X and $\phi : K \to \mathbb{R}$ be convex. If for $x^* \in K$ we have $\phi'(x^*) = 0$ then x^* is a minimal solution of ϕ on K.*

We summarize this situation in the following:

Theorem 9.1.11. *For convex problems every admissible extremaloid is a solution of the variational problem.*

In view of this theorem it turns out that the method of pointwise minimization can be extended to a much larger class of variational problems, where the integrand can be convexified by use of a suitable supplement (see Section 9.3).

The invariance property stating that equivalent problems have the same extremaloids, established in the subsequent theorem, leads to the following principle:

> *An extremaloid of a problem that is convexifiable is a solution of the original variational problem. In particular, the explicit convexification does not have to be carried out, instead the solvability of certain ordinary differential equations has to be verified*

(see below).

Theorem 9.1.12. *Every extremaloid for the Lagrangian L is an extremaloid for the supplemented Lagrangian*

$$\tilde{L} := L - \langle F_x, \dot{x} \rangle - F_t$$

and vice versa, where F is a supplement potential.

Proof. We have

$$\tilde{L}_{\dot{x}} = L_{\dot{x}} - F_x,$$

and

$$\tilde{L}_x = L_x - \dot{x}^T F_{xx} - F_{tx}.$$

Moreover

$$\frac{d}{dt} F_x(t, x(t)) = F_{xt}(t, x(t)) + \dot{x}(t)^T F_{xx}(t, x(t)).$$

If x satisfies the Euler–Lagrange equation in integral form w.r.t. L, i.e.

$$L_{\dot{x}} = \int_a^t L_x d\tau + c,$$

then there is a constant $\tilde{c}$ such that

$$\tilde{L}_{\dot{x}} = \int_a^t \tilde{L}_x d\tau + \tilde{c},$$

which we shall now show, using the continuity of F_x

$$\int_a^t \tilde{L}_x d\tau = \int_a^t (L_x - \dot{x}^T F_{xx} - F_{tx}) d\tau = L_{\dot{x}} - c - \int_a^t (\dot{x}^T F_{xx} + F_{xt}) d\tau$$
$$= L_{\dot{x}} - F_x + F_x(a, x(a)) - c = \tilde{L}_{\dot{x}} - \tilde{c}. \qquad \square$$

As equivalent variational problems have the same extremaloids Theorem 9.1.11 also holds for convexifiable problems. The following theorem can be viewed as a central guideline for our further considerations:

Theorem 9.1.13. *If, for a given variational problem, there exists an equivalent convex problem, then every admissible extremaloid is a minimal solution.*

9.2 Smoothness of Solutions

Theorem 9.2.1. *Let $L : U \to \mathbb{R}$ be such that L_q continuous. Let $x \in S$ be such that*

(a) $V_t := \{q \in \mathbb{R}^n \mid (x(t), q, t) \in U\}$ *is convex for all* $t \in [a, b]$.

(b) $L(x(t), \cdot, t)$ *be strictly convex on* V_t *for all* $t \in [a, b]$.

(c) $\lambda : [a, b] \to \mathbb{R}^n$ *where* $t \mapsto \lambda(t) := L_q(x(t), \dot{x}(t), t)$ *is continuous.*

Then x is smooth (i.e. $x \in C^1[a, b]^n$).

Proof. Let $q \mapsto \phi_t(q) := L(x(t), q, t) - \langle \lambda(t), q \rangle$, Then ϕ_t is strictly convex on V_t. We have

$$\phi_t'(\dot{x}(t)) = L_q(x(t), \dot{x}(t), t) - \lambda(t) = 0$$

for all $t \in [a, b]$, hence $\dot{x}(t)$ is the unique minimal solution of ϕ_t on V_t. Let (t_k) be a sequence in $[a, b]$ with $t_k \uparrow t$. Then there exists an interval $I_t := [\bar{t}, t)$ and $K \in \mathbb{N}$ such that $t_k \in I_t$ for $k > K$ and we have $L_q(x, \dot{x}, \cdot)$ and x are continuous on I_t. We obtain

$$0 = L_q(x(t_k), \dot{x}(t_k), t_k) - \lambda(t_k) \to L_q(x(t), \dot{x}(t-), t) - \lambda(t).$$

Hence $\dot{x}(t-)$ is minimal solution of ϕ_t on V_t. As ϕ_t is strictly convex, we finally obtain: $\dot{x}(t) = \dot{x}(t-)$ which proves the theorem. $\square$

Corollary 9.2.2. *Let x^* be an extremaloid, and let*

(a) $V_t := \{q \in \mathbb{R}^n \mid (x^*(t), q, t) \in U\}$ *be convex for all* $t \in [a, b]$

(b) $L(x^*(t), \cdot, t)$ *be strictly convex on* V_t.

Then x^* is an extremal.

Proof. $\lambda : [a, b] \to \mathbb{R}^n$ with $t \mapsto \lambda(t) := L_q(x^*(t), \dot{x}^*(t), t)$ is continuous (Weierstrass–Erdmann condition is satisfied). $\square$

The following example shows that the above theorem does not hold if the strict convexity of $L(x^*(t), \cdot, t)$ is violated:

Example 9.2.3. Let $L(p,q,t) := ((t-\frac{1}{2})_+)^2 q + \frac{1}{2}p^2$ for $t \in [0,1]$. We observe: L is convex but not strictly convex w.r.t. q.

We want to consider the corresponding variational problem for the boundary conditions $x(0) = 0, x(1) = 1$.

For the Euler–Lagrange equation we obtain

$$\frac{d}{dt}\left(\left(t-\frac{1}{2}\right)_+\right)^2 = x \Rightarrow x^*(t) = 2 \cdot \left(t - \frac{1}{2}\right)_+,$$

i.e. x^* is not smooth.

On the other hand $\lambda(t) = L_q(t) = ((t-\frac{1}{2})_+)^2$ is continuous and hence the Weierstrass–Erdmann condition is satisfied. Thus x^* is extremaloid of the convex variational problem and hence a minimal solution (Theorem 9.1.12) . The function ϕ_t is not strictly convex

$$\phi_t(q) = \left(\left(t-\frac{1}{2}\right)_+\right)^2 q + \frac{1}{2}p^2 - \left(\left(t-\frac{1}{2}\right)_+\right)^2 q = \frac{1}{2}p^2$$

i.e. ϕ_t is constant w.r.t. q, which means that every q is a minimal solution of ϕ_t. In particular the set of minimal solutions of ϕ_t is unbounded (beyond being non-unique).

Definition 9.2.4. Let $x^* \in \mathrm{RCS}^1[a,b]^n$ be an extremaloid, and let $L^0_{\dot{x}\dot{x}}(t) := L_{\dot{x}\dot{x}}(x^*(t), \dot{x}^*(t), t)$ satisfy the *strong Legendre–Clebsch condition*, i.e. $L^0_{\dot{x}\dot{x}}$ is positive definite on $[a,b]$, then x^* is called a *regular extremaloid*.

Remark 9.2.5. The *Legendre–Clebsch condition*, i.e. $L^0_{\dot{x}\dot{x}}$ positive semi-definite on $[a,b]$, is a classical necessary condition for a minimal solution of the variational problem (see [38]).

We would like to point out that a regular extremaloid is not always an extremal, as can be seen in the following example:

Example 9.2.6. Consider the variational problem on the interval $[-2\pi, 2\pi]$, with the boundary conditions $x^*(2\pi) = x^*(-2\pi) = 0$ given by the Lagrangian being defined by $(p,q,t) \mapsto L(p,q,t) := \cos q + \frac{t}{\gamma} \cdot q$ for $\gamma > 2\pi$ and for $(p,q,t) \in U = \mathbb{R} \times (-\frac{3}{2}\pi, \frac{3}{2}\pi) \times [-2\pi, 2\pi]$. Then $L_q = -\sin q + \frac{t}{\gamma}$ and $L_{qq} = -\cos q$, the latter being positive for $\frac{\pi}{2} < |q| < \frac{3}{2}\pi$. For the Euler–Lagrange equation we obtain

$$\frac{d}{dt}\left(-\sin \dot{x} + \frac{t}{\gamma}\right) = 0,$$

i.e.

$$-\sin \dot{x} + \frac{t}{\gamma} = c.$$

In particular, any solution of the above equation satisfies the Weierstrass–Erdmann condition.

Choosing x^* to be even (i.e. $\dot{x}^*$ odd), we obtain according to Equation (9.5)

$$c = \frac{1}{b-a}\int_a^b L_{\dot{x}}(x^*(t), \dot{x}^*(t), t)dt = \frac{1}{4\pi}\int_{-2\pi}^{2\pi}\Big(-\sin(\dot{x}^*(t)) + \frac{t}{\gamma}\Big)dt = 0,$$

and hence we have to solve

$$\sin q = \frac{t}{\gamma},$$

such that $\frac{\pi}{2} < |q| < \frac{3}{2}\pi$ in order to satisfy the strong Legendre–Clebsch condition. We obtain two different solutions:

1. For $2\pi \geq t > 0$ and $\frac{\pi}{2} < q < \frac{3}{2}\pi$ we obtain

$$q(t) = -\arcsin\left(\frac{t}{\gamma}\right) + \pi,$$

and for $-2\pi \leq t < 0$ and $-\frac{\pi}{2} > q > -\frac{3}{2}\pi$

$$q(t) = -\arcsin\left(\frac{t}{\gamma}\right) - \pi.$$

2. For $2\pi \geq t > 0$ and $-\frac{\pi}{2} > q > -\frac{3}{2}\pi$ we obtain

$$q(t) = -\arcsin\left(\frac{t}{\gamma}\right) - \pi,$$

and for $-2\pi \leq t < 0$ and $\frac{\pi}{2} < q < \frac{3}{2}\pi$

$$q(t) = -\arcsin\left(\frac{t}{\gamma}\right) + \pi.$$

Obviously, both solutions are discontinuous at $t = 0$. The extremals are then obtained via integration.

1. For $t > 0$

$$x^*(t) = t\Big(-\arcsin\left(\frac{t}{\gamma}\right) + \pi\Big) - \gamma\sqrt{1 - \left(\frac{t}{\gamma}\right)^2} - D_1,$$

and for $t < 0$

$$x^*(t) = t\Big(-\arcsin\left(\frac{t}{\gamma}\right) - \pi\Big) - \gamma\sqrt{1 - \left(\frac{t}{\gamma}\right)^2} - D_1,$$

where $D_1 = 2\pi(-\arcsin(\frac{2\pi}{\gamma}) + \pi) - \gamma\sqrt{1 - (\frac{2\pi}{\gamma})^2}$ according to the boundary conditions.

2. For $t > 0$

$$x^*(t) = t\left(- \arcsin\left(\frac{t}{\gamma}\right) - \pi\right) - \gamma\sqrt{1 - \left(\frac{t}{\gamma}\right)^2} - D_2,$$

and for $t < 0$

$$x^*(t) = t\left(- \arcsin\left(\frac{t}{\gamma}\right) + \pi\right) - \gamma\sqrt{1 - \left(\frac{t}{\gamma}\right)^2} - D_2,$$

where $D_2 = 2\pi(-\arcsin(\frac{2\pi}{\gamma}) - \pi) - \gamma\sqrt{1 - (\frac{2\pi}{\gamma})^2}$. Both extremaloids are admissible and both satisfy the strong Legendre condition.

Pointwise Minimization. The method of pointwise minimization (w.r.t. p, q for fixed t) leads directly to a global solution of the variational problem. For the linear supplement choose the *Dubois–Reymond form* $\lambda(t) = \int_{-2\pi}^{t} L_x d\tau + c = c$. Let $\Phi_t : \mathbb{R}\times(-\frac{3}{2}\pi, \frac{3}{2}\pi) \to \mathbb{R}$ with $\Phi_t(p,q) := L(p,q,t) + \lambda(t)q + \dot{\lambda}(t)p = \cos q + \frac{t}{\gamma}q + cq =: \phi_t(q)$, where $\phi_t : (-\frac{3}{2}\pi, \frac{3}{2}\pi) \to \mathbb{R}$. Choosing x^* to be even (i.e. $\dot{x}^*$ odd), we obtain as above (see Equation (9.5))

$$c = \frac{1}{b-a}\int_a^b L_{\dot{x}}(x^*(t), \dot{x}^*(t), t)dt = \frac{1}{4\pi}\int_{-2\pi}^{2\pi}\left(- \sin(\dot{x}^*(t)) + \frac{t}{\gamma}\right)dt = 0.$$

For $t > 0$ and $q > 0$ we obtain: $\phi_t(q) > -1$ and for $\phi_t(-\pi) = \cos(-\pi) - \frac{t}{\gamma}\pi$, hence we must look for minimal solutions on $(-\frac{3}{2}\pi, -\frac{\pi}{2})$. On this interval ϕ_t is convex.

The necessary condition yields

$$\sin q = \frac{t}{\gamma},$$

for which we obtain the following solution: let $q = -\pi + r$ where $r \in [-\frac{\pi}{2}, \frac{\pi}{2}]$ then $\sin q = \sin(-\pi + r) = -\sin r = \frac{t}{\gamma}$ hence $r = -\arcsin\frac{t}{\gamma}$.

For $t < 0$ for analogous reasons minimal solutions are found on $(\frac{\pi}{2}, \frac{3}{2}\pi)$. Hence, for $2\pi \geq t > 0$ and $-\frac{\pi}{2} > q > -\frac{3}{2}\pi$ we obtain

$$q(t) = -\arcsin\left(\frac{t}{\gamma}\right) - \pi,$$

and for $-2\pi \leq t < 0$ and $\frac{\pi}{2} < q < \frac{3}{2}\pi$

$$q(t) = -\arcsin\left(\frac{t}{\gamma}\right) + \pi.$$

As ϕ_t does not depend on p, every pair $(p, q(t))$ is a pointwise minimal solution of Φ_t on $\mathbb{R} \times (-\frac{3}{2}\pi, \frac{3}{2}\pi)$. We choose

$$x^*(t) = p(t) = \int_{-2\pi}^{t} q(\tau)d\tau.$$

x^* is in $\mathrm{RCS}^1[-2\pi, 2\pi]$, even, and satisfies the boundary conditions. According to the principle of pointwise minimization, x^* is the global minimal solution of the variational problem.

We would like to point out that Carathéodory in [17] always assumes that x^* is smooth.

9.3 Weak Local Minima

For our subsequent discussion that is based on a convexification of the Lagrangian we need the following

Lemma 9.3.1. *Let V be an open and A a compact subset of a metric space (X, d), and let $A \subset V$. Then there is a positive δ such that*

$$\bigcup_{x \in A} K(x, \delta) \subset V.$$

Proof. Suppose for all $n \in \mathbb{N}$ there exists $x_n \in X \setminus V$ and $a_n \in A$ such that $d(x_n, a_n) < \frac{1}{n}$. As A compact there is a convergent subsequence (a_k) such that $a_k \to \bar{a} \in A \subset V$. We obtain

$$d(x_k, \bar{a}) \leq d(x_k, a_k) + d(a_k, \bar{a}) \to 0,$$

a contradiction to $X \setminus V$ closed. □

Lemma 9.3.2. *Let $M \in L(\mathbb{R}^{2n})$ be a matrix of the structure*

$$M = \begin{pmatrix} A & C^T \\ C & D \end{pmatrix}$$

where $A, C, D \in L(\mathbb{R}^n)$ and D positive definite and symmetric. Then M is positive (semi-)definite if and only if $A - C^T D^{-1} C$ is positive (semi-)definite.

Proof. Let $f : \mathbb{R}^n \times \mathbb{R}^n \to \mathbb{R}$ defined by

$$f(p, q) := \begin{pmatrix} p \\ q \end{pmatrix}^T \begin{pmatrix} A & C^T \\ C & D \end{pmatrix} \begin{pmatrix} p \\ q \end{pmatrix} = p^T A p + 2q^T C p + q^T D q.$$

Minimization w.r.t. q for fixed p yields: $2Dq = -2Cp$ and hence

$$q(p) = -D^{-1}Cp.$$

By inserting this result into f we obtain

$$f(p, q(p)) = p^T Ap - 2p^T C^T D^{-1}Cp + p^T C^T D^{-1}Cp = p^T(A - C^T D^{-1}C)p.$$

Using our assumption $A - C^T D^{-1}C$ positive (semi-)definite it follows that $f(p, q(p)) > 0$ for $p \neq 0$ ($f(p, q(p)) \geq 0$ in the semi-definite case). For $p = 0$ and $q \neq 0$ obviously $f(p, q) > 0$.

On the other hand, let W be positive (semi-)definite. Then (0,0) is the only (a) minimal solution of f. Hence the function $p \mapsto f(p, q(p))$ has 0 as the (a) minimal solution, i.e. $A - C^T D^{-1}C$ is positive (semi-)definite. □

Definition 9.3.3. We say that the *Legendre–Riccati condition* is satisfied, if there exists a continuously differentiable symmetrical matrix function $W : [a, b] \to L(\mathbb{R}^n)$ such that for all $t \in [a, b]$ the expression

$$L^0_{xx} + \dot{W} - (L^0_{x\dot{x}} + W)(L^0_{\dot{x}\dot{x}})^{-1}(L^0_{\dot{x}x} + W) \tag{9.7}$$

is positive definite.

We would like to point out that Zeidan (see [116]) treats variational problems by discussing the Hamiltonian. She presents a corresponding condition for W w.r.t. the Hamiltonian in order to obtain sufficient conditions.

If the Legendre–Riccati condition is satisfied, we shall introduce a *quadratic supplement potential* based on the corresponding matrix W such that the supplemented Lagrangian is strictly convex (compare also [33], Vol. I, p. 251). Klötzler in [18] p. 325 uses a modification of the Hamiltonian, also leading to Riccati's equation, such that the resulting function is concave, in the context of extensions of field theory.

Let in the sequel U be open.

Theorem 9.3.4 (Fundamental Theorem). *Let $L : U \to \mathbb{R}$ be continuous and $L(\cdot, \cdot, t)$ twice continuously differentiable.*

An admissible regular extremaloid x^* is a weak local minimal solution of the given variational problem, if the Legendre–Riccati condition is satisfied.

Proof. Let a differentiable $W : [a, b] \to L(\mathbb{R}^n)$ be given, such that the Legendre–Riccati condition is satisfied for W. Then we choose the quadratic supplement potential

$$F : [a, b] \times \mathbb{R}^n \to \mathbb{R}$$

with $F(t,p) = -\frac{1}{2}p^T W(t)p$ which leads to an equivalent variational problem with the modified Lagrange function

$$\begin{aligned}\tilde{L}(p,q,t) &:= L(p,q,t) - \langle q, F_p(t,p)\rangle - F_t(t,p) \\ &= L(p,q,t) + \langle q, W(t)p\rangle + \frac{1}{2}\langle p, \dot{W}(t)p\rangle.\end{aligned}$$

We shall now show that there is a $\delta > 0$ such that for all $t \in [a,b]$ the function $(p,q) \mapsto \phi_t(p,q) := \tilde{L}(p,q,t)$ is strictly convex on $K_t := K((x^*(t), \dot{x}^*(t)), \delta)$: we have

$$M := \phi_t''(x^*(t), \dot{x}^*(t)) = \begin{pmatrix} L^0_{pp} + \dot{W} & L^0_{pq} + W \\ L^0_{qp} + W & L^0_{qq} \end{pmatrix}(t)$$

is positive definite using the Legendre–Riccati condition and Lemma 9.3.2 (note that $L^0_{qp} = {L^0_{pq}}^T$). Then $\phi_t''(p,q)$ is positive definite on an open neighborhood of $(x^*(t), \dot{x}^*(t))$. As the set $\{(x^*(t), \dot{x}^*(t)) \cup (x^*(t), \dot{x}^*(t-)) | t \in [a,b]\}$ is compact there is – according to Lemma 9.3.1 – a (universal) δ such that on K_t the function ϕ_t'' is positive definite, and therefore ϕ_t strictly convex on K_t, for all $t \in [a,b]$. As the RCS^1-ball with center x^* and radius δ is contained in the set

$$S_\delta := \{x \in \mathrm{RCS}^1[a,b]^n \mid (x(t), \dot{x}(t)) \in K_t \ \forall t \in [a,b]\},$$

we obtain that the extremal x^* is a (proper) weak local minimum. Thus we have identified a (locally) convex variational problem with the Lagrangian $\tilde{L}$ that is equivalent to the problem involving L (Theorems 9.1.11, 9.1.12) . □

Remark 9.3.5. The Legendre–Riccati condition is guaranteed if the *Legendre–Riccati Matrix Differential Equation*

$$L^0_{xx} + \dot{W} - (L^0_{x\dot{x}} + W)(L^0_{\dot{x}\dot{x}})^{-1}(L^0_{\dot{x}x} + W) = cI$$

for a positive $c \in \mathbb{R}$ has a symmetrical solution on $[a,b]$. Using the notation $R := L^0_{\dot{x}\dot{x}}$, $Q := L^0_{\dot{x}x}$, $P := L^0_{xx}$ and $A := -R^{-1}Q$, $B := R^{-1}$, $C := P - Q^T R^{-1} Q - cI$ then for $V := -W$ the above equation assumes the equivalent form (Legendre–Riccati equation)

$$\dot{V} + VA + A^T V + VBV - C = 0.$$

Definition 9.3.6. The first order system

$$\begin{aligned}\dot{Z} &= CY - A^T Z \\ \dot{Y} &= AY + BZ\end{aligned}$$

is called the *Jacobi equation in canonical form*. The pair of solutions (Z, Y) is called *self-conjugate* if $Z^T Y = Y^T Z$.

For the proof of the subsequent theorem we need the following

Lemma 9.3.7 (Quotient Rule).

$$\frac{d}{dt}(A^{-1}(t)) = -A^{-1}(t)\dot{A}(t)A^{-1}(t)$$

$$\frac{d}{dt}(B(t)A^{-1}(t)) = \dot{B}(t)A^{-1}(t) - B(t)A^{-1}(t)\dot{A}(t)A^{-1}(t)$$

$$\frac{d}{dt}(A^{-1}(t)B(t)) = A^{-1}(t)\dot{B}(t) - A^{-1}(t)\dot{A}(t)A^{-1}(t)B(t).$$

Proof. We have

$$0 = \frac{d}{dt}(I) = \frac{d}{dt}(A(t)A^{-1}(t)) = \dot{A}(t)A^{-1}(t) + A(t)\frac{d}{dt}(A^{-1}(t)).$$ □

Theorem 9.3.8. *If the Jacobi equation has a solution (Z, Y) such that $Y^T Z$ is symmetrical and Y is invertible then $V := ZY^{-1}$ is a symmetrical solution of the Legendre–Riccati equation*

$$\dot{V} + VA + A^T V + VBV - C = 0.$$

Proof. If $Y^T Z$ is symmetrical, then $V := ZY^{-1}$ is also symmetrical, as

$$(Y^T)^{-1}(Y^T Z)Y^{-1} = ZY^{-1} = V.$$

V is a solution of the Legendre–Riccati equation, because according to the quotient rule (Lemma 9.3.7) we have

$$\begin{aligned}\dot{V} &= \dot{Z}Y^{-1} - ZY^{-1}\dot{Y}Y^{-1}\\ &= (CY - A^T Z)Y^{-1} - ZY^{-1}(AY + BZ)Y^{-1}\\ &= C - A^T V - VA - VBV.\end{aligned}$$ □

Definition 9.3.9. Let x^* be an extremaloid and let A, B, C as in Remark 9.3.5. Let (x, y) be a solution of the Jacobi equation in vector form

$$\begin{aligned}\dot{z} &= Cy - A^T z\\ \dot{y} &= Ay + Bz,\end{aligned}$$

such that $y(a) = 0$ and $z(a) \neq 0$. If there is a point $t_0 \in (a, b]$ such that $y(t_0) = 0$ then we say t_0 is a *conjugate point* of x^* w.r.t. a.

For the next theorem compare Hartmann ([36], Theorem 10.2 on p. 388).

Theorem 9.3.10. *If the regular extremaloid x^* does not have a conjugate point in $(a, b]$ then there exists a self-conjugate solution (Z_1, Y_1) of the matrix Jacobi equation such that Y_1 is invertible on $[a, b]$.*

Furthermore, the Legendre–Riccati condition is satisfied.

Proof. For any solution (Z, Y) of the matrix Jacobi equation

$$\dot{Z} = CY - A^T Z$$
$$\dot{Y} = AY + BZ$$

it turns out that $\frac{d}{dt}(Z^T Y - Y^T Z) = 0$ using the product rule and the fact that the matrices B and C are symmetrical. Hence the matrix $Z^T Y - Y^T Z$ is equal to a constant matrix K on $[a, b]$. If we consider the following initial value problem:

$$Y(a) = I, \quad Y_0(a) = 0$$
$$Z(a) = 0, \quad Z_0(a) = I$$

then obviously for the solution (Z_0, Y_0) the matrix K is equal to zero, i.e. (Z_0, Y_0) is self-conjugate.

Apparently

$$\begin{pmatrix} Y & Y_0 \\ Z & Z_0 \end{pmatrix}$$

is a fundamental system. Hence any solution (y, z) can be represented in the following way: $y = Yc_1 + Y_0 c_2$ and $z = Zc_1 + Z_0 c_2$. If $y(a) = 0$ then $0 = y(a) = Ic_1 = c_1$ and $z(a) = Ic_2$, thus $y = Y_0 c_2$ and $z = Z_0 c_2$.

It turns out that $Y_0(t)$ is non-singular on $(a, b]$, for suppose there is $t_0 \in (a, b]$ such that $Y_0(t_0)$ is singular, then the linear equation $Y_0(t_0)c = 0$ has a non-trivial solution c_0. Let $y_0(t) := Y_0(t)c_0$ on $[a, b]$ then $y_0(a) = 0$ and $y_0(t_0) = 0$. Moreover for $z_0(t) := Z_0(t)c_0$ we have $z_0(a) = Ic_0 = c_0 \neq 0$, hence t_0 is a conjugate point of a, a contradiction.

We now use a construction that can be found in Hestenes: there is an $\varepsilon > 0$ such that x^* can be extended to a function $\bar{x}^*$ on $[a - \varepsilon, b]$ and $L_{\dot{x}\dot{x}}(\bar{x}^*, \dot{\bar{x}}^*, \cdot)$ remains positive definite on $[a - \varepsilon, b]$. If we insert $\bar{x}^*$ into $L_{x\dot{x}}$ and L_{xx}, the matrices A, B, C (using the notation of Remark 9.3.5) retain their properties and we can consider the corresponding Jacobi equation extended to $[a - \varepsilon, b]$. Then Lemma 5.1 in [38] on p. 129 yields an $a_0 < a$ such that $\bar{x}^*$ has no conjugate point w.r.t. a_0 on $(a_0, b]$. But then, using the same initial conditions at a_0 and the same argument as in the first part of the proof, we obtain a self-conjugate solution (Z_1, Y_1) on $[a_0, b]$ such that Y_1 is non-singular on $(a_0, b]$. The restriction of Y_1 is, of course, a solution of the Jacobi equation on $[a, b]$ that is non-singular there.

Using Theorem 9.3.8 it follows that the $Z_1 Y_1^{-1}$ is a symmetrical solution of the Legendre–Riccati equation. □

Theorem 9.3.11 (Fundamental Theorem of Jacobi–Weierstrass). *If the regular extremal x^* does not have a conjugate point on $(a, b]$ then x^* is a weak local minimal solution of the given variational problem.*

9.3.1 Carathéodory Minimale

Definition 9.3.12. A function $x^* \in \mathrm{RCS}^1[a,b]^n$ is called a *Carathéodory Minimale*, if for every $t_0 \in (a,b)$ there are $s, t \in (a,b)$ with $s < t_0 < t$ such that $x^*|_{[s,t]}$ is a weak local solution of the variational problem

$$\min \int_s^t L(x(\tau), \dot{x}(\tau), \tau) d\tau$$

$$\text{on } S_{s,t} := \{x \in \mathrm{RCS}^1[s,t]^n \mid (x(\tau), \dot{x}(\tau), \tau) \in U,\ \forall \tau \in [s,t],$$
$$x(s) = x^*(s), x(t) = x^*(t)\}.$$

As the consequence of the Theorem 9.3.4 we obtain (compare [17], p. 210) the

Theorem 9.3.13. *Every regular extremal is a Carathéodory Minimale.*

Proof. The matrix $W := r(t - t_0) \cdot I$ satisfies, for large r on the interval $[t_0 - 1/r, t_0 + 1/r]$, the Legendre–Riccati condition: in fact, for $p \in \mathbb{R}^n$ with $\|p\| = 1$ we obtain

$$p^T(P(t) + rI - (Q^T(t) + r(t - t_0)I)\, R^{-1}(t)(Q(t) + r(t - t_0)I))p$$
$$\geq r - (\|Q^T\| + 1)\, \|R^{-1}\|\, (\|Q(t)\| + 1) - \|P(t)\| > 0$$

for r large enough. □

9.4 Strong Convexity and Strong Local Minima

Definition 9.4.1. Let $\tau : \mathbb{R}_{\geq 0} \to \mathbb{R}_{\geq 0}$ be non-decreasing, $\tau(0) = 0$ and $\tau(s) > 0$ for $s > 0$. Then we call τ a module function.

The following two lemmata can be found in [59]:

Lemma 9.4.2. *Let X be a normed space, $U \subset X$ open, and $f : U \to \mathbb{R}$ differentiable and K a convex subset of U. Then the following statements are equivalent:*

(a) *$f : K \to \mathbb{R}$ is uniformly convex, i.e. f is convex and for a module function and all $x, y \in K$*

$$f\left(\frac{x+y}{2}\right) \leq \frac{1}{2} f(x) + \frac{1}{2} f(y) - \tau(\|x - y\|),$$

(b) *there is a module function* τ_1 *such that for all* $x, y \in K$

$$f(y) - f(x) \geq \langle y - x, f'(x) \rangle + \tau_1(\|x - y\|),$$

(c) *there is a module function* τ_2 *such that for all* $x, y \in K$

$$\langle y - x, f'(y) - f'(x) \rangle \geq \tau_2(\|x - y\|).$$

Proof. (a) $\Rightarrow$ (b):

$$\begin{aligned}\frac{1}{2}(f(x) + f(y)) &\geq f\left(\frac{x+y}{2}\right) + \tau(\|x - y\|) - f(x) + f(x)\\ &\geq \left\langle \frac{x+y}{2} - x, f'(x) \right\rangle + f(x) + \tau(\|x - y\|).\end{aligned}$$

Multiplication of both sides by 2 yields

$$f(y) - f(x) \geq \langle y - x, f'(x) \rangle + 2\tau(\|x - y\|).$$

(b) $\Rightarrow$ (a): Let $\lambda \in [0, 1]$ and let $z := \lambda x + (1 - \lambda)y$ then

$$\begin{aligned}\langle x - z, f'(z) \rangle &\leq f(x) - f(z) - \tau_1(\|z - x\|)\\ \langle y - z, f'(z) \rangle &\leq f(y) - f(z) - \tau_1(\|z - y\|).\end{aligned}$$

Multiplication of the first inequality by λ and the second by $1 - \lambda$ and subsequent addition yields

$$\begin{aligned}0 = \langle 0, f'(z) \rangle &= \langle \lambda x + (1 - \lambda)y - z, f'(z) \rangle\\ &\leq \lambda f(x) + (1 - \lambda) f(y)) - f(z) - \lambda\tau_1(\|z - x\|) - (1 - \lambda)\tau_1(\|z - y\|),\end{aligned}$$

showing that f is convex and we obtain for $\lambda = \frac{1}{2}$

$$f\left(\frac{x+y}{2}\right) \leq \frac{1}{2} f(x) + \frac{1}{2} f(y) - \tau_1\left(\frac{\|x - y\|}{2}\right).$$

(b) $\Rightarrow$ (c): we have

$$\begin{aligned}f(y) - f(x) &\geq \langle y - x, f'(x) \rangle + \tau_1(\|x - y\|),\\ f(x) - f(y) &\geq \langle x - y, f'(y) \rangle + \tau_1(\|x - y\|).\end{aligned}$$

Addition of both inequalities yields

$$\langle x - y, f'(x) - f(y) \rangle \geq 2\tau_1(\|x - y\|).$$

(c) $\Rightarrow$ (b): Let $x, y \in K$. First we show that f is convex. For $h := y - x$ we obtain by the mean value theorem an $\alpha \in (0, 1)$ such that

$$f(y) - f(x) = \langle h, f'(x + \alpha h)\rangle.$$

Using (c) we have

$$\langle \alpha h, f'(x + \alpha h) - f'(x)\rangle \geq 0,$$

hence

$$\langle h, f'(x)\rangle \leq \langle h, f'(x + \alpha h)\rangle = f(y) - f(x),$$

showing that f is convex by a similar argument as above.

Moreover by (c)

$$\left\langle \frac{x+y}{2} - x, f'\left(\frac{x+y}{2}\right) - f'(x)\right\rangle \geq \tau_2\left(\frac{\|x-y\|}{2}\right),$$

i.e.

$$\frac{1}{2}\left\langle y - x, f'\left(\frac{x+y}{2}\right)\right\rangle \geq \frac{1}{2}\langle y - x, f'(x)\rangle + \tau_2\left(\frac{\|x-y\|}{2}\right).$$

Therefore

$$\begin{aligned} f(y) - f(x) &= f(y) - f\left(\frac{x+y}{2}\right) + f\left(\frac{x+y}{2}\right) - f(x) \\ &\geq \left\langle y - \frac{x+y}{2}, f'\left(\frac{x+y}{2}\right)\right\rangle + \left\langle \frac{x+y}{2} - x, f'(x)\right\rangle \\ &\geq \frac{1}{2}\langle y - x, f'(x)\rangle + \tau_2\left(\frac{\|x-y\|}{2}\right) + \frac{1}{2}\langle y - x, f'(x)\rangle \\ &= \langle y - x, f'(x)\rangle + \tau_2\left(\frac{\|x-y\|}{2}\right). \end{aligned}$$

□

Lemma 9.4.3. *Let X be a normed space, $U \subset X$ open, and $g : U \to \mathbb{R}$ twice continuously differentiable. If for an $x^* \in U$ and $c > 0$ we have*

$$\langle g''(x^*)h, h\rangle \geq c\|h\|^2$$

for all $h \in X$, there exists a $\delta > 0$ such that g is strongly convex *on $K := K(x^*, \delta)$. In particular,*

$$g\left(\frac{x+y}{2}\right) \leq \frac{1}{2}g(x) + \frac{1}{2}g(y) - \frac{c}{8}\|x-y\|^2$$

for all $x, y \in K$.

Proof. As g'' is continuous, there is a $\delta > 0$ such that for all $x \in K(x^*, \delta)$: $\|g''(x) - g''(x^*)\| \leq \frac{c}{2}$ Hence for all $x \in K$ and all $h \in X$ we obtain

$$\begin{aligned}\langle g''(x)h, h\rangle &= \langle (g''(x) - g''(x^*))h, h\rangle + \langle g''(x^*)h, h\rangle \\ &\geq -\|g''(x) - g''(x^*)\|\|h\|^2 + c\|h\|^2 \geq \frac{c}{2}\|h\|^2.\end{aligned}$$

Let $x, y \in K$ then the mean value theorem yields

$$\begin{aligned}g(y) - g(x) - \langle y - x, g'(x)\rangle &= \langle y - x, g''(x + \alpha(y - x))(y - x)\rangle \\ &\geq \frac{c}{2}\|y - x\|^2.\end{aligned}$$

For $\tau_1(s) = \frac{c}{2}s^2$ in (b) of the previous lemma the assertion follows. □

Theorem 9.4.4 (Uniform Strong Convexity of the Lagrangian). *Let x^* be an extremal and let the Legendre–Riccati condition be satisfied, then there is a $\delta > 0$ and a $c > 0$ such that for all $(p, q), (u, v) \in K_t := K((x^*(t), \dot{x}^*(t)), \delta)$ and for all $t \in [a, b]$ we have*

$$\tilde{L}\left(\frac{p+u}{2}, \frac{q+v}{2}, t\right) \leq \frac{1}{2}\tilde{L}(p, q, t) + \frac{1}{2}\tilde{L}(u, v, t) - \frac{c}{8}(\|p - u\|^2 + \|q - v\|^2).$$

Proof. Let ϕ_t be as in Theorem 9.3.4. As the set $K_1 := \{(p, q) \in \mathbb{R}^{2n} \mid \|p\|^2 + \|q\|^2 = 1\}$ is compact and as $t \mapsto \phi_t''(x^*(t), \dot{x}^*(t))$ is continuous on $[a, b]$ there is a positive $c \in \mathbb{R}$ such that for all $t \in [a, b]$

$$\begin{pmatrix} p \\ q \end{pmatrix}^T \phi_t''(x^*(t), \dot{x}^*(t)) \begin{pmatrix} p \\ q \end{pmatrix} \geq c \tag{9.8}$$

on K_1, i.e. $t \mapsto \phi_t''(x^*(t), \dot{x}^*(t))$ is uniformly positive definite on $[a, b]$.

According to Lemma 9.3.1 there is a $\rho > 0$ such that on the compact set in $\mathbb{R}^{2n+1}$

$$\overline{\bigcup_{t \in [a,b]} K((x^*(t), \dot{x}^*(t), t), \rho)} \subset U$$

we have uniform continuity of $(p, q, t) \mapsto \phi_t''(p, q)$. Hence there is a $\delta > 0$ such that for all $t \in [a, b]$ and all $(u, v) \in K_t := K((x^*(t), \dot{x}^*(t)), \delta)$ we have

$$\|\phi_t''(u, v) - \phi_t''(x^*(t), \dot{x}^*(t))\| \leq \frac{c}{2},$$

and hence on that set

$$\begin{pmatrix} p \\ q \end{pmatrix}^T \phi_t''(u, v) \begin{pmatrix} p \\ q \end{pmatrix} \geq \frac{c}{2}.$$

Thus we obtain that $(p,q) \mapsto \tilde{L}(p,q,t)$ is uniformly strongly convex on K_t for all $t \in [a,b]$, i.e.

$$\tilde{L}\Big(\frac{p+u}{2},\frac{q+v}{2},t\Big) \le \frac{1}{2}\tilde{L}(p,q,t) + \frac{1}{2}\tilde{L}(u,v,t) - \frac{c}{8}(\|p-u\|^2 + \|q-v\|^2)$$

for all $(p,q),(u,v) \in K_t$ and all $t \in [a,b]$. □

For the corresponding variational functional the above theorem leads to the

Corollary 9.4.5 (Strong Convexity of the Variational Functional). *Let*

$$B(x^*,\delta) := \{x \in \mathrm{RCS}^1[a,b]^n \mid \|x(t)-x^*(t)\|^2 + \|\dot{x}(t)-\dot{x}^*(t)\|^2 < \delta \ \forall t \in [a,b]\},$$

then the variational functional $\tilde{f}$ belonging to the Lagrangian $\tilde{L}$ is uniformly strongly convex with respect to the Sobolev norm

$$\tilde{f}\Big(\frac{x+y}{2}\Big) \le \frac{1}{2}\tilde{f}(x) + \frac{1}{2}\tilde{f}(y) - \frac{c}{8}\|x-y\|_W^2.$$

Let $V := \{x \in B(x^*,\delta) \mid x(a)=\alpha, x(b)=\beta\}$ be the subset of those functions satisfying the boundary conditions then on V the original functional f is also uniformly strongly convex w.r.t. the Sobolev norm

$$f\Big(\frac{x+y}{2}\Big) \le \frac{1}{2}f(x) + \frac{1}{2}f(y) - \frac{c}{8}\|x-y\|_W^2.$$

Furthermore, every minimizing sequence converges to the minimal solution w.r.t. the Sobolev norm (strong solvability).

Proof. For the variational functional $\tilde{f}(x) := \int_a^b \tilde{L}(x(t),\dot{x}(t),t)dt$ we then obtain

$$\begin{aligned}\tilde{f}\Big(\frac{x+y}{2}\Big) &= \int_a^b \tilde{L}\Big(\frac{x(t)+y(t)}{2},\frac{\dot{x}(t)+\dot{y}(t)}{2},t\Big)dt \\ &\le \int_a^b \Big(\frac{1}{2}\tilde{L}(x(t),\dot{x}(t),t) + \frac{1}{2}\tilde{L}(y(t),\dot{y}(t),t)\Big)dt \\ &\quad - \frac{c}{8}\int_a^b (\|x(t)-y(t)\|^2 + \|\dot{x}(t)-\dot{y}(t)\|^2)dt \\ &= \frac{1}{2}\tilde{f}(x) + \frac{1}{2}\tilde{f}(y) - \frac{c}{8}\|x-y\|_W^2\end{aligned}$$

for all $x,y \in B(x^*,\delta)$. Thus $\tilde{f}$ is strictly convex on $B(x^*,\delta)$. On V the functional $f(x) := \int_a^b L(x(t),\dot{x}(t),t)dt$ differs from the functional $\tilde{f}$ only by a constant, hence has the same minimal solutions. On V the inequality

$$f\Big(\frac{x+y}{2}\Big) \le \frac{1}{2}f(x) + \frac{1}{2}f(y) - \frac{c}{8}\|x-y\|_W^2$$

is satisfied, i.e. f is strongly convex on V w.r.t. the Sobolev norm. □

Using Theorem 9.3.10 we obtain the following corollary (retaining the notation of the previous corollary):

Corollary 9.4.6 (Strong Convexity). *If the regular extremal x^* does not have a conjugate point then there is a $\delta > 0$ such that on $B(x^*, \delta)$ the modified variational functional $\tilde{f}$ is uniformly strongly convex with respect to the Sobolev norm: and on V the original functional f is also uniformly strongly convex w.r.t. the Sobolev norm.*

9.4.1 Strong Local Minima

We now investigate the question under what conditions we can guarantee that weak local minimal solutions are in fact strong local minimal solutions. It turns out that such a strong property can be presented and proved without the use of embedding theory. We will show that strong local minima require a supplement for the Lagrangian that is generated by the supplement potential

$$(t, p) \mapsto F(t, p) := \frac{1}{2}\langle p, W(t)p\rangle - \langle \lambda(t), p\rangle,$$

where $W(t)$ is symmetrical and the linear term is chosen in such a way that – along the extremal – the necessary conditions for optimality for $\hat{L}$ coincide with the fulfillment of the Euler–Lagrange equation and the quadratic term, with its convexification property, provides the sufficient optimality conditions.

The supplemented Lagrangian then has the following structure:

$$\begin{aligned}\hat{L}(p, q, t) &= L(p, q, t) + F_t + \langle F_p, q\rangle \\ &= L(p, q, t) + \frac{1}{2}\langle p, \dot{W}(t)p\rangle - \langle \dot{\lambda}(t), p\rangle + \langle W(t)p, q\rangle - \langle \lambda(t), q\rangle.\end{aligned}$$

If we define

$$\lambda(t) := L_q(x^*(t), \dot{x}^*(t), t) + W(t)x^*(t),$$

then – using the Euler–Lagrange equation for the original Lagrangian L, i.e. $\frac{d}{dt}L_q = L_p$, we obtain in fact the necessary conditions

$$\hat{L}_p(x^*(t), \dot{x}^*(t), t) = \hat{L}_q(x^*(t), \dot{x}^*(t), t) = 0.$$

In the subsequent theorem we make use of the following lemma.

Lemma 9.4.7. *Let $r > 0$ and the ball $\overline{K}(x_0, r) \subset \mathbb{R}^n$. Let $U \supset \overline{K}(x_0, r)$ open. Let $g : U \to \mathbb{R}$ be continuous, and let g be for all $x \in S(x_0, r)$ directionally differentiable with the property $g'(x, x - x_0) > 0$. Then every minimal solution of g on $\overline{K}(x_0, r)$ is in the interior of that ball. In particular, if g is differentiable then $g'(x^*) = 0$ for every minimal solution x^* in $K(x_0, r)$.*

Proof. Since $\overline{K}(x_0, r)$ is compact, g has a minimal solution x^* there. Suppose $\|x^* - x_0\| = r$ then the function $\varphi : [0,1] \to \mathbb{R}$ with

$$t \mapsto \varphi(t) := g(x^* + t(x_0 - x^*))$$

has a minimal solution at 0, hence $\varphi'(0) \geq 0$. Since

$$\begin{aligned}\varphi'(0) &= \lim_{t \downarrow 0} \frac{g(x^* + t(x_0 - x^*)) - g(x^*)}{t} \\ &= g'(x^*, x_0 - x^*) = -g'(x^*, x_0 - x^*) < 0,\end{aligned}$$

a contradiction. □

Theorem 9.4.8 (Strong Local Minimum). *Let x^* be an admissible, regular extremal and let the Legendre–Riccati condition be satisfied.*

Let there exist a $\kappa > 0$ such that for all $t \in [a,b]$ and all p with $\|p - x^*(t)\| < \kappa$ the set $V_{t,p} := \{q \in \mathbb{R}^n \mid (p,q,t) \in U\}$ is convex and the function $L(p,\cdot,t)$ is convex on $V_{t,p}$. Then x^* is a locally strong minimal solution of the variational problem, i.e. there is a positive d, such that for all $x \in K := \{x \in S \mid \|x - x^*\|_\infty < d\}$ we have

$$\int_a^b L(x^*(t), \dot{x}^*(t), t)dt \leq \int_a^b L(x(t), \dot{x}(t), t)dt.$$

Proof. In Theorem 9.4.4 we have constructed positive constants c and δ such that for all (p,v), $(u,v) \in K_t := K(x^*(t), \dot{x}^*(t), \delta)$ we have

$$\tilde{L}\left(\frac{p+u}{2}, \frac{q+v}{2}, t\right) \leq \frac{1}{2}\tilde{L}(p,q,t) + \frac{1}{2}\tilde{L}(u,v,t) - \frac{c}{8}(\|p-u\|^2 + \|q-v\|^2).$$

But $\hat{L}$ differs from $\tilde{L}$ only by the linear term

$$-\langle \dot{\lambda}(t), p\rangle - \langle \lambda(t), q\rangle.$$

In particular $\tilde{L}'' = \hat{L}''$, i.e. the convexity properties remain unchanged, and hence $\hat{L}(\cdot,\cdot,t)$ is strongly convex on $K_t := K((x^*(t), \dot{x}^*(t)), \delta)$ with a uniform constant c for all $t \in [a,b]$.

From Lemma 9.4.2 we obtain for this c that

$$\begin{aligned}\langle \hat{L}_q(x^*(t), q, t) - \hat{L}_q(x^*(t), \dot{x}^*(t), t), q - \dot{x}^*(t)\rangle &= \langle \hat{L}_q(x^*(t), q, t), q - \dot{x}^*(t)\rangle \\ &\geq \frac{c}{2}\|q - \dot{x}^*(t)\|^2\end{aligned}$$

(*strong monotonicity* of $\hat{L}_q(x^*(t), \cdot, t)$).

The uniform continuity of $\hat{L}_q$ guarantees that for $0 < \varepsilon < c \cdot \delta/4$ there is a $0 < d \le \min\{\delta/2, \kappa\}$ such that for all $t \in [a, b]$

$$\|\hat{L}_q(x^*(t), q, t) - \hat{L}_q(p, q, t)\| < \varepsilon$$

for all $p \in x^*(t) + K(0, d)$. We obtain

$$\begin{aligned}&\langle \hat{L}_q(x^*(t), q, t) - \hat{L}_q(p, q, t), q - x^*(t)\rangle \\ &\le \|\hat{L}_q(x^*(t), q, t) - \hat{L}_q(p, q, t)\|\|q - x^*(t)\| < \varepsilon\|q - x^*(t)\|.\end{aligned}$$

Let $\rho = \delta/2$, then on the sphere $\{q \mid \|q - \dot{x}^*(t)\| = \rho\}$ it follows that for all $p \in x^*(t) + K(0, d)$

$$\langle \hat{L}_q(p, q, t), q - \dot{x}^*(t)\rangle \ge c\|q - \dot{x}^*(t)\|^2 - \varepsilon\|q - x^*(t)\| = \frac{c}{2}\rho^2 - \varepsilon\rho > 0.$$

Hence from Lemma 9.4.7 we obtain for every $p \in x^*(t) + K(0, d)$ a minimal solution of $q^*(p) \in K(\dot{x}^*(t), \rho)$ such that

$$\hat{L}_q(p, q^*(p), t) = 0.$$

From the convexity of $\hat{L}(p, \cdot, t)$ we conclude that $q^*(p)$ is a minimal solution of $\hat{L}(p, \cdot, t)$ on $V_{t,p}$. (Note that $\hat{L}(p, \cdot, t)$ differs from $L(p, \cdot, t)$ only by an affine term in q.)

We shall now show that for all $t \in [a, b]$ we have that $(x^*(t), \dot{x}^*(t))$ is a minimal solution of $\hat{L}(\cdot, \cdot, t)$ on $W_t := \{(p, q) \in \mathbb{R}^n \times \mathbb{R}^n \mid (p, q, t) \in U \text{ and } \|p - x^*(t)\| < d\}$.

For, suppose there exists $(p, q) \in W_t$ such that

$$\hat{L}(x^*(t), \dot{x}^*(t), t) > \hat{L}(p, q, t).$$

As x^* is extremal , and as $\hat{L}(\cdot, \cdot, t)$ is convex on K_t we have $(x^*(t), \dot{x}^*(t))$ is minimal solution of $\hat{L}(\cdot, \cdot, t)$ on K_t by construction of $\hat{L}$. As $(p, q^*(p)) \in K_t$ we obtain

$$\hat{L}(x^*(t), \dot{x}^*(t), t) \le \hat{L}(p, q^*(p), t) \le \hat{L}(p, q, t) < \hat{L}(x^*(t), \dot{x}^*(t), t),$$

a contradiction.

For all $x \in K = \{x \in S \mid \|x^* - x\|_\infty < d\}$ we then have

$$\int_a^b \hat{L}(x^*(t), \dot{x}^*(t), t)dt \le \int_a^b \hat{L}(x(t), \dot{x}(t), t)dt$$

and, as the integrals differ on S only by a constant, we obtain the corresponding inequality also for the original Lagrangian

$$\int_a^b L(x^*(t), \dot{x}^*(t), t)dt \le \int_a^b L(x(t), \dot{x}(t), t)dt,$$

which completes the proof. □

The previous theorem together with Theorem 9.3.10 leads to the following

Corollary 9.4.9 (Strong Local Minimum). *Let x^* be an admissible and regular extremal without conjugate points.*

Let there exist a $\kappa > 0$ such that for all $t \in [a, b]$ and all p with $\|p - x^*(t)\| < \kappa$ the set $V_{t,p} := \{q \in \mathbb{R}^n \mid (p, q, t) \in U\}$ is convex and the function $L(p, \cdot, t)$ is convex on $V_{t,p}$. Then x^* is a locally strong minimal solution of the variational problem, i.e. there is a positive d, such that for all $x \in K := \{x \in S \mid \|x - x^*\|_\infty < d\}$ we have

$$\int_a^b L(x^*(t), \dot{x}^*(t), t)dt \leq \int_a^b L(x(t), \dot{x}(t), t)dt.$$

Remark 9.4.10. If in particular $U = U_1 \times U_2 \times [a, b]$, where $U_1 \subset \mathbb{R}^n$ open, $U_2 \subset \mathbb{R}^n$ open and convex, and $L(p, \cdot, t) : U_2 \to \mathbb{R}$ convex for all $(p, t) \in U_1 \times [a, b]$ then the requirements of the previous theorem are satisfied.

9.5 Necessary Conditions

We briefly restate the Euler–Lagrange equation as a necessary condition in the piecewise continuous case. The standard proof carries over to this situation (see Hestenes [38], Lemma 5.1, 70).

Theorem 9.5.1. *Let $L, L_x, L_{\dot{x}}$ be piecewise continuous. Let x^* be a solution of the variational problem such that its graph is contained in the interior of U. Then x^* is an extremaloid, i.e. there is a $c \in \mathbb{R}^n$ such that*

$$L_{\dot{x}}(x^*(t), \dot{x}^*(t), t) = \int_a^t L_x((x^*(\tau), \dot{x}^*(\tau), \tau)d\tau + c.$$

9.5.1 The Jacobi Equation as a Necessary Condition

A different approach for obtaining the Jacobi equation is to consider the variational problem that corresponds to the second directional derivative of the original variational problem:

Let x^* be a solution of the original variational problem, let

$$V := \{h \in \mathrm{RCS}^1[a, b]^n \mid h(a) = h(b) = 0\},$$

and let

$$\phi(\alpha) := f(x^* + \alpha h) = \int_a^b L(x^*(t) + \alpha h(t), \dot{x}^*(t) + \alpha \dot{h}(t), t)dt.$$

Then the necessary condition yields

$$0 \le \phi''(0) = f''(x^*, h) = \int_a^b \langle L^0_{\dot{x}\dot{x}}\dot{h}, \dot{h}\rangle + 2\langle L^0_{\dot{x}x}h, \dot{h}\rangle + \langle L^0_{xx}h, h\rangle dt$$
$$= \int_a^b \langle R\dot{h}, \dot{h}\rangle + 2\langle Qh, \dot{h}\rangle + \langle Ph, h\rangle dt,$$

using our notation in Remark 9.3.5. Then the (quadratic) variational problem

$$\text{Minimize } f''(x^*, \cdot) \text{ on } V$$

is called the accessory (secondary) variational problem. It turns out that the corresponding Euler–Lagrange equation

$$\frac{d}{dt}(R\dot{h} + Qh) = Q^T\dot{h} + Ph$$

or in matrix form

$$\frac{d}{dt}(R\dot{Y} + QY) = Q^T\dot{Y} + PY$$

yields the Jacobi equation in canonical form

$$\dot{Z} = CY - A^T Z$$
$$\dot{Y} = AY + BZ$$

by setting $Z := R\dot{Y} + QY$ and using the notation in Remark 9.3.5.

Lemma 9.5.2. *If $h^* \in V$ is a solution of the Jacobi equation in the piecewise sense then it is a minimal solution of the accessory problem.*

Proof. Let

$$\Omega(h, \dot{h}) := \langle R\dot{h}, \dot{h}\rangle + 2\langle Qh, \dot{h}\rangle + \langle Ph, h\rangle$$

h^* is extremal of the accessory problem, i.e. (in the piecewise sense)

$$\frac{d}{dt}(2R\dot{h}^* + 2Qh^*) = 2Ph^* + 2Q^T\dot{h}^*.$$

Hence

$$\Omega(h^*, \dot{h}^*) = \langle R\dot{h}^* + Qh^*, \dot{h}^*\rangle + \left\langle \frac{d}{dt}(R\dot{h}^* + Qh^*), h^*\right\rangle = \frac{d}{dt}(\langle R\dot{h}^* + Qh^*, h^*\rangle).$$

Using the Weierstrass–Erdmann condition for the accessory problem, i.e. $R\dot{h}^* + Qh^*$ continuous, we can apply the main theorem of differential and integral calculus

$$\int_a^b \Omega(h^*, \dot{h}^*)dt = \langle R\dot{h}^* + Qh^*, h^*\rangle|_a^b = 0,$$

as $h^*(a) = h^*(b) = 0$. Hence h^* is minimal solution as $f''(x^*, \cdot)$ is non-negative. □

Theorem 9.5.3 (Jacobi's Necessary Condition for Optimality). *If a regular extremal x^* is a minimal solution of the variational problem on S, then a has no conjugate point in (a,b).*

Proof. For otherwise, let (h^*, k^*) be a non-trivial solution of the Jacobi equation with $h(a) = 0$, and let $c \in (a,b)$ be such a conjugate point. Then $h^*(c) = 0$ and $k^*(c) \neq 0$. Because $k^*(c) = R\dot{h}^*(c) + Qh^*(c) = R\dot{h}^*(c)$ it follows that $\dot{h}^*(c) \neq 0$. We define $y(t) = h^*(t)$ for $t \in [a,c]$, and $y(t) = 0$ for $t \in [c,b]$. In particular $\dot{y}(c) \neq 0$. Obviously, y is a solution of the Jacobi equation (in the piecewise sense) and hence, according to the previous lemma, a minimal solution of the accessory problem, a contradiction to Corollary 9.2.2 on smoothness of solutions, as Ω is strictly convex w.r.t. $\dot{h}$. □

9.6 C^1-variational Problems

Let

$$S_1 := \{x \in C^1[a,b]^n \mid (x(t), \dot{x}(t), t) \in U, x(a) = \alpha, x(b) = \beta\},$$

and

$$V_1 := \{h \in C^1[a,b]^n \mid h(a) = 0, h(b) = 0\}.$$

As the variational problem is now considered on a smaller set ($S_1 \subset S$), sufficient conditions carry over to this situation.

Again, just as in the RCS^1-theory, the Jacobi condition is a necessary condition.

Theorem 9.6.1. *If a regular extremal $x^* \in C^1[a,b]^n$ is a minimal solution of the variational problem on S_1, then a has no conjugate point in (a,b).*

Proof. Let $c \in (a,b)$ be a conjugate point of a. We consider the quadratic functional

$$h \mapsto g(h) := \int_a^b \Omega(h, \dot{h})dt$$

for $h \in V_1$, where

$$\Omega(h, \dot{h}) := \langle R\dot{h}, \dot{h}\rangle + 2\langle Qh, \dot{h}\rangle + \langle Ph, h\rangle.$$

We show that there is $\hat{h} \in V$ with $g(\hat{h}) < 0$: for suppose $g(h) \geq 0$ for all $h \in V$ then, as in Lemma 9.5.2 every extremal h^* is a minimal solution of g, and we use the construction for y as in the proof of Theorem 9.5.3 to obtain a solution of the Jacobi equation in the piecewise sense (which is, again according to Lemma 9.5.2 a minimal solution of the accessory problem) but is not smooth, in contradiction to Corollary 9.2.2, as Ω is strictly convex w.r.t. $\dot{h}$. Hence there is $\hat{h} \in V$ with $g(\hat{h}) < 0$. Now we use the process of smoothing of corners as described in Carathéodory [17]. Thus we obtain a $\bar{h} \in C^1[a,b]^n$ with $g(\bar{h}) < 0$. Let $\alpha \mapsto \phi(\alpha) = f(x^* + \alpha\bar{h})$ then $g(\bar{h}) = \phi''(0) \geq 0$ as x^* is minimal solution of f on S_1, a contradiction. □

9.7 Optimal Paths

For a comprehensive view of our results we introduce the notion of an optimal path. Our main objective in this context is, to characterize necessary and sufficient conditions.

Definition 9.7.1. A function $x^* \in \mathrm{RCS}^1[a,b]^n$ is called an optimal path, if it is a weak local solution of the variational problem

$$\min \int_a^t L(x(\tau), \dot{x}(\tau), \tau) d\tau$$

on

$$S_t := \{x \in \mathrm{RCS}^1[a,t]^n \mid (x(\tau), \dot{x}(\tau), \tau) \in U,\ \forall \tau \in [a,t],\ x(a) = \alpha, x(t) = x^*(t)\}$$

for all $t \in [a,b)$.

Theorem 9.7.2. *Consider the variational problem*

$$\min \int_a^b L(x(\tau), \dot{x}(\tau), \tau) d\tau$$

on

$$S := \{x \in \mathrm{RCS}^1[a,t]^n \mid (x(\tau), \dot{x}(\tau), \tau) \in U,\ \forall \tau \in [a,b],\ x(a) = \alpha, x(t) = \beta\}.$$

Let $x^* \in \mathrm{RCS}^1[a,b]^n$ be a regular extremal, then the following statements are equivalent:

(a) *x^* is an optimal path.*

(b) *The variational problem has an equivalent convex problem in the following sense: for the original Lagrangian there is a locally convexified Lagrangian $\tilde{L}$ such that for every subinterval $[a,\tau] \subset [a,b)$ there is a $\delta > 0$ with the property that $\tilde{L}(\cdot,\cdot,t)$ is strictly convex on the ball $K(x^*(t), \dot{x}^*(t), \delta)$ for all $t \in [a,\tau]$.*

(c) *The variational problem has an equivalent convex problem (in the sense of (b)) employing a quadratic supplement.*

(d) *For every subinterval $[a,\tau] \subset [a,b)$ there is a $\delta > 0$ with the property that $(x^*(t), \dot{x}^*(t))$ is a pointwise minimum of $\tilde{L}(\cdot,\cdot,t)$ on the ball $K(x^*(t), \dot{x}^*(t), \delta)$ for all $t \in [a,\tau]$.*

(e) *The Legendre–Riccati condition is satisfied on $[a,b)$.*

(f) *The Jacobi matrix equation has a non-singular and self-conjugate solution on $[a,b)$.*

(g) *a has no conjugate point in (a,b).*

9.8 Stability Considerations for Variational Problems

For many classical variational problems (Brachistochrone, Dido problem, geometrical optics) the optimal solutions exhibit singularities in the end points of the definition interval. A framework to deal with problems of this kind is to consider these singular problems as limits of "well-behaved" problems. This view requires – aside from a specification of these notions – certainty about the question whether limits of the solutions of the approximating problems are solutions of the original problem.

We will now apply our stability principles to variational problems and at first only vary the restriction set.

Continuous convergence of variational functionals requires special structures. If e.g. the variational functionals are continuous and convex, then by Remark 5.3.16 the continuous convergence already follows from pointwise convergence.

We will now show that from standard assumptions on the Lagrangian L the continuity of the variational functional on the corresponding restriction set follows, where the norm on $C^{(1)}[a,b]^m$ is defined as the sum of the maximum norm of y and $\dot{y}$

$$\|y\|_{C^1} := \|y\|_{\max} + \|\dot{y}\|_{\max}.$$

Let in the sequel $U \subset \mathbb{R}^{2m+1}$ and let $M = \{x \in C^{(1)}[a,b]^m \mid (x(t), \dot{x}(t), t) \in U \forall t \in [a,b]\}$. Then the following theorem holds (compare [35]):

Theorem 9.8.1. *Let $L : U \to \mathbb{R}$ be continuous. Then the variational functional*

$$x \mapsto f(x) := \int_a^b L(x(t), \dot{x}(t), t)\, dt$$

is continuous on M.

Proof. Let the sequence $(x_n) \subset M$ and $x \in M$ and let

$$\lim_{n\to\infty} \|x_n - x\|_{C^1} = 0, \tag{9.9}$$

hence by definition $(x_n, \dot{x}_n) \to (x, \dot{x})$ uniformly on $C[a,b]^m$. Uniform convergence implies according to Theorem 5.3.6 continuous convergence, i.e. for $t_n \to t_0$ it follows that $(x_n(t_n), \dot{x}_n(t_n), t_n) \to (x(t_0), \dot{x}(t_0), t_0)$. From the continuity of L on U it then follows that

$$L(x_n(t_n), \dot{x}_n(t_n), t_n) \to L(x(t_0), \dot{x}(t_0), t_0).$$

Let now $\varphi_n := L(x_n, \dot{x}_n, \cdot) : [a,b] \to \mathbb{R}$ and $\varphi := L(x, \dot{x}, \cdot) : [a,b] \to \mathbb{R}$, then φ_n and φ are continuous on $[a,b]$ and we conclude: (φ_n) converges continuously to φ. Again by Theorem 5.3.6 uniform convergence follows. Therefore $L(x_n, \dot{x}_n, \cdot)$ converges uniformly to $L(x, \dot{x}, \cdot)$, whence $f(x_n) \to f(x)$. □

We obtain the following stability theorems for variational problems:

Theorem 9.8.2. *Let $V \subset \mathbb{R}^{2m}$ and $U := V \times [a,b]$ be open. Let $L : U \to \mathbb{R}$ be continuous. Let the sequence $(\alpha_n, \beta_n) \in \mathbb{R}^{2m}$ of boundary values converge to $(\alpha, \beta) \in \mathbb{R}^{2m}$.*

Let further

$$\begin{aligned} f(x) &:= \int_a^b L(x(t), \dot{x}(t), t)\, dt, \\ S &:= \{x \in M \mid x(a) = \alpha,\ x(b) = \beta\}, \\ S_n &:= \{x \in M \mid x(a) = \alpha_n,\ x(b) = \beta_n\}. \end{aligned}$$

Then every point of accumulation of minimal solutions of the variational problems "minimize f on S_n" is a minimal solution of the variational problem "minimize f on S".

Proof. According to theorems (Stability Theorem 5.3.19 and Theorem 9.8.1) we only have to verify

$$\lim S_n = S.$$

Let for each $n \in \mathbb{N}$

$$c_n := \frac{\alpha - \alpha_n + \beta_n - \beta}{b - a}, \quad d_n := \alpha_n - \alpha - c_n \cdot a,$$

then

$$t \mapsto v_n(t) := t \cdot c_n + d_n$$

converges in $C^{(1)}[a,b]^n$ (w.r.t. the norm $\|\cdot\|_{C^1}$) to 0. Let $x \in S$.

By Lemma 9.3.1 (see also [35], p. 28) we have for sufficiently large $n \in \mathbb{N}$

$$x_n := x + v_n \in M,$$

and hence by construction $x_n \in S_n$. Apparently $(x_n)_{n\in\mathbb{N}}$ converges to x. Therefore $\underline{\lim}_n S_n \supset S$.

On the other hand, uniform convergence also implies pointwise convergence at the end points, whence $\overline{\lim}_{n\to\infty} S_n \subset S$ follows. Since $\overline{\lim}_{n\to\infty} S_n \supset \underline{\lim}_n S_n$ we obtain

$$\lim_{n\to\infty} S_n = S.$$

The assertion follows using the stability Theorem 5.3.19 and Theorem 9.8.1. □

As an example we consider

9.8.1 Parametric Treatment of the Dido problem

We consider the following version of the Dido problem, which in the literature is also referred to as the special isoperimetric problem (see Bolza [12], p. 465). We choose Bolza's wording:

Dido problem. Among all ordinary curves of fixed length, connecting two given points P_1 and P_2, the one is to be determined which – together with the chord P_1P_2 – encloses the largest area.

This general formulation requires a more specific interpretation. In a similar manner as Bolza we put the points P_1, P_2 on the x-axis, symmetrical to zero and maximize the integral (Leibniz's sector formula, s. Bohn, Goursat, Jordan and Heuser)

$$f(x,y) = \frac{1}{2}\int_a^b (x\dot{y} - y\dot{x})dt \tag{9.10}$$

under the restriction

$$G(x,y) := \int_a^b \sqrt{\dot{x}^2 + \dot{y}^2}dt = \ell, \quad x(a) = -\alpha,\ x(b) = \alpha,\ y(a) = y(b) = 0,$$

where $\alpha, \ell \in \mathbb{R}_{>0}$ ($2\alpha < \ell$) are given. Along the chord P_1P_2 the integrand is equal to 0 ($y = \dot{y} = 0$), since $[P_1, P_2]$ lies on the x-axis. Hence also for the closed curve, generated through the completion of (x, y) by the interval $[P_1, P_2]$ we have

$$f(x,y) = \int_a^b (x\dot{y} - y\dot{x})\,dt + \int_{P_1}^{P_2} (x\dot{y} - y\dot{x})\,dt = \int_a^b (x\dot{y} - y\dot{x})\,dt.$$

Discussion of Leibniz's Formula. For arbitrary "ordinary" curves it is debatable, what the meaning of Formula (9.10) is. For all $(x,y) \in C^{(1)}[a,b]^2$ (resp. $(x,y) \in \mathrm{RCS}^{(1)}[a,b]^2$) the integral is defined (and for counterclockwise oriented boundary curves of normal regions in fact the value is the enclosed area) but we avoid this discussion and maximize f on the whole restriction set

$$S = \{(x,y) \in C^{(1)}[a,b]^n \mid x(a) = P_1, x(b) = P_2, G(x,y) = \ell\}.$$

As a first step we show that Formula (9.10) is translation invariant for closed curves.

Lemma 9.8.3 (Translation invariance). *Let piecewise continuously differentiable functions $(x,y) \in (\mathrm{RCS}^{(1)}[t_a,t_b])^2$ be given with $x(t_b) - x(t_a) = 0$ and $y(t_b) - y(t_a) = 0$.*

Let the translation vector $(r_1, r_2) \in \mathbb{R}^2$ *be given and* $(u, v) = (x + r_1, y + r_2)$ *the closed curve* (x, y) *shifted by* (r_1, r_2). *Then*

$$\int_{t_a}^{t_b} (x\dot{y} - y\dot{x})dt = \int_{t_a}^{t_b} (u\dot{v} - v\dot{u})dt.$$

Proof. Apparently $\dot{u} = \dot{x}, \dot{v} = \dot{y}$ and hence

$$\int_{t_a}^{t_b} (u\dot{v} + v\dot{u})dt = \int_{t_a}^{t_b} (x\dot{y} - y\dot{x})dt + r_1 \int_{t_a}^{t_b} \dot{y}dt - r_2 \int_{t_a}^{t_b} \dot{x}dt.$$

The curve is closed, i.e. $x(t_a) = x(t_b), y(t_a) = y(t_b)$, whence

$$\int_{t_a}^{t_b} \dot{y}dt = y(t_b) - y(t_a) = 0 \quad \text{and} \quad \int_{t_a}^{t_b} \dot{x}dt = x(t_b) - x(t_a) = 0,$$

and hence the assertion. □

Leibniz's sector formula is meant for closed curves. Therefore we have extended the curves under consideration (which will be traversed in counterclockwise direction) by the interval $(-\alpha, \alpha)$. The curve integral in Formula (9.10) is homogeneous w.r.t. $(\dot{x}, \dot{y})$ and hence independent of the parameterization.

For our further treatment we choose the parameterization w.r.t. the arc length. Since the total length of the closed curve is given by $\ell + 2\alpha$, we have to consider piecewise continuously differentiable functions $(x, y) \in (\text{RCS}^{(1)}[0, \ell + 2\alpha])^2$ with $\dot{x}^2 + \dot{y}^2 = 1$ on the interval $[0, \ell + 2\alpha]$.

On the subinterval $[\ell, \ell + 2\alpha]$ the desired curve runs along the x-axis and has the parameterization $y = 0$ and $t \mapsto x(t) := t - \ell - \alpha$.

Apparently

$$\int_0^{\ell+2\alpha} (x\dot{y} - y\dot{x})dt = \int_0^{\ell} (x\dot{y} - y\dot{x})dt$$

holds, since the integral vanishes on $[\ell, \ell + 2\alpha]$.

If the restriction set is shifted parallel to the x-axis by $\beta \in \mathbb{R}$, then due to Lemma 9.8.3 for $u := x$ and $v := y + \beta$

$$\int_0^{\ell+2\alpha} (u\dot{v} - v\dot{u})dt = \int_0^{\ell+2\alpha} (x\dot{y} - y\dot{x})dt.$$

On $[\ell, \ell + 2\alpha]$ we have $v = \beta$ and hence

$$\begin{aligned} \int_{\ell}^{\ell+2\alpha} (u\dot{v} - v\dot{u})dt &= -\beta \int_{\ell}^{\ell+2\alpha} \dot{u}dt \\ &= -\beta(u(\ell + 2\alpha) - u(\ell)) = -\beta(\alpha + \alpha) = -2\alpha\beta. \end{aligned}$$

Therefore the integrals

$$\int_0^{\ell+2\alpha} u\dot{v} - v\dot{u}dt$$

and

$$\int_0^{\ell} u\dot{v} - v\dot{u}dt$$

differ for all (u, v) only by the constant (independent of u and v) $2\alpha\beta$.

Therefore we obtain: if a minimal solution of f on the unshifted restriction set is shifted by $(0, \beta)$ then it is a minimal solution of f on the shifted restriction set S.

9.8.2 Dido problem

We will now deal with the Dido problem using a parametric supplement. We are looking for a curve, which – together with the shore interval $[P_1, P_2]$ – forms a closed curve and encloses a maximal area. Let ℓ be the length of the curve.

We want to determine a smooth (resp. piecewise smooth) parametric curve

$$(x, y) : [0, \ell] \to \mathbb{R}^2$$

with preassigned start $(\alpha, 0)$ and end point $(-\alpha, 0)$, which maximizes

$$f(x, y) := \int_0^{\ell} (x\dot{y} - \dot{x}y)dt, \tag{9.11}$$

where the parameterization is done over the length of the curve, i.e. for all $t \in [0, \ell]$ we have

$$\dot{x}^2(t) + \dot{y}^2(t) = 1. \tag{9.12}$$

Let $0 < \alpha < \frac{\ell}{2}$. The restriction set is given by

$$S := \{(x, y) \in C^{(1)}[0, \ell]^2 \mid \dot{x}^2 + \dot{y}^2 = 1, x(0) = \alpha, x(\ell) = -\alpha, y(0) = y(\ell) = 0\}.$$

We now apply the approach of pointwise minimization (see Theorem 9.1.7) to the supplemented functional $f + \Lambda$, with the supplement

$$\Lambda(x, y) = \int_0^{\ell} (r(\dot{x}^2 + \dot{y}^2) - \lambda\dot{x} - \eta\dot{y} - \dot{\lambda}x - \dot{\eta}y)\, dt, \tag{9.13}$$

for λ, η and $r \in C^{(1)}[0, \ell]$.

As soon as λ, η and r are specified, the supplement Λ is constant on S.

For the corresponding Euler–Lagrange equation we obtain

$$\frac{d}{dt}(2r\dot{x} - y) = \dot{y} \tag{9.14}$$

$$\frac{d}{dt}(2r\dot{y} + x) = -\dot{x}. \tag{9.15}$$

Together with Equation (9.12) we obtain for $r = \frac{\ell}{2\pi}$ and the boundary condition $\alpha = 0$ the parameterized circle as an admissible extremal

$$x(t) = -r\sin\left(\frac{t}{r}\right) \quad y(t) = r\left(1 - \cos\left(\frac{t}{r}\right)\right). \tag{9.16}$$

But the function $f + \Lambda$ is not convex. We shall try to treat this variational problem by use of successive minimization for a quadratic supplement. The minimization will essentially be reduced to the determination of the deepest point of a parabola. For a suitable supplement we choose a function $\mu \in C^{(1)}(0, \ell]$ and a $r \in \mathbb{R}_{\geq 0}$, leading to a quadratically supplemented functional $\tilde{f}$

$$\tilde{f}(x) = \int_0^\ell \dot{y}x - \dot{x}y + r(\dot{x}^2 + \dot{y}^2) + \mu(t)(x\dot{x} + y\dot{y}) + \frac{1}{2}\dot{\mu}(t)(x^2 + y^2)dt \tag{9.17}$$

(see Section 9.3). For the supplemented Lagrangian $\tilde{L}$ we then obtain

$$\tilde{L}(p, q, t) = p_1q_2 - p_2q_1 + r(q_1^2 + q_2^2) + \mu(t)(p_1q_1 + p_2q_2) + \frac{1}{2}\dot{\mu}(t)(p_1^2 + p_2^2). \tag{9.18}$$

Minimization w.r.t. $q = (q_1, q_2)$ for fixed p (successive minimization) of the convex quadratic function

$$q \mapsto r(q_1^2 + q_2^2) + (\mu(t)p_1 - p_2)q_1 + (p_1 + \mu(t)p_2)q_2, \tag{9.19}$$

setting the partial derivatives w.r.t. q_1 and q_2 to zero, leads to the equations

$$\mu(t)p_1 - p_2 + 2rq_1 = 0 \quad \text{i.e. } q_1 = \frac{p_2 - \mu(t)p_1}{2r}. \tag{9.20}$$

and

$$p_1 + 2rq_2 + \mu(t)p_2 = 0 \quad \text{i.e. } q_2 = -\frac{p_1 + \mu(t)p_2}{2r}. \tag{9.21}$$

Insertion into Equation (9.18) yields

$$\begin{aligned}\Psi(p) :&= \frac{-(\mu(t)p_1 - p_2)^2 - (p_1 + \mu(t)p_2)^2}{4r} + \frac{1}{2}\dot{\mu}(t)(p_1^2 + p_2^2) \\ &= -\frac{(\mu^2(t) + 1)(p_1^2 + p_2^2)}{4r} + \frac{1}{2}\dot{\mu}(t)(p_1^2 + p_2^2) \\ &= \frac{1}{2}\left(\dot{\mu}(t) - \frac{(\mu^2(t) + 1)}{2r}\right)(p_1^2 + p_2^2).\end{aligned}$$

If now μ is chosen such that the following ODE

$$\dot{\mu} = \frac{\mu^2 + 1}{2r} \tag{9.22}$$

is satisfied on all of $(0, \ell)$, then Ψ is constant equal to zero and every $p \in \mathbb{R}^2$ is a minimal solution of Ψ. The Differential Equation (9.22) (ODE with separated variables) has a general solution of the form

$$\mu(t) = \tan\left(\frac{t - C}{2r}\right),$$

where C is a constant.

We can try to choose $C \in \mathbb{R}, r \in \mathbb{R}_{\geq 0}$ in a suitable way and put $C := \frac{\ell}{2}$. However, if we choose $r := \frac{\ell}{2\pi}$, then

$$\mu(t) = \tan\left(\frac{1}{2}\left(\frac{t}{r} - \pi\right)\right) \tag{9.23}$$

is a solution of Equation (9.22) on the open interval $(0, \ell)$ but μ has a singularity at $t = \ell$, i.e. for the original problem the convexification fails (at least in this manner).

We now consider the situation for $r > \frac{\ell}{2\pi}$. Using the formulae

$$s \mapsto \tan\left(\frac{s}{2}\right) = \frac{\sin s}{1 + \cos s} = \frac{1 - \cos s}{\sin s},$$

and $x(t) = p_1, y(t) = p_2, \dot{x} = q_1$ and $\dot{y} = q_2$ we obtain with Equations (9.20) and (9.21) by direct insertion the following candidate for an optimal solution:

$$x^*(t) = r \sin\left(\frac{t - C}{r}\right) \quad \text{and} \quad y^*(t) = r\left(1 + \cos\left(\frac{t - C}{r}\right)\right).$$

If we succeed in choosing the constants C and r in such a way that (x^*, y^*) is inside the restriction set and μ is defined on the whole interval $[0, \ell]$, then we have found the optimal solution.

We want to make the domain of definition for μ as large as possible, and in order to achieve this we choose its center for $C := \frac{\ell}{2}$. For the boundary condition w.r.t. x we then obtain the equations

$$r \sin\left(-\frac{\ell}{2r}\right) = -\alpha \quad \text{and} \quad r \sin\left(\frac{\ell}{2r}\right) = \alpha.$$

For $0 < \alpha < \frac{\ell}{2}$, there is always a unique $r(\alpha) > \frac{\ell}{2\pi}$, to satisfy these conditions, since for $r = \frac{\ell}{2\pi}$ we have $r \sin\left(\frac{\ell}{2r}\right) = 0$ and for $r > \frac{\ell}{2\pi}$ the mapping

$$r \mapsto r \sin\left(\frac{\ell}{2r}\right) =: \phi(r)$$

is strictly increasing on $[\frac{\ell}{2\pi}, \infty)$. In order to compute the limit for $r \to \infty$, we introduce the new variable $R := \frac{1}{r} > 0$. Using the rule of de L'Hospital we obtain

$$\lim_{R \to 0} \frac{\sin(\frac{R\ell}{2})}{R} = \lim_{R \to 0} \frac{\frac{\ell}{2}\cos(\frac{R\ell}{2})}{1} = \frac{\ell}{2}.$$

Apparently $\lim_{\alpha\to 0} r(\alpha) = \frac{\ell}{2\pi}$, due to the strict monotonicity and continuity of ϕ.

Let $\beta(\alpha) := y^*(0) = r(\alpha)(1+\cos(\frac{-\frac{\ell}{2}}{r(\alpha)}))$.

According to the principle of pointwise minimization (x^*, y^*) is a minimal solution of the variational functional f on the restriction set, shifted by $(0, \beta(\alpha))$

$$\tilde{S} = \{(x,y)\in C^{(1)}[0,\ell]^2 \mid \dot{x}^2 + \dot{y}^2 = 1, x(0) = \alpha, x(\ell) = -\alpha, y(0) = y(\ell) = \beta(\alpha)\}.$$

Since shifting of the restriction set leads to optimal solutions (see introductory remark to Leibniz's formula) $(x^*, y^*) - (0, \beta(\alpha))$ is a solution of the problem for $\alpha > 0$. Furthermore

$$\beta(\alpha) \xrightarrow{\alpha\to 0} \frac{\ell}{2\pi}(1+\cos(\pi)) = 0$$

holds. Altogether we observe that the solutions for $\alpha > 0$ converge for $\alpha \to 0$ to the Solution (9.16) in $C^1[0,\ell]^2$: in order to see this, let (α_n) be a sequence tending to zero with $\alpha_n \in (0,\ell)$ for all $n \in \mathbb{N}$ and let (x_n, y_n) be solutions for α_n, then using the notation $r_n := r(\alpha_n)$ we have

$$x_n(t) = r_n \sin\left(\frac{t-\ell/2}{r_n}\right) \quad \text{and} \quad y_n(t) = r_n\left(1+\cos\left(\frac{t-\ell/2}{r_n}\right)\right) - \beta(\alpha_n).$$

$$\begin{aligned}
|x_n(t) - x(t)| &= \left|\left(\frac{\ell}{2\pi} - r_n\right)\sin\left(\frac{t}{r_n}\right) - \frac{\ell}{2\pi}\left(\sin\frac{t}{r_n} - \sin\frac{t}{\ell/2\pi}\right)\right. \\
&\quad \left. + r_n\left(\cos\frac{\ell}{2r_n} + 1\right)\sin\left(\frac{t}{r_n}\right) - r_n \sin\frac{\ell}{2r_n}\cos\frac{t}{r_n}\right| \\
&\le \left|\left(\frac{\ell}{2\pi} - r_n\right)\right| + \frac{\ell}{2\pi}\left|\sin\frac{t}{r_n} - \sin\frac{t}{\ell/2\pi}\right| \\
&\quad + r_n\left|\cos\frac{\ell}{2r_n} + 1\right| + r_n\left|\sin\frac{\ell}{2r_n}\right|.
\end{aligned}$$

Since $|\sin\frac{t}{r_n} - \sin\frac{t}{\ell/2\pi}| \le \ell|\frac{1}{r_n} - \frac{1}{\ell/2\pi}|$ it follows that

$$\max_{t\in[0,\ell]} |x_n(t) - x(t)| \xrightarrow{n\to\infty} 0.$$

Correspondingly we obtain

$$\max_{t\in[0,\ell]} |\dot{x}_n(t) - \dot{x}(t)| \le \ell\left|\frac{1}{r_n} - \frac{1}{\ell/2\pi}\right| + \left|\cos\frac{\ell}{2r_n} + 1\right| + \left|\sin\frac{\ell}{2r_n}\right| \to 0,$$

and in a similar manner

$$\begin{aligned}
|y_n(t) - y(t)| &\le 2\left|\frac{\ell}{2\pi} - r_n\right| + r_n\left|\sin\frac{\ell}{2r_n}\right| + \frac{\ell^2}{2\pi}\left|\frac{1}{r_n} - \frac{1}{\ell/2\pi}\right| \\
&\quad + \left|r_n\cos\frac{\ell}{2r_n} + 1\right| + \beta(\alpha_n) \\
&\to 0,
\end{aligned}$$

as well as

$$\max_{t\in[0,\ell]} |\dot{y}_n(t) - \dot{y}(t)| \le \left|\cos\frac{\ell}{2r_n} + 1\right| + \left|\sin\frac{\ell}{2r_n}\right| \ell \left|\frac{1}{r_n} - \frac{1}{\ell/2\pi}\right| \to 0.$$

According to stability Theorem 9.8.2 it follows that (9.16) is a solution of the Dido problem.

Remark. The above stated solution (x, y) of the Dido problem is, because of the negative orientation of the curve a minimal solution (with negative 'area'). The boundary conditions of the Dido problem do not determine a unique solution (contrary to the approximating problems for $\alpha > 0$, the solutions of which also have a negative orientation), since fixing a point on the circle does not determine its position. Interchanging e.g. x and y also leads to a solution of the Dido problem (rotated circle by 90°), since the 'area' is the same.

If one requires in addition $y \ge 0$, then the opposite circle with parametric representation $x(t) = r\sin(\frac{t}{r})$ and $y(t) = r(1-\cos(\frac{t}{r}))$ is a (maximal)-solution, if we choose the supplement

$$\Lambda(x, y) = \int_0^\ell -r(\dot{x}^2 + \dot{y}^2) - \lambda\dot{x} - \eta\dot{y} - \dot{\lambda}x - \dot{\eta}y\, dt. \tag{9.24}$$

For the corresponding Euler–Lagrange equations we then obtain

$$\frac{d}{dt}(2r\dot{x} + y) = -\dot{y} \tag{9.25}$$

$$\frac{d}{dt}(2r\dot{y} - x) = \dot{x}. \tag{9.26}$$

9.8.3 Global Optimal Paths

For many variational problems the end points of the curve to be determined play a particular role. Here we often encounter singularities. Let for $V \subset \mathbb{R}^{2m}$ open and $\alpha, \beta \in \mathbb{R}^m$ the restriction set be defined as

$$S := \{x \in C^{(1)}[a, b]^n \mid x(a) = \alpha, x(b) = \beta, (x(t), \dot{x}(t)) \in V \text{ for all } t \in [a, b]\}.$$

On each subinterval of the interval (a, b) the curves we are looking for are mostly well-behaved.

In order to avoid difficulties (singularities) in the end points, we will now introduce the following notion of optimality.

Definition 9.8.4. A $x^* \in S$ is called a *global optimal path*, if for each $\tau \in [0, \frac{b-a}{2})$ x^* is a solution of the following variational problems

$$\text{Minimize} \int_{a+\tau}^{b-\tau} L(x(t), \dot{x}(t), t)dt$$

on the restriction set S_τ, defined by

$$S_\tau := \{x \in C^{(1)}[a,b]^n \mid x(a+\tau) = x^*(a+\tau), x(b-\tau) = x^*(b-\tau),\\ (x(t), \dot{x}(t)) \in V \; \forall t \in [a+\tau, b-\tau]\}.$$

Then the following stability theorem holds:

Theorem 9.8.5. *Let $L : V \times [a,b]$ be continuous. Then every global optimal path is a minimal solution of the variational problem*

$$\text{Minimize } f(x) = \int_a^b L(x(t), \dot{x}(t), t)dt \text{ on } S.$$

Proof. Let $(\tau_n)_{n\in\mathbb{N}}$ be a sequence tending to zero in $(0, \frac{b-a}{2})$ and $x \in S$.

Let $a_n = x^*(a+\tau_n)$, $b_n = x^*(b-\tau_n)$, $u_n(t) = (t-a)\frac{b_n-\beta-v_n(b)}{b-a}$ and $v_n(t) = (t-b+\tau_n)\frac{a_n-a}{a-b+\tau_n}$.

Since $x(a) = x^*(a)$ and $x(b) = x^*(b)$ the function sequences

$$(u_n)_{n\in\mathbb{N}}, (v_n)_{n\in\mathbb{N}}, (\dot{u}_n)_{n\in\mathbb{N}} \text{ and } (\dot{v}_n)_{n\in\mathbb{N}} \text{ converge uniformly to 0.}$$

For an $n_0 \in \mathbb{N}$ we have by Lemma 9.3.1

$$x_n := x + u_n + v_n \in S_{\tau_n}.$$

This means $\underline{\lim}\, S_{\tau_n} \supset S$. By Theorem 5.3.20 the assertion follows. □

9.8.4 General Stability Theorems

So far we have in our stability considerations only varied the restriction set. We will now consider varying the variational functional. We obtain the first general stability assertion in this context for monotone convergence (see Theorem 5.2.3).

Theorem 9.8.6. *Let $V \subset \mathbb{R}^m \times \mathbb{R}^m$, $L : V \times [a,b] \to \mathbb{R}$ continuous and for all $n \in \mathbb{N}$ $L_n : V \times [a,b] \times \mathbb{R}$ continuous, where $(L_n)_{n\in\mathbb{N}}$ converges pointwise monotonically to L. Let*

$$W := \{x \in C^{(1)}[a,b]^m \mid (x(t), \dot{x}(t)) \in V \; \forall t \in [a,b]\},$$

$$f : W \to \mathbb{R},\; x \mapsto f(x) := \int_a^b L(x(t), \dot{x}(t), t)\, dt,$$

and

$$f_n : W \to \mathbb{R},\; x \mapsto f_n(x) := \int_a^b L_n(x(t), \dot{x}(t), t)\, dt$$

be defined. For a set $S \subset W$ and the sequence $(S_n \subset W)_{n\in\mathbb{N}}$ let $\underline{\lim}_{n\to\infty} S_n \supset S$ be satisfied.

Then each point of accumulation of the sequence of minimal solutions of the variational problems "minimize f_n on S_n", which is in S, is a minimal solution of the variational problem "minimize f on S".

Proof. Let $x \in W$. The sequence of continuous functions $L_n(x(\cdot), \dot{x}(\cdot), \cdot)_{n\in\mathbb{N}}$ converges pointwise monotonically on $[a, b]$ to the continuous function $L(x(\cdot), \dot{x}(\cdot), \cdot)$. The corresponding functionals are by Theorem 9.8.1 also continuous. According to the theorem of Dini for real-valued functions on $[a, b]$ the convergence is uniform. Therefore $f_n(x) \to f(x)$. The convergence of the values of the functionals is also monotone. By the theorem of Dini in metric spaces (see Theorem 5.2.2) this convergence is continuous. The Stability Theorem 5.3.20 then implies the assertion. □

We also have the

Theorem 9.8.7. *Let $U \subset \mathbb{R}^{2m+1}$, and let the sequence of continuous functions $(L_n : U \to \mathbb{R})_{n\in\mathbb{N}}$ converge continuously to the continuous function $L : U \to \mathbb{R}$. Let*

$$R := \{x \in C^{(1)}[a, b]^m \mid (x(t), \dot{x}(t), t) \in U \text{ for all } t \in [a, b]\}.$$

Then the sequence of functionals

$$f_n(x) = \int_a^b L_n(x(t), \dot{x}(t), t)\, dt$$

converges continuously to

$$f(x) = \int_a^b L(x(t), \dot{x}(t), t)\, dt$$

on R.

Let $S \subset R$ and $(S_n)_{n\in\mathbb{N}}$ be a sequence of subsets of the restriction set of R with

$$\underline{\lim}_{n} S_n \supset S. \tag{9.27}$$

Let for each $n \in \mathbb{N}$ x_n be a minimal solution of f_n on S_n.

Then each point of accumulation of the sequence $(x_n)_{n\in\mathbb{N}}$, lying in S, is a minimal solution of f on S.

Proof. Let $x_n \to x$ be convergent w.r.t. $\|\cdot\|_{C^1}$.

We show that the sequence $\varphi_n := L_n(x_n(\cdot), \dot{x}_n(\cdot), \cdot) : [a, b] \to \mathbb{R}$ converges uniformly to $\varphi := L(x(\cdot), \dot{x}(\cdot), \cdot) : [a, b] \to \mathbb{R}$. As a composition of continuous functions φ is continuous. It suffices to verify continuous convergence of $(\varphi_n)_{n\in\mathbb{N}}$, since this implies uniform convergence.

Let $t_n \in [a,b]$ be a to $t \in [a,b]$ convergent sequence, then

$$(x_n(t_n), \dot{x}_n(t_n), t_n) \to (x(t), \dot{x}(t), t).$$

Since the functions $L_n : U \to \mathbb{R}$ converge continuously to $L : U \to \mathbb{R}$, we have

$$L_n(x_n(t_n), \dot{x}_n(t_n), t_n) \xrightarrow{n\to\infty} L(x(t), \dot{x}(t), t).$$

Uniform convergence of $(\varphi_n)_{n\in\mathbb{N}}$ to φ implies continuous convergence of the sequence of functionals $(f_n)_{n\in\mathbb{N}}$ to f.

The remaining part of the assertion follows from Stability Theorem 5.3.20. □

Remark 9.8.8. Let R be open and convex. If for every $t \in [a,b]$ the subsets $U_t := \{(p,q) \in \mathbb{R}^{2n} \mid (p,q,t) \in U\}$ are convex, $L : U_t \to \mathbb{R}$ is convex and the sequence of functions $(L_n(\cdot,\cdot,t) : U_t \to \mathbb{R})_{n\in\mathbb{N}}$ is convex, then continuous convergence already follows from pointwise convergence of the functionals

$$f_n(x) := \int_a^b L_n(x(t), \dot{x}(t), t)\, dt$$

to

$$f(x) := \int_a^b L(x(t), \dot{x}(t), t)\, dt$$

(see Remark 5.3.16).

More generally:

Remark 9.8.9. A pointwise convergent sequence $(L_n : U \to \mathbb{R})_{n\in\mathbb{N}}$ converges continuously to $L : U \to \mathbb{R}$, if U is open and convex and for every $n \in \mathbb{N}$ the Lagrangian L_n is componentwise convex. The convexity in the respective component can be replaced by requiring concavity (see section on convex operators in the chapter on stability).

By Theorem 9.8.7 and Remark 9.8.9 we obtain

Theorem 9.8.10. *Let $U \subset \mathbb{R}^{2n+1}$ open and convex. Let the sequence of continuous functions $(L_n : U \to \mathbb{R})_{n\in\mathbb{N}}$ be componentwise convex and pointwise convergent to $L : U \to \mathbb{R}$.*

Let

$$R := \{x \in C^{(1)}[a,b]^m \mid (x(t), \dot{x}(t), t) \in U \textit{ for all } t \in [a,b]\}.$$

The sequence of functionals

$$f_n(x) = \int_a^b L_n(x(t), \dot{x}(t), t)\, dt$$

converges continuously to

$$f(x) = \int_a^b L(x(t), \dot{x}(t), t)\, dt$$

on R.

Let $S \subset R$ *and* $(S_n)_{n\in\mathbb{N}}$ *be a sequence of subsets of the restriction set* R *with*

$$\underline{\lim}_n S_n \supset S.$$

Let for every $n \in \mathbb{N}$ x_n *be a minimal solution of* f_n *on* S_n.

Then each point of accumulation of the sequence $(x_n)_{n\in\mathbb{N}}$, *which is in* S, *is a minimal solution of* f *on* S.

As an example for the above stability theorem we consider

9.8.5 Dido problem with Two-dimensional Quadratic Supplement

We treat the problem: minimize the functional (which represents the 2-fold of the negative Leibniz formula)

$$f(x) = \int_0^{2\pi} -x_1\dot{x}_2 + \dot{x}_1 x_2 dt \tag{9.28}$$

on

$$S = \{x = (x_1, x_2) \in C^{(1)}[0, 2\pi]^2 \mid \|\dot{x}\| = 1, x_1(0) = x_1(2\pi) = 1, \\ x_2(0) = x_2(2\pi) = 0\}.$$

As supplement we choose a sum of a parametric and a (two-dimensional) linear supplement, i.e. for a positive $\rho \in C^{(1)}[0, 2\pi]$ and $\lambda \in C^{(1)}[0, 2\pi]^2$ let

$$\Lambda(x) = \int_0^{2\pi} \left(\frac{1}{2}\rho\,\|\dot{x}\|^2 - \lambda^T\dot{x} - x^T\dot{\lambda}\right) dt, \tag{9.29}$$

which is equal to the constant $2\pi \int_0^{2\pi} \rho(t)dt - \lambda_1(2\pi) + \lambda_1(0)$ on S.

According to the approach of pointwise minimization we have to minimize the function below for fixed $t \in [0, 2\pi]$

$$(p, q) \mapsto \varphi^t(p, q) := -p_1q_2 + p_2q_1 + \frac{1}{2}\rho(t)(q_1^2 + q_2^2) - \lambda^T(t)q - p^T\dot{\lambda}(t) \tag{9.30}$$

on $\mathbb{R}^4$. A necessary condition is that the partial derivatives vanish. This results in the equations

$$-q_2 - \dot{\lambda}_1(t) = 0; \qquad q_1 - \dot{\lambda}_2(t) = 0 \tag{9.31}$$

$$p_2 + \rho(t)q_1 - \lambda_1(t) = 0; \qquad -p_1 + \rho(t)q_2 - \lambda_2(t) = 0. \tag{9.32}$$

Just as Queen Dido, we expect the unit circle as the solution with the parameterization

$$x_1^*(t) = \cos(t) \quad \text{and} \quad x_2^*(t) = \sin(t)$$

and we obtain the following equations for specification of ρ and λ

$$\dot{\lambda}_1(t) = -\cos(t), \qquad \dot{\lambda}_2(t) = -\sin(t) \tag{9.33}$$

$$\sin(t) - \rho(t)\sin(t) = \lambda_1(t), \qquad -\cos(t) + \rho(t)\cos(t) = \lambda_2(t). \tag{9.34}$$

For $\rho \equiv 2$ and $\lambda(t) = (-\sin(t), \cos(t))$ these equations are satisfied. But the function $f + \Lambda$ is not convex and the Conditions (9.31) and (9.32) do not constitute sufficient optimality conditions.

In order to obtain those, we will now employ the approach of convexification by a quadratic supplement.

For $C = \begin{pmatrix} 0 & -1 \\ 1 & 0 \end{pmatrix}$ and a $\rho \in \mathbb{R}_{>0}$ we will now (for fixed $t \in [0, 2\pi]$) try to convexify the function

$$\phi^t(p, q) = p^T Cq + \frac{1}{2}\rho\|q\|^2$$

using a quadratic supplement.

In other words, we attempt to find a continuously differentiable matrix function

$$Q : [0, 2\pi] \to \mathbb{R}^2 \times \mathbb{R}^2$$

such that for all $t \in [0, 2\pi]$ the function

$$\mathbb{R}^4 \ni (p, q) \mapsto \psi(p, q) = -p_1 q_2 + q_1 p_2 + \frac{1}{2}\rho\,\|q\|^2 + q^T Qp + \frac{1}{2} p^T \dot{Q} p,$$

is convex, where $p = (p_1, p_2)$ and $q = (q_1, q_2)$.

The quadratic supplement function $E : \mathbb{R}^2 \times \mathbb{R}^2 \times [0, 2\pi] \to \mathbb{R}$ we choose to be of the form (compare Section 9.3)

$$(p, q, t) \mapsto E(p, q, t) = q^T Q(t)p + \frac{1}{2} p^T \dot{Q}(t)p,$$

where $Q = \mu I$ with a $\mu \in C^{(1)}[0, 2\pi]$ and I is the 2×2-unit matrix.

Then $\dot{Q} = \dot{\mu} I$ and for all $t \in [0, 2\pi]$

$$\phi^t(p, q) + E(p, q, t) = p^T Cq + \frac{1}{2}\rho q^T q + q^T Q(t)p + \frac{1}{2} p^T \dot{Q}(t)p.$$

This function is convex on $\mathbb{R}^2 \times \mathbb{R}^2$, if and only if for all $(p, q) \in \mathbb{R}^4$ the Hessian matrix

$$H^t(p,q) = \begin{pmatrix} \dot{Q}(t) & C+Q \\ C^T+Q & \rho I \end{pmatrix}$$

is positive semi-definite.

According to Lemma 9.3.2 this is the case, if and only if

$$M_t := \dot{Q}(t) - \frac{1}{\rho}(C+Q(t))^T(C+Q(t))$$

is positive semi-definite. We have

$$\begin{aligned} M_t &= \begin{pmatrix} \dot{\mu}(t) & 0 \\ 0 & \dot{\mu}(t) \end{pmatrix} - \frac{1}{\rho}\begin{pmatrix} \mu(t) & 1 \\ -1 & \mu(t) \end{pmatrix}\begin{pmatrix} \mu(t) & -1 \\ 1 & \mu(t) \end{pmatrix} \\ &= \begin{pmatrix} \dot{\mu}(t) & 0 \\ 0 & \dot{\mu}(t) \end{pmatrix} - \frac{1}{\rho}\begin{pmatrix} 1+\mu^2(t) & 0 \\ 0 & 1+\mu^2(t) \end{pmatrix} \\ &= \begin{pmatrix} \dot{\mu}(t) - \frac{1}{\rho}(1+\mu^2(t)) & 0 \\ 0 & \dot{\mu}(t) - \frac{1}{\rho}(1+\mu^2(t)) \end{pmatrix}. \end{aligned}$$

This matrix is positive semi-definite, if

$$\dot{\mu}(t) - \frac{1}{\rho}(1+\mu^2(t)) \geq 0$$

holds.

The differential equation of Riccati-type

$$\dot{\mu} = \frac{1}{\rho}(1+\mu^2)$$

has for $\rho > 2$ on all of $[0, 2\pi]$ the solution $\mu_\rho(t) = \tan(\frac{t-\pi}{\rho})$.

But for $\rho = 2$ we observe that $\mu_2(t) = \tan(\frac{t-\pi}{2})$ is only defined on the open interval $(0, 2\pi)$.

For a complete solution of the isoperimetric problem we need the following stability consideration:

Let $(\rho_n)_{n\in\mathbb{N}}$ be a sequence in $(2, \infty)$, converging to 2. Based on the above considerations we will establish that for every $n \in \mathbb{N}$ the variational problem

$$\text{Minimize } f_n(x) = \int_0^{2\pi} (-x_1\dot{x}_2 + x_2\dot{x}_1 + \frac{1}{2}\rho_n(\dot{x}_1^2 + \dot{x}_2^2))dt \qquad (P_n)$$

on

$$S_n = \left\{ x \in C[0,2\pi] \cap C^{(1)}(0,2\pi) \,\middle|\, \|\dot{x}\| = 1,\ x(0) = (1,0), \right.$$
$$\left. x(2\pi) = \left(\cos\left(\frac{4\pi}{\rho_n}\right), \sin\left(\frac{4\pi}{\rho_n}\right)\right) \right\}$$

has a minimal solution

$$t \mapsto x_n(t) = \left(\cos\left(\frac{2t}{\rho_n}\right), \sin\left(\frac{2t}{\rho_n}\right)\right).$$

At first we observe that x_n is an extremal of (P_n), since for the Euler–Lagrange equations we have

$$\rho_n \ddot{x}_1 + \dot{x}_2 = -\dot{x}_2 \tag{9.35}$$

$$\rho_n \ddot{x}_2 - \dot{x}_1 = \dot{x}_1. \tag{9.36}$$

The extremal is a solution, because the problem (P_n) has an equivalent convex variational problem w.r.t. the supplement potential

$$F_n(t,p) = p^T Q_n p,$$

where

$$Q_n(t) = \begin{pmatrix} \mu_n(t) & 0 \\ 0 & \mu_n(t) \end{pmatrix},$$

and $\mu_n(t) = \tan(\frac{t-\pi}{\rho_n})$.

Moreover, for each $x \in S$ the function

$$t \mapsto y_n(t) = x(t) + \frac{t}{2\pi}\left(-1 + \cos\left(\frac{4\pi}{\rho_n}\right), \sin\left(\frac{4\pi}{\rho_n}\right)\right)$$

is an element of S_n.

Apparently $\lim_{n\to\infty} y_n = x$ in $C^1[0,2\pi]^2$ and hence $\underline{\lim}_n S_n \supset S$. Let

$$f_0(x) := \int_0^{2\pi} (-x_1\dot{x}_2 + x_2\dot{x}_1 + \dot{x}_1^2 + \dot{x}_2^2)dt.$$

Apparently f and f_0 only differ by a constant on S. The functions f_n and f_0 are by Theorem 9.8.1 continuous. The corresponding Lagrangians are componentwise convex and pointwise convergent. The stability Theorem 9.8.10 yields that $x(t)$ is solution of the original problem, since the sequence (x_n) converges in $C^1[0,2\pi]^2$ to x.

9.8.6 Stability in Orlicz–Sobolev Spaces

Definition 9.8.11. Let $T \subset \mathbb{R}^r$ be an open set and μ the Lebesgue measure and let $\mu(T) < \infty$, let Φ be a Young function. Then the *Orlicz–Sobolev space* $W^m L^\Phi(\mu)$ consists of all functions u in $L^\Phi(\mu)$, whose weak derivatives (in the distributional sense) $D^\alpha u$ also belong to $L^\Phi(\mu)$. The corresponding norm is given by

$$\|u\|_{m,\Phi} := \max_{0\le|\alpha|\le m} \|D^\alpha u\|_{(\Phi)}.$$

W.r.t. this norm $W^m L^\Phi(\mu)$ is a Banach space (see [2]).

Theorem 9.8.12. *Let $T \subset \mathbb{R}^r$ be an open set and μ the Lebesgue measure and let $\mu(T) < \infty$, let Φ be a Young function, satisfying the Δ_2^∞-condition. Let J be set of all mappings*

$$L : T \times \mathbb{R} \times \mathbb{R}^n \to \mathbb{R}_{\geq 0}$$

satisfying the following properties:

(a) *for all $(w, z) \in \mathbb{R} \times \mathbb{R}^n$ the mapping $t \mapsto L(t, w, z) : T \to \mathbb{R}_{\geq 0}$ is measurable*

(b) *for all $t \in T$ the mapping $(w, z) \mapsto L(t, w, z) : \mathbb{R} \times \mathbb{R}^n \to \mathbb{R}_{\geq 0}$ is convex*

(c) *there is a $M \in \mathbb{R}_{\geq 0}$ such that for all $(t, w, z) \in T \times \mathbb{R} \times \mathbb{R}^n$*

$$L(t, w, z) \leq M\left(1 + \Phi(w) + \sum_{i=1}^{n} \Phi(z_i)\right)$$

holds.

Then the set

$$F := \left\{ f \,\middle|\, \exists L \in J : f : W^1 L^\Phi(\mu) \to \mathbb{R}_{\geq 0}, u \mapsto f(u) = \int_T L(t, u(t), \nabla u(t)) d\mu(t) \right\}$$

is equicontinuous.

Proof. Due to (b) F is a family of convex functions. For each $f \in F$ and each $u \in W^1 L^\Phi(\mu)$

$$0 \leq f(u) \leq \int_T M(1 + \Phi(u) + \sum_{i=1}^{n} \Phi(D_i u) d\mu = M\left(\mu(T) + f^\Phi(u) + \sum_{i=1}^{n} f^\Phi(D_i u)\right)$$

holds. If u is in the $W^1 L^\Phi(\mu)$-ball with radius R, then $\|u\|_{(\Phi)}$ and $\|D_i u\|_{(\Phi)} \leq R$ for $i = 1, \ldots, n$. The modular f^Φ is because of Theorem 6.3.10 bounded on bounded subsets of $L^\Phi(\mu)$, i.e. there is a $K \in \mathbb{R}$ such that

$$f(u) \leq M(\mu(T) + (n + 1)K).$$

Therefore F is uniformly bounded on every ball of $W^1 L^\Phi(\mu)$ and according to Theorem 5.3.8 equicontinuous. □

Existence results for variational problems in Orlicz–Sobolev spaces can be found in [8].

9.9 Parameter-free Approximation of Time Series Data by Monotone Functions

In this section we treat a problem that occurs in the analysis of Time Series: we determine a parameter-free approximation of given data by a smooth monotone function in order to detect a monotone trend function from the data or to eliminate it in order to facilitate the analysis of cyclic behavior of difference data ("Fourier analysis"). In the discrete case, this type of approximation is known as *monotone regression* (see [13], p. 28 f.).

In addition to the data themselves, our approximation also takes derivatives into account, employing the mechanism of variational calculus.

9.9.1 Projection onto the Positive Cone in Sobolev Space

$$\text{Minimize} \int_a^b (v-x)^2 + (\dot{v}-\dot{x})^2 dt \text{ on } S,$$

where $S := \{v \in \mathrm{RCS}^1[a,b] \,|\, \dot{v} \geq 0\}$.

Linear Supplement. Let $F(t,p) = \eta(t)\cdot p$, then

$$\frac{d}{dt}F(\cdot, v(\cdot)) = F_t + F_p \cdot \dot{v} = \dot{\eta} v + \eta \dot{v}.$$

Using the *transversality conditions* for free endpoints: $\eta(a) = \eta(b) = 0$, we obtain that the supplement is constant on S

$$\int_a^b \dot{\eta} v + \eta \dot{v} dt = [\eta(t)v(t)]_a^b = 0.$$

Let $\tilde{L}(p,q) := L(p,q) - \dot{\eta}p - \eta q = \frac{1}{2}(x-p)^2 - \dot{\eta}p + \frac{1}{2}(\dot{x}-q)^2 - \eta q$. Pointwise minimization of $\tilde{L}$ w.r.t. p and q is broken down into two separate parts:

1. $\min\{\frac{1}{2}(x-p)^2 - \dot{\eta}p \,|\, p \in \mathbb{R}\}$ with the result: $\dot{\eta} = p - x$
2. $\min\{\frac{1}{2}(\dot{x}-q)^2 - \eta q \,|\, q \in \mathbb{R}_{\geq 0}\}$.

In order to perform the minimization in 2. we consider the function

$$\psi(q) := \frac{1}{2}(\dot{x}-q)^2 - \eta q.$$

Setting the derivative equal to zero, we obtain for $c := \dot{x} + \eta$

$$\psi_q = q - \dot{x} - \eta = q - c = 0.$$

For $c \geq 0$ we obtain a solution of the parabola ψ.

For $c < 0$ we have $q = 0$ because of the monotonicity of the parabola to the right of the global minimum.

For $c \geq 0$ we obtain the linear inhomogeneous system of linear differential equations

$$\begin{pmatrix} \dot{\eta} \\ \dot{v} \end{pmatrix} = \begin{pmatrix} 0 & 1 \\ 1 & 0 \end{pmatrix} \begin{pmatrix} \eta \\ v \end{pmatrix} + \begin{pmatrix} -x \\ \dot{x} \end{pmatrix}.$$

For $c(t) = \dot{x}(t) + \eta(t) < 0$, this inequality holds, because of the continuity of c, on an interval I, i.e. $\dot{v}(t) = 0$ on I, hence. $v(t) = \gamma$ there. We obtain $\dot{\eta}(t) = \gamma - x(t)$ on I, i.e.

$$\eta(t) - \eta(t_1) = \int_{t_1}^{t} (\gamma - x(\tau))d\tau = \gamma(t - t_1) - \int_{t_1}^{t} x(\tau)d\tau.$$

Algorithm (Shooting Procedure)

Choose $\gamma = v(a)$, notation: $c(t) = \dot{x}(t) + \eta(t)$, note: $\eta(a) = 0$.
Start: $c(a) = \dot{x}(a)$.
If $\dot{x}(a) \geq 0$ then set $t_1 = a$, goto 1.
If $\dot{x}(a) < 0$ then set $t_0 = a$, goto 2.

1. Solve initial value problem

$$\begin{pmatrix} \dot{\eta} \\ \dot{v} \end{pmatrix} = \begin{pmatrix} 0 & 1 \\ 1 & 0 \end{pmatrix} \begin{pmatrix} \eta \\ v \end{pmatrix} + \begin{pmatrix} -x \\ \dot{x} \end{pmatrix}$$

with the initial values $\eta(t_1), v(t_1)$, for which we have the following explicit solutions (see below):

$$\eta(t) = (v(t_1) - x(t_1))\sinh(t - t_1) + \eta(t_1)\cosh(t - t_1)$$
$$v(t) = x(t) + (v(t_1) - x(t_1))\cosh(t - t_1) + \eta(t_1)\sinh(t - t_1).$$

Let t_0 be the first root with change of sign of $c(t)$ such that $t_0 < b$, goto 2.

2. $\eta(t) = v(t_0)(t - t_0) - \int_{t_0}^{t} x(\tau)d\tau + \eta(t_0)$. Let t_1 be the first root with change of sign of $c(t)$ such that $t_1 < b$, goto 1.

For given $\gamma = v(a)$, this algorithm yields a pair of functions (η_γ, v_γ). Our aim is, to determine a γ such that $\eta_\gamma(b) = 0$. In other words: let $\gamma \mapsto \phi(\gamma) := \eta_\gamma(b)$, then we have to solve the (1-D)-equation $\phi(\gamma) = 0$.

Solution of Differential Equation in 2. Let $U(t)$ a fundamental system then the solution of the (inhomogeneous) system is given by

$$V(t) = U(t) \int_{t_1}^{t} U^{-1}(\tau) b(\tau) d\tau + U(t) D,$$

where $D = U^{-1}(t_1) V(t_1)$. In our case we have the fundamental system

$$U(t) = \begin{pmatrix} \cosh t & \sinh t \\ \sinh t & \cosh t \end{pmatrix},$$

and hence

$$U^{-1}(t) = \begin{pmatrix} \cosh t & -\sinh t \\ -\sinh t & \cosh t \end{pmatrix},$$

which yields (using corresponding addition theorems)

$$U(t)U^{-1}(\tau) = \begin{pmatrix} \cosh(t-\tau) & \sinh(t-\tau) \\ \sinh(t-\tau) & \cosh(t-\tau) \end{pmatrix}.$$

We obtain

$$\begin{aligned} V(t) = \int_{t_1}^{t} &\begin{pmatrix} \cosh(t-\tau) & \sinh(t-\tau) \\ \sinh(t-\tau) & \cosh(t-\tau) \end{pmatrix} \begin{pmatrix} -x(\tau) \\ \dot{x}(\tau) \end{pmatrix} d\tau \\ &+ \begin{pmatrix} \cosh(t-t_1) & \sinh(t-t_1) \\ \sinh(t-t_1) & \cosh(t-t_1) \end{pmatrix} \begin{pmatrix} \eta(t_1) \\ v(t_1) \end{pmatrix} \end{aligned}$$

i.e.

$$\begin{aligned} \eta(t) &= \int_{t_1}^{t} \dot{x}(\tau)\sinh(t-\tau) - x(\tau)\cosh(t-\tau) d\tau \\ &\quad + \eta(t_1)\cosh(t-t_1) + v(t_1)\sinh(t-t_1) \\ v(t) &= \int_{t_1}^{t} \dot{x}(\tau)\cosh(t-\tau) - x(\tau)\sinh(t-\tau) d\tau \\ &\quad + \eta(t_1)\sinh(t-t_1) + v(t_1)\cosh(t-t_1). \end{aligned}$$

The integrals can be readily solved using the product rule

$$\begin{aligned} \eta(t) &= [x(\tau)\sinh(t-\tau)]_{t_1}^{t} + \eta(t_1)\cosh(t-t_1) + v(t_1)\sinh(t-t_1) \\ v(t) &= [x(\tau)\cosh(t-\tau)]_{t_1}^{t} + \eta(t_1)\sinh(t-t_1) + v(t_1)\cosh(t-t_1). \end{aligned}$$

We finally obtain the following explicit solutions

$$\begin{aligned} \eta(t) &= (v(t_1) - x(t_1))\sinh(t-t_1) + \eta(t_1)\cosh(t-t_1) \\ v(t) &= x(t) + (v(t_1) - x(t_1))\cosh(t-t_1) + \eta(t_1)\sinh(t-t_1). \end{aligned}$$

Existence. If we minimize the strongly convex functional $f(v) := \int_a^b (v - x)^2 + (\dot{v} - \dot{x})^2 dt$ on the closed and convex subset S of the Sobolev space $W^1_{2,1}[a, b]$, where $S := \{v \in W^1_{2,1}[a, b] \mid \dot{v} \geq 0\}$, then, according to Theorem 3.13.5 f has a unique minimal solution in S. But the proof of Theorem 9.9.3 carries over to the above problem, yielding that both v and η are in $C^1[a, b]$.

9.9.2 Regularization of Tikhonov-type

In practice it turns out that fitting the derivative of v to the derivative $\dot{x}$ of the data has undesired effects. Instead we consider the following problem:

$$\text{Minimize } f_\alpha(v) := \frac{1}{2}\int_a^b \alpha(v - x)^2 + (1 + \alpha)\dot{v}^2 dt$$

on the closed and convex subset S of the Sobolev space $W^1_{2,1}[a, b]$, where

$$S := \{v \in W^1_{2,1}[a, b] \mid \dot{v} \geq 0\}.$$

Let $\tilde{L}(p, q) := L(p, q) - \dot{\eta}p - \eta q = \alpha\frac{1}{2}(x - p)^2 - \dot{\eta}p + \frac{1}{2}(1 + \alpha)q^2 - \eta q$. Pointwise minimization of $\tilde{L}$ w.r.t. p and q is broken down into two separate parts:

1. $\min\{\alpha\frac{1}{2}(x - p)^2 - \dot{\eta}p \mid p \in \mathbb{R}\}$ with the result: $\dot{\eta} = \alpha(p - x)$
2. $\min\{\frac{1+\alpha}{2}q^2 - \eta q \mid q \in \mathbb{R}_{\geq 0}\}$.

In order to perform the minimization in 2. we consider the function

$$\psi(q) := \frac{1 + \alpha}{2}q^2 - \eta q.$$

Setting the derivative equal to zero, we obtain for $c := \eta$

$$\psi_q = (1 + \alpha)q - \eta = (1 + \alpha)q - c = 0.$$

For $c \geq 0$ we obtain a minimal solution of the parabola ψ.

For $c < 0$ we have $q = 0$ because of the monotonicity of the parabola to the right of the global minimum.

For $c \geq 0$ we obtain the inhomogeneous system of linear differential equations

$$\begin{pmatrix} \dot{\eta} \\ \dot{v} \end{pmatrix} = \begin{pmatrix} 0 & \alpha \\ \frac{1}{1+\alpha} & 0 \end{pmatrix} \begin{pmatrix} \eta \\ v \end{pmatrix} + \begin{pmatrix} -\alpha x \\ 0 \end{pmatrix}.$$

For $c(t) = \eta(t) < 0$, this inequality holds, because of the continuity of c, on an interval I, i.e. $\dot{v}(t) = 0$ on I, and we obtain $v(t) = \gamma$ there. Hence $\dot{\eta}(t) = \alpha(\gamma - x(t))$ on I, i.e.

$$\eta(t) - \eta(t_1) = \alpha \int_{t_1}^t (\gamma - x(\tau))d\tau = \alpha\gamma(t - t_1) - \alpha \int_{t_1}^t x(\tau)d\tau.$$

Algorithm (Shooting Procedure)

Start: Choose $\gamma = v(a) \neq x(a)$, notation: $c(t) = \eta(t)$, note: $c(a) = \eta(a) = 0$. Put $t_1 = a$, if $\gamma > x(a)$ goto 1, else goto 2.

1. Solve initial value problem

$$\begin{pmatrix} \dot{\eta} \\ \dot{v} \end{pmatrix} = \begin{pmatrix} 0 & \alpha \\ \frac{1}{1+\alpha} & 0 \end{pmatrix} \begin{pmatrix} \eta \\ v \end{pmatrix} + \begin{pmatrix} -\alpha x \\ 0 \end{pmatrix}$$

with the initial values $\eta(t_1), v(t_1)$, for which we have the following solutions:

$$\begin{aligned} \eta(t) &= -\alpha \int_{t_1}^{t} x(\tau) \cosh \beta(t-\tau) d\tau \\ &\quad + \eta(t_1) \cosh \beta(t-t_1) + v(t_1)\beta \sinh \beta(t-t_1) \\ v(t) &= -\frac{\alpha}{\beta} \int_{t_1}^{t} x(\tau) \sinh \beta(t-\tau) d\tau \\ &\quad + \frac{1}{\beta}\eta(t_1) \sinh \beta(t-t_1) + v(t_1) \cosh \beta(t-t_1)), \end{aligned}$$

where $\beta := \sqrt{\frac{\alpha}{1+\alpha}}$ (see below).

Let t_0 be the first root with change of sign of $c(t)$ such that $t_0 < b$, goto 2.

2. $\eta(t) = \alpha v(t_0)(t-t_0) - \alpha \int_{t_0}^{t} x(\tau) d\tau + \eta(t_0)$. Let t_1 be the first root with change of sign of $c(t)$ such that $t_1 < b$, goto 1.

For given $\gamma = v(a)$, this algorithm yields a pair of functions (η_γ, v_γ). Our aim is, to determine a γ such that $\eta_\gamma(b) = 0$. In other words: let $\gamma \mapsto \phi(\gamma) := \eta_\gamma(b)$, then we have to solve the 1-D equation $\phi(\gamma) = 0$ (see discussion below).

Solution of the Differential Equation in 2. Let $U(t)$ be a fundamental system then the solution of the (inhomogeneous) system is given by

$$V(t) = U(t) \int_{t_1}^{t} U^{-1}(\tau) b(\tau) d\tau + U(t) D,$$

where $D = U^{-1}(t_1) V(t_1)$. For the eigenvalues of

$$\begin{pmatrix} 0 & \alpha \\ \frac{1}{1+\alpha} & 0 \end{pmatrix}$$

we obtain $\beta_{1,2} := \pm \sqrt{\frac{\alpha}{1+\alpha}}$ resulting in the following fundamental system:

$$U(t) = \begin{pmatrix} \beta \cosh \beta t & \beta \sinh \beta t \\ \sinh \beta t & \cosh \beta t \end{pmatrix}$$

and hence

$$U^{-1}(t) = \frac{1}{\beta}\begin{pmatrix} \cosh \beta t & -\beta \sinh \beta t \\ -\sinh \beta t & \beta \cosh \beta t \end{pmatrix}$$

which yields (using corresponding addition theorems)

$$U(t)U^{-1}(\tau) = \begin{pmatrix} \cosh \beta(t-\tau) & \beta \sinh \beta(t-\tau) \\ \frac{1}{\beta}\sinh \beta(t-\tau) & \cosh \beta(t-\tau) \end{pmatrix}.$$

We obtain

$$\begin{aligned} V(t) = &\int_{t_1}^{t} \begin{pmatrix} \cosh \beta(t-\tau) & \beta \sinh \beta(t-\tau) \\ \frac{1}{\beta}\sinh \beta(t-\tau) & \cosh \beta(t-\tau) \end{pmatrix} \begin{pmatrix} -\alpha x(\tau) \\ 0 \end{pmatrix} d\tau \\ &+ \begin{pmatrix} \cosh \beta(t-t_1) & \beta \sinh \beta(t-t_1) \\ \frac{1}{\beta}\sinh \beta(t-t_1) & \cosh \beta(t-t_1) \end{pmatrix} \begin{pmatrix} \eta(t_1) \\ v(t_1) \end{pmatrix} \end{aligned}$$

i.e.

$$\begin{aligned} \eta(t) &= -\alpha \int_{t_1}^{t} x(\tau) \cosh \beta(t-\tau) d\tau \\ &\quad + \eta(t_1) \cosh \beta(t-t_1) + v(t_1)\beta \sinh \beta(t-t_1) \\ v(t) &= -\frac{\alpha}{\beta} \int_{t_1}^{t} x(\tau) \sinh \beta(t-\tau) d\tau \\ &\quad + \frac{1}{\beta}\eta(t_1) \sinh \beta(t-t_1) + v(t_1) \cosh \beta(t-t_1)). \end{aligned}$$

Discussion of Existence, Convergence, and Smoothness. We minimize the convex functional

$$f_\alpha(v) := \frac{1}{2}\int_a^b \alpha(v-x)^2 + (1+\alpha)\dot{v}^2 dt$$

on the closed and convex subset S of the Sobolev space $W_{2,1}^1[a,b]$, where $S := \{v \in W_{2,1}^1[a,b] \mid \dot{v} \geq 0\}$. For f_α we have the following representation:

$$f_\alpha(v) := \frac{1}{2}\int_a^b (\alpha(v^2 + \dot{v}^2 - 2vx + x^2) + \dot{v}^2)dt.$$

The convex functional

$$f(v) := \frac{1}{2}\int_a^b (v^2 + \dot{v}^2 - 2vx + x^2)dt$$

is apparently uniformly convex on $W_{2,1}^1[a,b]$, since

$$f\left(\frac{u+v}{2}\right) = \frac{1}{2}(f(u) + f(v)) - \frac{1}{4}\|u - v\|^2_{W_{2,1}^1[a,b]}.$$

Let further $g(v) := \frac{1}{2}\int_a^b \dot{v}^2 dt$, then apparently $f_\alpha = \alpha f + g$. By the Regularization Theorem of Tikhonov-type 8.7.2 we obtain the following statement: Let now α_n be a positive sequence tending to zero and $f_n := \alpha_n f + g$. Let finally v_n be the (uniquely determined) minimal solution of f_n on S, then the sequence (v_n) converges to the (uniquely determined) minimal solutions of f on $M(g, S)$, where $M(g, S)$ apparently consists of all constant functions (more precisely: of all those functions which are constant a.e.). On this set f assumes the form

$$f(v) = \frac{1}{2}\int_a^b (v^2 - 2vx + x^2)dt = \frac{1}{2}\int_a^b (v - x)^2 dt.$$

The minimal solution of f on $M(g, S)$ is apparently the mean value

$$\mu = \frac{1}{b-a}\int_a^b x(t)dt.$$

We now turn our attention to solving the (1-D)-equation $\phi(\gamma) = 0$.
We observe for $t > a$

$$\frac{1}{\beta}\sinh\beta(t-a) = \int_a^t \cosh\beta(t-\tau)d\tau,$$

hence for $\gamma = v(a)$ and $t_1 = a$:

(a) If $\gamma > (1+\alpha)\max x(t)$ then $\eta(t) > 0$ for all $t > a$, since

$$\eta(t) = \alpha\int_a^t \left(\frac{\beta^2}{\alpha}\gamma - x(\tau)\right)\cosh\beta(t-\tau)d\tau,$$

and $\frac{\beta^2}{\alpha} = \frac{1}{1+\alpha}$. We conclude $\phi(\gamma) = \eta(b) > 0$.

(b) If $\gamma < \min x(t)$ then $\eta(t) < 0$ since

$$\eta(t) = \alpha\int_a^t (\gamma - x(\tau))d\tau$$

for all $t > a$, hence $\phi(\gamma) = \eta(b) < 0$.

The decision whether 1. or 2. is taken in the above algorithm can be based on the equation $\dot{\eta} = \alpha(v - x)$, in particular: $\dot{\eta}(a) = \alpha(\gamma - x(a))$.

In order to establish the continuity of the function ϕ we consider the following initial value problem for the non-linear ODE system

$$\dot{\eta} = \alpha(v - x)$$

$$\dot{v} = \frac{1}{1+\alpha}\frac{\eta + |\eta|}{2}$$

for $\eta(a) = 0$ and $v(a) = \gamma$, which – for non-negative η – corresponds to the above linear system, whereas for negative η the second equation reads $\dot{v} = 0$. The right-hand side apparently satisfies a global Lipschitz condition. Due to the theorem on continuous dependence on initial values (see [59]) the solution exists on the whole interval $[a, b]$, both v and η belong to $C^1[a, b]$ and depend continuously on the initial conditions, in particular $\eta(b)$ depends continuously on γ. Since $\eta(b)$ changes sign between $\gamma > (1 + \alpha) \max x(t)$ and $\gamma < \min x(t)$ there is a γ_0 such that for $v(a) = \gamma_0$ we have $\eta(b) = 0$, where γ_0 can be determined e.g. via the bisection method.

Remark 9.9.1. The above smoothness result is essentially a special case of Theorem 9.9.3.

The question remains, how a positive trend in the data x can be detected. We define:

Definition 9.9.2. *x does not contain a positive trend*, if the monotone regression, i.e. the minimal solution of $h(v) := \frac{1}{2}\int_a^b (v - x)^2 dt$ on S is the constant function identical to the mean value μ.

Apparently, in this case the minimal solution of $f_\alpha(v) = \frac{1}{2}\int_a^b \alpha(v - x)^2 + (1 + \alpha)\dot{v}^2 dt$ on S is also equal to μ for all $\alpha > 0$.

In other words: the existence of a positive trend can be detected for each positive α. The corresponding solutions can also be obtained via appropriate numerical methods applied to the above non-linear ODE system.

An Alternative Approach

We minimize the convex functional

$$g_\alpha(v) := \frac{1}{2}\int_a^b (1 + \alpha)(v - x)^2 + \alpha\dot{v}^2 dt$$

on the closed and convex subset S of the Sobolev space $W_{2,1}^1[a, b]$, where $S := \{v \in W_{2,1}^1[a, b] \mid \dot{v} \geq 0\}$ under the assumption that $h(v) := \frac{1}{2}\int_a^b (v - x)^2 dt$ has a minimal solution on S, the monotone regression. For g_α we have the following representation:

$$g_\alpha(v) := \frac{1}{2}\int_a^b (\alpha(v^2 + \dot{v}^2 - 2vx + x^2) + (v - x)^2) dt.$$

As was pointed out above, the convex functional

$$f(v) := \frac{1}{2}\int_a^b (v^2 + \dot{v}^2 - 2vx + x^2) dt$$

is uniformly convex on $W_{2,1}^1[a, b]$. Let (α_n) be a positive sequence tending to zero, then due to Theorem 8.7.2 the sequence (x_n) of minimal solutions of $g_n := g_{\alpha_n}$ converges to the monotone regression.

9.9.3 A Robust Variant

We can modify the functional f_α in the spirit of robust statistics by introducing a differentiable and strictly convex Young function Φ

$$f_\alpha(v) := \int_a^b \alpha\Phi(v-x) + \frac{1}{2}\dot{v}^2 dt$$

using the same linear supplement as above we obtain the following initial value problem:

$$\dot{\eta} = \alpha\Phi'(v-x)$$
$$\dot{v} = \frac{\eta + |\eta|}{2}$$

for $\eta(a) = 0$ and $v(a) = \gamma$. If Φ is twice continuously differentiable with bounded second derivative, then the right-hand side satisfies a global Lipschitz condition. In particular this results in a continuous dependence on initial conditions.

Let $\gamma > \max x(t)$. Since $\dot{v} \geq 0$ we have $v \geq \gamma$ hence $\eta(t) = \alpha \int_a^t \Phi'(v(\tau) - x(\tau))d\tau > 0$ for all $a < t \leq b$, in particular $\eta(b) > 0$.

If $\gamma < \min x(t) =: \rho$, then $\dot{\eta}(a) = \alpha\Phi'(\gamma - x(a)) < 0$. Hence there is a non-empty interval $(0, r)$ such that $\dot{\eta}(t) < 0$ on $(0, r)$. Let r_0 be the largest such r and suppose $r_0 < b$. Then $\eta(t) < 0$ on $(0, r_0)$, thus $\dot{v} = 0$ on $(0, r_0)$, i.e. $v(t) = \gamma$ on $[0, r_0)$. But then $\dot{\eta}(t) = \alpha\Phi'(\gamma - x(t)) \leq \alpha\Phi'(\gamma - \rho) < 0$ for all $t \in [0, r_0)$, hence $\dot{\eta}(r_0) < 0$ and there is an $\varepsilon > 0$ such that $\dot{\eta}(r_0 + \varepsilon) < 0$, a contradiction.

We conclude: $\eta(b) < 0$ for $\gamma < \min x(t)$. Due to the continuous dependence on initial values the intermediate value theorem again guarantees the existence of a γ_α such that $\eta(b) = 0$.

Now we consider the convergence question for $\alpha \to 0$:

Since $\gamma_\alpha \in [\min x(t), \max x(t)]$ there is a convergent subsequence $\gamma_n := \gamma_{\alpha_n} \to \gamma_0$. Let $y_n := (v_n, \eta_n)^T$ be the sequence of solutions obtained for each α_n, then according to the theorem (Folgerung B.2 in [59])

$$\|y_n(t) - y_0(t)\| \leq |\gamma_n - \gamma_0| e^{L(t-a)} + \frac{\alpha_n}{L} \sup_{[a,b]} |\Phi'(v_0(t) - x(t))| (e^{L(t-a)} - 1)$$

where $y_0 = (v_0, \eta_0)^T$ is the solution of

$$\dot{\eta} = 0$$
$$\dot{v} = \frac{\eta + |\eta|}{2}$$

for the initial conditions $\eta(a) = 0$ and $v(a) = \gamma_0$. Apparently $\eta_0(t) = 0$ for all $t \in [a, b]$, hence $\dot{v}_0 = 0$ on $[a, b]$ and thus $v_0(t) = \gamma_0$. Hence $(v_n, \eta_n)^T$ converges to $(\gamma_0, 0)^T$ in $(C[a, b])^2$.

But $v_n \to v_0$ also in $C^1[a,b]$: we have $\eta_n(t) = \alpha_n \int_a^t \Phi'(v_n(\tau) - x(\tau))d\tau$ and hence

$$|\dot{v}_n(t) - \dot{v}_0(t)| \leq \alpha_n \int_a^t |\Phi'(v_n(\tau) - x(\tau))|d\tau \leq \alpha_n \int_a^b |\Phi'(v_n(\tau) - x(\tau))|d\tau \leq \alpha_n C,$$

since $\{|\Phi'(v_n - x)|\}$ is bounded.

Now, our theorem on second stage solutions 5.6.2 yields that v_0 is a minimal solution of

$$f(v) := \int_a^b \Phi(v(\tau) - x(\tau))d\tau \quad \text{on } M(g,S),$$

with $g(v) := \frac{1}{2}\int_a^b (\dot{v})^2 dt$. Apparently $M(g,S)$ consists of all constant functions. Since Φ is strictly convex, the strictly convex function $\lambda \mapsto \int_a^b \Phi(\lambda - x(\tau))d\tau$ has a uniquely determined solution λ_0. Hence $\lambda_0 = \gamma_0$, and this holds for every convergent subsequence, thus $\gamma_\alpha \to_{\alpha\to 0} \lambda_0$ and $v_\alpha \to v_0 = \lambda_0$ in $C^1[a,b]$.

We have thus proved the following existence and convergence theorem:

Theorem 9.9.3. *Let Φ be a strictly convex and twice continuously differentiable Young function with bounded second derivative. Then the problem*

$$\textit{Minimize } f_\alpha(v) := \int_a^b \alpha\Phi(v - x) + \frac{1}{2}\dot{v}^2 dt \textit{ on } S := \{v \in C^1[a,b] \mid \dot{v} \geq 0\}$$

has a unique solution $v_\alpha \in S$ for each $\alpha > 0$.

For $\alpha \to 0$ we obtain $v_\alpha \to v_0 = \lambda_0$ in $C^1[a,b]$, where λ_0 is the uniquely determined minimal solution of the function $\lambda \mapsto \int_a^b \Phi(\lambda - x(\tau))d\tau$.

Remark 9.9.4. The proof above also shows that the corresponding η_α of the linear supplement belongs to $C^1[a,b]$.

9.10 Optimal Control Problems

In contrast to variational problems we cannot assume that the controls u are in $C(T)$. For convex-concave Lagrangians we obtain the following assertion about equicontinuity of the corresponding functionals:

Theorem 9.10.1. *Let (T,Σ,μ) be a finite measure space, where T is a compact set. Let J be a family of mappings $L : \mathbb{R}^n \times \mathbb{R}^m \to \mathbb{R}$, satisfying the following properties:*

(a) *$L(p,\cdot) : \mathbb{R}^m \to \mathbb{R}$ is convex for all $p \in \mathbb{R}^n$*

(b) *$L(\cdot,r) : \mathbb{R}^n \to \mathbb{R}$ is concave for all $r \in \mathbb{R}^m$*

(c) *$L(p,r) \leq M(1 + \sum \Phi(r_i))$ for all $p \in \mathbb{R}^n, r \in \mathbb{R}^m$*

where Φ is a Young function, satisfying the Δ_2^∞-condition. Let V be an open convex subset of $(C(T))^n$ and U an open convex subset of $(L^\Phi(\mu))^m$ then the family F of all functions $f : V \times U \to \mathbb{R}$ with

$$f(x,u) := \int L(x,u)d\mu \quad \text{with } L \in J$$

is equicontinuous, provided that F is pointwise bounded from below.

Proof. Let $U_0 \subset U$ be open and bounded. Let $x \in V$ and $u \in U_0$, then (see Theorem 6.3.10) there is a constant K, such that

$$\begin{aligned} f(x,u) &= \int_T L(x,u)d\mu \leq M \int_T \left(1 + \sum_{i=1}^m \Phi(u_i)\right) d\mu \\ &\leq M\mu(T) + M \sum_{i=1}^m \int_T \Phi(u_i)d\mu \\ &= M\mu(T) + M \sum_{i=1} f^\Phi(u_i) \leq M\mu(T) + m \cdot K. \end{aligned}$$

In particular $f(x,\cdot)$ is continuous for all $x \in V$. Moreover, the family F is pointwise bounded. On the other hand $-L(\cdot,r) : \mathbb{R}^n \to \mathbb{R}$ is convex and hence continuous according to Theorem 5.3.11 for all $r \in \mathbb{R}^n$. Let now $x_n \to x$ in $(C^1(T))^n$, then as in Theorem 9.8.1 $L(x_n,u) \to L(x,u)$ uniformly, hence $\int_T L(x_n,u)d\mu \to \int_T L(x,u)d\mu$, i.e. $f(\cdot,u)$ continuous on V for all $u \in U$. With Theorem 5.4.9 the assertion follows. □

For convex-concave Lagrangians pointwise convergence already implies pointwise convergence of the corresponding functionals using the theorem of Banach–Steinhaus:

Theorem 9.10.2. *Let (T,Σ,μ) be a finite measure space, where T is a compact set and let Φ be a Young function, satisfying the Δ_2^∞-condition.*

Let the sequence $(L_k)_{k\in\mathbb{N}}$ of mappings $L_k : \mathbb{R}^n \times \mathbb{R}^m \to \mathbb{R}$ have the following properties:

(a) *$L_k(p,\cdot) : \mathbb{R}^m \to \mathbb{R}$ is convex for all $p \in \mathbb{R}^n$*

(b) *$L_k(\cdot,r) : \mathbb{R}^n \to \mathbb{R}$ is concave for all $r \in \mathbb{R}^m$*

(c) *$L_k(p,r) \leq M(1 + \sum \Phi(r_i))$ for all $p \in \mathbb{R}^n, r \in \mathbb{R}^m$*

(d) *$L_k(p,r) \to_{k\to\infty} L_0(p,r)$ for all $(p,r) \in \mathbb{R}^n \times \mathbb{R}^m$.*

Let V be an open subset of $(C(T))^n$ and U an open subset of $(L^\Phi(\mu))^m$ then the sequence $(f_k : V \times U \to \mathbb{R})_{k\in\mathbb{N}}$ with $f_k(x,u) := \int_T L_k(x,u)d\mu$ converges pointwise to $f_0(x,u) := \int_T L_0(x,u)d\mu$.

Proof. Let $u \in (L^\Phi(\mu))^m$ be a step function, i.e. $u := \sum_{i=1}^s a_i \chi_{T_k}$ with $\bigcup_{i=1}^s T_k = T$ and $T_i \cap T_j = \emptyset$ for $i \neq j$, then

$$f_k(x,u) = \sum_{k=1}^s \int_{T_i} L_k(x(t), a_i) d\mu(t).$$

Let $g_{k,i} := -\int_{T_i} L_k(x(t), a_i) d\mu(t)$, then

$$g_{k,i} \xrightarrow{k\to\infty} g_i = -\int_{T_i} L_0(x(t), a_i) d\mu(t)$$

holds, since $-L_k(\cdot, a_i)$ is convex and hence according to Theorem 5.3.6 $-L_k(\cdot, a_i)$ $\to_{k\to\infty} -L_0(\cdot, a_i)$ continuously convergent and therefore $(-L_k(x(\cdot), a_i))$ on T_i also uniformly convergent.

In all we obtain $f_k(x,u) \to f_0(x,u)$. Using the theorem of Banach–Steinhaus 5.3.17 the assertion follows. □

9.10.1 Minimal Time Problem as a Linear L^1-approximation Problem

A *minimal time problem* is obtained if the right end point of the time interval can be freely chosen, and we look for the shortest time τ_0, such that via an admissible control the given point $c \in \mathbb{R}^n$ with $c \neq 0$ is reached from the starting point.

If the function $f(x) = \|x(b) - c\|^2$ is viewed in dependence of b, we look for the smallest b, such that a minimal solution with minimal value 0 exists. In particular we encounter optimal controls which are of bang-bang type. This is specified more precisely in Theorem 9.10.5.

An important class of such problems can be treated by methods of linear L^1-approximation. Let $\mathrm{RS}[0,\tau]$ denote the set of all piecewise continuous functions, such that the right-sided limit exists. Let $K_\tau := \mathrm{RCS}^{(1)}[0,\tau]^n$, $X_\tau := \mathrm{RS}[0,\tau]^m$ and

$$Q_\tau := \{u \in X_\tau \mid |u_i(t)| \leq 1,\ i \in \{1,\dots,m\}, t \in [0,\tau]\}. \tag{9.37}$$

Let $A \in \mathrm{RS}[0,\infty)^{n\times n}$, $B \in \mathrm{RS}[0,\infty)^{n\times m}$. Let further

$$\begin{aligned} R_\tau := \{u \in X_\tau \mid &\exists x \in K_\tau \forall t \in [0,\tau] : \dot{x}(t) = A(t)x(t) + B(t)u(t), \\ &x(0) = 0,\ x(\tau) = c\}. \end{aligned} \tag{9.38}$$

The minimal time problem (MT) is now posed in the following form:

$$\text{Minimize } \tau \text{ under the condition } Q_\tau \cap R_\tau \neq \emptyset. \tag{9.39}$$

If the vector space X_τ is equipped with the norm

$$\|u\|_\tau = \sup\{|u_i(t)| \mid i \in \{1,\dots,m\},\ t \in [0,\tau]\} \tag{9.40}$$

then Q_τ is the unit ball in the normed space $(X_\tau, \|\cdot\|_\tau)$. In particular for those τ with $Q_\tau \cap R_\tau \neq \emptyset$ we have

$$w(\tau) := \inf\{\|u\|_\tau \mid u \in R_\tau\} \leq 1, \tag{9.41}$$

and for τ with $Q_\tau \cap R_\tau = \emptyset$ we obtain for all $u \in R_\tau$ $\|u\|_\tau > 1$ and hence $w(\tau) \geq 1$. We will now attempt to prove that for the minimal time τ_0

$$w(\tau_0) = 1 \tag{9.42}$$

holds.

A sufficient criterion will be supplied by Theorem 9.10.6.

Let H be a fundamental matrix of the ODE in (9.38), where $H(0)$ is the n-th unit matrix. Then the Condition (9.38) using $Y(t) := H(\tau)H^{-1}(t)B(t)$, $i \in \{1, \ldots, n\}$ can be expressed in the form of n linear equations

$$\int_0^\tau Y_i(t)u(t)dt = c_i \quad \text{for } i \in \{1, \ldots, n\}, \tag{9.43}$$

where Y_i is the i-th row of Y.

Using the above considerations we arrive at the following problem:

Find a pair (τ, u^*) with the following properties:

$$\begin{aligned} &u^* \text{ is an element of minimal norm in} \\ S_\tau &:= \{u \in X_\tau \mid u \text{ satisfies Equation (9.43)}\}, \end{aligned} \tag{9.44}$$

and

$$\|u^*\|_\tau = 1. \tag{9.45}$$

In order to draw the connection to the above mentioned approximation in the mean, we will reformulate for each τ the optimization problem given by Equations (9.43) and (9.44). Since $c \neq 0$, there is a $c_{i_0} \neq 0$. In the sequel let w.l.o.g. $i_0 = n$. Then Equation (9.43) can be rewritten using the functions

$$Z := Y_n/c_n \quad \text{and} \quad Z_i := Y_i - \frac{c_i}{c_n}Y_n \quad \text{for } i \in \{1, \ldots, n-1\} \tag{9.46}$$

as

$$\int_0^\tau Z_i(t)u(t)dt = 0, \quad i \in \{1, \ldots, n-1\}, \quad \int_0^\tau Z(t)u(t)dt = 1. \tag{9.47}$$

In addition to the problem

$$\text{Minimize } \|u\|_\tau \text{ under the Conditions (9.47)} \tag{D1}$$

we consider

$$\text{Maximize } \psi_\tau(u) := \int_0^\tau Z(t)u(t)dt \tag{D2}$$

under the conditions

$$\int_0^\tau Z_i(t)u(t)dt = 0 \quad \text{for } i \in \{1,\dots,n-1\} \text{ and } \|u\|_\tau = 1. \tag{9.48}$$

Remark 9.10.3. One can easily see that problems (D1) and (D2) are equivalent in the following sense:

If W is the maximal value of (D2), then $1/W$ is the minimal value of (D1) and a u_0 is a solution of (D2), if and only if u_0/W is a solution of (D1).

We will now establish that problem (D2) can be viewed as a dual problem of the following non-restricted problem in $\mathbb{R}^{n-1}$:

$$\text{Minimize } \varphi_\tau(\alpha_1,\dots,\alpha_{n-1}) := \int_0^\tau \left\| Z^T(t) - \sum_{i=0}^{n-1} \alpha_i Z_i^T(t) \right\|_1 dt \tag{P1}$$

on $\mathbb{R}^{n-1}$, where $\|\cdot\|_1$ denotes the norm $\rho \mapsto \sum_{i=1}^m |\rho_i|$ in $\mathbb{R}^m$.

Apparently we have for all $\rho, \sigma \in \mathbb{R}^m$

$$\left| \sum_{i=1}^m \rho_i \sigma_i \right| \le \left(\max_{1\le i\le m} |\sigma_i| \right) \sum_{i=1}^m |\rho_i|. \tag{9.49}$$

Remark 9.10.4. (P1) is a problem of approximation in the mean, treated in Chapter 1 for $m = 1$.

The following theorem holds:

Theorem 9.10.5. *The problems* (P1) *and* (D2) *are weakly dual and* (P1) *is solvable. If α^* is a solution of* (P1) *and if each component of $h = (h_1,\dots,h_m)^T := Z^T - \sum_{i=1}^{n-1} \alpha_i^* Z_i^T$ has only finitely many zeros in $[0,\tau]$, then a solution u^* of* (D2) *can be constructed in the following way:*

Let $i \in \{1,\dots,m\}$ fixed, and let $\{t_1 < \dots < t_k\}$ the zeros with change of sign of the i-th component h_i of h. Let $t_0 := 0$, $t_{k+1} := \tau$ and

$$\varepsilon_i := \begin{cases} 1, & \text{if } h_i \text{ on } [0,t_1) \text{ non-negative} \\ -1, & \text{otherwise.} \end{cases}$$

Then the following componentwise defined function

$$u_i^*(t) = \varepsilon_i(-1)^{j-1}, \quad \text{for } t \in [t_{j-1},t_j),\ j \in \{1,\dots,k+1\} \tag{9.50}$$

is a maximal solution of (D2), *and $\psi_\tau(u^*) = \varphi_\tau(\alpha^*)$ holds, i.e. the problem* (D2) *is dual to* (P1).

Proof. Let $\alpha = (\alpha_1, \ldots, \alpha_{n-1}) \in \mathbb{R}^{n-1}$ and let u satisfy Equation (9.48). With Equation (9.49) we obtain

$$\begin{aligned}\psi_\tau(u) &= \int_0^\tau Z(t)u(t)dt = \int_0^\tau \Big(Z(t) - \sum_{i=1}^{n-1} \alpha_i Z_i(t)\Big)u(t)dt \\ &\leq \int_0^\tau \Big|\Big(Z(t) - \sum_{i=1}^{n-1} \alpha_i Z_i(t)\Big)u(t)\Big| dt \\ &\leq \int_0^\tau \Big\|Z^T(t) - \sum_{i=1}^{n-1} \alpha_i Z_i^T(t)\Big\|_1 \|u\|_\tau dt \\ &\leq \int_0^\tau \Big\|Z^T(t) - \sum_{i=1}^{n-1} \alpha_i Z_i^T(t)\Big\|_1 dt = \varphi_\tau(\alpha), \end{aligned} \tag{9.51}$$

and hence weak duality has been established. According to the theorem of Weierstrass (P1) has a minimal solution α^*. Due to the Duality Theorem of Linear Approximation (see Theorem 3.12.4) problem (D2), extended to the space $(L^\infty[0,\tau]^m, \|\cdot\|_\tau)$, has a solution $\bar{u}$ with $\|\bar{u}\|_\tau = 1$ and

$$\int_0^\tau Z_i(t)\bar{u}(t) = 0 \quad \text{for all } i \in \{1, \ldots, n-1\}.$$

Apparently $(L^\infty[0,\tau]^m, \|\cdot\|_\tau)$ is the dual space of $(L^1[0,\tau]^m, \int_0^\tau \|x(t)\|_1 dt)$ (see Theorem 7.6.3). For $\bar{u}$ the inequality corresponding to (9.51) is satisfied as an equality. In particular

$$\begin{aligned}\psi_\tau(\bar{u}) &= \int_0^\tau h^T(t)\bar{u}(t)dt = \int_0^\tau \Big(\sum_{i=1}^{m} |h_i(t)|\Big)dt \\ &= \int_0^\tau \Big\|Z^T(t) - \sum_{i=1}^{n-1} \alpha_i^* Z_i^T(t)\Big\|_1 dt = \varphi_\tau(\alpha^*). \end{aligned} \tag{9.52}$$

From the above equality it follows that $\bar{u} = u^*$ a. e. , for

$$\int_0^\tau h(t)^T \bar{u}(t)dt = \int_0^\tau \Big(\sum_{i=1}^{m} |h_i(t)|\Big)dt = \int_0^\tau h(t)^T u^*(t)dt.$$

For assume there is $T_0 \subset [0,\tau]$ with $\mu(T_0) > 0$ with $\bar{u}(t) \neq u^*(t)$ for all $t \in T_0$. Then there is

$$h_{i_0}(t)\overline{u_{i_0}}(t) < |h_{i_0}(t)| = h_{i_0}(t)u_{i_0}^*(t)$$

for all $t \in T_0 \setminus N_{i_0}$, where N_{i_0} denotes the zeros of $h_{i_0}(t)$. Hence

$$\int_{T_0} h_{i_0}(t)\overline{u_{i_0}}(t)dt < \int_{T_0} |h_{i_0}(t)|dt,$$

leading to a contradiction, since $|h(t)^T \bar{u}(t)| \leq \sum_{i=1}^m |h_i(t)|$ for all $t \in [0,\tau]$.

Therefore u^* satisfies the Condition (9.48). Since $u^* \in \mathrm{RS}[0,\tau]$, u^* is a solution of (D2), for which $\psi_\tau(u^*) = \varphi_\tau(\alpha^*)$ holds, i.e. (D2) is dual to (P1). □

We obtain the following sufficient criterion for minimal time solutions:

Theorem 9.10.6. *Let τ_0 be such that a minimal solution α^* of* (P1) *has the value $\varphi_{\tau_0}(\alpha^*) = 1$ and each component of $h := Z^T - \sum_{i=1}^{n-1} \alpha_i^* Z_i^T$ has only finitely many zeros. Then τ_0 solves the minimal time problem and the function u^* defined by* (9.50) *is a time-optimal control. By*

$$x^*(t) := \int_0^t Y(s)u^*(s)ds \quad \textit{for all } t \in [0,\tau_0] \tag{9.53}$$

a time-optimal state-function is defined.

Proof. Let $\tau < \tau_0$. Then

$$r := \varphi_\tau(\alpha^*) = \int_0^\tau \|h(s)\|_1 ds < \int_0^{\tau_0} \|h(s)\|_1 ds = 1$$

holds. Let α_τ be a minimal solution of φ_τ, then $r_\tau := \varphi_\tau(\alpha_\tau) \le r$. Let $u^0 = u^*|_{[0,\tau]}$ be a solution of (D2), then according to Theorem 9.10.5 we obtain

$$\psi_\tau(u^0) = \varphi_\tau(\alpha_\tau) = r_\tau < 1.$$

By Remark 9.10.3 we recall u^0 is a solution of (D2) with value r_τ if and only if $\frac{1}{r_\tau}u^0$ is a solution of (D1) with value $\frac{1}{r_\tau}$. More explicitly: for all $u \in R_\tau$

$$1 < 1/r_\tau = \|1/r_\tau u^0\|_\tau \le \|u\|_\tau,$$

but then $1/r_\tau u^0 \notin Q_\tau$, and hence $\tau < \tau_0$ cannot be a minimal solution of (MT). □

Example 9.10.7. We consider the control of a vehicle on rails. We want to arrive at the point $c = (1,0)^T$ in the shortest possible time. The ODE system in this situation is given by

$$\dot{x}_1 = x_2 \quad \text{and} \quad \dot{x}_2 = u, \tag{9.54}$$

i.e. $A = \left(\begin{smallmatrix} 0 & 1 \\ 0 & 0 \end{smallmatrix}\right)$ and $B = \left(\begin{smallmatrix} 0 \\ 1 \end{smallmatrix}\right)$. A fundamental matrix H with $H(0) = E$ is immediately obtained by $H(t) = \left(\begin{smallmatrix} 1 & t \\ 0 & 1 \end{smallmatrix}\right)$. For the inverse $H^{-1}(t) = \left(\begin{smallmatrix} 1 & -t \\ 0 & 1 \end{smallmatrix}\right)$ holds. This leads to

$$Y(t) = \begin{pmatrix} 1 & \tau_0 \\ 0 & 1 \end{pmatrix} \begin{pmatrix} 1 & -t \\ 0 & 1 \end{pmatrix} \begin{pmatrix} 0 \\ 1 \end{pmatrix} = \begin{pmatrix} 1 & \tau_0 - t \\ 0 & 1 \end{pmatrix} \begin{pmatrix} 0 \\ 1 \end{pmatrix} = \begin{pmatrix} \tau_0 - t \\ 1 \end{pmatrix},$$

i.e.

$$Y_1(t) = \tau_0 - t \quad \text{and} \quad Y_2(t) = 1. \tag{9.55}$$

Using (9.45) and (9.46) we have $Z = Y_1$ and $Z_1 = Y_2$. For problem (P1) we obtain in this case *minimize* φ_{τ_0} *on* $\mathbb{R}$, where

$$\varphi_{\tau_0}(\alpha) := \int_0^{\tau_0} |\tau_0 - t - \alpha| dt.$$

According to the characterization theorem of convex optimization it is necessary and sufficient that for all $h \in \mathbb{R}$

$$0 \leq \varphi'_{\tau_0}(\alpha, h) = h \int_0^{\tau_0} \operatorname{sign}(\tau_0 - t - \alpha) dt$$

holds. For $\alpha = \tau_0/2$ we have $\int_0^{\tau_0} \operatorname{sign}(\tau_0/2 - t)dt = 0$. Hence $\alpha = \tau_0/2$ is a minimal solution of (P1) with value $W = \tau_0^2/4$. From requirement $W = 1$ the minimal time $\tau_0 = 2$ follows.

The function $\tau_0 - t - \tau_0/2 = \tau_0/2 - t$ changes sign at $t = 1$. By (9.50) the control

$$u^*(t) = \begin{cases} 1 & \text{for } t \in [0, 1) \\ -1 & \text{for } t \in [1, 2] \end{cases} \tag{9.56}$$

is time-optimal. Using (9.53), (9.55), and (9.56) it follows that

$$x^*(t) := \int_0^t \begin{pmatrix} 2 - s \\ 1 \end{pmatrix} u^*(s) ds.$$

Due to Theorem 9.10.6 the pair (x^*, u^*) is the solution to our problem.

Bibliography

[1] Achieser, N. I.: Vorlesungen über Approximationstheorie, Akademie Verlag, Berlin, 1953.

[2] Adams, R. A.: Sobolev Spaces, Academic Press, New York, 1975.

[3] Akimovich, B. A.: On the uniform convexity and uniform smoothness of Orlicz spaces. Teoria Functii Functional Anal & Prilozen, 15 (1972), 114–120.

[4] Asplund, E., Rockafellar, R. T.: Gradients of Convex Functions, Trans. Amer. Math. Soc. 139 (1969), 443–467.

[5] Attouch, H.: Variational Convergence for Functions and Operators. Applicable Mathematics Series, Pitman Advanced Publishing Program, Boston, London, Melbourne, 1984.

[6] Bellmann, R. E.: Dynamic Programming, Princeton University Press, Princeton, N. J., 1957.

[7] Bennet, C., Sharpley, R.: Interpolation of Operators, Academic Press, Vol. 129 in Pure and Appl. Math., 1988.

[8] Bildhauer, M.: Convex Variational Problems, Lecture Notes in Mathematics 1818, Springer, Heidelberg, N. Y., 2003.

[9] Birnbaum, Z. W., Orlicz, W.: Über die Verallgemeinerung des Begriffes der zueinander konjugierten Potenzen, Studia Math. 3 (1931), 1–64.

[10] Bliss, G. A.: The Problem of Lagrange in the Calculus of Variations, American Journal of Mathematics, Vol. LII (1930), 673–744.

[11] Blum, E., Oettli, W.: Mathematische Optimierung, Springer, Berlin, Heidelberg, New York, 1975.

[12] Bolza, O.: Vorlesungen über Variationsrechnung, Teubner Leipzig u. Berlin, 1909.

[13] Borg, I., Lingoes, J.: Multidimensional Similarity Structure Analysis. Springer, Berlin, Heidelberg, New York, 1987.

[14] Boyd, D. W.: The Hilbert Transform on rearrangement-invariant Spaces, Canad. J. of Math. 19 (1967), 599–616.

[15] Browder, F.: Problèmes Nonlinéaires, Univ. of Montreal Press, 1966.

[16] Browder, F. E.: Nonlinear operators and non-linear equations of evolution in Banach Spaces, Proc. Symp. Nonlinear Funct. Analysis, Chicago Amer. Math. Soc. 81, pp. 890–892, 1975.

[17] Carathéodory, C.: Variationsrechnung und partielle Differentialgleichungen erster Ordnung. Teubner, Leipzig u. Berlin, 1935.

[18] Carathéodory, C.: Variationsrechnung und partielle Differentialgleichungen erster Ordnung: Variationsrechnung. Herausgegeben, kommentiert und mit Erweiterungen zur Steuerungs- u. Dualitätstheorie versehen von R. Klötzler, B. G. Teubner, Stuttgart, Leipzig, 1994.

[19] Cesari, L.: Optimization Theory and Applications, Springer, New York, Heidelberg, Berlin, 1983.

[20] Chen, S.: Geometry of Orlicz Spaces, Dissertationes Math. 356, 1996.

[21] Chen, S., Huiying, S.: Reflexive Orlicz Spaces have uniformly normal structure, Studia Math. 109(2) (1994), 197–208.

[22] Cheney, E. E.: Introduction to Approximation Theory, McGraw-Hill, 1966

[23] Choquet,G.: Lectures on Analysis I, II Amsterdam, New York 1969.

[24] Day, M. M.: Normed Linear Spaces (3rd Edn.), Ergebnisse der Mathematik u. ihrer Grenzgebiete, Bd 21., Springer, 1973.

[25] Descloux, I.: Approximation in L^p and Tschebyscheff approximation. SIAM J. Appl. Math. 11 (1963), 1017–1026.

[26] Diaz, J. B., Metcalf, F. T.: On the structure of the set of subsequential limit points of successive approximations, Bull. Amer. Math. Soc. 73 (1967), 516–519.

[27] Dominguez, T., Hudzik, H.,Lopez, G., Mastylo, M., Sims, B.: Complete Characterization of Kadec–Klee Properties in Orlicz Spaces, Houston Journal of Mathematics 29(4) (2003), 1027–1044.

[28] Dunford, N. Schwartz, J. T.: Linear Operators, Part I, Wiley – Interscience, 1958.

[29] Fan, K., Glicksberg, I.: Some geometric properties of the spheres in a normed linear space, Duke Math. J. 25 (1958), 553–568.

[30] Figiel, T.: On moduli of convexity and smoothness, Studia Math 56 (1976), 121–155.

[31] Fleming, W. H., Rishel, R. W.: Deterministic and Stochastic Optimal Control, Springer, Berlin, Heidelberg, New York, 1975.

[32] Garrigos, G., Hernandez, E., Martell, J.: Wavelets, Orlicz Spaces, and greedy Bases, submitted Jan. 2007.

[33] Giaquinta, M., Hildebrand, S.: Calculus of Variations I and II, Springer, Berlin, Heidelberg, New York, 1996.

[34] Gustavson, S. A.: Nonlinear systems in semi-infinite programming, in: Numerical Solution of Systems of Nonlinear Algebraic Equations, Eds. G. D. Byren and C. A. Hall, Academic Press, N. Y., London, 1973.

[35] Harms, D.: Optimierung von Variationsfunktionalen, Dissertation, Kiel, 1983.

[36] Hartmann, Ph.: Ordinary Differential Equations, Wiley and Sons, New York, 1964.

[37] Hebden, M. D.: A bound on the difference between the Chebyshev Norm and the Hölder Norms of a function, SIAM J. of Numerical Analysis 8 (1971), 271–277.

[38] Hestenes, M. R.: Calculus of Variations and Optimal Control Theory, J. Wiley and Sons, New York, 1966.

[39] Heuser, H.: Funktionalanalysis, B. G. Teubner, Stuttgart, 1975.

[40] Hewitt, E., Stromberg, K.: Real and Abstract Analysis. Springer, 1965.

[41] Hirzebruch, F., Scharlau, W.: Einführung in die Funktionalanalysis; BI Bd. 296, Mannheim 1971.

[42] Holmes, R. B.: A Course on Optimization and Best Approximation; Lecture Notes in Math. 257, Springer 1972.

[43] Hudzik, H., Maligandra, L.: Amemiya norm equals Orlicz norm in general, Indag. Math. 11 (2000), 573–585.

[44] Ioffe, A. D., Tichomirov, V. M.: Theorie der Extremalaufgaben, Deutscher Verlag der Wissenschaften, Berlin, 1979.

[45] James, R. C.: Weakly compact sets, Trans. Amer. Math. Soc. 113 (1964), 129–140.

[46] Kaminska, A.: On uniform rotundity of Orlicz spaces, Indag. Math. A85 (1982), 27–36.

[47] Karlovitz, L.: Construction of nearest points in the L_p, p even, and L^∞ norms, J. Approx. theory 3 (1970), 125–127.

[48] Klötzler, R., Pickenhain, S.: Pontrjagin's Maximum Principle for multidimensional control problems, Optimal Control (Freiburg 1991), 21–30, Internat. Ser. Numer. Math. 111, Birkhäuser, Basel, 1993.

[49] Köthe, G.: Topologische lineare Räume I. Springer, Berlin, Göttingen, Heidelberg 1966.

[50] Kosmol, P., Flache Konvexität und schwache Differenzierbarkeit, Manuscripta Mathematica 8 (1973), 267–270.

[51] Kosmol, P.: Über Approximation stetiger Funktionen in Orlicz-Räumen, Journ. of Approx. Theory 8 (1973), 67–83.

[52] Kosmol,P.: Flache Konvexität von Orliczräumen, Colloq. Math. XXXI, pp. 249–252, 1974.

[53] Kosmol, P.: Nichtexpansive Abbildungen bei Optimierungsverfahren, First Symposium on Operations Research, University of Heidelberg, Sept. 1–3, 1976, in Operations Research Verfahren/ Methods of Operations Research XXV, 88–92, R. Henn et al. (Ed.), Verlag Anton Hain, Meisenheim, 1976.

[54] Kosmol, P.: Optimierung konvexer Funktionen mit Stabilitätsbetrachtungen, Dissertationes Mathematicae CXL, pp. 1–38, 1976.

[55] Kosmol, P.: Regularization of Optimization Problems and Operator Equations, Lecture Notes in Economics and Mathematical Systems 117, Optimization and Operations Research, Oberwolfach 1975, pp. 161–170, Springer 1976.

[56] Kosmol, P., Wriedt, M.: Starke Lösbarkeit von Optimierungsaufgaben, Math. Nachrichten 83 (1978), 191–195.

[57] Kosmol, P.: Bemerkungen zur Brachistochrone, Abh. Math. Univ. Sem. Hamburg 54 (1984), 91–94.

[58] Kosmol, P.: Ein Algorithmus für Variationsungleichungen und lineare Optimierungsaufgaben, Deutsch-französisches Treffen zur Optimierungstheorie, Hamburg, 1986.

[59] Kosmol, P.: Optimierung und Approximation. de Gruyter Lehrbuch, Berlin, New York, 1991. 2. überarbeitete und erweiterte Auflage 2010.

[60] Kosmol, P.: Methoden zur numerischen Behandlung nichtlinearer Gleichungen und Optimierungsaufgaben. B. G. Teubner Studienbücher, Stuttgart, zweite Auflage, 1993.

[61] Kosmol, P: An Elementary Approach to Variational Problems, Proceedings of the 12th Baikal International Conference on Optimization Methods and their Applications, Section 2. Optimal Control, Irkutsk, Baikal, pp. 202–208, June 24–July 1, 2001.

[62] Kosmol, P.: Projection Methods for Linear Optimization, J. of Contemporary Mathematical Analysis (National academy of Sciences of Armenia) XXXVI(6) (2001), 49–56.

[63] Kosmol, P., Pavon, M.: Solving optimal control problems by means of general Lagrange functionals, Automatica 37 (2001), 907–913.

[64] Kosmol, P.: Über Anwendungen des matrixfreien Newtonverfahrens, Berichtsreihe des Mathematischen Seminars der Universität Kiel (Kosmol 2005B), 2005.

[65] Kosmol, P., Müller-Wichards, D.: Homotopy Methods for Optimization in Orlicz Space, Proceedings of the 12th Baikal Int. Conf. on Optimization Methods and their Applications, Vol. 1, pp. 224–230, June 24–July 1, 2001.

[66] Kosmol, P., Müller-Wichards, D.: Homotopic Methods for Semi-infinite Optimization, J. of Contemporary Mathematical Analysis, National academy of Sciences of Armenia, XXXVI(5) (2001), 35–51.

[67] Kosmol, P., Müller-Wichards, D.: Pointwise Minimization of supplemented Variational Problems, Colloquium Mathematicum, 101(1) (2004), 15–49.

[68] Kosmol, P., Müller-Wichards, D.: Stability for Families of Nonlinear Equations, Isvestia NAN Armenii. Matematika, 41(1) (2006), 49–58.

[69] Kosmol, P., Müller-Wichards, D.: Optimierung in Orlicz-Räumen, Manuscript, University of Kiel, 2008.

[70] Kosmol, P., Müller-Wichards, D.: Strong Solvability in Orlicz-Spaces, J. of Contemporary Mathematical Analysis, National academy of Sciences of Armenia, 44(5) (2009), 271–304. Isvestia NAN Armenii. Matematika, No 5 (2009), 28–67.

[71] Krabs, W.: Optimierung und Approximation, Teubner, Stuttgart, 1972.

[72] Krasnosielskii, M. A., Ruticki, Ya, B.: Convex Functions and Orlicz Spaces. P. Noordhoff, Groningen, 1961.

[73] Kripke, B.: Best Approximation with respect to nearby norms, Numer. Math. 6 (1964), 103–105.

[74] Krotov, W. F., Gurman, W. J.: Methoden und Aufgaben der optimalen Steuerung (russ.), Moskau, Nauka, 1973.

[75] Kurc, W.: Strictly and Uniformly Monotone Musielak–Orlicz Spaces and Applications to Best Approximation, J. Approx. Theory 69 (1992), 173–187.

[76] Levitin, E. S., Polyak, B. T.: Convergence of minimizing sequences in conditional extremum problems, Doklady 5 (1966/1968), 764–767.

[77] Lindenstrauss, J.: On the modulus of smoothness and divergent series in Banach spaces. Mich. Math. J. 10 (1963), 241–252.

[78] Lovaglia, A. R.: Locally uniform convex Banach spaces, Trans. Amer. Math. XVIII (1955), 225–238.

[79] Luenberger, D. G.: Optimization by vector space methods, Wiley and Sons, NY, 1969.

[80] Luxemburg, W. A. J.: Banach function Spaces, Doctoral Thesis, Delft, 1955.

[81] Luxemburg, W. A., Zaanen, A.: Conjugate Spaces of Orlicz Spaces, Indag. Math. XVIII (1956), 217–228.

[82] Luxemburg, W. A. J., Zaanen, A. C.: Riesz spaces I, Amsterdam, London 1971.

[83] Maleev, R. P., Troyanski, S. L.: On the moduli of convexity and smoothness in Orlicz spaces. Studia Math, 54 (1975), 131–141.

[84] Meyer, Y.: Wavelets and operators. Cambridge Studies in Advanced Mathematics 37, 1992.

[85] Milnes, H. W.: Convexity of Orlicz Spaces, Pacific J. Math. VII (1957), 1451–1483.

[86] Müller-Wichards, D.: Über die Konvergenz von Optimierungsmethoden in Orlicz-Räumen. Diss. Uni. Kiel, 1976

[87] Müller-Wichards, D.: Regularisierung von verallgemeinerten Polya Algorithmen. Operations Research Verfahren, Verlag Anton Hain, Meisenheim, XXIV, pp. 119–136, 1978.

[88] Musielak, J.: Orlicz Spaces and Modular Spaces, Lecture Notes in Mathematics, 1034, 1983.

[89] Natanson, L. P.: Theorie der Funktionen einer reellen Veränderlichen; Akademie Verlag Berlin, 1961.

[90] Peetre, I.: Approximation of Norms, J. Approx. Theory 3 (1970), 243–260.

[91] Phelps, R. R.: Convex Functions, Monotone Operators and Differentiability, Lecture Notes in Mathematics, Vol. 1364, Springer, 1989.

[92] Rao, M. M.: Linear functionals on Orlicz Spaces, Nieuw. Arch. Wisk. 12 (1964), 77–98.

[93] Rao, M. M.: Smoothness of Orlicz spaces, Indagationes Mathematicae 27 (1965), 671–689.

[94] Rao, M. M.: Linear functionals on Orlicz Spaces: general theory, Pac. J. Math. 25 (1968), 553–585.

[95] Rao, M. M., Ren, Z. D.: Theory of Orlicz Spaces, Pure and Applied Mathematics, Marcel Dekker, New York, 1991.

[96] Rao, M. M., Ren, Z. D.: Applications of Orlicz Spaces, Pure and Applied Mathematics, Marcel Dekker, New York, 2002.

[97] Reid, W. T.: A Matrix Differential Equation of Riccati Type, American Journal of Mathematics LXVIII (1946).

[98] Rice, J. R.: Approximation of Functions. Vol. I and II, Addison Wesley, 1964 and 1969.

[99] Rockafellar, R. T.: Characterization of the subdifferentials of convex functions, Pac. J. Math. 17 (1966), 497–510.

[100] Rockafellar, R. T.: Convex Analysis, Princeton, 1970.

[101] Roubicek, T.: Relaxation in Optimization Theory and Variational Calculus. De Gruyter Series in Nonlinear Analysis and Applications 4, 1997.

[102] Soardi, P.: Wavelet bases in rearrangement invariant function spaces. Proc. Amer. Math. Soc. 125(12) (1997), 3669–3673.

[103] Temlyakov, M. M.: Greedy algorithms in Banach spaces, Advances in Computational Mathematics, 14 (2001), 277–292.

[104] Temlyakov, M. M.: Convergence of some greedy algorithms in Banach spaces, The Journal of Fourier Analysis and Applications, Vol. 8(5) (2002), 489–505.

[105] Temlyakov, M. M.: Nonlinear Methods of Approximation, Foundations of Computational Mathematics, 3 (2003), 33–107.

[106] Tsenov, I. V.: Some questions in the theory of functions. Mat. Sbornik 28 (1951), 473–478 (Russian).

[107] Turett, B.: Fenchel–Orlicz spaces. Diss. Math. 181, 1980.

[108] Tychonoff, A. N.: Solution of incorrectly formulated problems and the regularization method. Soviet Math. Dokl. 4 (1963), 1035–1038.

[109] Vainberg, M. M.: Variational Method and Method of monotone Operators in the Theory of non-linear Equations. Wiley and Sons, 1973.

[110] Völler, J.: Gleichgradige Stetigkeit von Folgen monotoner Operatoren, Diplomarbeit am Mathematischen Seminar der Universität Kiel, 1978.

[111] Werner, D.: Funktionalanalysis, Springer, 6. Auflage, 2007.

[112] Werner, H.: Vorlesungen über Approximationstheorie. Lect. Notes in Math. 14, Springer, 1966.

[113] Wloka J.: Funktionalanalysis und Anwendungen, W. de Gruyter, Berlin, 1971.

[114] Zaanen, A. C.: Integration, North Holland, p. 372, 1967.

[115] Zang, I.: A Smoothing-Out Technique for Min-Max-Optimization, Mathematical Programming, pp. 61–77, 1980.

[116] Zeidan, V.: Sufficient Conditions for the Generalized Problem of Bolza, Transactions of the AMS 275 (1983), 561–586.

[117] Zeidan, V.: Extended Jacobi Sufficient Conditions for Optimal Control, SIAM Control and Optimization 22 (1984), 294–301.

[118] Zeidan, V.: First and Second Order Sufficient Conditions for Optimal Control and calculus of Variations, Applied Math. and Optimiz. 11 (1984), 209–226.

[119] Zeidler, E.: Nonlinear Functional Analysis and its Applications III, Springer, Berlin, Heidelberg, New York, 1985.

List of Symbols

C^0	polar of the set C
C^{00}	bipolar of the set C
X^*	dual space of X
$x_n \rightharpoonup x$	weak convergence of sequence (x_n) to x in space X
$S(X)$	unit sphere in X
$U(X)$	unit ball in X
$K(x_0, r)$	open ball with center x_0 and radius r
$\mathrm{conv}(A)$	convex hull of set A
$E_p(S)$	extreme points of set S
$\mathrm{Epi}(f)$	epigraph of the function f
$\mathrm{Dom}(f)$	set where function f is finite
$S_f(r)$	level set of function f at level r
$\partial f(x_0)$	subgradient of function f at x_0
$M(f, K)$	set of minimal solutions of function f on set K
(f, K)	minimize function f on set K
(g_1, g_2, K)	minimize function g_1 on $M(g_2, K)$
f^*	convex conjugate of function f
f^{**}	biconjugate of function f
g^+	concave conjugate of function g
χ_A	characteristic function of set A
$x\|_A$	function x restricted to set A
Φ	Young function
Ψ	conjugate of Φ
f^Φ	modular
(T, Σ, μ)	measure space
$L^\Phi(\mu)$	Orlicz space
$M^\Phi(\mu)$	subspace of $L^\Phi(\mu)$: closure of step functions
$C^\Phi(\mu)$	Orlicz class: $\mathrm{Dom}(f^\Phi)$
$H^\Phi(\mu)$	subspace of $L^\Phi(\mu)$ contained in $C^\Phi(\mu)$
ℓ^Φ	Orlicz sequence space
m^Φ	subspace of ℓ^Φ: closure of step functions
c^Φ	Orlicz class: $\mathrm{Dom}(f^\Phi)$ in ℓ^Φ

h^{Φ}	subspace of ℓ^{Φ} contained in c^{Φ}
ℓ^{∞}	space of bounded sequences
$\|\cdot\|_{(\Phi)}$	Luxemburg norm
$\|\cdot\|_{\Phi}$	Orlicz norm
Δ_2-condition	growth condition for Φ
Δ_2^{∞}-condition	growth condition for Φ in neighborhood of ∞
Δ_2^{0}-condition	growth condition for Φ in neighborhood of 0
δ-condition	convexity condition for Φ
δ^{∞}-condition	convexity condition for Φ in neighborhood of ∞
δ^{0}-condition	convexity condition for Φ in neighborhood of 0
ρ_X	module of smoothness for uniformly differentiable spaces X
δ_X	module of convexity for uniformly convex spaces X
δ_X	module of convexity for uniformly convex spaces X
τ_{x,x^*}	module of convexity for locally uniformly convex functions
$\mathrm{RS}[a,b]$	space of piecewise continuous functions on $[a,b]$
$\mathrm{RCS}^1[a,b]^n$	space of piecewise continuously differentiable functions on $[a,b]$ with values in $\mathbb{R}^n$
$C^1[a,b]^n$	space of continuously differentiable functions on $[a,b]$ with values in $\mathbb{R}^n$
$C(T)$	space of continuous functions on T
$W^m L^{\Phi}(\mu)$	Orlicz–Sobolev space

Index

U

V

W

Y

Z